1st International Conference on Advances in Engineering and Technology (ICAET-2018)

IOP Conference Series: Materials Science and Engineering
Volume 414

Quetta, Pakistan
2 - 3 April 2018

Editor:

Muhammad Najam Khan
Abdul Malik Rehan Abbasi
Syed Mobasher Aftab
Faisal Mushtaq
Rehanul Haq Siddiqui
Muhammad Oasim Siddiqi
Saqib Siddiqui
Syed Abid Hussain
Syed Mohammad Ali Shah
Maqsood Ahmad
Mohammad Tariq

ISBN: 978-1-5108-7196-0
ISSN: 1757-8981

Printed from e-media with permission by:

Curran Associates, Inc.
57 Morehouse Lane
Red Hook, NY 12571

Some format issues inherent in the e-media version may also appear in this print version.

Printed by Curran Associates, Inc. (2018)

For permission requests, please contact the Institute of Physics at the address below.

Institute of Physics
Dirac House, Temple Back
Bristol BS1 6BE UK

Phone: 44 1 17 929 7481
Fax: 44 1 17 920 0979

techtracking@iop.org

Additional copies of this publication are available from:

Curran Associates, Inc.
57 Morehouse Lane
Red Hook, NY 12571 USA
Phone: 845-758-0400
Fax: 845-758-2633
Email: curran@proceedings.com
Web: www.proceedings.com

Table of contents

Volume 414

1st International Conference on Advances in Engineering and Technology (ICAET-2018)

2–3 April 2018, Baleli, Quetta 87300, Pakistan

Accepted papers received: 9 August 2018
Published online: 13 September 2018

Preface

Papers

IOP Conf. Series: Materials Science and Engineering **414** (2018) 011001 doi:10.1088/1757-899X/414/1/011001

1st International Conference on Advances in Engineering and Technology-2018

Conference Proceedings

1st International Conference on Advances in Engineering and Technology (ICAET-2018) "Sharing the Ideas"

02 April 2018 - 03 April 2018
BUITEMS, Quetta, Pakistan

Balochistan University of Information Technology, Engineering and Management Sciences, Quetta, Pakistan.

IOP Conf. Series: Materials Science and Engineering **414** (2018) 011001 doi:10.1088/1757-899X/414/1/011001

1st International Conference on Advances in Engineering and Technology-2018

Table of Contents

IOP Conf. Series: Materials Science and Engineering **414** (2018) 011001 doi:10.1088/1757-899X/414/1/011001

1st International Conference on Advances in Engineering and Technology-2018

Preface

This conference proceedings contains contributions from the Engineers, Scientists and Institutional Educationists coming from many countries across the world who have decided to participate in the 1st International Conference on Advances in Engineering and Technology (ICAET-2018). The conference was held at Balochistan University of Information Technology, Engineering and Management Sciences, (BUITEMS) Quetta, Pakistan from 2nd to 3rd of April 2018. The papers included were evaluated/peer reviewed by a number of colleagues from Pakistan and worldwide. The purpose is to guarantee quality, credibility and accuracy of scientific contents for the different sessions during the Conference. Out of 100 papers received, a total of 45 papers were accepted for for publication. A large number of contributions covers the manifold interests of the Engineers, Scientists and the interdisciplinary community. This is in view of the dedication and efforts made in facing new challenges arising in our times. It is the time in which new technologies, resources and various novel issues are emerging in developed and under developed countries; while at the same time solutions of the same are being found. The Conference proceedings and all the other scientific presentations related to the ICAET-2018, must be attributed to the extraordinary efforts made by the conference Scientific Publication Committee, chaired by Dr. Syed Mobasher Aftab.The tremendous dedication of Dr. Muhammad Najam Khan and Dr. Faiza Akhter and all senior colleagues are also appreciated, whose engagement and endeavors to actualize this concept of the conference started a long time ago. We also wish to thank the organizers of ICAET-2018, who followed cautiously the entire process of printing of this book of abstracts.

Editorial Team,
ICAET 2018,
BUITEMS,Quetta, Pakistan.

IOP Conf. Series: Materials Science and Engineering **414** (2018) 011001 doi:10.1088/1757-899X/414/1/011001

1st International Conference on Advances in Engineering and Technology-2018

Aim and Scope of the Confernce

The idea of the 1st International Conference on Advances in Engineering and Technology-2018, is to bring together the scientists, scholars, engineers and students from all over the world to present the ongoing research activities and to promote research relations. This conference provides opportunities for the participants to exchange new ideas, applications, problems, issues, solutions and experiences and to establish research relations.

Experts from Pakistan and abroad participated in ICAET-2018. We are delighted to have you as one of the honorable participants in the conference by presenting your research work. The scope of the conference includes, but is not limited to the following topics:

- Civil Engineering
- Geological Engineering and Geosciences
- Mining Engineering
- Petroleum and Gas Engineering
- Chemical Engineering
- Textile Engineering and Technology
- Mechanical and Material Engineering
- Environmental Engineering and Management
- Architecture and Town Planning
- Interdisciplinary

IOP Conf. Series: Materials Science and Engineering **414** (2018) 011001 doi:10.1088/1757-899X/414/1/011001

1st International Conference on Advances in Engineering and Technology-2018

Organizing Committees

Conference General Chair	Engr. Ahmed Farooq Bazai (SI)	
Conference Chair	Dr. Muhammad Najam Khan Malghani	
Committee	**Chair**	**Co-chair**
Secretary	Dr. Faisal Mushtaq	Dr. Saeedullah Jan Mandokhail
Publication	Dr. Syed Mobasher Aftab	Dr. Saqib Siddiqui
Technical Program	Dr. Muhammad Qasim Siddiqui	Engr. Arshad Ali
Information Technology	Engr. Muhammad Anwar Khan	Dr. Abdul Malik Rehan Abbasi
Registration and Exhibit	Dr. Syed Mohammad Ali Shah	Dr. Faiza Akhtar
Local Arrangement	Dr. Syed Kamran Sami	Ar. Jalal Faisal
Finance	Dr. Syed Zameer ul Hassan	Dr. Ali Nawaz Mengal

IOP Conf. Series: Materials Science and Engineering **414** (2018) 011001 doi:10.1088/1757-899X/414/1/011001

1st International Conference on Advances in Engineering and Technology-2018

Editorial Board

1st International Conference on Advances in Engineering and Technology-2018

Email ID: syed.ali@buitms.edu.pk

10. **Dr. Maqsood Ahmad**
Professor, Department of Environmental
Management and Policy,
BUITEMS, Quetta, Pakistan
Email ID: maqsood@buitms.edu.pk

11. **Dr. Mohammad Tariq**
Professor, Department of Petroleum and Gas Engineering,
BUITEMS, Quetta, Pakistan
Email ID: mohammad.tariq@buitms.edu.pk

IOP Conf. Series: Materials Science and Engineering **414** (2018) 011001 doi:10.1088/1757-899X/414/1/011001

1st International Conference on Advances in Engineering and Technology-2018

Keynote Speakers

1. **Dr. Jiří Militký**
Professor, Faculty of Textile Engineering,
Technical University of Liberec,
Czech Republic.

2. **Dr. Farid Nasir Bin Ani**
Research Professor, Faculty of Mechanical Engineering,
Department of Thermo-fluids,
Universiti Teknologi Malaysia, Malaysia.

3. **Dr. Pakamas Chetpattananondh**
Associate Professor, Chemical Engineering Department,
Prince of Songkla University, Songkhla, Thailand.

4. **Dr. Arfan Ghani**
Senior Lecturer, Electrical/Electronic Engineering,
School of Computing, Electronics and Maths,
Coventry University, Coventry, UK.

5. **Dr. Muhammad Kashif Khan**
Lecturer, Aerospace Engineering,
School of Mechanical, Aerospace and Automotive Engineering,
Coventry University, Coventry, UK.

IOP Conf. Series: Materials Science and Engineering **414** (2018) 011001 doi:10.1088/1757-899X/414/1/011001

1st International Conference on Advances in Engineering and Technology-2018

Invited Speakers

1. **Dr. Mohammad Ahmad Farooqui**
   Professor of Geology and Program Director,
   Dual Degree Program/Head, Academics and Research,
   COMSATS Institute of Information Technology,
   Lahore, Pakistan.

2. **Dr. Asad Ullah Khan**
   Professor, Head of the Department/Chairman,
   Department of Chemical Engineering,
   COMSATS Institute of Information Technology,
   Lahore, Pakistan.

3. **Dr. Arshad Hussain**
   Professor, Chemical Engineering Department,
   School of Chemical and Materials Engineering,
   National University of Science and Technology,
   Islamabad, Pakistan.

4. **Dr. Abdul Waheed Bhutto**
   Professor and Dean, Faculty of Engineering,
   Dawood University of Engineering and Technology,
   Karachi, Pakistan.

5. **Dr. Amanullah Marri**
   Professor, Department of Civil Engineering,
   NED University of Engineering and Technology,
   Karachi, Pakistan.

6. **Dr. Suleman Aziz Lodhi**
   Professor, University of Sargodha,
   Lahore Campus, Lahore, Pakistan.

7. **Dr. Mahmood Saleem**
   Professor, Department of Chemical Engineering,
   ICET, University of the Punjab, Lahore, Pakistan.

8. **Dr. Saeed Badshah**
   Associate Professor and Chairman,
   Department of Mechanical Engineering,
   International Islamic University,
   Islamabad, Pakistan.

9. **Dr. Zulfiqar Ali**

1st International Conference on Advances in Engineering and Technology-2018

Associate Professor and Chairman,
Department of Mining Engineering,
University of Engineering and Technology,
Lahore, Pakistan.

10. **Dr. Yasir Nawab**
Assistant Professor, Textile Composite Materials
Research Group, National Textile University,
Faisalabad, Pakistan.

11. **Engr. Muhammad Owais Naseem**
Senior Engineer (Drilling), Drilling & Well
Engineering Department,
Pakistan Petroleum Limited, Karachi, Pakistan.

IOP Conf. Series: Materials Science and Engineering **414** (2018) 011001 doi:10.1088/1757-899X/414/1/011001

1st International Conference on Advances in Engineering and Technology-2018

Summary of Oral Schedule

Day	Date	Timing	Session Name	Session Coding	Venue
1st	2nd April	12:00-02:00 PM	Civil and Construction Materials	1-A	Orange Hall
			Textile Engineering	1-B	Green Hall
			Mechanical and Material Engineering	1-C	Pink Hall
			Chemical Engineering	1-D	Purple Hall
			Interdisciplinary- General	1-E	Training Hall
		03:00-06:20 PM	Civil and Construction Materials	2-A	Orange Hall
		03:00-05:00 PM	Textile Engineering	2-B	Green Hall
		03:00-05:20 PM	Mechanical and Material Engineering	2-C	Pink Hall
		03:00-06:00 PM	Chemical Engineering	2-D	Purple Hall
		03:00-04:40 PM	Interdisciplinary-Engineering	2-E	Training Hall
2nd	3rd April	10:00-12:20 PM	Environmental Engineering and Management	3-A	Pink Hall
		10:00-11:40 AM	Geological Engineering	3-B	Purple Hall
		10:00-12:00 Noon	Petroleum and Gas Engineering	3-C	Orange Hall
		10:00-12:20 PM	Architecture and Town Planning	3-D	Green Hall
		12:00-02:00 PM	Environmental Engineering and Management	4-A	Pink Hall
			Geological Engineering	4-B	Purple Hall
			Mining Engineering	4-C	Orange Hall

IOP Conf. Series: Materials Science and Engineering **414** (2018) 011001 doi:10.1088/1757-899X/414/1/011001

1st International Conference on Advances in Engineering and Technology-2018

IOP Conf. Series: Materials Science and Engineering **414** (2018) 011001 doi:10.1088/1757-899X/414/1/011001

1st International Conference on Advances in Engineering and Technology-2018

IOP Conf. Series: Materials Science and Engineering **414** (2018) 011002 doi:10.1088/1757-899X/414/1/011002

Peer review statement

All papers published in this volume of *IOP Conference Series: Materials Science and Engineering* have been peer reviewed through processes administered by the proceedings Editors. Reviews were conducted by expert referees to the professional and scientific standards expected of a proceedings journal published by IOP Publishing.

Effect of tempered glass fines in concrete at elevated temperature

Iffat Siddique[1], Momina Rauf[1], Sadia Umer Khayam[1], Muhammad Alamgeer Shams[1], Husnain Faisal[1]

[1]University of Engineering and Technology, Lahore.

Email. mominababri@gmail.com

Abstract. Millions of tons of waste glass are produced every year. Solid waste management is one of the biggest environmental problems. Also, there is a need to discover alternative materials that can compensate for the depletion of natural resources. The purpose of this research is to assess the feasibility of waste tempered glass fines (TGF) as partial replacement of sand in concrete mix. Concrete cubes with four different replacement percentages (0%, 5%, 10%, 15% & 20%) and three different sizes of tempered glass fines (TGF) were tested for compressive strength at ambient and elevated temperature (790°C). The mass loss and elastic modulus were also inspected. The results indicated that waste TGF could be used as an efficient sand replacement material. An enhanced compressive strength was achieved for all of concrete mix design samples. Among all the mixes, 5% by weight replacement of sand with TGF (particle size= 1.7-4.75mm) gave the maximum enhancement of compressive strength. The same trend was observed for mass loss and modulus of elasticity.
Keywords. Temperature glass fines, concrete, solid waste management.

1. Introduction

Concrete is one of the most widely used construction material throughout the world because of its durability, strength, ease of fabrication and noncombustible nature. The rapidly growing use of concrete thus, puts an increased demand of the natural resources while leading to the depletion of the resources at the same time. Therefore, for the sustainability of the construction industry, new innovative, non-conventional cost-effective materials need to be employed which can compensate for the lack of such resources along with enhancement of the behavior of concrete structures under different loading conditions.

Moreover, waste reduction is also a global issue that needs to be addressed. Waste glass is a growing burden worldwide. Despite its ideal properties for recycling, it is not suitable for landfill [1]. Recycling of waste glass not only helps conserve earth's natural resources but also reduces demand of landfill spaces, while saving energy and money at the same time [2]. Glass can be used as a pozzolanic/cementitious material [3]. Various past researches have been dedicated to the use of waste glass as an aggregate and/or cement replacement material [4-7].

Glass has same physical properties as sand but it shows a lower rate of water absorption and therefore, is an interesting material for concrete development [8]. It has been reported in many past researches that with the inclusion of glass aggregates the compressive strength reduces [9,10]. Several other studies conducted in the last decade show that along with volume of glass aggregates used, particle size plays an important role in alkali silica reaction (ASR) [11,12]. The results indicate that ASR increases with the increased particle size. Expansion and subsequent cracking of concrete due to ASR can be eliminated if less than 20% of waste glass aggregates of particle size less than 1.18mm are used. Also reduced particle size increases pozzolanicity of the concrete mix. [13].

Fire resistance is an important natural attribute of concrete. When concrete is subjected to elevated temperature, strength and stiffness of concrete reduces due to several reasons such as evaporation of water from voids, aggregate decomposition and dehydration of cement paste etc. [14,15]. Also, higher temperature and faster cooling rate produce negative impact on residual mechanical properties of concrete and strength reduces to almost zero at 600°C [16]. The volume and size of aggregates is another important factor that greatly influences the behavior of concrete at elevated temperatures [17,18].

Different materials can be employed to assess the fire-resistant properties of concrete. Poutos et al. proved that glass aggregates exhibit better temperature stability owing to its lower specific heat compared to its natural sand and due to pozzolanic activity of glass [19].

Tempering is a thermal process of glass strengthening by creating compressive stresses [20]. Tempered glass is a special glass type with enhanced features of high toughness and thermal resistance. It is extensively used in sliding doors and partitions in houses and offices, facades, car windshields etc. For a glass to be tempered, it is cut to required size and shape, before it is toughened and it cannot be re-toughened. Thus, broken tempered glass and wrongly fabricated sizes serve as a potential waste material that can be utilized in concrete. Moreover, the enhanced thermal resistance can lead to a better behavior when concrete is exposed to fire [21].

The described research program is aimed at improving the strength properties of concrete at normal and elevated temperature (790°C) on one hand. While to check the feasibility of using this material as sand replacement on the other hand. In this study, tempered glass fines (TGF) were used in concrete as partial replacement of fine aggregate and effect of size and dosage of TGF were studied.

2. Experimental Program

2.1. Materials

The concrete mix used in this study was prepared from Ordinary Portland Cement-Type I (OPC), natural coarse aggregates, locally available clean river sand (max. particle size 4.75mm), tap water and tempered glass fines. The bulk densities of cement, sand and coarse aggregate were 1440 kg/m^3, 1463 kg/m^3 and 1455 kg/m^3 respectively. Tempered glass fines (TGF) were incorporated in the concrete mix as a partial replacement of sand in various sizes and percentages. The acronym of three types of TGF along with their respective sizes are listed in table 2.

2.2. Mix Proportion

A total of thirteen different concrete mixes were prepared with varying amount and sizes of TGF and a minimum compressive strength of 20 MPa. In all the mixes a constant W/C ratio of 0.42 was taken. Sand was replaced with TGF in four different amounts by weight as 5%, 10%, 15% and 20% respectively. The quantities of the constituent materials for all mix designs are summarized in table 1.

Table 1. Proportions of concrete mix designs

Mix Type	Cement (kg/m^3)	Sand (kg/m^3)	Aggregate (kg/m^3)	TGF (kg/m^3)
Control	205.71	418.12	831.77	0.00
CGF-5	205.71	397.21	831.77	19.51
CGF-10	205.71	376.31	831.77	39.02
CGF-15	205.71	355.40	831.77	58.53
CGF-20	205.71	334.49	831.77	78.04
MGF-5	205.71	397.21	831.77	17.52
MGF-10	205.71	376.31	831.77	35.04
MGF-15	205.71	355.40	831.77	52.56
MGF-20	205.71	334.49	831.77	70.08
FGF-5	205.71	397.21	831.77	14.44
FGF-10	205.71	376.31	831.77	28.88
FGF-15	205.71	355.40	831.77	43.32
FGF-20	205.71	334.49	831.77	57.76

Note: The numbers (5,10,15,20) in designation of specimen, %age replacement of sand by TGF

IOP Conf. Series: Materials Science and Engineering **414** (2018) 012001 doi:10.1088/1757-899X/414/1/012001

Table 2. Properties of TGF

TGF Type	Acronym	Passing sieve No.	Retaining sieve No.
Coarse Glass Fines	CGF	4	12
Medium Glass Fines	MGF	12	40
Fine Glass Fines	FGF	40	200

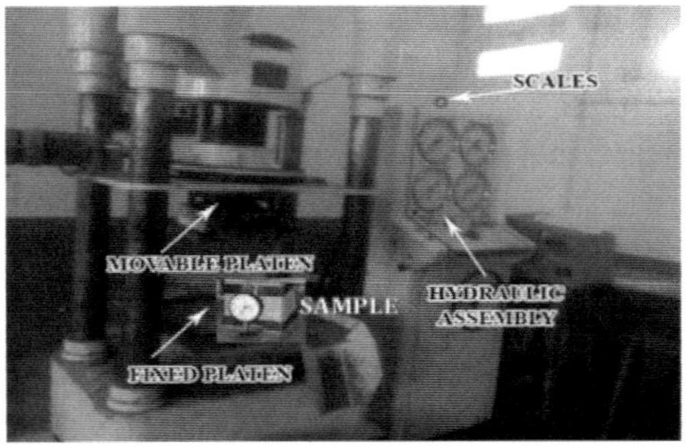

Figure 1: Test setup and test sample

2.3. Specimen Detail & Testing Methodology
According to British Standard, 100 mm x 100 mm cubic specimens were prepared for compression test. Six specimens were prepared of each concrete mix. Three of them were tested at ambient temperature and three were tested after exposure to an elevated temperature of 790^{0}C. Nabertherm LV 15/11 furnace was used to attain the 790^{0}C temperature. The specimens were demolded after 24 hours of casting, cured for 28 days and were allowed to dry. Once the specimens were fully dried, half of them were one by one exposed to 790°C in a furnace at an approximate rate of 6.5^{o}C/ temperature rise per minute. These samples were allowed to naturally cool down to room temperature before they were tested. The compression test was performed on Denison Universal Testing Machine of capacity 300Ton, figure 1. All the specimens were loaded at the rate of 5 Ton per minute up to the failure of the sample in Universal Testing Machine to get the compressive strength. The average of three tested samples has been reported in table 3. The strains were calculated from the readings of the dial gauge attached to the sample, Figure 1. Mass of each specimen was measured before and after exposing it to high temperature to find the mass loss.

3. Results and Discussion
3.1 Compressive Strength
The results of the compressive strength test carried out at ambient and elevated temperatures are presented in table 3 and figure 2a &2b.

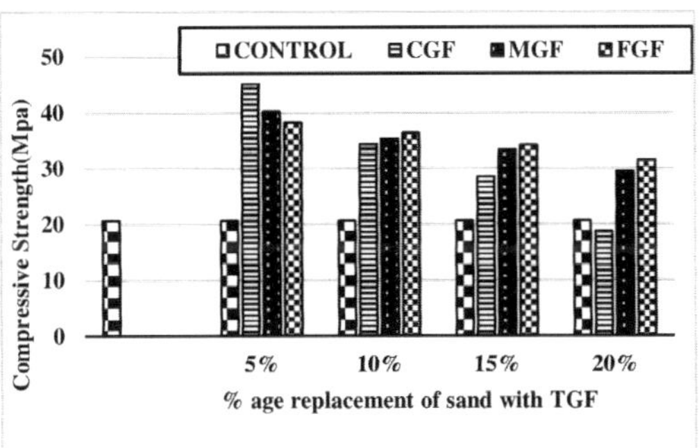

Figure 2a. Compressive strength at ambient temperature

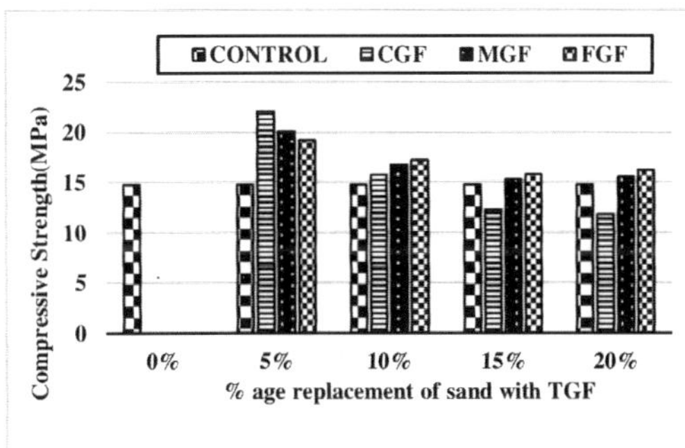

Figure 2b. Compressive strength at elevated temperature

It can be seen from the figure 2a & 2b that among the four different replacements of sand with TGF; 5% replacement of sand came out to be most efficient in improving the compressive strength. At the ambient temperature, CGF-5 mix delivered an increase of 119% (compared to control) in the ultimate compressive strength. In case of MGF-5 and FGF-5, this enhancement was up to 95% and 86% respectively. A similar trend was observed at the elevated temperature. Samples of CGF-5, MGF-5 and

FGF-5 mix showed a 49.9%, 16.6% and 10% enhanced ultimate compressive strength as compared to the samples of control mix (0% TGF). A decreasing trend was observed for the increased inclusion of TGF as replacement of sand. The specimens with 10%, 15% and 20% partial replacement of sand with TGF exhibited lesser enhancement in the respective ultimate compressive strengths as compared to the 5% replacement. But still the strength remained greater than control samples except for CGF-20.

This implies that although the coarser particle results in the maximum increase in compressive strength, but with the increased amount of CGF, the drop-in strength is also much more noticeable than the other two particle sizes. The reason is that the coarser TGF retain their original structure. The inherent surface compression in the structure of TGF, helps to resist much more compressive loads. On the other hand, the increased amount of CGF, being coarser, creates more voids and debonding in the concrete matrix. Such that the dominant advantage of the surface compression starts reducing. However, when the size of the particles is reduced, as the TGF structure is disintegrated and only the filling effect of the particles remains resulting in a more consolidated structure. And hence the compressive strength is increased but the effect is less pronounced for MGF and CGF concrete mixes.

Further, it can be said that use of TGF up to an optimum level results in enhanced compressive strength even at elevated temperatures. This highlights the advantage of TGF over annealed glass fines. The increase can be attributed to the improved thermal resistance, better crystalline structure, mechanical strength and surface compression, Table (1).

At elevated temperature of 790^{0}C the moisture is dried up from the pores leaving voids, the hydration products of cement paste and C-S-H gel are considerably disintegrated. And as coarser particles have a lesser water absorption than the finer particles, CGF mixes show a lesser reduction in strength at the same replacement level compared to the MGF and FGF.

Table 3. Ultimate Compressive Strength

Mix Type	**Ultimate Compressive Strength (MPa)**		
	Ambient Temperature	**790°C**	**%age Reduction**
Control	20.60	14.72	29
CGF-5	45.13	22.07	51.1
CGF-10	34.35	15.72	54.2
CGF-15	28.45	12.25	56.9
CGF-20	18.64	11.77	36.9
MGF-5	40.22	20.07	50.1
MGF-10	35.32	16.77	52.5
MGF-15	33.37	15.28	54.2
MGF-20	29.39	15.50	47.3
FGF-5	38.26	19.19	49.8
FGF-10	36.38	17.21	52.7
FGF-15	34.23	15.78	53.9
FGF-20	31.51	16.21	48.6

3.2 Mass Loss

All the dried/cooled specimens were weighed before and after exposure to elevated temperature. The mass loss due to exposure to elevated temperature is reported in figure 3.

Mass loss is lesser in concrete specimens with smaller sized TGF as compared to the coarser TGF. This exactly complies with the expected behavior. Because coarser particles have more voids. And so for CGF, more porous concrete matrix is obtained as compared to the MGF and FGF mixes. The elevated temperature dries out the unreacted water and squeezes out the entrapped air. Fine particles produce denser concrete matrix with less voids in them. And lesser mass loss was witnessed.

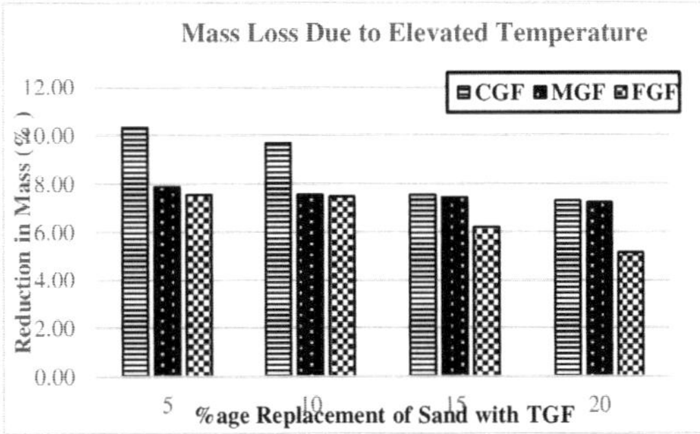

Figure 3. Mass loss due to elevated temperature

3.3 Modulus of Elasticity

Figures 4a & 4b show the effect of TGF on modulus of elasticity of concrete. At both the temperatures, the trend was similar as that of the compressive strength. The maximum improved strength was obtained for 5% replacement and coarser particles. The reason behind is that the higher MOE of TGF resulted in the higher MOE of the mix. Also, though elevated temperature caused a decrease in the MOE. But in case of concrete with TGF, the reduction was lower. Because between 600-800^{0}C the melting and re-solidification of glass occurs. This process fills up the micro cracks and so the reduction in the MOE is lowered.

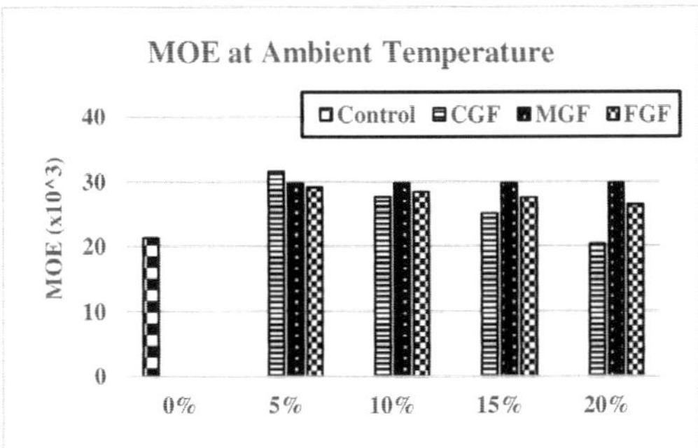

Figure 4a. MOE at ambient temperature

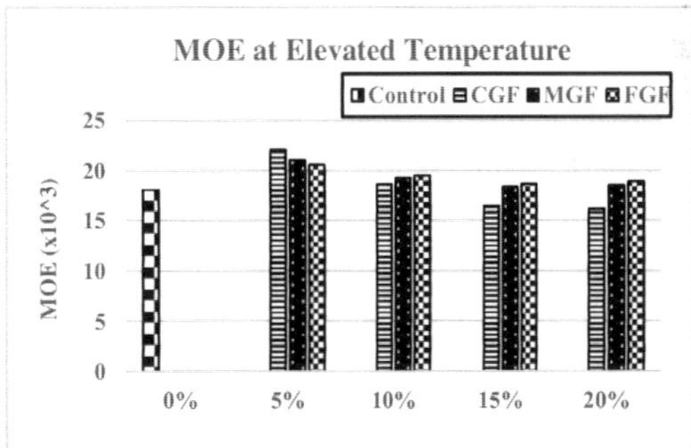

Figure 4b. MOE at elevated temperature

3.4 Mode of Failure

The samples failed by bulging and usual inclined cracks at ambient temperature. The bulging was more visible in case of CGF samples because of the weak bonding at the interface, figure 5. Spalling was pronounced in specimens after they were exposed to elevated temperature. The effect increased with the increased size and amount of TGF, figure 6. Hair like cracking was also visible in heated specimens before the application of any compressive load in figure 7. This is the sample that was exposed to heating prior to testing.

Figure 5. Bulging during Compression Test

Figure 6. Spalling compression test

Figure 7. Hair Line Cracks before Compression Test

4. Conclusion
The behavior of concrete with tempered glass was studied in the presented work. Based on the experimental investigation, following conclusions can be drawn.

- TGF can be used as an efficient sand replacement material with additional benefit of enhanced compressive strength. This will be beneficial from the solid waste management point of view as well as the exhaustion of natural resources.
- Among all the replacement percentages 5% replacement was found to be the most efficient in enhancing the compressive strength. Hence, it is the optimum dosage for the sand replacement.
- CGF-5 proves to be more efficient than MGF-5 and FGF-5 (12.2% and 17.9% respectively) in the improvement of compressive strength but at the same time the coarser particles result in greater mass loss i.e. 10.78%, 7.87% and 7.56% for CGF-5, MGF-5 and FGF-5 respectively. Therefore, it is recommended to carefully select the particle size especially when the concrete is frequently exposed to elevated temperature.
- TGF caused considerable increase in Modulus of Elasticity at ambient temperature and a lesser reduction in concrete stiffness at elevated temperature provided that the amount of TGF replacement does not exceed 10%. Therefore, it is an efficient sand replacement material especially for those structures having probability of being frequently exposed to elevated temperature and/or to fire.

5. Acknowledgements
The experimental work was conducted in Concrete Testing Laboratory of UET Lahore, Pakistan.

References
[1] Heriyanto 2018 From waste glass to building materials – An innovative solution for waste glass. *Journal of Cleaner Production* 191:192-206.
[2] Hogland W 2002 Remediation of an old landsfill site *Environmental Science and Pollution Research* 9(S1):49-54.
[3] Jani Y, Hogland W. 2014 Waste glass in the production of cement and concrete – A review *Journal of Environmental Chemical Engineering* 2(3):1767-1775.
[4] Pike R, Hubbard D. Physicochemical studies of the destructive alkali-aggregate reaction in concrete. Journal of Research of the National Bureau of Standards. 1957;59(2):127.
[5] Ducman V, Mladenovič A, Šuput J. 2002 Lightweight aggregate based on waste glass and its alkali–silica reactivity. *Cement and Concrete Research* 32(2):223-226.
[6] Tan K, Du H. Use of waste glass as sand in mortar: Part I – Fresh, mechanical and durability properties. *Cement and Concrete Composites*. 2013;35(1):109-117.
[7] Shao Y, Lefort T, Moras S, Rodriguez D. Studies on concrete containing ground waste glass. *Cement and Concrete Research*. 2000;30(1):91-100.
[8] Ismail Z, AL-Hashmi E. Recycling of waste glass as a partial replacement for fine aggregate in concrete. *Waste Management*. 2009;29(2):655-659.
[9] Taha B, Nounu G. Properties of concrete contains mixed colour waste recycled glass as sand and cement replacement. *Construction and Building Materials*. 2008;22(5):713-720.
[10] Topçu İ, Canbaz M.2004 Properties of concrete containing waste glass. *Cement and Concrete Research*. 34(2):267-274.
[11] Takata R., Sato S., Nonaka T., Ogata H., Hattori K.2004 Investigation on alkali-silica reaction utilizing waste glass in concrete and suppression effect by natural zeolite. *29th Conference on our world in concrete and structures*.
[12] Corinaldesi V, Gnappi G, Moriconi G, Montenero A. Reuse of ground waste glass as aggregate for mortars. *Waste Management*. 2005;25(2):197-201.
[13] Idir R, Cyr M, Tagnit-Hamou A. Use of fine glass as ASR inhibitor in glass aggregate mortars. *Construction and Building Materials*. 2010;24(7):1309-1312.
[14] Nassif A. 2002 Postfiring stress-strain hysteresis of concrete subjected to various heating and cooling regimes *Fire and Materials* 26(3):103-109.
[15] Nassif A. 2006 Postfire full stress–strain response of fire-damaged concrete *Fire and Materials* 30(5):323-332.
dos Santos CC, Rodrigues JPC 2016 Calcareous and granite aggregate concretes after fire. *J Build Eng* [Internet]. 8:231–42. *Available from: http://dx.doi.org/10.1016/j.jobe.2016.09.009*
[16] Sawa A, Manita P, Sideris KK. 2005 Influence of elevated temperatures on the mechanical properties of blended cement concretes prepared with limestone and siliceous aggregate *Cement and Concrete Composites* 27(2): 239–248.
[17] Sideris KK. 2007 Mechanical characteristics of self-consolidating concretes exposed to elevated temperatures. *Journal of Materials of Civil Engineering ASCE* 19(8): 648–654.
[18] Poutos K, Alani A, Walden P, Sangha C.2008 Relative temperature changes within concrete made with recycled glass aggregate *Construction and Building Materials* 22(4):557-

[19] To Q.D, He.Q.C, Cossavella M,Morcant K., Panait A., Yvonnet J. The tempering of glass and the failure of tempered glass plates with pin-loaded joints: Modelling and simulation.
[20] Reynolds S. 2015 The Benefits of Glass Tempering and Strengthening. Swift Glass. *Archived from the original on 2017-12-14. Retrieved 2017-12-13*

Thermodynamic investigation of M-cycle assisted open-cycle desiccant air conditioning systems

Hassan Niaz[1], Muhammad Sultan[1], Zahid Mahmood Khan[1], Muhammad Hamid Mahmood[1] and Yasir Niaz[2]

[1]Department of Agricultural Engineering, Faculty of Agricultural Sciences & Technology, Bahauddin Zakariya University, Bosan Road, 60800 Multan, Pakistan
[2]Baluchistan University of Information Technology, Engineering & Management Sciences, Quetta, Pakistan

Email: muhammadsultan@bzu.edu.pk

Abstract. There is a necessity of low cost air conditioning (AC) systems for the agricultural sector of Pakistan e.g. product storage/preservation, greenhouse growing and thermal comfort of animals etc. Solid desiccant air-conditioning (DAC) system can be a handy solution in this regard. The present study gives a detail overview of DAC assisted with the Maisotsenko cycle (M-cycle) system and its applicability for the agriculture sector and livestock applications. Ideal humidity and temperature requirements for the agricultural and livestock applications have been represented on psychometric charts. Comparison between M-cycle assisted DAC and conventional DAC systems has been given which shows the significance of DAC technology in the AC sciences and also represented this difference as graphically. Desiccant AC systems are getting lots of attention in order to control the humidity in various air conditioning applications e.g. product storage, greenhouses and thermal comfort for the livestock. Different materials and arrangements of the desiccant air conditioning system are checked whether which arrangement is suitable for Pakistan situation. Additionally, some analysis has been made to investigate the DAC system assisted with M-cycle and checked the feasibility of system for ambient conditions of Multan. Results show that the DAC system can be low cost heat driven air conditioning system for the human comfort as well as agricultural sector of Pakistan.
Keywords: desiccant; air-conditioning; M-cycle; agriculture; product storage; Pakistan.

1. Introduction

Desiccant air conditioning (DAC) will be a definite solution for air conditioning instead of vapor compression air conditioning (VAC) as it has no effect on global warming (GW) and has zero ozone depletion potential (ODP) due to non-appearance of harmful refrigerants used [1]. Furthermore, individual evaporating cooling systems cannot be used individually for the regions where the controlled temperature as well as relative humidity is required [2, 3]. DAC system can handle the both temperature as well as relative humidity at the same time discretely. Desiccants are used to control relative humidity and evaporative coolers are used for control of temperature and these can be operated at low regeneration temperatures or low grade heat sources [4, 5]. For need of the humidity and temperature control, desiccant air conditioning systems and desiccants are becoming eye catching technology for human as well as non-human thermal comfort mostly in humid and hot climates [1, 4, 6, 7]. System schematics are shown in Figure 1, which show the system with integration of any evaporative cooler (M-cycle) and the system without integration of evaporative cooling system (M-cycle). M-cycle (advance indirect evaporative cooling system) is innovative thermodynamic conception in the field of evaporative cooling systems that can lower down temperature up to dew point [8, 9]. Therefore, a set of equations are used for the analysis of desiccant wheel and performance evaluation of M-cycle which are developed by Sultan et al. and Jurinak et al., respectively [9, 10].

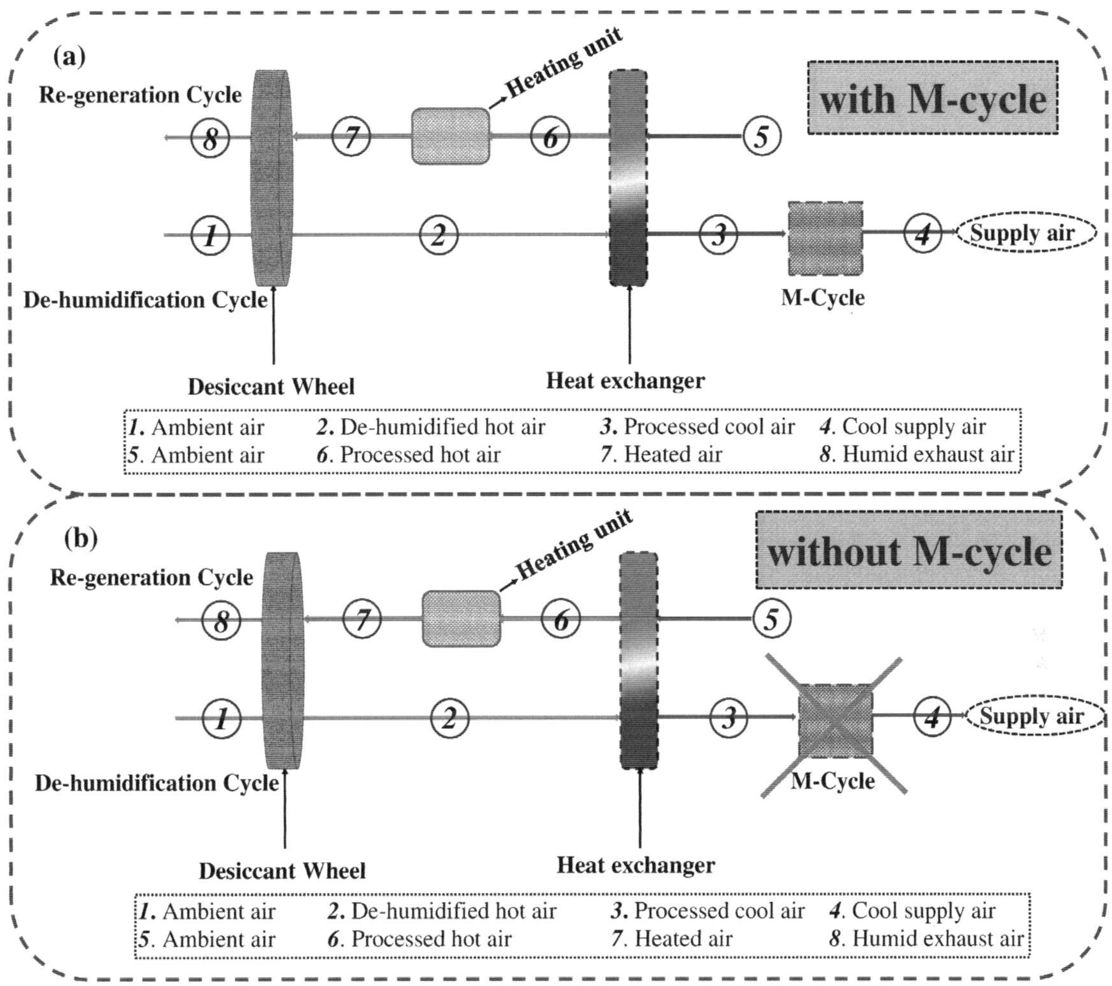

Figure 1: Typical desiccant air conditioning systems: **(a)** Desiccant air conditioning system with integration of M-cycle; **(b)** Desiccant air conditioning system without integration of M-cycle.

2. Research methodology

For the analysis of DAC system, grain type silica gel is used in desiccant wheel. Silica gel is cheap, easily available material and has a high capability of water uptake [2, 7, 11, 12]. Desiccant wheel is evaluated under static conditions. Unit mas flow rate is considered for system analysis. The schematics of the systems have shown in Figure 1. The systems mainly contain a desiccant wheel, flat plate heat exchanger, heating unit and M-cycle unit. The effectiveness of the heat exchanger is considered as 0.90, the outdoor air is passed through the system from 1 to 4 to be air conditioned for required space. Therefore, conditions of air associated with DAC, HX and M-cycle are estimated for same flow rate (unit mass flow rate) using the following equations (1) to (4). Equation (1) and (2) are used for the desiccant wheel performance [7, 12] and equation (3) and (4) are for the heat exchanger and M-cycle performance calculations [9, 13], respectively. Furthermore, COP of the system is calculated by using equation (5) [14].

$$F_{1,ip} = \frac{-2865}{(T_{ip}+273.15)^{1.49}} + 4.344\left(\frac{w_{ip}}{1000}\right)^{0.8624} \qquad (1.1)$$

$$F_{2,ip} = \frac{(T_{ip}+273.15)^{1.49}}{6360} - 1.127\left(\frac{w_{ip}}{1000}\right)^{0.07969} \qquad (1.2)$$

$$\eta_{F1} = \frac{F_{1,2}-F_{1,1}}{F_{1,8}-F_{1,1}} \quad (2.1)$$

$$\eta_{F2} = \frac{F_{2,2}-F_{2,1}}{F_{2,8}-F_{2,1}} \quad (2.2)$$

where F_1 and F_2 are combined potentials, η_{F1}, η_{F2} are the efficiencies in terms of the combined potentials and (η_{F1}, η_{F2}) = (0.05, 0.95) for high performance desiccant wheel and '*ip*' indicates the state of air in the system (1, 2, and 8).

$$T_3 = T_{2,db} - \varepsilon_{hx}\left(T_{2,db} - T_{1,db}\right) \quad (3)$$

where 'ε' is the efficiency of the HX whose value is considered as 0.9

$$T_{out} = 6.70 + 0.2630\ (T_{in}) + 0.5298\ (w_{in}) \quad (4)$$

$$COP = \frac{Cooling\ capacity}{Heat\ input} = \frac{|h_1-h_4|}{|h_7-h_6|} \quad (5.1)$$

Thermal COP can be calculated in terms of enthalpy at different stages.

$$h = 1.006T + w(2501 + 1.86T) \quad (5.2)$$

T is temperature in °C, w is humidity ratio in g/kg-DA.

3. Results and discussions

In this study, performance of desiccant wheel is evaluated for ambient conditions of Multan. Moreover, analysis based on monthly and hourly data as shown in Figure 2 and **Error! Reference source not found.**, respectively. The results show the feasibility of material for air conditioning purpose as relative humidity and temperature are concerned. A very low amount of temperature and relative humidity can be obtained by using both systems discussed in above headings, whereas M-cycle assisted DAC system shows better results in terms of controlled relative humidity and temperature requirement for the air conditioning space. Furthermore, the temperature and relative humidity requirements for fruits, vegetables, buildings, livestock and product's storage can also be obtained by DAC system while the size of the desiccant wheel will be changed [2, 6, 11]. Analysis of monthly data shows that DAC system is valid for tropical regions and also for subtropical regions in winter or monsoon season when the relative humidity is very much high. Whereas analysis of hourly data shows that DAC system works better in night conditions rather than day conditions because of higher relative humidity.

IOP Conf. Series: Materials Science and Engineering **414** (2018) 012002 doi:10.1088/1757-899X/414/1/012002

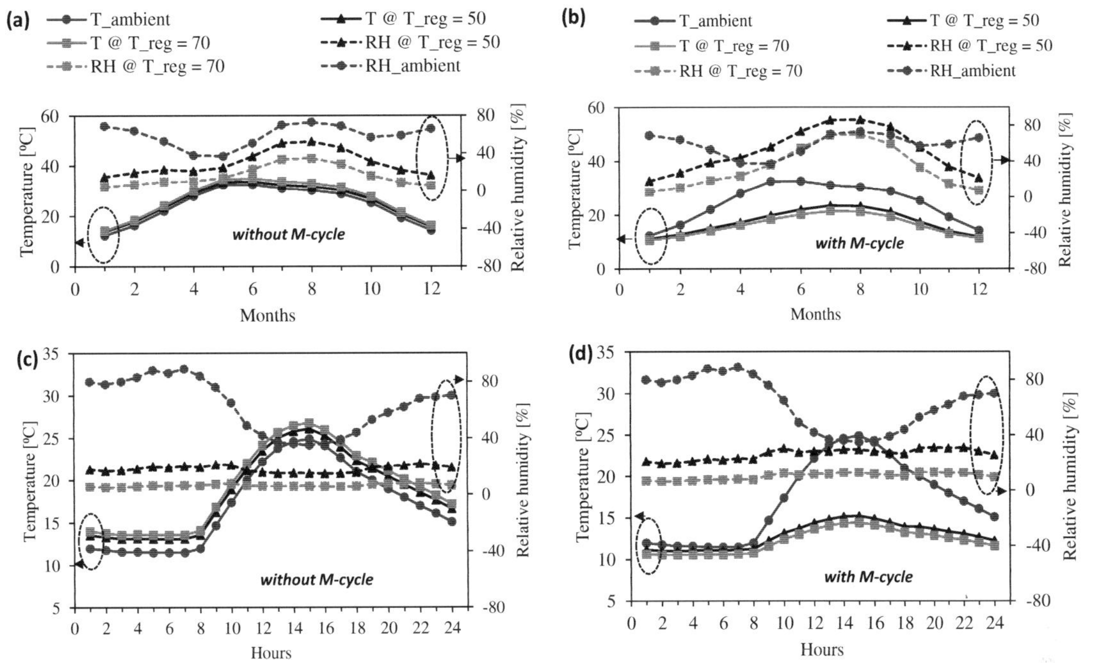

Figure 2: Monthly and hourly data of ambient air conditions which are at T_{reg}=70 °C and T_{reg}=50 °C for Multan: **(a)** monthly data for system without M-cycle; **(b)** monthly data for system with M-cycle; **(c)** hourly data for system without M-cycle and **(d)** hourly data for system with M-cycle.

Data analysis shows that DAC system integrated with M-cycle provides better air conditions than that of system without M-cycle in both monthly conditions and also in hourly conditions as shown in the Figure .

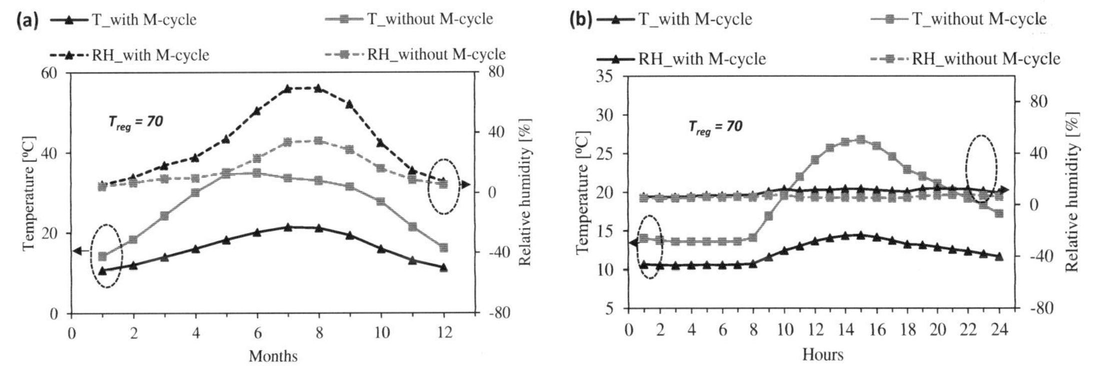

Figure 3: Comparison of systems at T_{reg}=70 °C for Multan condition: (a) comparison on monthly basis and (b) comparison on hourly basis.

Furthermore, there are no. of parameters involved in the performance evaluation of DAC system [2, 4, 5, 11] such as;

IOP Conf. Series: Materials Science and Engineering **414** (2018) 012002 doi:10.1088/1757-899X/414/1/012002

- Air flow rate of supply air and exhaust air.
- Rotational speed of the desiccant wheel.
- The regeneration temperature of the desiccant wheel.

These parameters have a great influence on the performance of the desiccant wheel. If regeneration temperature is increased to a high level, the COP of the system decreased. Results show that how regeneration temperature influences on the COP and how the trend line moves with the temperature variation as shown in the Figure 4.

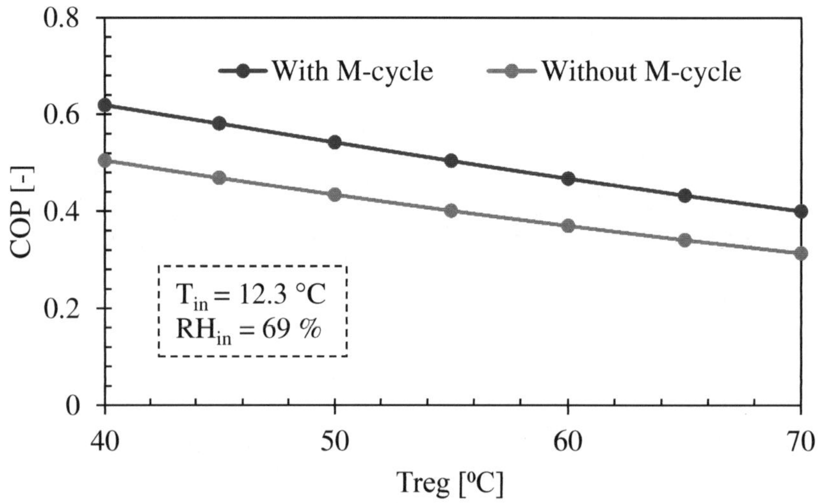

Figure 4: Calculation of COP of the DAC system (with and without integration of M-cycle) by changing the regeneration temperature at specific ambient conditions for the month of January

As the regeneration temperature decreases, the value of COP of the DAC system is increases due to less ambient temperature. Results also shows that the system assisted with M-cycle has better COP than that of the system without the integration of M-cycle. Furthermore, both systems are applicable depends upon the ambient conditions of the space need to be air conditioned. As the regeneration temperature will be increased, the COP of the system will be decreased. So, regeneration temperature depends upon the material properties. Every material has an optimum regeneration value at which the system has maximum COP [2, 4].

4. Conclusions

The aim of this study is to check the feasibility of the desiccant air conditioning system with and without integration of M-cycle. DAC system without integration of M-cycle can be used for the climatic conditions where the temperature is low or controlled relative humidity is required e.g. northern areas in Pakistan, while DAC system with integration of M-cycle can be used where both temperature and relative humidity control is required. A set of equations are used for the analysis and performance evaluation of DAC system as well as M-cycle developed in the literature. Analysis have shown the validity of the systems to satisfy the actual cooling loads for different applications such as; product storage, greenhouses and thermal comfort for the livestock. The air flow rate, regeneration temperatures and the size of the desiccant wheel depends upon the cooling load requirements which vary for different applications and also for different climatic conditions. A considerable working range is achieved by using M-cycle assisted DAC system for different agricultural applications. Furthermore, DAC system can be designed for any specific purpose by using analysis and set of equations used in this study.

References

[1] Kashif M, Sultan M and Khan Z M 2017 Alternative air-conditioning options for developing countries Eur. J. Eng. Res. Sci. **2** 76-79.

[2] Mahmood M H, Sultan M, Miyazaki T and Koyama S 2016 Desiccant air-conditioning system for storage of fruits and vegetables: Pakistan preview Evergreen Joint J. Novel Carbon Res. Sci. Green Asia Strategy **3** 12-17.

[3] Sultan M, El-Sharkawy I I, Miyazaki T, Saha B B and Koyama S 2015 An overview of solid desiccant dehumidification and air conditioning systems Renewable Sustainable Energy Rev. **46** 16-29.

[4] Sultan M and Miyazaki T 2017 Energy-efficient air-conditioning systems for nonhuman applications In Refrigeration. IntechOpen 97-117.

[5] Sultan M, Miyazaki T, Saha B B, Koyama S and Maisotsenko V S 2015 Steady-state analysis on thermally driven adsorption air-conditioning system for agricultural greenhouses. Procedia Eng. **118** 185-192.

[6] Yamaguchi S and Saito K 2013 Numerical and experimental performance analysis of rotary desiccant wheels Int. J. Heat Mass Transfer **60** 51-60.

[7] Panaras G, Mathioulakis E, Belessiotis V and Kyriakis N 2010 Experimental validation of a simplified approach for a desiccant wheel model Energy Build. **42** 1719-1725.

[8] Mahmood M H, Sultan M, Miyazaki T, Koyama S and Maisotsenko V S 2016 Overview of the Maisotsenko cycle–a way towards dew point evaporative cooling Renewable Sustainable Energy Rev. **66** 537-555.

[9] Sultan M, Miyazaki T, Mahmood M H and Khan Z M 2018 Solar assisted evaporative cooling based passive air-conditioning system for agricultural and livestock applications J. Eng. Sci. Technol. **13** 693-703.

[10] Jurinak J J 1983 Open cycle solid desiccant cooling component models and system simulations Ph.D. Thesis.

[11] Sultan M, Miyazaki T and Koyama S 2018 Optimization of adsorption isotherm types for desiccant air-conditioning applications Renewable Energy **121** 441-450.

[12] Panaras G, Mathioulakis E and Belessiotis V 2011 Solid desiccant air-conditioning systems-e design parameters Energy **36** 2399-2406.

[13] Panaras G, Mathioulakis E, Belessiotis V and Kyriakis N 2010 Theoretical and experimental investigation of the performance of a desiccant air-conditioning system Renewable Energy **35** 1368-1375.

[14] Handbook A F 2009 American Society of Heating, Refrigerating and Air-conditioning Engineers. Inc.: Atlanta, GA, USA.

IOP Conf. Series: Materials Science and Engineering **414** (2018) 012003 doi:10.1088/1757-899X/414/1/012003

Fluidized bed heat exchange capacity of Alumina, coal-char and bio-char solids

F Mushtaq[1], I Ullah[1,2], I Khan[1], S K Sami[1], W Alam[1], M Sadiq[1], Z Tariq[1] and A Jaan[1]

[1]Faculty of Engineering and Architecture, Department of Chemical Engineering, Balochistan University of Information Technology, Engineering and Management Sciences (BUITEMS), Airport Road, Baleli, Quetta, Balochistan, Pakistan.
[2]National Engineering Research, Center of Chemical Fertilizer Catalyst, College of Chemical Engineering, Fuzhou University, Fuzhou, PR China.

Email. faisal.mushtaq@buitms.edu.pk

Abstract. Fluidized bed heat exchanger (FBHx) units have significant applications in various industrial processes. Heat transfer coefficient, pressure drop and superficial gas velocity are the most salient parameters used to characterize the behavior of solid particles in FBHx. This research identifies the pressure drop and surface heat transfer coefficient of Alumina, coal-char and bio-char solids in FBHx. The pressure drop and heat transfer coefficient were estimated at fixed bed height of 6cm by maintaining air flow rate of 8-100 Liters per minute (LPM). Digital temperature sensors are used to measure the bed temperature, air inlet temperature, and heater surface temperature. The results of this study suggest that the effects on pressure drop and surface heat transfer coefficient in FBHx are influenced by bed materials. Among the tested materials, pressure drop and heat transfer coefficient were found of order Alumina>coal-char>bio-char.
Keywords. Fluidized bed heat exchanger, pressure drop, superficial gas velocity, surface heat transfer coefficient, carbon solids.

1. Introduction

Fluidization bed units share the characteristics flow of fluid, where gas or liquid moved through bed of solid particles. A fluid like behavior is attained when the force exerted by the liquid or gas on solid particles in the direction of force and buoyant forces minimizes the downward gravitational forces acting on solid particles. Owning to the corresponding motion, fluidization of the solids can be achieved using gas or liquid phase. Fluidized beds are frequently used as reactors where the gas or liquid is contacted with suspended solids usually used as a catalyst [1-3]. Fluidized bed reactors (FBR) are considered valuable source in industrial applications because they offer high rate of mass transfer and less pressure drop [4]. In addition, FBR presents high rate of mixing and capacity to operate in continuous phase [5]. Fluidized bed heat exchange (FBHx) units offers much complicated heat and mass transport mechanism, where the heat carried by the gas or liquid is used to exchange heat to free flowing solid to raise the solids temperature to desired conditions, and later the reaction carried out at the surface of solids eventually transport mass back [6, 7]. FBHx can be categorized into four classes; i) catalytic gas reactor, where the reactants and products are in the gas phase, and the reaction takes places on the solid surface, either the reactant gases are brought to or solids are pre-heated and maintained to desired temperature ii) reaction gases which utilizes solids as a source of heat energy to bring the desired reaction conditions, where the reactants and the products are in the gas phase but solids used as heat carrier source iii) gas-solid combination, where the reactants and products are in the gas and solid phase iv) fluidization with no chemical reaction, where the product gases are dried over solid surface [8].

Gas-solids fluidization regimes follows the sequence stationary bed> bubbling bed> slugging bed> turbulent bed >fast moving bed> pneumatic conveying bed, where stationary bed regime refer to minimum fluidization velocity (U_{mf}) that does not tends to move the solid bed. However, with the increase in superficial gas velocity (U_g), the system approaches bubbling bed regime, where the bubbles initiate mixing of solids [8]. U_{mf} is considered as one of the most significant property when describing the fluidization behavior in FBHx [9]. Several methods described in literature are used to estimate U_{mf} in fluidized bed by; i) heat transfer method, ii) pressure drop method and iii) voidage method. In first method, wall heat transfer coefficient is estimated with the increase in gas velocity and the point is noted where heat transfer coefficient increase significantly [8, 10]. This method of finding U_{mf} is quite costly and requires developed experimental setup [11], where heat transfer data are estimated by maintaining steady state conditions. In the second method, pressure drop over the bed and U_g are plotted and point of alteration between stationary and bubbling bed regime are indentified. In third method, U_{mf} is identified when the free space inside the bed begin to increase the bed extension as the U_g increases. Although voidage method is most generally used, however it is difficult to position out the point where bed extension actually begins. In general, U_{mf} mainly depends on physico-chemical properties of solids, liquid and gas phase and bed geometry [10].

Fluidization of bio-char and coal-char solids produced from various industrial applications provides a prospect to develop and utilize by-product carbon rich solids. Several researcher have found widespread applications of bio-char and coal-char as a source of carbon material, such as metallurgical and mineral processing [12], biomass valorization [13], bio-solid handling [14], soil decontamination [15], carbon catalyzed gas-phase reactions [16]. Among these, gas-phase reactions catalyzed over carbon in FBHx are gaining importance over the recent decade [17-20]. The key concerns of using carbon in FBHx are non-uniform nature and variation in density associated with carbons, such as bio-char and coal-char solids which offers complex multiphase system. The effects of significant properties on FBHx capacity needs to be investigated for improved system performance and design. The primary aim of this study is to investigate the effect of air flow rate on pressure drop, U_{mf} and surface heat transfer coefficient of Alumina, bio-char and coal-char solids.

2. Materials and Methods

The coal samples in the form of lumps were collected from the Machh Bolan Mine Company located at Quetta, Pakistan. Sawdust waste biomass samples were collected from local carpenter shop. The coal and sawdust samples of 3 kg were kept in air tight plastic bags and labeled for further processing. The coal and sawdust samples were subjected to combustion at temperature of 550°C and 350°C, respectively for char production. The maximum combustion temperature of coal and sawdust were selected and based on the studies [21, 22]. The coal-char and bio-char samples were sieved to obtain uniform particle of 14 and 20 mesh size, respectively and were used without physico-chemical modification. Granular fused Alumina (white Aluminum Oxide) type V of density 3770 kg/m^3 and 80 grit size (average particle size 177 μm) and pour density of approximately 1620 kg/m^3 was used as a reference material. SOLTEQ® fluidization and fluid bed heat transfer Unit (Model: HE 162) apparatus was used as shown in Figure 1.

The unit comprised of glass cylinder was carefully loaded with 1.25 kg of tested sample and leveled. The glass chamber is coupled with air distributor chamber that supports the bed material at the bottom. The air distributor chamber provided was designed to keep uniform airflow distribution with minimal pressure drop. The air passes through the bed material and vent to atmosphere through an air filter at the top. The bed temperature was measured by thermocouple labeled as T_1. Compressed air was supplied through filter and pressure regulator assembly and airflow rotameters equipped with low flow rate (V) 6-50 liters per minutes (LPM) and high flow 40-440 LPM through air flow control valves before entering the distribution chamber. The compressed air temperature was measured with thermocouple (T_2). Variable transformer provided with electrical heater power of 250 W and surface area of 22.8 cm^2 was used to control input heater power (Q). The heating element fitted with thermocouple was used to determine the heater surface temperature (labeled as T_3). The pressure drop across the bed at various levels was measured by liquid filled manometer upto 300mm H_2O pressure drop. For excessive pressure in chamber, the chamber is fitted with pressure relief valve.

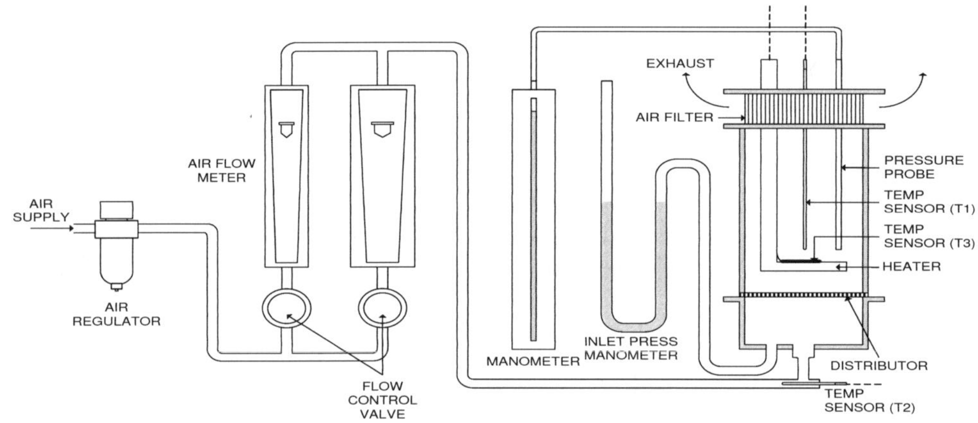

Figure 1. Schematic Diagram of Fluidization and Fluid Bed Heat Exchanger (Model: HE 162) used in this study.

The pressure drop across the bed Δp was measured and is equal to the effective weight of the bed per unit area.

$$\Delta p = M / \rho_p \, Sb \, (\rho_p - \rho_f) \, g \qquad (1)$$

Where mass M of particles of density ρ_p charged in the bed of cross-sectional area Sb fluidized by a fluid of density ρ_f and g gravitational acceleration.

The U_{mf} under ambient conditions was measured by charging 1.25 kg of particles into the chamber to form a bed. Initially, the bed is vigorously fluidized for 1-2 minutes to breakdown particle interlocking. The pressure drop under these conditions was found decreasing as the fluidizing gas velocity was reduced. Several step measurements of the pressure drop across the bed as function of reducing fluidizing gas velocity were noted. The results were plotted as of bed pressure drop and superficial gas velocity. The U_{mf} is derived from the velocity profile, where the intersection between the raising and smoothing of the profile identified. It was expected that the pressure drop tends to have value equal to or less than (because of bed column particle-wall interaction) bed weight per unit area. However, the increase in U_{mf} linearly increases the pressure drop across the bed until the bed approaches fluidization, a condition where gas flow through bed is considered laminar.

The surface heat transfer coefficient is calculated using Equation 2.

$$h = Q / A_h \, \Delta T_m \qquad (2)$$

Where Q is the heater power in W in W/m^2.K, A_h is surface area of the heater in m^2 and ΔT_m is the mean temperature difference in K and calculated from

$$\Delta T_m = \Delta T_1 - \Delta T_2 \qquad (3)$$

Where ΔT_1 is the temperature difference between heater surface (T_3) and air inlet (T_2) and ΔT_2 is the temperature difference between heater surface (T_3) and bed material (T_1).

The following are the details of calculations:

For superficial gas velocity

a. Material= Al_2O_3, mean particle diameter=d_p=177 μm (80grit), particle density= ρ_p=3770kg/m^3, mass of the particles=m=1.25kg, Air flow through bed, V_b= (V/60)x(T_1/T_2) and superficial gas velocity , $U_g=V_b/(1000 \times A)$, where, cross sectional area of the bed= A= 0.008659 m^2

b. Material= Coal-char, mean particle diameter=d_p=1410 μm (14 mesh), particle density= ρ_p=0.0009kg/m^3, mass of the particles=m=1.25kg, Air flow through bed, V_b= (V/60)x(T_1/T_2) and superficial gas velocity , $U_g=V_b/(1000 \times A)$, where, cross sectional area of the bed= A= 0.008659 m^2

For surface heat transfer coefficient

a. Material= Al_2O_3, mean particle diameter=d_p=177μm (80grit), particle density= ρ_p=3770kg/m^3, mass of the particles=m=1.25kg, Air flow through bed, V_b= (V/60)x(T_1/T_2) and superficial gas velocity , $U_g=V_b/(1000 \times A)$, where cross sectional area of the bed= A= 0.008659 m^2, surface heat transfer coefficient=h= $Q/A_h \times \Delta T_m$, $\Delta T_m=\Delta T_1 - \Delta T_2$ (where $\Delta T_1= T_3 - T_2$ and $\Delta T_2= T_3 - T_1$) and A_h= surface heat transfer area of heater= 0.00228 m^2

b. Material= Coal-char, mean particle diameter=d_p=1410 μm (14 mesh), particle density= ρ_p=0.0009kg/m^3, mass of the particles=m=1.25kg, Air flow through bed, V_b= (V/60)x(T_1/T_2) and superficial gas velocity , $U_g=V_b/(1000 \times A)$, where cross sectional area of the bed= A= 0.008659 m^2. Surface heat transfer coefficient=h= $Q/A_h \times \Delta T_m$, $\Delta T_m=\Delta T_1 - \Delta T_2$ (where $\Delta T_1= T_3 - T_2$ and $\Delta T_2= T_3 - T_1$) and A_h= surface heat transfer area of heater= 0.00228 m^2

c. Material= Bio-char, mean particle diameter=d_p=841 μm (20 mesh), particle density= ρ_p=0.0003 kg/m^3, mass of the particles=m=1.25kg, Air flow through bed, V_b= (V/60)x(T_1/T_2) and superficial gas velocity , $U_g=V_b/(1000 \times A)$, where cross sectional area of the bed= A= 0.008659 m^2. Surface heat transfer coefficient=h= $Q/A_h \times \Delta T_m$, $\Delta T_m=\Delta T_1 - \Delta T_2$ (where $\Delta T_1= T_3 - T_2$ and $\Delta T_2= T_3 - T_1$) and A_h= surface heat transfer area of heater= 0.00228 m^2

3. Results and Discussion

3.1. Effects of change in bed height, bed pressure drop and upward air flow velocity of Al_2O_3 granular bed

The results of change in bed height, pressure drop with increasing and decreasing air flow rate is given in Table 1 and Table 2 and redrawn as Figure 2 and Figure 3 as a function of superficial gas velocity, bed height and pressure drop. As can be observed from Table 1 and 2 that pressure drop is directly proportional to the superficial gas velocity. At low air flow rate, drastic increase in pressure drop can be related to lesser penetration of air within the bed material. However, pressure drop was observed fairly constant during later stages of fluidization. This is mainly due to the fact that particles are kept separated ensuing higher free space and increase in viodage. The U_{mf} calculated of 0.04 m/s at 97 mm of H_2O/mm of bed height from Figure 3.

Table 1. Effects bed height, pressure drop and air flow velocity of Al_2O_3 bed with increasing air flow rate.

Air flow rate (*LPM*)	T_1(*K*)	T_2(*K*)	H (*mm*)	V_b (*L/s*)	U_g (*m/s*)	*ΔP/bed height (mm of H_2O/mm)*
8	307.5	306.5	60	0.13	0.015	55.3
12	309.3	306.5	61	0.20	0.023	76.4
16	309.1	306.4	69	0.27	0.031	102.8
20	309.6	306.5	73	0.34	0.039	124.7
24	311.6	306.5	81	0.41	0.047	96.6
30	313.9	306.4	85	0.51	0.059	97.4
40	310.7	307.6	89	0.67	0.078	99.7
60	310.4	306.6	93	1.01	0.117	101.6
80	311.2	306.7	97	1.35	0.156	104.8
100	311.3	307	115	1.69	0.195	108.6

IOP Conf. Series: Materials Science and Engineering **414** (2018) 012003 doi:10.1088/1757-899X/414/1/012003

Table 2. Effects of bed height, pressure drop and air flow velocity of Al_2O_3 bed with decreasing air flow rate.

Air flow rate (*LPM*)	T_1(*K*)	T_2(*K*)	H (*mm*)	V_b *(L/s)*	U_g (*m/s*)	Δ*P/bed height* (*mm of* H_2O/mm)
100	316.6	299.5	80	1.76	0.20	111.4
80	315.4	299.2	73	1.41	0.16	106.6
60	315.9	298.6	70	1.06	0.12	103.8
40	306.5	298.1	68	0.69	0.08	101.2
30	304.2	297.8	65	0.51	0.06	100.5
24	296.9	297.3	62	0.40	0.05	99.8
20	296.3	296.9	60	0.33	0.04	95.2
16	296.1	296.2	59	0.27	0.03	80.2
12	295.9	295.7	59	0.20	0.02	58.5
8	295.4	295.4	59	0.13	0.02	41.8

3.2. Effects of change in bed height, bed pressure drop and upward air flow velocity of coal-char bed

The results of change in bed height, pressure drop with increasing and decreasing air flow rate is given in Table 3 and 4 and redrawn as Figure 4 and Figure 5 as a function of superficial gas velocity, bed height and pressure drop. As can be observed from Table 3 and 4, pressure drop increases with increase in superficial gas velocity at low to moderate air flow rate, which can be related to lesser penetration of air within the compacted bed material. However, once the bed is fluidized, pressure drop was observed fairly constant. The U_{mf} determined was equal to 0.116 m/s from Figure 5. The maximum pressure drop with coal-char bed was observed 33.12 mm of H_2O/mm of bed height, where Al_2O_3 bed of the same mass produced 111.4 mm of H_2O/mm of bed height. U_{mf} was estimated at 0.116 m/s at 21.5 mm of H_2O/mm of bed height pressure drop. This difference can be correlated to large difference is density.

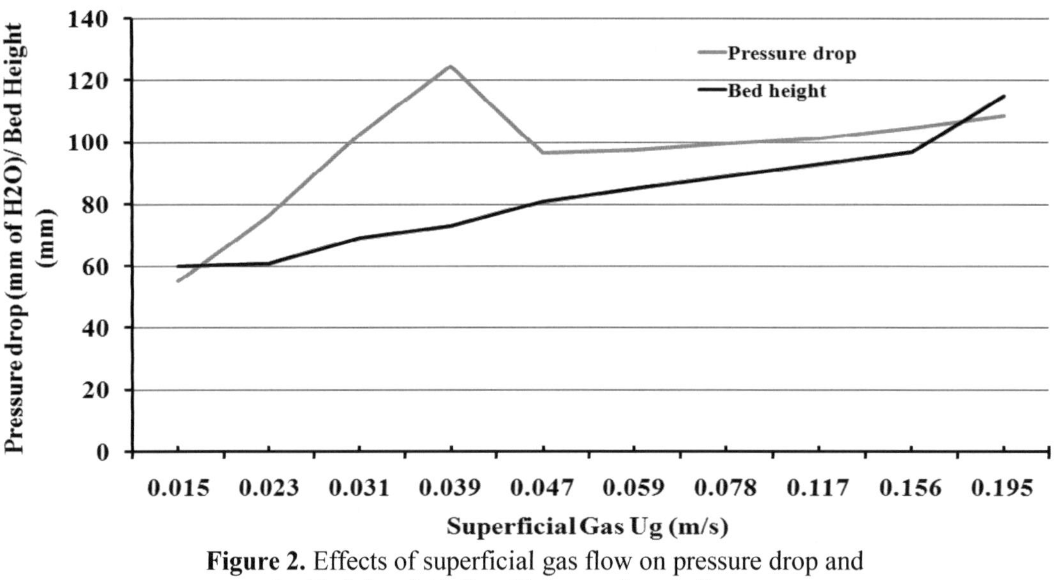

Figure 2. Effects of superficial gas flow on pressure drop and bed height of Al_2O_3 with increasing air flow rate

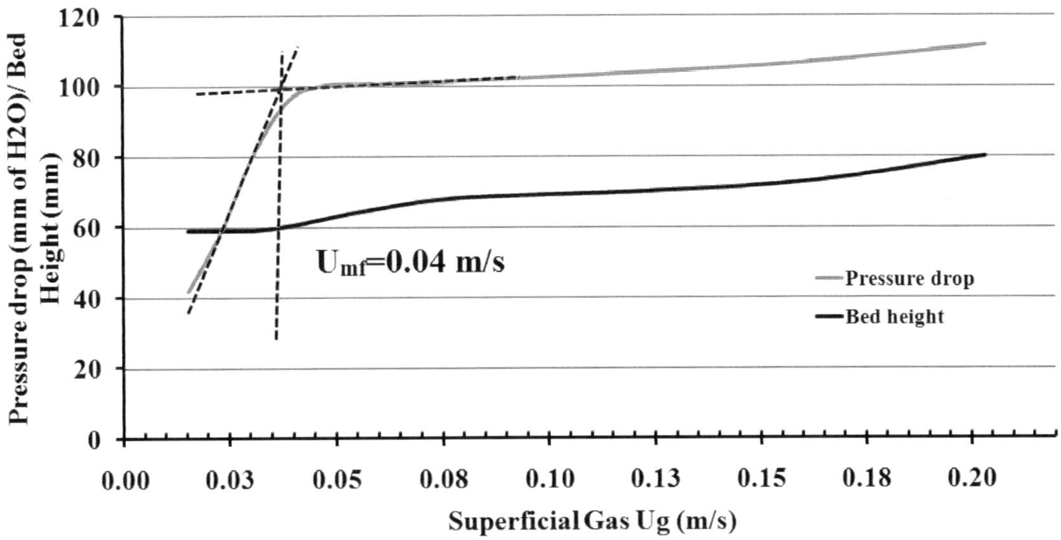

Figure 3. Effects of superficial gas flow on pressure drop and bed height of Al_2O_3 with decreasing air flow rate

Table 3. Effects of change in bed height, pressure drop of coal-char bed with increasing air flow rate.

Air flow rate (*LPM*)	T_1(*K*)	T_2(*K*)	H (*mm*)	V_b *(L/s)*	U_g (*m/s*)	*ΔP/bed height* (*mm of* *H_2O/mm*)
8	307.5	306.5	60	0.13	0.015	7.62
12	309.3	306.5	61	0.20	0.023	25.4
16	309.1	306.4	69	0.26	0.031	32.02
20	309.6	306.5	73	0.33	0.038	30.48
24	311.6	306.5	81	0.39	0.045	31.00
30	313.9	306.4	85	0.49	0.056	38.1
40	310.7	307.6	89	0.66	0.076	33.12
60	310.4	306.6	93	0.99	0.114	33.02
80	311.2	306.7	97	1.31	0.152	22.94
100	311.3	307	115	1.64	0.190	27.94

Table 4. Effects of change in bed height, pressure drop of coal-char bed with decreasing air flow rate.

Air flow rate (*LPM*)	T_1(*K*)	T_2(*K*)	H (*mm*)	V_b *(L/s)*	U_g (*m/s*)	*ΔP/bed height* (*mm of H_2O/mm*)
100	311.3	307	115	1.89019	0.218	27.94
80	311.2	306.7	100	1.64257	0.190	22.94
60	311.2	306.6	93	1.52709	0.176	33.02
40	310.7	307.6	87	1.43553	0.166	33.12
30	313.9	306.4	84	1.36655	0.158	38.1
24	311.6	306.5	81	1.3279	0.153	25.86
20	309.6	306.5	73	1.20448	0.139	30.48
16	309.1	306.4	70	1.15648	0.134	32.02
12	309.3	306.5	62	1.02398	0.118	25.4
8	307.5	306.5	60	0.99675	0.115	7.62

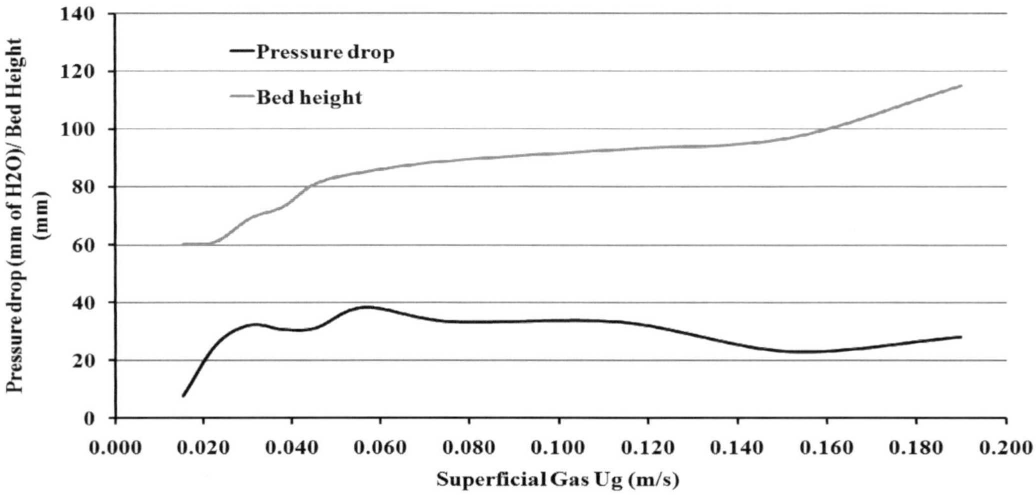

Figure 4. Effects of superficial gas velocity bed height and pressure drop of coal-char with increasing air flow rate

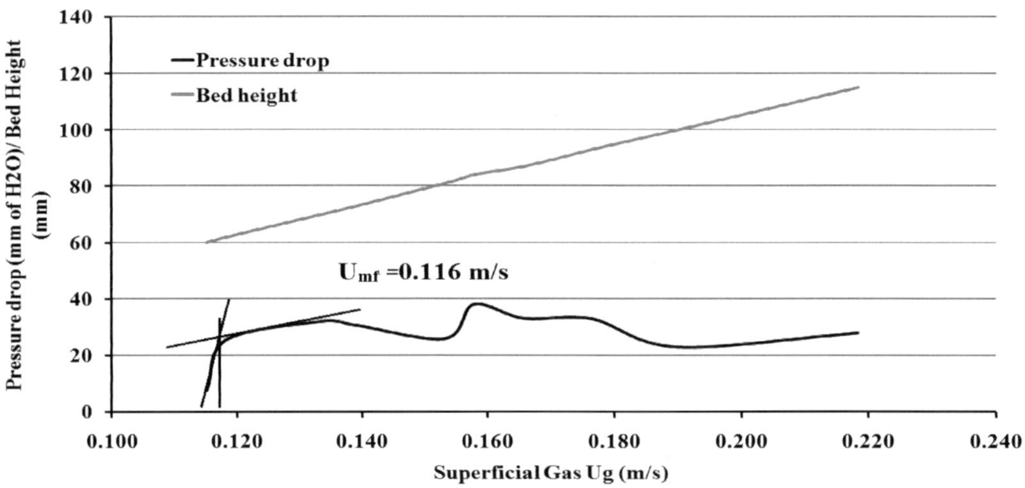

Figure 5. Effects of superficial gas velocity bed height and pressure drop of coal-char with decreasing air flow rate

3.3. Effects of heater power, superficial gas velocity on surface heat transfer coefficient of Al_2O_3, coal-char and bio-char bed

The results of change in heater power and superficial gas velocity on surface heat transfer coefficient of Al2O3, coal-char and bio-char are presented in Table 5, Table 6 and Table 7, respectively and compared in Figure 6. It can be observed that increase in heater power and superficial gas velocity increases the surface heat transfer coefficient with all the tested material. The highest *h* was observed with Al_2O_3 of 717.70 W/m^2.K and lowest with bio-char of 43.76 W/m^2.K. Among the tested material, the surface heat transfer coefficient were observed of order Al_2O_3 >coal-char>bio-char. This difference can be mainly correlated to the difference is physico-chemical properties of the material. The heat exchange capacity in fluidized system of metals and carbon based material can be influenced greatly where the bed is fluidized.

Table 5. Effects of heater power and superficial gas velocity on surface heat transfer coefficient of Al_2O_3

Height of heater (mm)	Height of T_1 Probe (mm)	Air flow rate (LPM)	T_1(K)	T_2(K)	T_3(K)	Heater power Q (W)	V_b (L/s)	Superficial gas Velocity U_g (m/s)	ΔT_m	Surface heat transfer coefficient h (W/m^2.K)
40	40	8	326.4	307.3	356.2	**4.3**	0.142	**0.016**	19.100	98.74
40	40	12	325.7	307.1	356.4	**4.6**	0.212	**0.024**	18.600	108.47
40	40	16	324.4	306.3	356.2	**4.9**	0.282	**0.033**	18.100	118.74
40	40	20	323.2	306.5	356.7	**5.5**	0.351	**0.041**	16.700	144.45
40	40	26	323	307	356.8	**5.7**	0.456	**0.053**	16.000	156.25
40	40	30	321.6	306.4	357	**6.3**	0.525	**0.061**	15.200	181.79
40	40	40	318.3	306.3	356.9	**7.5**	0.693	**0.080**	12.000	274.12
40	40	60	316.5	306.5	357	**9.2**	1.033	**0.119**	10.000	403.51
40	40	80	316.7	306.4	358.5	**11.6**	1.378	**0.159**	10.300	493.95
40	40	100	314.2	306.5	358.3	**12.6**	1.709	**0.197**	7.700	717.70

Table 6. Effects of heater power and superficial gas velocity on surface heat transfer coefficient of coal-char

Height of heater (mm)	Height of T_1 Probe (mm)	Air flow rate (LPM)	T_1(K)	T_2(K)	T_3(K)	Heater power Q (W)	V_b (L/s)	Superficial gas Velocity U_g (m/s)	ΔT_m	Surface heat transfer coefficient h (W/m^2.K)
40	40	8	340.9	305.5	329.8	**4.3**	0.149	**0.017**	35.4	53.28
40	40	12	339.6	305.3	330	**4.6**	0.222	**0.026**	34.3	58.82
40	40	16	338.2	304.5	329.8	**4.9**	0.296	**0.034**	33.7	63.77
40	40	20	338	304.7	330.3	**5.5**	0.370	**0.043**	33.3	72.44
40	40	26	337.8	305.2	330.4	**5.7**	0.480	**0.055**	32.6	76.69
40	40	30	335.4	304.6	330.6	**6.3**	0.551	**0.064**	30.8	89.71
40	40	40	334.1	304.5	329.5	**7.5**	0.731	**0.084**	29.6	111.13
40	40	60	331.3	304.7	330.6	**9.2**	1.087	**0.126**	26.6	151.70
40	40	80	330.5	304.6	334.1	**11.6**	1.447	**0.167**	25.9	196.44
40	40	100	330	306.7	333.9	**12.6**	1.793	**0.207**	23.3	237.18

Table 7. Effects of heater power and superficial gas velocity on surface heat transfer coefficient of bio-char

Height of heater (mm)	Height of T_1 Probe (mm)	Air flow rate (LPM)	T_1(K)	T_2(K)	T_3(K)	Heater power Q (W)	V_b (L/s)	Superficial gas Velocity U_g (m/s)	ΔT_m	Surface heat transfer coefficient h (W/m^2.K)
40	40	8	340.9	305.5	329.8	**4.3**	0.149	**0.017**	35.4	53.28
40	40	12	339.6	305.3	330	**4.6**	0.222	**0.026**	34.3	58.82
40	40	16	338.2	304.5	329.8	**4.9**	0.296	**0.034**	33.7	63.77
40	40	20	338	304.7	330.3	**5.5**	0.370	**0.043**	33.3	72.44
40	40	26	337.8	305.2	330.4	**5.7**	0.480	**0.055**	32.6	76.69
40	40	30	335.4	304.6	330.6	**6.3**	0.551	**0.064**	30.8	89.71
40	40	40	334.1	304.5	329.5	**7.5**	0.731	**0.084**	29.6	111.13
40	40	60	331.3	304.7	330.6	**9.2**	1.087	**0.126**	26.6	151.70
40	40	80	330.5	304.6	334.1	**11.6**	1.447	**0.167**	25.9	196.44
40	40	100	330	306.7	333.9	**12.6**	1.793	**0.207**	23.3	237.18

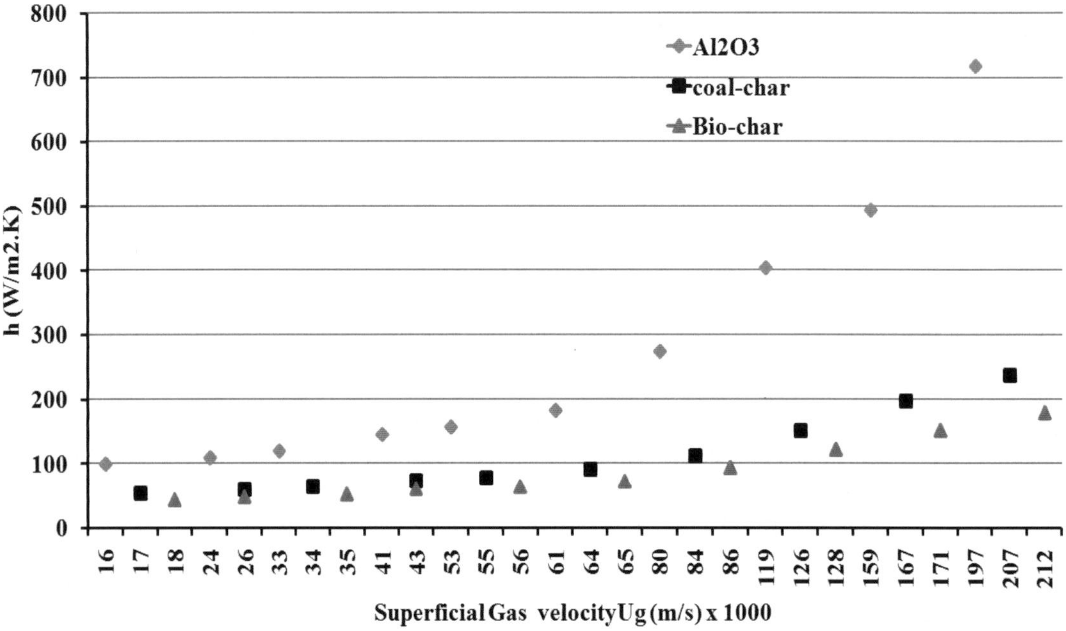

Figure 6. Comparison of surface heat transfer coefficient of the tested material

4. Conclusion

This study investigated the effects of change in bed height, bed pressure drop and upward air flow velocity on superficial gas velocity and fluidization behavior. Further, the effect of superficial gas velocity on surface heat transfer coefficient of Alumina, coal-char and bio-char were tested. The results of this study suggests that pressure drop increases with increase in superficial gas velocity at low to moderate air flow rate, which can be correlated to lesser penetration of air within the compacted bed material. However, once the bed is fluidized, pressure drop was observed fairly constant which is mainly due to the increase in free viodage space. In addition, increase in heater power and superficial gas velocity increases the surface heat transfer coefficient with all the tested material. Among the material tested, the surface heat transfer coefficient were observed of order Al_2O_3 >coal-char>bio-char. The intermediate heat exchange properties of coal-char suggests potential application as slow raise in bed solid temperature and controls over final process temperature when tested as catalyst in bench scale fluidized bed reactor.

Acknowledgement

Authors gratefully acknowledge the support of Balochistan University of Information Technology, Engineering and Management Sciences (BUITEMS), Quetta, Balochistan, Pakistan.

References

[1] Aho A, Kumar N, Eränen K, Salmi T, Hupa M and Murzin D Y 2007 Catalytic pyrolysis of biomass in a fluidized bed reactor: influence of the acidity of H-beta zeolite *Process Safety and Environmental Protection* **85** 473-80

[2] Asadullah M, Ito S-i, Kunimori K, Yamada M and Tomishige K 2002 Biomass gasification to hydrogen and syngas at low temperature: novel catalytic system using fluidized-bed reactor *Journal of Catalysis* **208** 255-9

[3] Zhang H, Xiao R, Huang H and Xiao G 2009 Comparison of non-catalytic and catalytic fast pyrolysis of corncob in a fluidized bed reactor *Bioresource Technology* **100** 1428-34

[4] Karmakar M K, Loha C, De S and Chatterjee P K 2018 *Coal and Biomass Gasification*: Springer) pp 93-114
[5] Natarajan E, Nordin A and Rao A 1998 Overview of combustion and gasification of rice husk in fluidized bed reactors *Biomass and bioenergy* **14** 533-46
[6] Kim S D and Kang Y 1997 Heat and mass transfer in three-phase fluidized-bed reactors—an overview *Chemical Engineering Science* **52** 3639-60
[7] Behjat Y, Shahhosseini S and Hashemabadi S H 2008 CFD modeling of hydrodynamic and heat transfer in fluidized bed reactors *International Communications in Heat and Mass Transfer* **35** 357-68
[8] Kunii D and Levenspiel O 2013 *Fluidization engineering*: Elsevier
[9] Molerus O and Wirth K-E 2012 *Heat transfer in fluidized beds* vol 11: Springer Science & Business Media
[10] Wen C and Yu Y 1966 A generalized method for predicting the minimum fluidization velocity *AIChE Journal* **12** 610-2
[11] Gupta C K and Sathiyamoorthy D 1998 *Fluid bed technology in materials processing*: CRC press
[12] Ma Z, Bai J, Li W, Bai Z and Kong L 2013 Mineral transformation in char and its effect on coal char gasification reactivity at high temperatures, Part 1: Mineral transformation in char *Energy & Fuels* **27** 4545-54
[13] Trivedi N S, Mandavgane S A and Chaurasia A 2018 Characterization and valorization of biomass char: a comparison with biomass ash *Environmental Science and Pollution Research* **25** 3458-67
[14] Pang S 2016 *Fuel Flexible Energy Generation*: Elsevier pp 241-69
[15] Menéndez J, Arenillas A, Fidalgo B, Fernández Y, Zubizarreta L, Calvo E G and Bermúdez J M 2010 Microwave heating processes involving carbon materials *Fuel Processing Technology* **91** 1-8
[16] Ma Z, Bai J, Bai Z, Kong L, Guo Z, Yan J and Li W 2014 Mineral transformation in char and its effect on coal char gasification reactivity at high temperatures, part 2: char gasification *Energy & Fuels* **28** 1846-53
[17] Sun L, Luo K and Fan J 2018 Production of synthetic natural gas by CO methanation over Ni/Al2O3 catalyst in fluidized bed reactor *Catalysis Communications* **105** 37-42
[18] Carnevali D, Guévremont O, RIgamonti M G, Stucchi M, Cavani F and Patience G S 2018 Gas-phase fructose conversion to furfural in micro-fluidized bed reactor *ACS Sustainable Chemistry & Engineering* **6** 5580-7
[19] Mattisson T, Keller M, Linderholm C, Moldenhauer P, Rydén M, Leion H and Lyngfelt A 2018 Chemical-looping technologies using circulating fluidized bed systems: Status of development *Fuel Processing Technology* **172** 1-12
[20] Zhu Y, Chen K, Yi C, Mitra S and Barat R 2018 Dry reforming of methane over palladium–platinum on carbon nanotube catalyst *Chemical Engineering Communications* **7** 1-9
[21] Aqsha A, Mahinpey N, Mani T, Salak F and Murugan P 2011 Study of sawdust pyrolysis and its devolatilisation kinetics *The Canadian Journal of Chemical Engineering* **89** 1451-7
[22] Mushtaq F, Qadeer A, Mir F and Aftab A 2012 Coal fired power generation potential of Balochistan *Petroleum & Coal* **54** 132-42

IOP Conf. Series: Materials Science and Engineering **414** (2018) 012004 doi:10.1088/1757-899X/414/1/012004

Investigation of carbon based adsorbents for the development of thermally-driven adsorption cooling systems

Faizan Shabir[1], Muhammad Sultan[1], Redmond R. Shamshiri[2], Zahid M. Khan[1]

[1]Department of Agricultural Engineering, Bahauddin Zakariya University, Bosan Road, Multan 60800, Pakistan
[2]Adaptive Agro-Tech Research Group International, 948 Via Verde, Del Rey oaks, California. 93955. USA

Email: muhammadsultan@bzu.edu.pk

Abstract. Adsorption cooling systems are considered as energy efficient and sustainable technologies from the perspective of environmental safety and thermal energy utilization. These systems possess zero potentials of ozone depletion and global warming. In adsorption cooling processes, knowledge of adsorbent-refrigerant pairs (e.g. adsorption equilibrium, kinetics and heat) is important. The system performance is directly related to interactions between the adsorbent and refrigerant. Thus, the overall thermodynamic performance of the system can be improved accordingly. In this study, numerous carbon based adsorbents are explored in detail with different types of refrigerants (e.g. ethanol, methanol, CO_2, R134A etc.). In order to select the optimum adsorbent-refrigerant pair. The analyses in the study are based on the experimental data on various adsorbent-refrigerant pairs available in the literature. Various adsorption isotherms models including: Dubinin-Astakhov, Tóth, Freundlich etc. present adsorption equilibrium data. Consequently, overall system analyses have been conducted by means of pressure-temperature-adsorption equilibrium (P-T-W) diagram. The P-T-W diagram is also drawn from the ideal cycle analysis in order to explain the performance of adsorption cooling systems. The coefficient of performance of the system has been calculated accordingly for the studied adsorbent-refrigerant pairs.
Keyword: adsorption cooling; activated carbon; refrigerant; optimization

1. Introduction

Thermally driven adsorption cooling systems are gaining much attention for the reason that, these systems are energy efficient and environmentally friendly. As the driving source of these systems could be solar energy or low grade waste heat [1]. Therefore, have a huge potential for adoption in those regions where there are more number of solar hours per year. Cooling load can be reduced in summer by utilizing adsorption cooling systems, where the vapor compression systems (VCS) cost a huge amount of energy. Also, thermally driven systems cause less deterioration of the ecosystem. As the refrigerant, they utilize (water, ethanol, CO2 and HFCs etc.) gives a low threat for ozone depletion in contrast to the conventional refrigerants (CFCs) of VCS. Adsorbent-refrigerant pairs for the development of adsorption cooling systems are studied experimentally by many researchers. Some of them are activated carbon-CO_2 [2-3], activated carbon-ammonia [4], activated carbon fiber-ethanol [5], activated carbon-R134a [6], Silica gel-water [7-8], Zeolite-water [9], MOF-ethanol [10] and polymer-water [11-12]. Apart of cooling adsorption system can be utilized for other applications like water desalination [13], carbon capturing [14] and ice production [15] etc.

In this study, two carbon based adsorbent-ethanol pairs (Supersorbon HS4-ethanol and Dezorex DB1-ethanol) are selected in order to analyze their potential for the development of adsorption cooling system. Thermodynamic performance of cooling cycle is evaluated through the ideal cycle analysis. Hence Specific cooling energy (SCE) and coefficient of performance (COP) are evaluated for both assorted pairs.

IOP Conf. Series: Materials Science and Engineering **414** (2018) 012004 doi:10.1088/1757-899X/414/1/012004

2. Working of adsorption cooling system

A typical adsorption cooling system which clearly indicates that these systems are like vapor compression cooling system (VCS) except the compressor part is presented in Figure 1. In vapor compression systems, the compressor is electric driven, thus incurs high input cost for cooling and air conditioning. While in adsorption cooling system prime mover is the thermal compressor, therefore, the system economics are totally depending on low-cost thermal energy source. Thermal compressor usually consists of two beds packed with adsorbent material named as (i) adsorption bed and (ii) desorption bed. In adsorption cooling cycle, low-pressure refrigerant from the evaporator enters the adsorption bed, where the refrigerant gets adsorbed on to the adsorbent at a pressure equal to evaporator pressure. The exothermic nature of adsorption process causes adsorbent to release heat during the adsorption process. The heat of adsorption is removed by cooling water cycle around the adsorption bed. When the adsorbent becomes fully saturated with refrigerant the bed is switched to desorption mode by heating bed through the hot water cycle. At the end of desorption process, high-temperature and high-pressure refrigerant vapors will move toward the condenser. In the condenser, the temperature of refrigerant gets reduced by exchanging the heat through cooling water cycle. Also, the change in phase will occur, vapor refrigerant will be converted into liquid refrigerant. Hereafter it will pass through an expansion valve, where the drop-in pressure gives low pressure and low-temperature liquid refrigerant. That low-pressure refrigerant offers the required cooling effect, which will exchange by the evaporator and converted into the vapor phase. For continuous operation of adsorption cooling system, the adsorption and desorption beds are switched, after a specified switching time, by changing the cooling and heating water supplies through the valve accordingly.

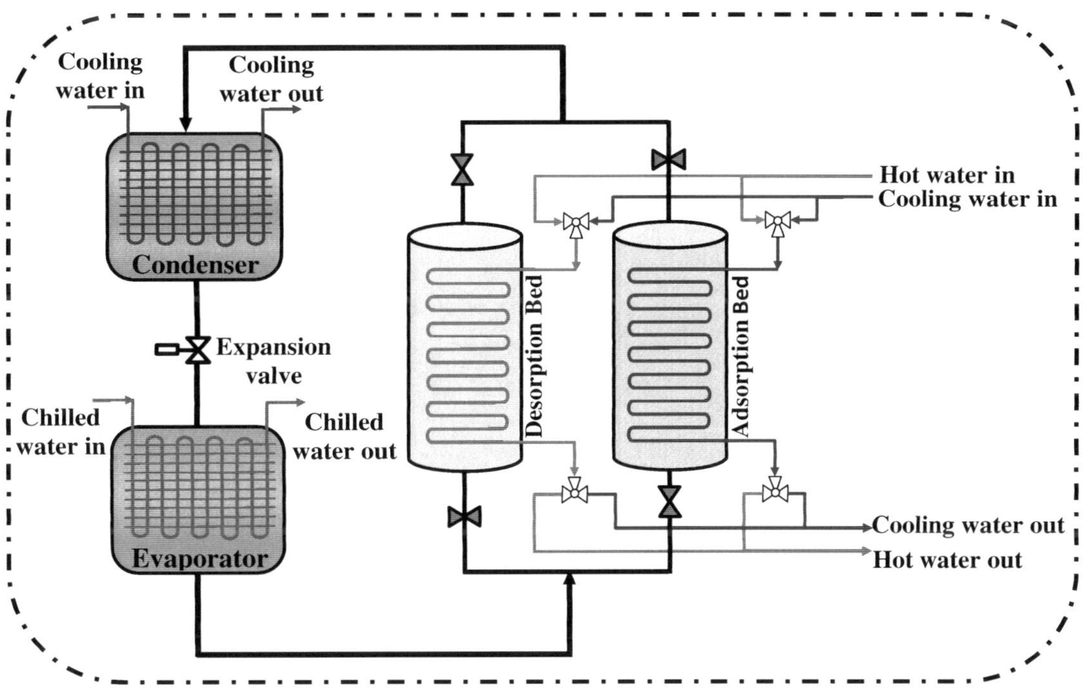

Figure 1. Schematic of a typical adsorption cooling system.

3. Thermodynamic evaluation of adsorption cooling cycle

Adsorption isotherms of the two assorted pairs Supersorbon HS4-ethanol and Dezorex DB1-ethanol are reproduced by using Dubinin–Astakhov (D–A) adsorption equilibrium model (Eq.1) [16]. Values of adsorption isotherm parameters are listed in. The equation of the D-A isotherm model is given as:

$$W = W_0 \exp\left\{-\left(\frac{A}{E}\right)^n\right\} \tag{1}$$

Where W, W_0, E, n and A are the equilibrium adsorption uptake, maximum adsorption uptake, Characteristic energy, structural heterogeneity parameter and adsorption potential, respectively. Adsorption potential of the absorbent-refrigerant pair can be evaluated as:

$$A = RT \ln\left(\frac{P_s}{P}\right) \tag{2}$$

Where P_s denotes the saturated pressure of refrigerant for a given temperature, P is the equilibrium pressure, T is the adsorption temperature and R is the gas constant.

Table 1. D-A adsorption isotherm model parameters

Adsorbent-Refrigerant Pair	D-A model Parameters			Reference
	W_0 [cm^3/kg]	E [kJ/mol]	n	
Supersorbon HS4-ethanol	426	8	2.4	[17]
Dezorex DB1-ethanol	508	7.8	1.2	[17]

Adsorption isotherms are outlined for different adsorption temperatures from 20-100°C with the difference of 10°C. The ideal adsorption Cooling cycle is traced along with these isotherms in Figure 2 for (a) Supersorbon HS4-ethanol and (b) Dezorex DB1-ethanol. The operating conditions for the ideal cycle are fixed at a regeneration temperature of 100°C. While evaporator and adsorption temperatures are set at 5°C and 30°C respectively.

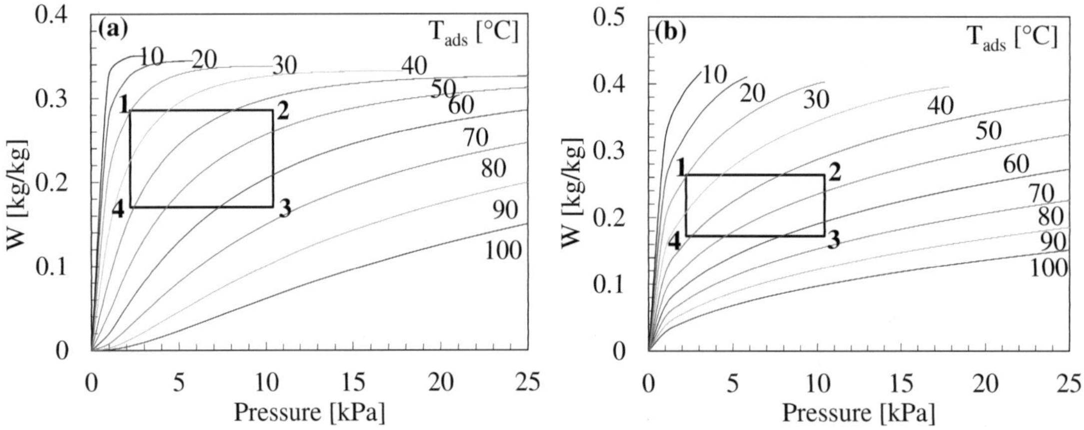

Figure 2. Ethanol adsorption isotherms produced by D-A adsorption isotherm model for: (a) Supersorbon HS4, and (b) Dezorex DB1, reproduced from [17].

Performance of ideal cooling cycle can also be evaluated by dühring diagram or pressure-temperature-concentration (P-T-W) diagram. P-T-W diagram of adsorbent-refrigerant pair expresses a thermodynamic relation between equilibrium pressure, adsorbent temperature and equilibrium adsorbed concentration of refrigerant. Isosteric lines in the P-T-W lines are produced by simplifying the D-A isotherm equation (Eq.1) in the following equations;

$$\ln\left(\frac{W}{W_0}\right) = -\left(\frac{A}{E}\right)^n \tag{3}$$

$$A = E\left(-\ln\left(\frac{W}{W_0}\right)\right)^{\frac{1}{n}} \tag{4}$$

$$RT\ln\left(\frac{P_S}{P}\right) = E\left(-\ln\left(\frac{W}{W_0}\right)\right)^{\frac{1}{n}} \tag{5}$$

$$P = P_s/\exp\left[\frac{E}{RT}\left(-\ln\left(\frac{W}{W_0}\right)\right)^{\frac{1}{n}}\right] \tag{6}$$

The adsorption cooling cycle consists of four consecutive processes 1-2 an isosteric heating process (pre-heating), 2-3 isobaric heating (desorption), 3-4 isosteric cooling process (precooling) and 4-1 isobaric cooling process (adsorption) Figure 3 [18]. P-T-W diagram of Supersorbon HS4-ethanol and Dezorex DB1-ethanol are shown in Figure 3 (a) and (b), respectively. The ideal cycle of both pairs is constructed for regeneration, evaporator and condenser temperatures of 100°C, 5°C and 30°C, respectively. To evaluate the performance of ideal cycle a thermodynamic model was employed, which is time independent [19]. Thus SCE can be defined as:

$$SCE = (W_{max} - W_{min})\left[LH_{T_e} - \int_{T_e}^{T_c} C_{P_{ref}}\, dT\right] \tag{7}$$

Whereby W_{max} is the maximum uptake of refrigerant evaluated for evaporation pressure and adsorption temperature and W_{min} is minimum uptake measured corresponding to the desorption temperature and condenser pressure. LH_{Te} is the vaporization enthalpy at evaporator temperature and $C_{P_{ref}}$ shows the specific heat capacity of refrigerant. Heat added to the adsorbent for the increase in temperature from T_1 and T_3. is given as:

$$Q_{ads} = \int_{T_1}^{T_3} C_{P_{ads}}\, dT \tag{8}$$

Where $C_{P_{ads}}$ is the specific heat capacity of adsorbent. Heat added to the refrigerant is given by:

$$Q_{ref} = W_{max}\int_{T_1}^{T_2} C_{P_{ref}}\, dT + \int_{T_2}^{T_3} W.C_{P_{ref}}\, dT + \int_{W_{min}}^{W_{max}} q_{st}\, dW \tag{9}$$

In Eq. (9), the first and second terms are sensible heats added to the refrigerant during pre-heating and desorption processes, respectively. Whilst third term is the heat of adsorption. Here the average values of q_{st} for Supersorbon HS4-ethanol and Dezorex DB1 ethanol Pairs are used 1126.05 kJ/kg and 1038.68 kJ/kg, respectively [17].

Thus, COP of the cycle can be described as:

$$COP = \frac{SCE}{Q_{ads}+Q_{ref}} \tag{10}$$

SCE and COP as a function of desorption temperature are found to increase with the increase in desorption temperature. SCP reach the maximum value of 180.1 and 128.8 kJ/kg for Supersorbon HS4-ethanol and Dezorex DB1-ethanol Pairs, respectively, at a regeneration temperature of 100°C. Minimum desorption temperature was found to be 58°C for corresponding evaporator temperature of 5°C. COP of Dezorex DB1-ethanol and Supersorbon HS4-ethanol pair, for the regeneration temperature range of 80-

IOP Conf. Series: Materials Science and Engineering **414** (2018) 012004 doi:10.1088/1757-899X/414/1/012004

0°C, was about 0.57 and 0.64, respectively. Therefore, these adsorbents could be an option for the utilization of low-grade waste heat.

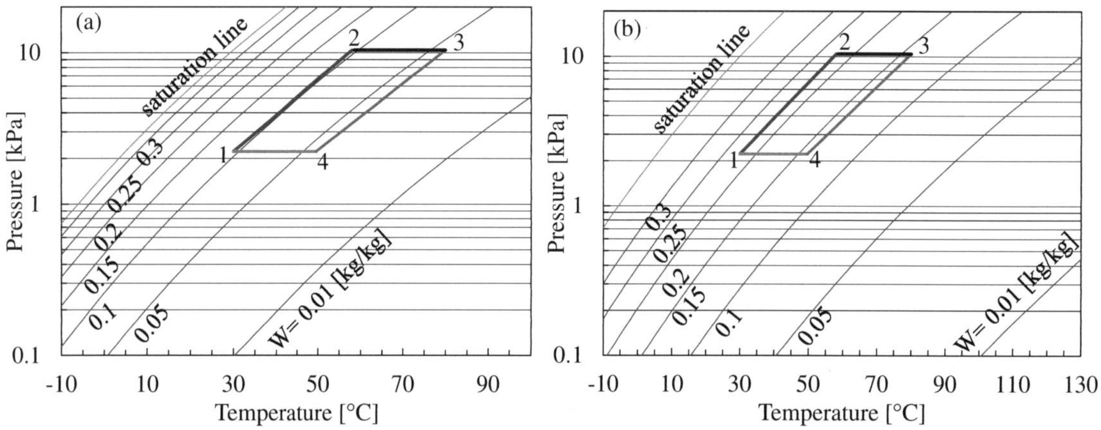

Figure 3. P-T-W diagram of (a) Supersorbon HS4-ethanol and (b) Dezorex DB1-ethanol Pairs

4. Conclusions

Adsorption isotherms of two assorted pairs reveal the higher adsorption uptakes for the Dezorex DB1-ethanol Pair as compared to Supersorbon HS4-ethanol pair. Ideal cycle analysis of adsorption cooling system of both activated carbon-ethanol pair is led at an evaporator temperature of 5°C and regeneration temperature of 100°C. It shows a higher specific cooling effect for Supersorbon HS4 is180.1, while Dezorex DB1 give relatively lower values for SCE of 128.8. Similarly, the coefficient of performance of Dezorex DB1-ethanol and Supersorbon HS4-ethanol pairs are 0.57 and 0.64, respectively, for the regeneration temperature ranging from 80°C to 90°C. At an evaporator temperature of 5°C, minimum desorption temperature was found i.e. 58°C, which proves the applicability of these adsorbents for thermally driven adsorption cooling systems.

Nomenclature

VCS	Vapor compression systems
W	Equilibrium adsorption uptake [kg/kg]
W_0	Maxium adsorption uptake [kg/kg]
E	Chahracteristic energy [kJ/mol]
n	Structural heterogenity parameter [-]
A	Adsorption potential [kJ/kg]
P_s	Saturated pressure of refrigerant for given temperature [kPa]
P	Equilibrium pressure [kPa]
T	Adsorption temperature [°C]
R	Gas constant.[kJ/kg.K]
LH	Vaporization enthalpy [kJ/kg]
C_p	Specific heat capacity [kJ/kg.K]
Q	Heat energy [kJ/kg]
q_{st}	Isosteric heat of adsorption [kJ/kg]
SCE	Specific cooling effect [kJ/kg]
COP	Coefficient of performance [-]

Subscripts

max	Maximum
min	Minimum
c	Condenser
e	Evaporator
ref	Refrigerant
ads	Adsorption

References

[1] Wang R Z and Oliveira R G 2006 Adsorption refrigeration-an efficient way to make good use of waste heat and solar energy *Prog. Energy. Combust. Sci.* **32** 424-458
[2] Saha B B, Jribi S, Koyama S and El-Sharkawy I I 2011 Carbon dioxide adsorption isotherms on activated carbons J. Chem. Eng. Data. 56 1974-81
[3] Jribi S, Miyazaki T, Saha B B, Pal A, Younes M M, Koyama S and Maalej A 2017 Equilibrium and kinetics of CO2 adsorption onto activated carbon *Int. J. Heat. Mass. Transfer.***108** 1941-6
[4] Tamainot-Telto Z, Metcalf S J, Critoph R E, Zhong Y and Thorpe R 2009 Carbon–ammonia pairs for adsorption refrigeration applications: ice making, air conditioning and heat pumping *Int. J. Refrig.* **32** 1212-29
[5] El-Sharkawy I I, Kuwahara K, Saha B B, Koyama S and Ng K C 2006 Experimental investigation of activated carbon fibers/ethanol pairs for adsorption cooling system application *Appl. Therm. Eng.* **26** 859-65
[6] Saha B B, Habib K, El-Sharkawy I I and Koyama S 2009 Adsorption characteristics and heat of adsorption measurements of R-134a on activated carbon *Int J. Refrig.* **32** 1563-9
[7] Chua H T, Ng K C, Chakraborty A, Oo N M and Othman M A 2002 Adsorption characteristics of silica gel+ water systems *J. Chem. Eng. Data.* **47** 1177-81
[8] Sultan M, Miyazaki T, Saha B B and Koyama S 2016 Steady-state investigation of water vapor adsorption for thermally driven adsorption based greenhouse air-conditioning system *Renewable. Energy.* **86** 785-95
[9] Wang D C, Xia Z Z and Wu J Y 2006 Design and performance prediction of a novel zeolite–water adsorption air conditioner *Ener. Convers. Manage.* **47** 590-610
[10] Saha B B, El-Sharkawy I I, Miyazaki T, Koyama S, Henninger S K, Herbst A and Janiak C 2015 Ethanol adsorption onto metal organic framework: Theory and experiments *Energy.* **79** 363-70
[11] Sultan M, El-Sharkawy I I, Miyazaki T, Saha B B, Koyama S, Maruyama T, Maeda S and Nakamura T 2015 Insights of water vapor sorption onto polymer based sorbents *Adsorpt.* **21** 205-15
[12] Sultan M, El-Sharkawy I I, Miyazaki T, Saha B B, Koyama S, Maruyama T, Maeda S and Nakamura T 2016 Water vapor sorption kinetics of polymer based sorbents: Theory and experiments *Appl. Therm. Eng.* **106** 192-202
[13] Ng K C, Thu K, Kim Y, Chakraborty A and Amy G 2013 Adsorption desalination: an emerging low-cost thermal desalination method *Desalin.* **308** 161-179
[14] Sah R P, Choudhury B and Das R K 2016 A review on low grade heat powered adsorption cooling systems for ice production *Renewable. Sustainable. Energy. Rev.* **62** 109-120.
[15] Yu C H, Huang C H and Tan C S 2012 A review of CO_2 capture by absorption and adsorption *Aerosol. Air. Qual. Res.* **12** 745-769
[16] Dubinin M M 1967 Adsorption in micropores 1967 *J. Colloid. Interface. Sci.* **23** 487-499
[17] Besedová E and Bobok D 2003 Adsorption of ethanol on activated carbon. *Chem. Pap.* **57** 27-34
[18] Askalany A A, Salem M, Ismael I M, Ali A H H, Morsy M G and Saha B B 2013 An overview on adsorption pairs for cooling *Renew. Sustain. Energy. Rev.* **19** 565–572.
[19] El-Sharkawy I I, Hassan M, Saha B B, Koyama S and Nasr M M 2009 Study on adsorption of methanol onto carbon based adsorbents. *Int. J. Refrig.* **32** 1579-1586

Effect of Brick Dust on Strength and Workability of Concrete

Muhammad Nasir Ayaz Khan[1], Nabeel Liaqat [2], Ibrar Ahmed [1], Abdul Basit [3], Muhammad Umar [4] and Muhammad Aftab Khan[2]

[1]University of Engineering and Technology Taxila, Pakistan
[2]National University of Science and Technology, Pakistan
[3]Swedish College of Engineering & Technology Wah Cantt, Pakistan
[4] Military College of Engineering Risalpur, Pakistan

Email: Engr.khannasir@gmail.com

Abstract. The aim of this research is to utilize the waste generated in concrete. Brick dust is lavish material which on dumping not only occupy land but also it has environmental problems which is hazardous to livings. This waste is generated in brick kilns, brick masonry construction sites and during transportation. By recycling brick dust the problem could be solved up to some extent. In this research, brick dust was used in plain cement concrete to check its fresh and hardened properties. Brick dust was used to check the workability and strength of concrete, using the water-cement ratio of 0.55 which was kept constant during research. Three samples were cast for each 3, 7, 14, 28 and 56 days with 0%, 5%, 10% 15% and 20% incorporation of brick dust. The test results reveals that replacing cement with brick dust shows higher workability than control sample for 5%, 10% and 15%, of which 15% was greater of all. However, strength results were quite competitive, replacing cement with 15% brick dust shows higher compressive strength. The split tensile test were also conducted, which shows high tensile strength by replacing cement with 15% brick dust. This research shows that cement can be replaced with brick dust.
Keywords: brick dust, compressive strength, workability, split tensile

1. Introduction:

Brick dust is lavish material which generates as waste in brick kilns and construction sites. This material goes into dumping and used as landfills which concerns our environment. Researchers around the globe use many innovative and waste materials in concrete to overcome the environmental and economic problems. These waste materials give higher or almost same concrete properties.

A report [1] stated that brick kilns generates most of the brick dust wastes where largest brick producing country is China and India lies second in list. The report further state that there are more than 2, 20,000 number of bricks unit in both countries with 80,000 brick units alone in China. Ahmed [2] concluded that in Pakistan, there are approximately 10,000 brick units which contributes about 1.5% GDP to country. These brick Kilns generates brick wastes in form of brick dust which not occupy lands but also hazardous to health and environment. Cement is most costly material in concrete, so its economical use should be adopted. In this way, many innovative and waste materials has been used by researchers to overcome the cost of concrete. These materials include fly ash, silica fume, marble dust, stone dust, and many others.

Researchers are using different materials as a replacement of cement to check the properties of concrete. The world is experiencing great expansion in concrete construction especially in developing countries. This leads to emission of carbon di oxide in atmosphere. The Cement Sustainability Initiative report [3] state that approximately 1 Ton of CO_2 is produced in preparation of 1 Ton of cement. Sayem et al [4] used waste plastic and stone dust in concrete. Properties like compressive strength, split tensile and unit weight were enhanced by using these wastes. Aliabdo et al [5] used crushed clay bricks in concrete and tested against non-traditional tests like XRD, thermo-gravimetric analysis (TGA) and microstructural analysis. Based on experimental observation they concluded that strength is reduced while thermal resistance, cost effects, and environmental aspects may be considered. Heidari and Hasanpour [6] reported that brick dust contain pozzolanic properties which can be used in concrete to replace cement. They further concluded that using brick dust has minor effects on strength loss. Khan et al [7] analyzed brick dust and marble powder as cement replacement. Results shows that workability is increased by using both ingredients in certain proportion, compressive strength is reduced by increasing brick dust content beyond 10% replacement. Duby and Porwal [8] reported that fly ash and brick dust mixed in concrete as admixture can save cost of cement up to 20% by providing the same strength. Bharti and Patel [9] reported that replacing cement with 15% silica fume and 10% brick kiln dust shows maximum compressive and tensile strength using 0.50 w/c ratio.

2. Brick Dust:

Brick dust occurs from loading or unloading, construction sites and brick kilns. This dust is used in dumping and filling. There are thousands Ton of brick waste generated each year around the world which goes in unplanned way. Pozzolanic materials such as brick dust and other ceramics powder has been used in concrete since ancient times. In ancient times the brick dust was used according to experiences and experiments as they were unaware of the properties of brick dust.

Bricks are made up of different types of clays and other materials like sand. Clay composed up of 20-30% Alumina, 50-60% Silica and other carbonates and oxides. Clay is responsible for the pozzolanic behavior of brick. Clay itself has no pozzolanic properties but when fired together with lime during brick making process it gains pozzolanic nature. [9]

2.1. Chemical Reaction of Brick Dust in Concrete:

Brick dust as pozzolana reacts with lime in presence of water to form hydraulic compounds. Calcium carbonate and water is produced when carbon dioxide reacts with calcium hydroxide. The chemical reactions are following:

Portland cement + Water → Calcium Silicate Hydrate

$Ca(OH)_2 + CO_2 \rightarrow CaCO_3 + H_2O$

Extra amount of hydraulic cement is formed when reacts with lime. The reaction is following:

Pozzolana + $Ca(OH)_2$ + water → C-H-S (Glue)

The former reaction of Portland cement with water is fast reaction which provides early strength to concrete where the later reaction of pozzolana with liberated lime in presence of water is slow reaction which effect early age strength. But after some time, the brick dust provides extra amount of C-H-S which contribute to strength of concrete. [10]

3. Materials Used in Research

3.1. Portland cement

Portland cement of Bestway Cement was used which was according to ASTM type 1. Cement was clean, free from slumps and impurities.

3.2. Fine Aggregates

Sand was obtained from Lawrencepur, Punjab. The sand was passed through sieve no 4. The fineness modulus of fine aggregate was determined according to ASTM C 136-93, which was 2.589.

3.3. Course Aggregates

Margalla crush was used in this research. Coarse aggregate sieve analysis was determined according to ASTM C 136.

3.4. Brick Dust

Brick Dust was obtained from Ring Road Kiln Peshawar. To attain the approximate fineness of cement, brick dust was further crushed in PCSIR Laboratories, Peshawar in grinding mill with 26 Nos cylindrical rods of length 3 feet each. Rotation of drum was set on 32 revolutions per minute. Brick dust retained about 2.6% on 90μm sieve. The physical properties of brick dust are presented in table 1.

Table 1. Physical Properties of Brick Dust

Properties	Brick Dust
Moister Content (%)	4.7
Specific Gravity (unit)	2.54
Fineness modulus	2.08

EDX Test was conducted in University of Peshawar Physics department to determine the chemical composition of brick dust which is presented in Figure 1. Brick dust is golden in colour as shown in figure 2.

Figure 1. Chemical Composition of Brick Dust

Figure 2. Brick Dust

4. Experimental Investigation

Various experiments have been conducted to study the properties of brick dust and concrete.

4.1. Preparation of concrete samples and testing

Total of five concrete mixes with brick dust were prepared designated as M_0, M_5, M_{10}, M_{15} and M_{20}, which represents 0%, 5%, 10%, 15% and 20% replacement of cement with brick dust respectively. Water-cement ratio 0.55 was used in this research. The details of mixes are shown below in figure 3.

S.No	Mix ID	Cement (g)	Brick Dust Gram	Brick Dust %	Sand (g)	Course Aggregate	Water (mL)	w/c
1	M_0	2557	0	0	5960	6900	1406	0.55
2	M_5	2430	127	5	5960	6900	1406	0.55
3	M_{10}	2302	255	10	5960	6900	1406	0.55
4	M_{15}	2174	383	15	5960	6900	1406	0.55
5	M_{20}	2045.6	511.4	20	5960	6900	1406	0.55

Figure 3. Mix Design

IOP Conf. Series: Materials Science and Engineering **414** (2018) 012005 doi:10.1088/1757-899X/414/1/012005

4.1.1. Slump Test:

Workability of the above mixes were measured with slump test according to ASTM C 143 [11] Standard slump cone with 12-inch diameter with 8 inches top opening diameter and 4 inches bottom opening diameter and base plate was used. Cone was filled with concrete in three layers by tamping 25 blows each layer. For each mix, three trials were prepared and values were noted.

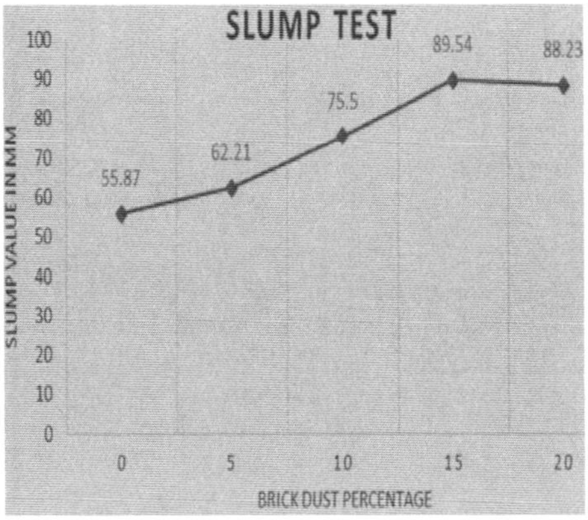

Figure 4. Slump Variation

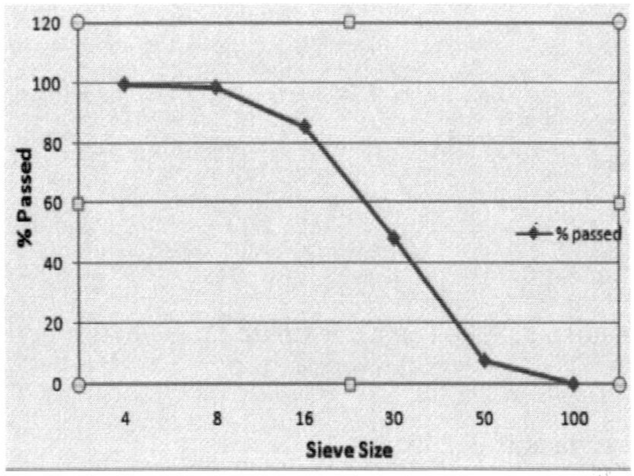

Figure 5. Sieve Analysis of Fine Aggregates

4.1.2. Compressive Strength Test:

Compressive strength test was conducted for 3 days, 7 days, 14 days, 28 days and 56 days according to ASTM C39 [12] Standard mold with 12 inches length and 6 inches diameter were properly oiled and filled with concrete in three layers. Each layers were propped with 25 blows for proper compaction. After initial curing molds were removed and cured in water tank for 3, 7, 14, 28 and 56 days. The load noted on UTM was load at which specimen were failed. Compressive strength value was taken as average of three samples value.

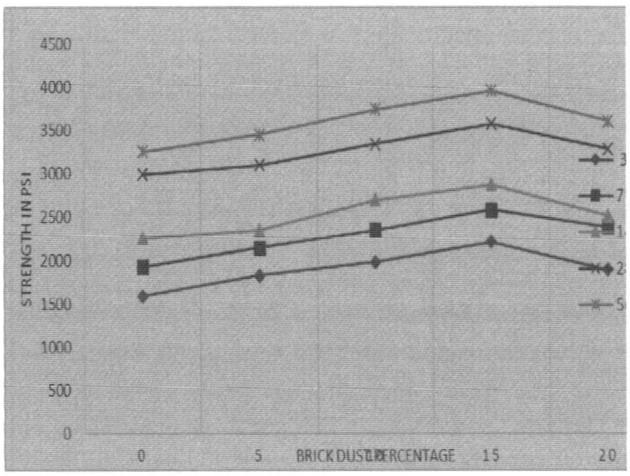

Figure 6. Compressive Strength at All Ages

Figure 7. Compressive Strength Test

4.1.3. Split tensile

Split tensile test was conducted for 28 days and 56 days according to ASTM C 496/C 496M – 04 [13] Total 30 specimens were cast for split tensile test, average of each 3.

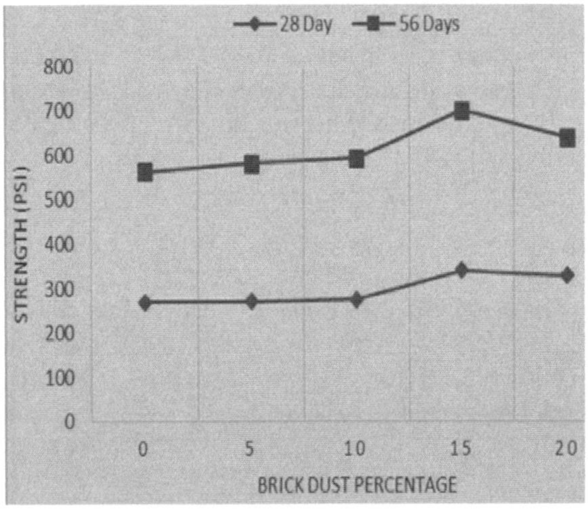

Figure 8. Split Tensile Strength at 28 days and 56 days

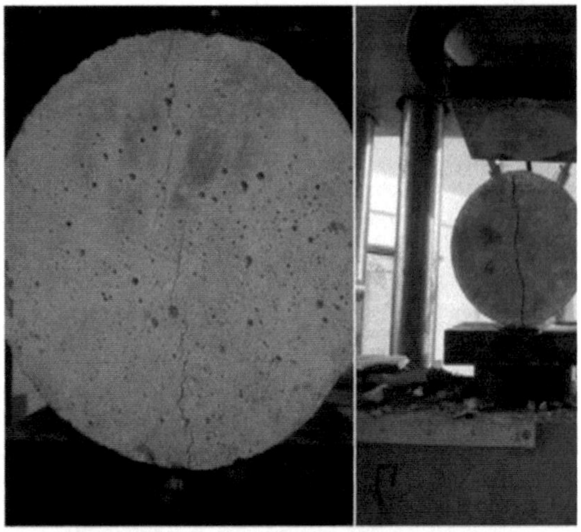

Figure 9. Split Tensile Strength Test

5. Results and Discussion

5.1. Slump test

Slump values tends to increase by increasing brick dust content in concrete samples. This increase in slump value may be due to spherical particles of brick dust which lubricate the mix and gives ball bearing effect to mixes. Beyond 15% replacement of cement with brick dust absorb more water hence slump value decreases. The trend of slump is shown in figure 3.

5.2. Compressive strength

It was observed that compressive strength increases with increasing brick dust content at all ages as shown in figure 6. Replacing cement with brick dust beyond 15% replacement tends to decrease strength. Maximum strength value was noted at 15% replacement. As brick dust is natural pozzolan so it produces more C-S-H gel which gives more strength [14] Secondly, the finer particles of brick dust forms a compact mass which fills the pores of concrete and results in more strength than control sample. Compressive strength setup is shown in figure 7.

5.3. Split tensile test

Split tensile shows similar strength gain pattern that is increasing up to 15% replacement of cement with brick dust and then tends to decrease as shown in figure 8. The increase may be due to finer particles of brick dust which contributes compact matrix. The test setup is shown in figure 9.

6. Conclusion of Research

In this research brick dust was incorporated in concrete by partial replacing of cement. Based on the result, the following conclusions can be drawn.

- The maximum slump was noted at 15% replacement after that it goes on decreasing. The minimum value of slump was observed at 20% replacement.
- Compressive strength tends to increase by increasing brick dust content up to 15%. Beyond this replacement the compressive strength shows inferior behavior at 20% replacement.
- Split tensile strength increases by increasing brick dust content up to 15%, beyond this replacement the split tensile strength tends to decrease. .
- Brick dust can be used as plasticizer as it decreases water demand because workability is increasing

IOP Conf. Series: Materials Science and Engineering **414** (2018) 012005 doi:10.1088/1757-899X/414/1/012005

References

[1] Kamyotra JS. 2015 *Brick Kilns in India*. **57**. Available from: http://www.cseindia.org/docs/aad2015/11.03.2015 Brick Presentation.pdf Another reference
[2] Available from : http://www.sdpi.org/sdc/12thsdc/panel_state_of_brick_kilns.html
[3] World Bussiness Council for Sustainable Development 2012;(June). The Cement Sustainability Initiative. Agenda for Action.
[4] Paper C. A 2014 Study of Green Concrete using Waste Plastic and Stone Dust.
[5] Aliabdo AA, Abd-Elmoaty AEM, Hassan HH. Utilization of crushed clay brick in concrete industry. Alexandria Eng J.;53(1):151–68
[6] Heidari A, Hasanpour B 2013 Effects of waste bricks powder of gachsaran company as a pozzolanic material in concrete. Asian J Civ Eng. 14(5):755–63
[7] Khan MM, Panchal S, Sharma A, Bharti AA 2017 Innovative Use of Brick Powder and Marble Dust As a Mineral Admixture in Concrete. Int J Civ Eng Technol. 8(81):987–90.
[8] Dubey S, Porwal RJ 2014 Studies on Effect of Cement Concrete Strength by using Waste Materials Likes Fly Ash and Brick Kiln Dust.3(6):317–9.
[9] Rogers SB 2011 Evaluation and Testing of Brick Dust as a Pozzolanic Additive to Lime Mortars for Architectural Conservation.
[10] Uddin MK 2004 Use of Brick Dust in Concrete as Mineral Admixture and Partial Replacement of Cement. J Civ Eng. 32(1):69–78.
[11] ASTM C143/C143M. Standard Test Method for Slump of Hydraulic-Cement Concrete. Astm C143. 2015;(1):1–4
[12] ASTM C39. Standard Test Method for Compressive Strength of Cylindrical Concrete Specimens. Am Soc Test Mater. 2016;i:1–7.
[13] ASTM C 496/C 496M – 04. Standard Test Method for Splitting Tensile Strength of Cylindrical Concrete Specimens. Annu B ASTM Stand Vol 0402 . 2011;1–5
[14] Paper C, Wales S, Wales S 2016 engineering properties of concrete.

Study the effects of epoxy/tri functional mercaptan resins structure based PDLC films

M Ellahi[1], H Ullah[1], M F Ali[2] and N N Panezai[1]

[1]Department of Chemistry, Faculty of Arts & Basic Sciences, Balochistan University of Information Technology, Engineering and Management Sciences (BUITEMS), Quetta 87100 Pakistan.
[2]Beijing University of Chemical Technology, Beijing 100029, People's Republic of China

Email. mujtaba.ellahi@gmail.com

Abstract. The present investigation is focused on to find out the effects of the structure of epoxy/tri functional mercaptan resins based polymer dispersed liquid crystal (PDLC) films on the morphology of polymer matrix via heat curing system. The interaction between epoxy reaction and polythiol group (-SH) resins turn played an essential role in step-growth polymerization reaction for the preparation of PDLC films. In addition to this, the polymer morphologies in this heat curing system have been regulated from polymer beads to porous polymer matrix by changing the liquid crystals (LCs) contents, chemical structures of epoxy tri functional mercaptan resins, and functionality of thiol hardeners. On the other hand, the thiol group (-SH) fractured the PDLC films continuously with changing driving voltage radically. With the enhancing content of the epoxy resins, the LC domain size decreased, while the thiol had a various effect on the LC domain size. It was of great importance for the optimisation and the possible applications of the PDLC films.
Keywords. epoxy/tri, mercaptan resins, PDLC films

1. Introduction

Polymer dispersed liquid crystal (PDLC) are a kind of functional macro molecular composite, that consist of haphazardly dispersed liquid crystal (LC) domain size entrenched in a polymer network [1,2]. In the off state, PDLC films have a milk-white form when the LC droplets with a +ve dielectric anisotropy are haphazardly leaning in the films. Electronic or charismatic field, the LC domain size arranges in a line direction parallel (||) to the electric or magnetic field, and the PDLCs films will switch to a clear state just about matches the polymer matrix refractive index (np) [3]. Based on this E-O property, their possible application as the active substrate in displays has been first projected [4], other E-O properties are important for Bathrooms/Shower enclosures, Projection displays, Hurricane resistant windows, and other devices which is increasing regularly [5-8].Usually, there are three properties required for PDLC film (smart glass) based tool applications: a low threshold voltage for the transition between on/off states, a high on/off contrast in transmittance and a fast response time for the transition between on/off states. These E-O properties depend mainly on the structures of the heat curing epoxy monomers [9-13]; the physical characteristics of the LC such as birefringence [14] and viscosity the interface and compatibility of LC and epoxy monomers; experimental conditions [15]; the electric field conditions [16]; and fastening energy of the domain size in PDLCs system (smart glass) [17, 18].

In current years, a number of studies on the influence of the dielectric anisotropy ($\Delta\varepsilon$) of the LC E-O PDLCs system have been done[19, 20].Since the $\Delta\varepsilon$ of the LC is free of the polymer, results point out that driving voltage is a falling function of the $\Delta\varepsilon$ of the LC, as in the case of pure LC in cells[21].Furthermore, it seems natural that a higher viscosity of the LC should increase both response times , other than the right viscosity of the LC inside the composite is hard to assess and it depends on a lot on the nature of the polymer, because of dissolved oligomers. In summary, LC as one main

PDLC films constituent plays a significant role in the growth of the PDLCs system E-O characteristics. In addition, polar CN- group joined by means of on the side substitutes like F atoms in various combinations generates high dipole moments (μ) which provide the increase to Δε standards that are extra functional for dropping the threshold voltage of the PDLC films.

At nearby, despite the fact that many studies have focused on the reactive polymer catalyst to fabricate PDLC films, and a small number of researches have focal point on belongings of the LC mixture physical parameters, such as Δn, Δε on the E-O characteristics of PDLCs systems. However, no research has been reported on effects of the LC molecular structures on the E-O characteristics of PDLCs system. In this article, NLCs has been selected due to its Characteristics for the PDLC films such as strong binding, optical clarity, strong adhesion, toughness, and flexibility effects on the terminal chain length of PDLC films. It is important to mention that adding the percentage of the thiol group (-SH) hardener, fractured the PDLC films continuously with best possible preparation condition was 40% LC, curing time 5.0 hrs at 70°c temperature. The effects of epoxy/tri functional mercaptan resins structure & E-O characteristics consequential PDLCs system cautiously examined. Similarly, the E-O substituting performance was straight dependent on LC domain. We think this study helps us not only to comprehend the relationship between the LC molecules & E-O characteristics of PDLCs system but also to find a new way to optimize the epoxy/tri functional thiol behaviour visions in structure & enhancing the E-O characteristics in curing thin glass system.

2. Materials and Methods

2.1 Materials

Fig. 1 shows the chemical structures of the materials used. In this study, the LC used contained of NLC (SLC-1717, Shijiazhuang ChengzhiYonghuaDisplay Material Co., Ltd., Shijiazhuang, P.R.China). The curable epoxy resins used wereDiglycidyl ether ofbisphenol A (DGEBA)(Sigma-Aldrich Company), Ethylene glycol diglycidyl ether (EGDE) resin (XY 669, Anhui Hengyuan Chemical Co., Ltd.), and Tri-methylol propane tris (3mercaptopropionate), Sigma-Aldrich Company). 3S-Thiol used as hardener for epoxy resins. Without cleansing, we used the compounds. The compositions of heating resins/3S-Thiol /NLC, as shows in Table 1.

2.2 Sample preparation

Ready samples contained of resins, tri- functional poly thiol hardener, and a constant amount of nematic liquid crystals (NLCs, 40.0wt %). First, carefully mixed in the particular mol ratio till a same blend was created. After that, based on rising liquid, the blend was packed in to the NLC cell whose internal appearances had been layered with ITO. Film thickness was controlled by a 15.0 ± 1.0 μm PET spacer. Then, prepared samples A1-A5 according to the mol ratios listed in Table 1.

IOP Conf. Series: Materials Science and Engineering **414** (2018) 012006 doi:10.1088/1757-899X/414/1/012006

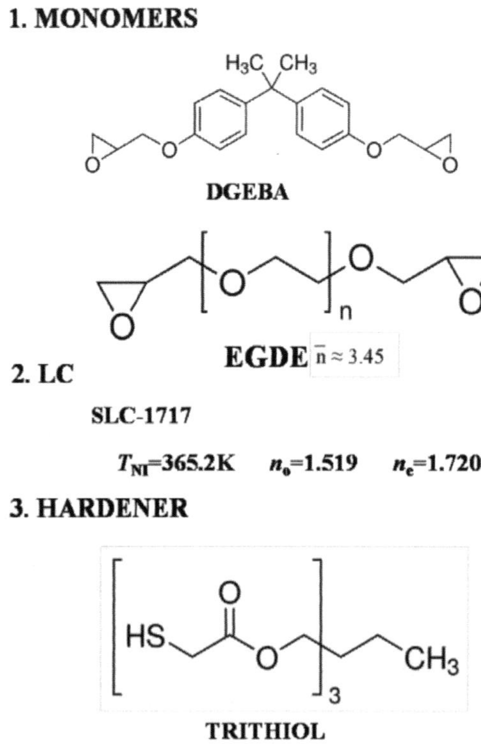

Figure 1. Chemical structures, names, and abbreviations of the materials

Table1. The arrangements of the samples A1-A5 studied for PDLCs system

Sample	Epoxy Resins/3S-Thiol (60 wt%) (5.0 hours at 70 ^{0}C)	SLC-1717/wt%
	EGDE/TRITHIOL/DGEBA/mol%	
A1	1.0/1.0/1.0	40
A2	2.0/1.0/2.0	40
A3	3.0/1.0/3.0	40
A4	4.0/1.0/4.0	40
A5	5.0/1.0/5.0	40

2.3 Analysis technique

LCD parameter tester (LCT-5016C, Changchun) used to require the E-O characteristics of the samples at ambient temperature. An electric field square wave (110 Hz) was functional and the space b/w the PDLCs system and photodiode was 350 mm. A schematic diagram of the E-O characteristics shown in scheme 1.

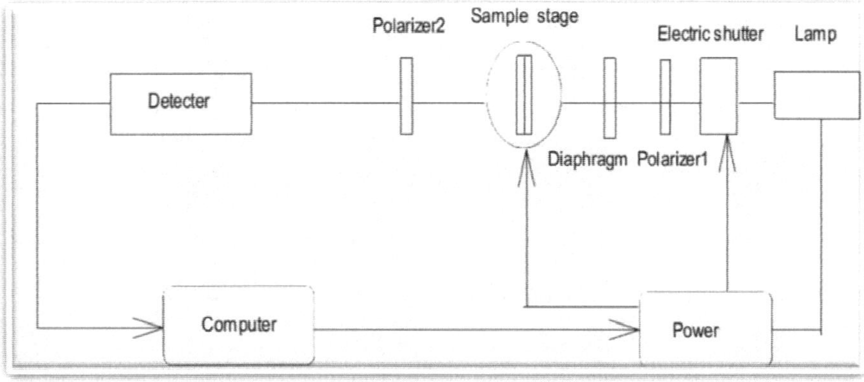

Scheme 1: A schematic diagram of the E-O properties

SEM(S-4800, Hitachi,) used to monitor the inner structures of the PDLCs System. The samples were initial to cut into small pieces and then soaked in n-hexane (C6H12) for 8 days at 250C to examine the LC from the polymer network. Afterward, the samples were dried under vacuum for 12 h. Then the surface of the polymer matrix was sputtered with gold and SEM to observe polymer crosslinked matrix.

3. Experimental Results and Discussion

3.1 Morphologies of the PDLCs System at 5.0 hrs at 70°C

With the purpose of appropriately quantity the droplet structure, Figure 2 shown the morphology of SEM pictures of the contract monomer network of A1-A5. The morphology of PDLCs system are reliant on the feed %age, heating situations, chemical arrangements, and on the particulars of the thin system development method [22]. The arrangement mol% of A1-A5 through 40% NLC contented & 700C curing heats is recorded in table 1. As the Figure 3 shown that the oval type droplet of the polymer matrix of PDLCs system A1-A5. 3S-Thiol responds through the epoxy group to arrange polymeric morphology scheme. Initially, the (-OH) gp: unlocked & improves another oxirane ring to arrange cross linked network chain of the PDLCs system. Furthermore, the increasing mol ratio % of the rigid chain segment and di functional epoxy resin increasing the cross-linking density consistently in all.

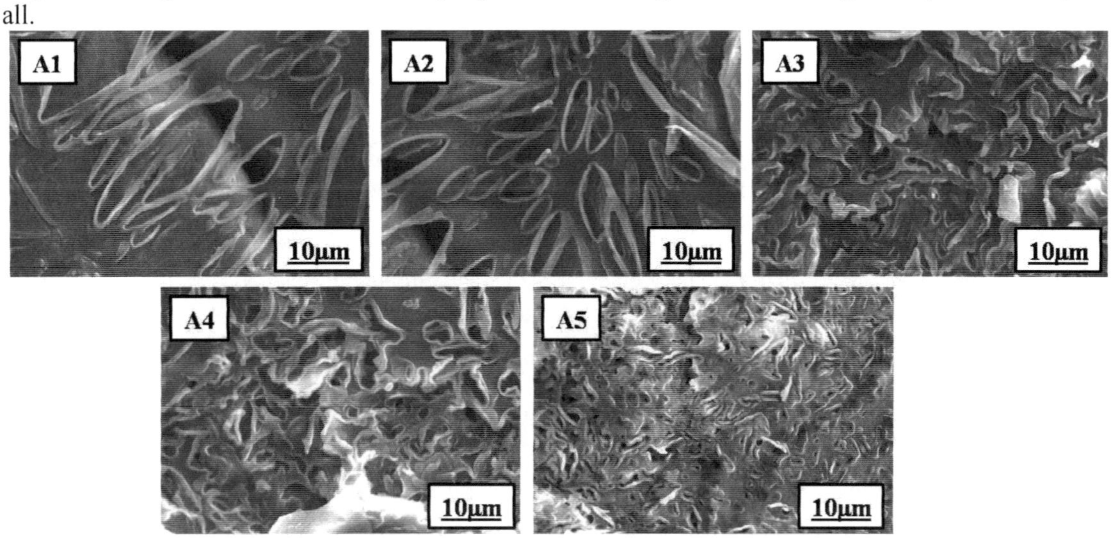

Figure 2. SEM micrographs of the polymer networks of the samples A1- A5

3.2. E-O Characteristics of PDLCs system at 5.0 hrs at 70°C

Gradually, additional oxirane and 3S-Thiolreacted shape domain size & NLCs divided from droplets, making extra and bigger droplet; before the PDLCs system became to develop easy to steer. A1 prepare with 1% each in EGDE, DGEBA, and poly thiol hardener respectively to take 5.0 hrs to acquire a PDLCs system through high-quality reaction current. At the same time, 3S-Thiol & 5.0 hrs: heating bands of samples A2, A3, A4, and A5 with constant mol ratio 1%. It shows that the samples heating for 5.0 hrs: have different E-O characteristics as sample A1-A5, As a result; it signifying to 3S-Thiol, the heating method essential merely 5.0 hrs.

The E-O effect in NLCs with an even director is not an essential condition in the presence of a warped director. Our findings have revealed a change in domain shape on the driving voltage of a PDLCs shutter with bipolar domain and investigated their effects on the polymer network, E-O characteristics, and conductivity the PDLCs system. Hence, thin film can be further successfully applied in display industries for the enhancement of E-O contrast with the specifically change of ON-scattering at minimum time. The transmission-applied voltage bands of A1-A5 are shows Figure 3.The enhancing the applied voltage, the transmission A1-A5 decreases progressively due to the increasing mol% of rigid chain segment of DGEBA.

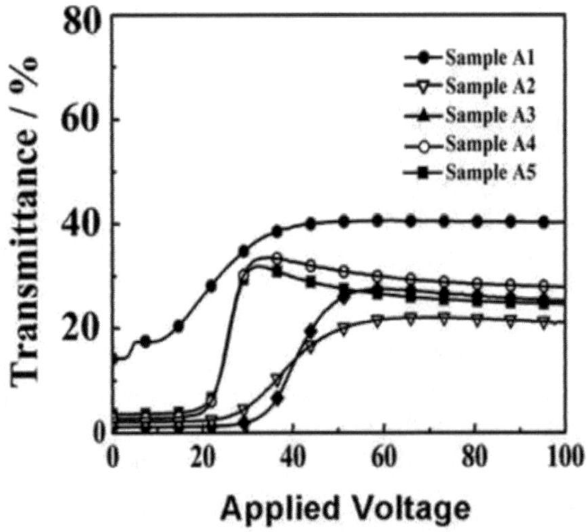

Figure 3. Applied voltage dependence of transmittance for samples A1-A5.

Gradually, additional oxirane and 3S-Thiolreacted shape domain size & NLCs divided from droplets, making extra and bigger droplet; before the PDLCs system became to develop easy to steer. A1 prepare with 1% each in EGDE, DGEBA, and poly thiol hardener respectively to take 5.0 hrs to acquire a PDLCs system through high-quality reaction current. At the same time, 3S-Thiol & 5.0 hrs: heating bands of samples A2, A3, A4, and A5 with constant mol ratio 1%. It shows that the samples heating for 5.0 hrs: have different E-O characteristics as sample A1-A5, As a result; it signifying to 3S-Thiol, the heating method essential merely 5.0 hrs.

The E-O effect in NLCs with an even director is not an essential condition in the presence of a warped director. Our findings have revealed a change in domain shape on the driving voltage of a PDLCs shutter with bipolar domain and investigated their effects on the polymer network, E-O characteristics, and conductivity the PDLCs system. Hence, thin film can be further successfully applied in display industries for the enhancement of E-O contrast with the specifically change of ON-scattering at minimum time. The transmission-applied voltage bands of A1-A5 are shows Figure 3.The enhancing the applied voltage, the transmission A1-A5 decreases progressively due to the increasing mol% of rigid chain segment of DGEBA.

IOP Conf. Series: Materials Science and Engineering **414** (2018) 012006 doi:10.1088/1757-899X/414/1/012006

$$V_{th} = \frac{d}{3a} \times \left(\frac{\rho p + 2}{\rho Lc}\right) \times \left(\frac{K(l^2 - 1}{\Delta\epsilon\, \epsilon_0}\right)^{1/2} \quad (1)$$

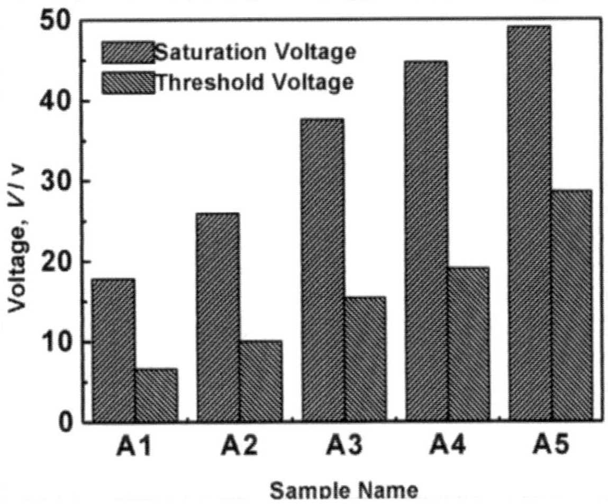

Figure 4. Saturation and Threshold voltages for samples A1-A5.

The curves of transmissibility-voltage and transmissibility-time can be obtained by the system. Analyzing the curves, the threshold voltage, saturation voltage, contrast ratio, steepness, and response time can be calculated from them as figure.4. The eq: (1) specifies the E-O characteristics of PDLCs system can be organised by LC domain size and other factors, such as film thickness, resistivity and dielectric anisotropy of LC. Hence, as shows in figure 4; V_{th} and V_{sat} increased with decreasingthe LC domain size. As for samples A1-A5, increasing the rigid chain segment and increasing the cross linking density altering the values of V_{th} and V_{sat}, which changes from 6.593V to 28.571V and 17.826V to 48.979V; correspondingly.

4. Conclusions

In this study, the optimized condition for fabricating smart glass thin PDLC films containing class nematic liquid crystals (NLCs) linear chain polymer mixtures. We have been examined two PDLCs system. At 5.0 hrs at 70°C using heating oxirane resins with heat heating processes mol%, composition feed ratio, numerous molecular structures and at different mol feed ratio hardener Tri-methylol propane tris (3mercaptopropionate), which affects the E-O characteristics of the PDLC systems. Moreover, an effect on of the smart glass thin films systems was powerfully prejudiced through rigid chain segment (DGEBA), while prejudiced morphology smart films system. The outcome of the combination and thermos tropic hotness on the domain size of smart glass films were studied. Additionally, using constant mol% 3S-Thiol variable driving voltages were optimized. Our findings have revealed a change in droplet shape on the driving voltage of a PDLC shutter with bipolar droplets and investigated their effects on the morphology, E-O properties, and conductivity of the PDLC thin films. The thiol group (-SH) fractured the PDLC films continuously with changing driving voltage radically. With the enhancing content of the epoxy resins, the LC domain size decreased, while the thiol had a various effect on the LC droplet size. Hence, this thin film can be further successfully applied in display industries for the enhancement of E-O and contrast with the specifically change of ON-scattering at minimum time.

IOP Conf. Series: Materials Science and Engineering **414** (2018) 012006 doi:10.1088/1757-899X/414/1/012006

5. Acknowledgment

This work was supported by the HEC fund for BUITEMS Funded Research Project No. 01/ ORIC/ BUITEMS/2011/610.The authors wish to express appreciation to the State Key Laboratory of Chemical Resource Engineering (BUST) and BUITEMS for laboratory experiment support and technical assistance.

References

[1] Drzaic, P.S. Polymer dispersed nematic liquid crystal for large area displays and light valves. *J. Appl.Phys.*1986 **60** 2142-48.

[2] Kurihara, S.; Masumoto, K.; Nonaka, T. Optical shutter driven photo chemically from anisotropic polymer network containing liquid crystalline and azo benzene molecules. *Appl. Phys. Lett.* 1998 **73** 160-62.

[3] Doane, J.W.; Vaz, N.A.; Wu, B.G.; Žumer, S. Field controlled light scattering from nematic micro droplets.*Appl. Phys. Lett.* 1986 **48** 269-71.

[4] Cheng, Z.; Wang, T.; Li, X.; Zhang, Y.; Yu, H.F. NIR-Vis-UV Light-Responsive Actuator Films of Polymer- Dispersed Liquid Crystal/ Graphene Oxide Nano composites. *ACS Appl. Mater. Interfaces.* 2015 **7** 27494-01.

[5] Yu, L.; Cheng, Z.; Dong, Z.; Zhang, Y.; Yu, H.F. Photo mechanical Response of Polymer-Dispersed Liquid Crystals/Graphene Oxide Nano composites. *J. Mater. Chem. C.* 2014 **2** 8501-06.

[6] Kim,S.H.;Heo,C.P.;Park,K.S.;Kim,B.K. Effect of pre polymer structure on the electro-optic performance of polymer dispersed liquid crystals. *Polym. Int.* 1998 **46** 143-149.

[7] Malik,P.;Raina,K.K.Droplet orientation and optical properties of polymer dispersed liquid crystal composite films. *Opt. Mater.* 2004 **27** 613-17.

[8] Liu, F.; Cao, H.; Mao, Q.; Song, P.; Yang, H. Effects of monomer structure on the morphology of polymer networks and the electro-optical properties of polymer-dispersed liquid crystal films. *Liq. Cryst.* 2012 **39** 419-24.

[9] Mujtaba Ellahi, Fang Liu, Ping Song, Yanzi Gao, Hui Cao, M.Y.Rafique, Murad Ali Khaskheli, M. Zubair Iqbal, Huai Yang. Influence of the multi-functional epoxy monomers structure on the electro-optical properties and morphology of polymer-dispersed liquid crystal films.*Polym. Bull.*2013 **70** 2967-80.

[10] Mujtaba Ellahi, Fang Liu, Ping Song, Yanzi Gao, M.Y.Rafique, DilFaraz Khan, Hui Cao, Huai Yang. Characterization and Morphology of Polymer-Dispersed Liquid Crystal Films. *Soft Mater.*2014 **12** 339-45.

[11] Mucha, M. Polymer as an important component of blends and composites with liquid crystals.*Prog.Polym. Sci.* 2003 **28** 837-73.

[12] Humaira Masood Siddiqi, Adeel Afzal, SamiaSajid ,Zareen Akhter. Synthesis, characterization and thermal oxidative stability of rigid epoxy polymers cured from aromatic mono and di-amines. *Journ. Of Polym. Reserh.* 2013 **20** 41-52.

[13] Tomohisa, G;Hideya, M. Preparation and characteristics of new reverse mode film of polymer dispersed liquid crystal type. *Appl. Phy. Lettr.* 1992 **60** 392-94.

[14] Mujtaba Ellahi, M.Y.Rafique, ShaguftaIshtiaque,M.Furqan Ali, JameelMemon, .Study on the effects of epoxy resin based PDLC films using polythiol group (-SH) as hardener and catalyst , *J. of Mater. Focus.* 2015 **4** 197-01.

[15] Mujtaba Ellahi, et al, .Study on the effects of isotropic cross-linked pristine morphology and electro-optical properties of PDLC films, *Polym. Bull.* 2015 **72** 2917-30.

[16] Farzana A, Muhammad J, Jae JY, et al. Comparative study on the electrooptical properties ofpolymer-dis- persed liquid crystal films with different mixtures of monomers and liquid crystals. *Appl. Polym. Sci.*2011 **121** 1424-30.

[17] Yu, H.F.; Liu, H.; Kobayashi, T. Fabrication and Photo response of Supra molecular Liquid-Crystalline Micro particles. *ACS Appl. Mater. Interfaces.* 2011 **3** 1333-40.

[18] H.Murai and T.Gotoh, Stabilization of Epoxy-Based Polymer-Dispersed Liquid Crystal Films by Addition of Excess Hardener. *Mol, Crys. & Liq. Cryl.* 1993 **226** 13-26.

[19] Song P, Cao H, Wang F, Liu F, Wang J, Ellahi M, Li F,Yang H, .The UV polymerisation temperature dependence of polymer-dispersed liquid crystals based on epoxies/acrylates hybrid polymer matrix components,*Liq. Crysl* 2012 **39** 1131-40.

[20] Mujtaba Ellahi, et al, .Effects of di and tetra functional epoxy monomers structure on the morphology and the electro-optical properties of polymer-dispersed liquid crystal films, *Inter. J. of Scien. & Tech. Reseh.* 2013 **2** 127-31.

[21] Senyurt, A.F.; Warren, G.; Whitehead, J.B.; Hoyle, C.E. Matrix physical structure effect on the electro-optic characteristics of thiol-ene based H-PDLC films. *Polym.* 2006 **47** 2741-49.

[22] Mujtaba Ellahi, YanziGao and M. Y. Rafique,. Influence of enhanced curing temperature of epoxy monomers structure on the electro-optical properties and morphology of polymer-dispersed liquid crystal films, *J. of Amer. Engg. Reserh.* 2013 **2** 01-06.

Use of slate in making lightweight concrete.

Muhammad Aftab Khan[1], Ibar Ahmed[2], Abdul Basit[1], Muhammad Umar[1], Nabeel Liaqat[1] and Muhammad Nasir Ayaz Khan[2]

[1]Swedish college of engineering and technology Wah Cantt, Punjab, Pakistan
[2]University of Engineering and Technology Taxila, Pakistan

Email. aftabkhanengineer@gmail.com

Abstract. This research was implemented to develop and to determine that how much self-weight of concrete decreases with the increase of Lightweight Aggregate and what its effect on the strength. Laboratory tests have been carried out to investigate the properties of fresh normal concrete as well as lightweight concrete and the strength development of hardened concrete. Five compositions of the concrete mix design were prepared and tested 0% of the Light Weight Aggregate, 10%, 20%, 30% and 40%. The workability was determined for each and every batch separately. The compressive strength test was carried out at the ages of 3, 7, and 28 days to examine the strength development of hardened concrete (normal + lightweight) mixes. Three Standard concrete cylinders were cast comprising of mixed, cured and tested, for analyzing the compressive strength and weight. The results show that water-cement ratio increases with increase of Light Weight Aggregate in concrete mix and their strength gradually decreases after the 10% use of LWA. It has been observed with different tests that up to 10% mix of Light Weight Aggregate in normal concrete, the strength increases 12.8% compared to normal concrete. It is practically concluded that addition of Light Weight Aggregate in concrete mix, the dead load reduces about 2.5% with the increase of every 10% batch. The trend of lightweight concrete uses is increases due to their lower dead load, strong thermal, acoustic, environmental, fire-retardant qualities and also reduced the building time.
Keywords. Slate, concrete, aggregate, fire-retardant

1. Introduction

Lightweight concrete is a type of concrete which contain an expanding agent that it increases the volume of the concrete mixture while giving additional qualities such as decreasing the dead weight. Light weight aggregate is that type of coarse aggregate which is used in the preparation of light weight concrete and such concrete then used in concrete blocks, structural concrete and pavements. Light weight aggregate is produced form shale, clay, or slate. Slate is a sedimentary rock grey in color, which is composed of clay or volcanic ash.

Fahad et al reported that using recycle plastic aggregate showed about 13% reduction in chloride penetration where compressive strength tends to decrease [1]. Emon et al reported that using brick chips as light weight aggregate along with low cost steel wire fibers enhance the mechanical performance of concrete such as compressive and split tensile strength [2]. Tang et al concluded that sedimentary light weight aggregate density is lower than normal aggregate and provide 29-35% lighter concrete when compared to normal density concrete [3]. Reddy et al performed experiments on preparation of fly ash pellet and finally he concluded that round shape of fly ash pellet gives better workability compared to the angular natural gravel [4]. Mo et al reported that using light weight aggregate in addition to fibers improve the bond properties of light weight concrete [5].

Alengaram et al reported that using oil palm kernel shell as light weight aggregate increase water absorption, low abrasion, increase compressive and split tensile strength, increase creep value and thermal conductivities was lower than normal weight concrete [6]. Hassanpour et al concluded that using fibers into light weight concrete reduces the workability, increase the density, increases the compressive strength and significantly higher splitting tensile strength compare to normal weight concrete [7]. Al-Khaiat and Haque reported that Lytag coarse and fine aggregate light weight concrete compressive

IOP Conf. Series: Materials Science and Engineering **414** (2018) 012007 doi:10.1088/1757-899X/414/1/012007

strength is less sensitive to initial curing but water penetration and durability is more sensitive to initial curing [8]. Kakade and Dhawale reported that coconut shell aggregate has good absorbance to shock, compressive strength is according to requirement of structure light weight concrete [9]. Akcaozoglu et al investigated the use of waste poly-ethylene Terephthalate (PET) bottle in concrete as light weight aggregate [10].

2. Experimental work

2.1 Data collection about LWA:

Tracing of lightweight aggregate was carried out throughout Pakistan. The 1st lightweight aggregate pumice aggregate was found, which is available mostly in Baluchistan desert area. Pumice aggregate is the residue of Lava and access to this material is quite difficult. Finally, the slate aggregate was found, and all the data and chemical composition of slate aggregate (lightweight) was collected from Engr. Muhammad Gul (ex-employee of the PCSIR). Lightweight aggregate in raw form was collected from PCSIR lab (Peshawar).

2.2 Collection of Materials:

Different materials have been collected from different sources. The lightweight aggregate is collected from the PCSIR laboratory (Peshawar) and from Manki Sharif source Nowshehra. The normal aggregate and sand are collected from the petty source of Model town (Main Source Margla) and (Main Source Nizampur). The Askari cement was used having compressive strength of 10000 Psi.

2.3 Tests on Materials:

2.3.1. Tests on Cement: The following tests were performed on cement.

Consistency test, initial & final setting time of cement, fineness of cement, and Soundness test on cement.

2.3.2. Tests on Sand: Sieve Analysis of Sand and Bulk Specific Gravity and Water absorption.

Fineness Modulus = Σ Cumulative retained / 100= 2.64

$$\text{Bulk Specific gravity} = A/(B+D-C) = 2.64$$

$$\text{Absorption \%} = (D-A)/A \times 100 = 2\%$$

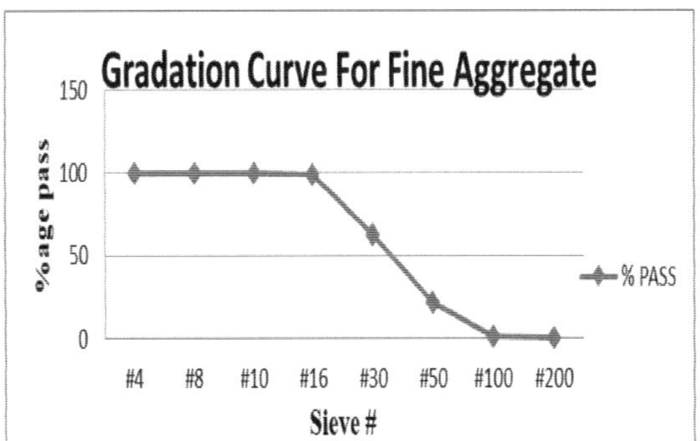

Figure 1: Gradation curve of Fine aggregate

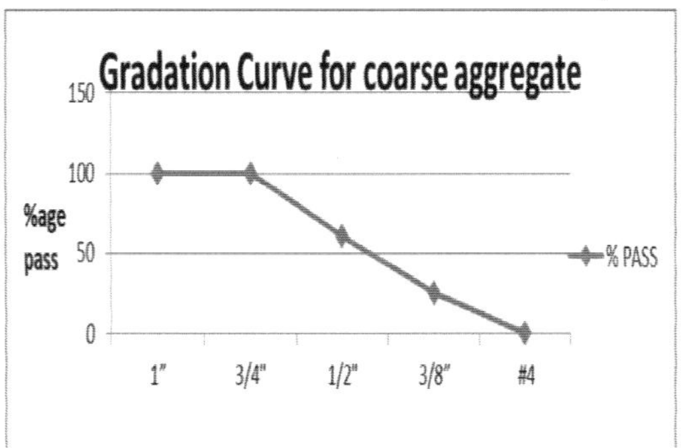

Figure 2: Gradation curve of coarse aggregate

The gradation curve in figure 1 shows that for fine aggregate the material passed 100% from Sieve#4 to Sieve#16 but did not passed from Sieve#100 and onward.

The gradation curve in figure 2 shows that for coarse aggregate the material passed 100% from 1inch Sieve to 3/4inch Sieve but did not passed from Sieve#4 and onward.

2.3.3. Tests on Light Weight Aggregate:

2.3.3.1. Specific Gravity and Water Absorption Test on Lightweight aggregate:
This test was performed according to ASTM-C 127
Specific gravity of aggregate =W3/(W2-W1) = 1721/(1755.2-720)= 1.63

Absorption of aggregate= W2-W3/W3=1755.2-1721/1720= 2.1

2.3.3.2. Sieve Analysis of Lightweight Aggregate:
This test was performed according to ASTM-C136 [12]

2.3.3.3 Tests on Normal Aggregate:

2.3.3.4. Specific Gravity and Water Absorption Test:
This test was performed according to ASTM-C 127 [11]

Specific gravity of aggregate =W3/(W2-W1) = 1930.9/(1944.9-1195) = 2.577
Absorption of aggregate= W2-W3/W3=1944.9-1930.9/1930.9= 0.72

2.3.3.5. Sieve Analysis of Normal Aggregate:
This test was performed according to ASTM-C136

2.4 Selection of samples:
Different sample were selected to continue research work i.e. 0%, 10%, 20%, 30% and 40% mix of Light Weight Aggregate with Normal Weight Aggregate.

2.5 Slump test for Concrete:
This test was performed according to ASTM-C143 [13]

Table 1. W/C and slump value

%age of LWA in Concrete	W/C ratio	Increase by % in W/C ratio	Slump value(inch)
0%	0.55	0	1.5
10%	0.55	0	2
20%	0.57	3.64	2
30%	0.59	7.27	1
40%	0.62	12.73	1.5

The detail of water cement ratio and slump value for different samples i.e. 0% 10% 20% 30% 40% mix is shown table 1. As the LWA addition was increased the w/c ratio was gradually increased but there is variation in slump value i.e. increase and decrease.

IOP Conf. Series: Materials Science and Engineering **414** (2018) 012007 doi:10.1088/1757-899X/414/1/012007

2.6 Compression test:

Concrete's compressive-strength mostly depends on the mix design. But it is affected by several other factors. Such as mixing of concrete, placing of concrete and curing of concrete as well as quality of concrete ingredients. For knowing compressive-strength of concrete, generally, concrete-cylinder is tested at laboratory.

Procedures of Making Concrete Cylinder:

Making concrete-cylinder specimen is simple and it is done in three simple steps.

1. Cleaning & fixing mould.
2. Mixing, Placing, Compacting & Finishing concrete.
3. Curing.

Figure 3. Compression test of concrete cylinder under observation

3. Results and discussions

The properties of the LWC mix samples determined by the following test: fineness test, workability test and Water cement ratio, and Compressive strength.

3.1 Fineness test

The fineness tests were performed on fine aggregate and course aggregate, and the result are listed in graph form shown in figure 1 and figure 2. The gradation curve in figure 1 shows that for fine aggregate the material passed 100% from Sieve#4 to Sieve#16 but did not passed from Sieve#100 and onward. The gradation curve in graph of figure 2 shows that for coarse aggregate the material passed 100% from 1inch Sieve to 3/4inch Sieve but did not passed from Sieve#4 and onward.

3.2 Workability and Water cement ratio

As the LWA addition was increased the w/c ratio was gradually increased and the test results are indicated in table 1. The concrete mixes workability was measured by using the slump cone test. The results listed in table 1 shows that most of the LWC mixes has a slump values ranging from 1inch to 2inch immediately after mixing. To be within the scope of this investigation, the workability of all the LWC mixes cast, was kept almost the same by changing the w/c ratio whenever needed.

3.3 Compression test:

The test results are shown in Table 2 with figure 4. These results show the comparison of normal weight aggregate with different type of LWA mixes using locally available natural lightweight aggregates, LWA. The concrete mixture tested for 7days, 14 days and 28days compressive strength, figure 3.

Table 2: Total summary table of compressive strength of different samples

Days	NC Weight (kg)	10%LWC Weight (kg)	20%LWC Weight (kg)	30%LWC Weight (kg)	40%LWC Weight (kg)
7d	13.29	12.86	12.58	12.34	11.84
7d	13.17	12.96	12.8	12.21	11.74
7d	13.23	12.91	12.69	12.28	11.79
14d	13.16	13.01	12.81	12.71	11.45
14d	13.48	13.05	12.86	12.72	11.92
14d	13.32	13.03	12.84	12.72	11.69
28d	13.31	13.14	13.01	12.6	11.94
28d	13.5	12.98	12.91	12.02	11.81
28d	13.41	13.06	12.96	12.31	11.88

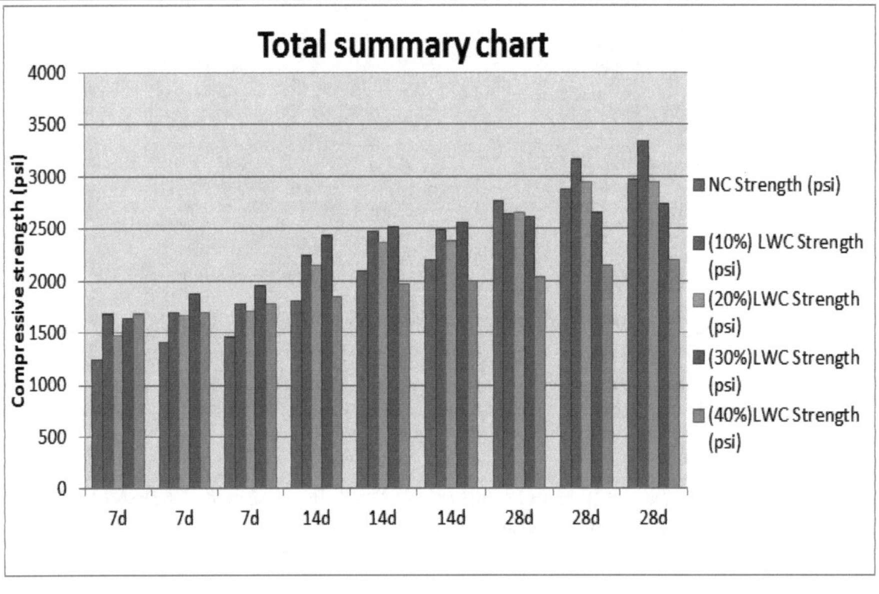

Figure 4: Total summary chart of compressive strength.

The graph of figure 4 shows the increase in strength of different samples with increase of LWA which were tested for 7days, 14days and 28days respectively. After 28days strength the 10% mix of LWA shows a tremendous increase in its compressive strength.

3.4 Difference between Weights of NWC & LWC is shown in table 3:

Table 3: Total summary table of weight difference of different samples

Days	Control Strength (psi)	(10%) LWC Strength (psi)	(20%)LWC Strength (psi)	(30%)LWC Strength (psi)	(40%)LWC Strength (psi)
7d	1239	1681	1480	1642	1685
7d	1411	1695	1660	1877	1691
7d	1460	1772	1701	1948	1773
14d	1798	2245	2151	2441	1841
14d	2089	2479	2370	2521	1964
14d	2201	2488	2379	2558	2001
28d	2766	2647	2655	2607	2037
28d	2875	3163	2936	2649	2144
28d	2964	3344	2942	2736	2198

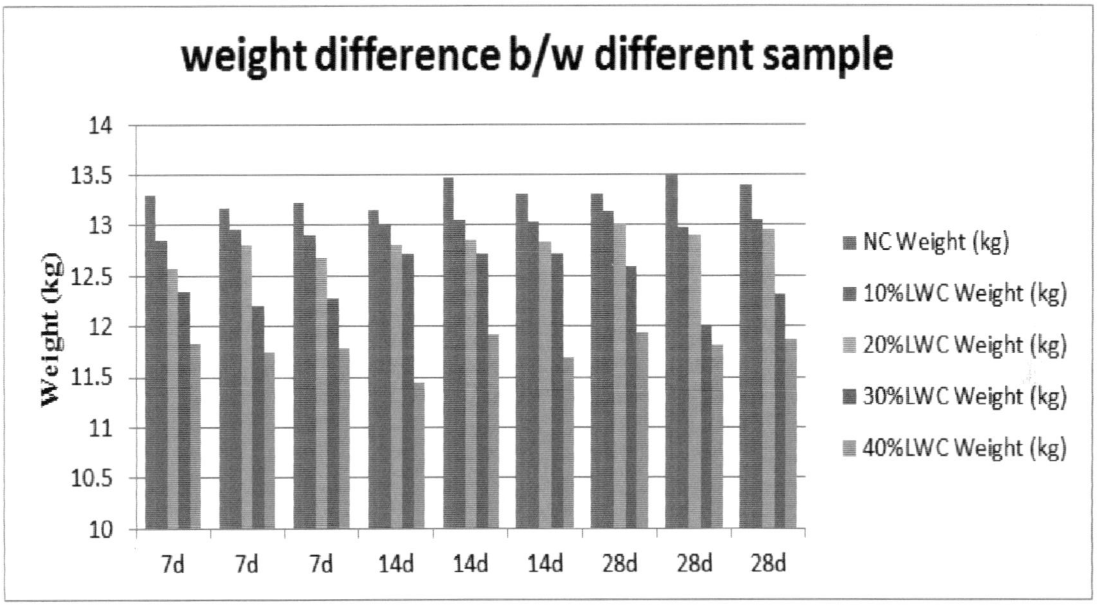

Figure 5. Graph showing the detail about the weight difference

The weight difference of sample 10%, 20%, 30% and 40% mix are shown in the graph of figure 5. The variation in weight of various samples is on regular basis i.e. there is no sudden increase or sudden decrease.

4. Conclusions

a. Incorporating 10% slate as light weight aggregate in concrete shows maximum strength at all ages.
b. The strength of light weight aggregate concrete increases by 12.8% at 28 days when 10% light weight aggregate is added.

c. Incorporating 40% light weight aggregate shows decline in strength by 25%.
d. The addition of Light Weight Aggregate in concrete mix, the dead load reduces about 2.5% with the increase of every 10% batch which makes the total reduction of dead load by 10% when 40% light weight aggregate is added.
e. The workability of light weight aggregate concrete tends to increase, the maximum value for slump was noted at 40% replacement which was 12.73% compared to control mix.

5. Recommendations

a. The Super plasticizer should be used to reduce the water cement ratio of the lightweight concrete and increase also their compressive strength.
b. The strength of LWAC should be increases with the use steel fiber with the specific ratio.
c. The strength of the Light Weight Aggregate Concrete should be enhancing with the use of bagasse ash with different trials.

References

[1] Alqahtani Fahad, Ghataora Gurmel, Khan Iqbal, Dirar, Samir Kioul, Azzedine Al-Otaibi, Mansour, Lightweight concrete containing recycled plastic aggregates. 2018;
[2] Bashar A, Manzur T, Yazdani N. Improving performance of light weight concrete with brick chips using low cost steel wire fiber. Constr Build Mater 2016;106:575–83.
[3] Tang C, Chen H, Wang S, Spaulding J. Cement & Concrete Composites Production of synthetic lightweight aggregate using reservoir sediments for concrete and masonry. Cem Concr Compos 2011;33(2):292–300.
[4] Reddy MVS, Nataraja MC, Sindhu K. Performance of Light Weight Concrete using Fly Ash Pellets as Coarse Aggregate Replacement. 2016;9(2):95–104.
[5] Mo KH, Alengaram UJ, Jumaat MZ. Bond properties of lightweight concrete – A review. Constr Build Mater 2016;112:478–96.
[6] Alengaram UJ, Abdullah B, Muhit A, Zamin M. Utilization of oil palm kernel shell as lightweight aggregate in concrete – A review. Constr Build Mater 2013;38:161–72.
[7] Hassanpour M, Shafigh P, Bin H. Lightweight aggregate concrete fiber reinforcement – A review. Constr Build Mater 2012;37:452–61.
[8] H. Al-Khaiat and M.N. Haque1, effect of initial curing on early strength and physical. 1998;28(6):859–66.
[9] Kakade SA, Dhawale AW. light weight aggregate concrete by using coconut shell. 2015;3(3):127–9.
[10] Atis CD. An investigation on the use of shredded waste PET bottles as aggregate in lightweight concrete. 2010;30:285–90.
[11] ASTM C 127, Specific Gravity, and Absorption test of coarse aggregate.
[12] ASTM C 136, Standard test for sieve analysis of fine aggregates.
[13] ASTM C 143, Standard test for slump of hydraulic cement concrete.

Effect of brick dust and lime powder on the performance of plain cement concrete

Abdul Basit[1], M Aftab Khan[2], Ibrar Ahmed[2], M Nasir Ayaz Khan[2] and M Umar[3]

[1]Swedish College of Engineering & Technology, Wah Cantt, Pakistan
[2]University of Engineering and Technology, Taxila, Pakistan[1]
[3] Military College of Engineering Risalpur, Pakistan

Email. basit.ukhan@gmail.com

Abstract. This paper illustrates the partial replacement of cement by brick dust and lime powder. The environmental condition is getting worse around the globe. One of the major causes of environment is waste generated from factories and working sites. The waste goes into dumping which not only occupy land but also cause ill effects to living. This unruly can be solved up to some degree by using these wastes in concrete as a cement replacement. An effort is made in this research to use lime powder and brick dust in plain concrete as cement replacement. The lime powder and brick dust was first used separately in concrete and then used a combination of brick dust and lime powder. Lime powder and brick dust was used separately in concrete up to 20% replacement with 5% interval. Than 5% brick dust was used with 15% lime powder, 10% brick dust with 10% lime powder, 15% brick dust with 5% lime powder. Using 20% brick dust separately in concrete gives good workability but less compressive strength compared to control sample. Replacing cement by 20% lime powder enhanced 3% compressive strength and split tensile strength compared to control mix. Whereas replacing cement with 15% lime powder and 5% brick dust gives optimum compressive strength and tensile strength compared to control mix. The results shows that lime powder increase the compressive strength and tensile strength compared to concrete made with brick dust. Using 20% lime and brick dust can save cost of cement in range of 7.2%- 12.5% which ultimately save the cost of a project.
Keywords: Lime powder, Brick dust, waste, workability, compressive strength, tensile strength.

1. Introduction

Many researchers are using waste materials which replace cement in concrete. Such materials provides competitive or more strength than conventional concrete. These materials properties must be evaluated in order to use them in feasible amount in concrete to get better results.

According to United Nation numbers, every year around two billion tons of wastes are generated each year in which least are recycled. There are many efforts in process to minimize waste generation around the world. Zero-waste companies are philosophy in which zero or neglecting amount wastes are generated. The waste generated is reuse in the same company. Likewise, many researchers are offering their services to reuse waste materials in concrete in order to get the same properties of concrete. [1]

Zhu et al [2], concluded that using waste of clay bricks decreases the followability and flexural strength of concrete, where compressive strength is changed slightly and chloride penetration was found good at 27% replacement. Aliabdo et al [3] reported that using crushed clay bricks tends to improve cement paste structure checked by X-ray Diffraction analysis (XRD), reduces number of pores and increases the density of structure. Aliabdo et al [4] used crushed bricks in cellular concrete and concluded that up to 50% replacement the compressive strength tends to increase while unit weight and thermal conductivity is reduced, however, porosity is increased. Dang et al [5] reported

IOP Conf. Series: Materials Science and Engineering **414** (2018) 012008 doi:10.1088/1757-899X/414/1/012008

that waste clay bricks possess pozolancity which attribute to increase in strength where microscopic analysis confirms the presence of SiO_2 and Al_2O_3 in waste clay bricks.

Burroughs et al [6] concluded that using lime powder decreases mixing time, reduce cost and CO_2. Turkel and Altuntas [7] studied that 30% replacement of cement with lime powder have positive effect on fresh properties, however, certain amount of lime powder, fly ash and silica fume can improve the workability together. Kumar et al [8] studied the comparison between intergrinding and blended lime stone and concluded that chemistries are same as for ordinary Portland cement. Liu et al [9] concluded that flexural strength increases by increasing limestone powder, compressive strength slightly changes where viscosity significantly decreases. Kanellopoulos et al [10] concluded that recycled lime powder is good alternative to use as cement replacement.

This work investigates the effect of using brick dust and lime powder in plain cement concrete. The trend of variation in workability, compressive strength and tensile strength of all the mixes with different percentages of cement replaced with brick dust and lime powder were observed. The mixes were prepared by replacing cement with bricks dust and lime powder separately as well as simultaneously. The results observed show an increasing trend in workability compare to control mix for both brick dust and lime powder. The compressive and tensile strength for introducing brick dust in concrete tends to reduce while lime powder alone has an enhancing effect on strength. Using brick dust and lime powder simultaneously in concrete give us a comparable strength concrete which will also decrease the cost of concrete.

2. Mix proportion

Different mixes were prepared with brick dust designated M0, MB1, MB2, MB3 and MB4 shows 0%, 5%, 10%, 15% and 20% brick dust respectively. Likewise, mixes prepared with lime powder M0, ML1, ML2, ML3 and ML4 shows 0%, 5%, 10%, 15% and 20% lime powder respectively. Also, different mixes MBL0, MBL1, MBL2, MBL3, MBL4 shows mixes prepared with 0% brick dust and lime powder, 15% brick dust and 5% lime powder, 10% brick dust and 10% lime powder, 7% brick dust and 13% lime and 5% brick dust and 15% lime powder respectively. Mix proportioning of three standard cylindrical samples by weight is shown in table 1.

Table 1. Proportioning of concrete mixes

S No	Mix	W/C	Brick Dust		Powder Lime		Cement (Kg)	Sand (Kg)	Coarse aggregate (Kg)
			%	Kg	%	Kg			
1	Mo	0.6	0	0	0	0	5.57	11.14	22.28
2	MB5	0.6	5	0.27	0	0	5.3	11.14	22.28
3	MB10	0.6	10	0.55	0	0	5.01	11.14	22.28
4	MB15	0.6	15	0.83	0	0	4.74	11.14	22.28
5	MB20	0.6	20	1.14	0	0	4.45	11.14	22.28
6	ML5	0.6	0	0	5	0.27	5.3	11.14	22.28
7	ML10	0.6	0	0	10	0.55	5.01	11.14	22.28
8	ML15	0.6	0	0	15	0.83	4.74	11.14	22.28
9	ML20	0.6	0	0	20	1.14	4.45	11.14	22.28
10	MBL1	0.6	15	0.83	5	0.27	4.45	11.14	22.28
11	MBL2	0.6	10	0.55	10	0.55	4.45	11.14	22.28
12	MBL3	0.6	7	0.38	13	0.78	4.45	11.14	22.28
13	MBL4	0.6	5	0.27	15	0.83	4.45	11.14	22.28

IOP Conf. Series: Materials Science and Engineering **414** (2018) 012008 doi:10.1088/1757-899X/414/1/012008

3. Experimental Program

3.1. Material testing

To determine the quantity of different chemical compounds in brick dust EDX (Energy Dispersive X-rays Spectroscopy) test was conducted in physics department lab, University of Peshawar. The results in percentage are shown in figure 1.

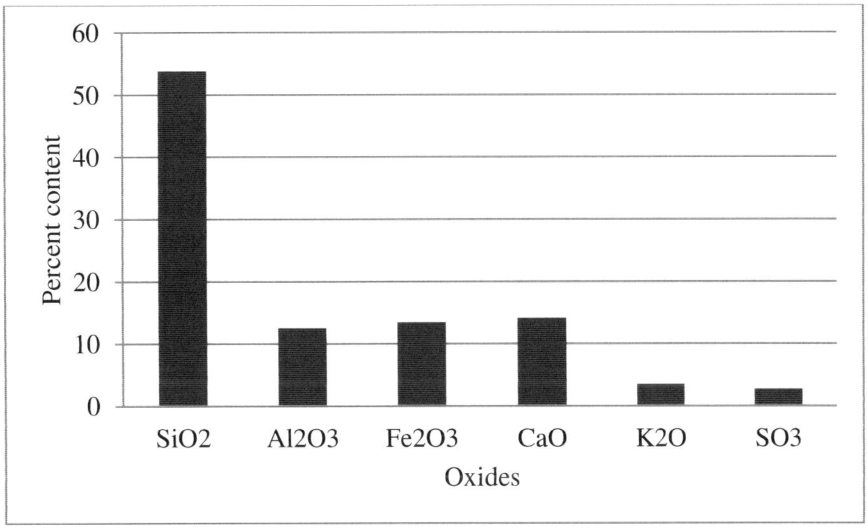

Figure 1. Oxides Composition of Brick Dust.

Lawrencepur sand was passed through sieves to calculate fineness modulus which was 2.41 according to ASTM C-136. Coarse aggregates obtained from Margalla were sieved according to ASTM C-136. Concrete cylinders having 6-inch diameter and 12 inches length are cast according to ASTM C-31. Kohat Cement of ASTM Type 1 was used in this research.

3.2. Workability evaluation

Slump test was performed on all above-mentioned mixes according to ASTM C-143. The results are graphically shown in Figure 2.

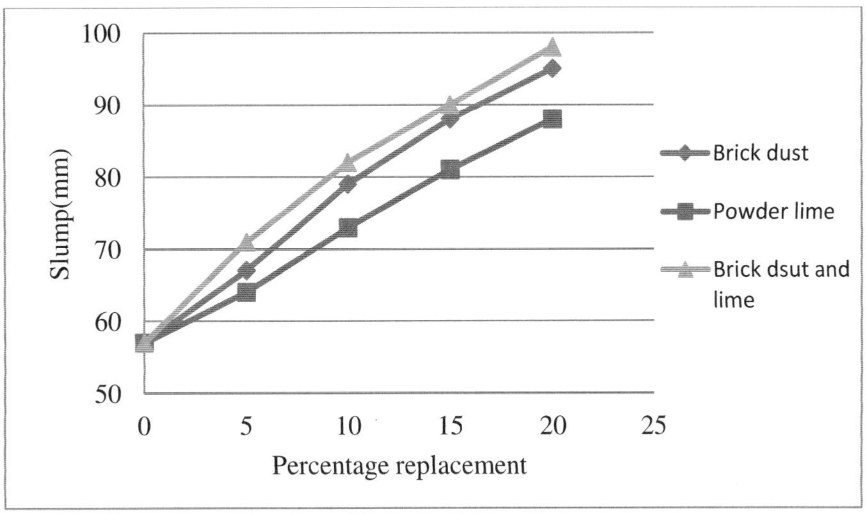

Figure 2. Slump test.

3.3. Compressive strength test

Compressive Strength for the mixes was checked according to ASTM C-139. Total 45 cylinders having size 12 inches in length and 6 inches in dia. the concrete moulds were properly oiled before filling concrete. The concrete cylinders were cured for 28 days according to ASTM C-192. The results were compared with control sample which are graphically represented in figure 3.

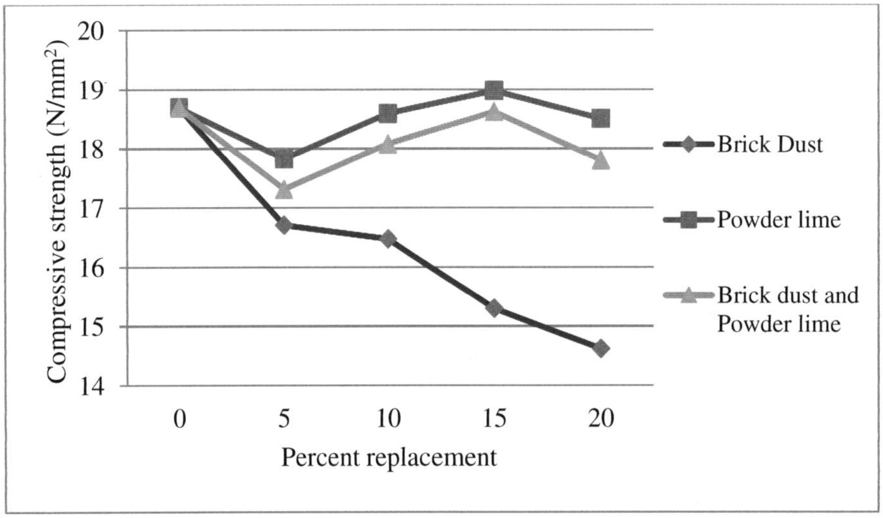

Figure 3. Twenty Eight Days Compressive Strength.

3.4.Split tensile strength test

Tensile Strength for the mixes was checked according to ASTM C-406. Total 45 cylinders having size 12 inches in length and 6 inches in diameter were tested. The concrete molds were properly oiled before filling concrete. The concrete cylinders were cured for 28 days according to ASTM C-192. The results were compared with control sample which are graphically represented in figure 4.

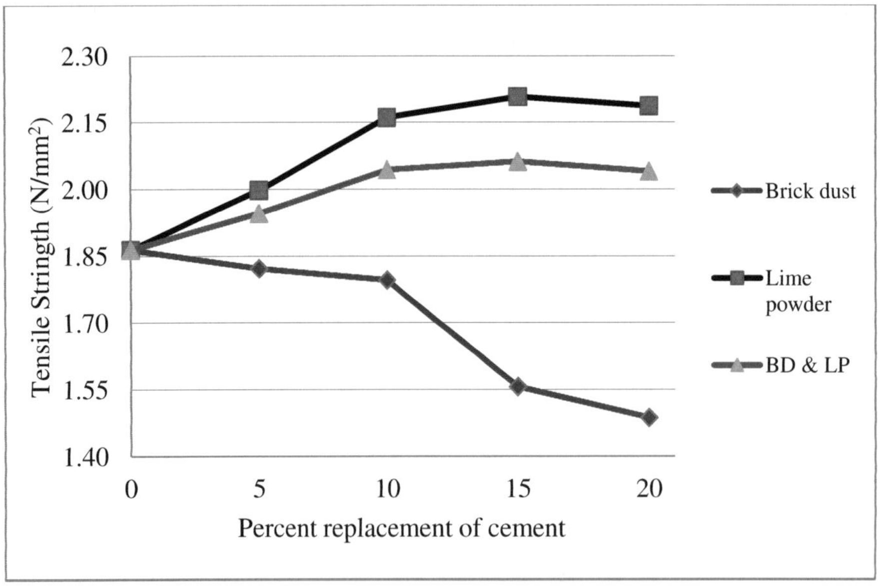

Figure 4. Twenty Eight Days Split Tensile Strength.

IOP Conf. Series: Materials Science and Engineering **414** (2018) 012008 doi:10.1088/1757-899X/414/1/012008

4. Results and Discussions

4.1 Workability

Concrete mix prepared with brick dust shows gradual increase in workability by increasing brick dust compared to control mix with no brick dust. The reason explained for the increment is due to the spherical particles of brick dust which can move on each other easily in mixing. The smoothness tends to decrease friction between particles. From the result, workability is increasing by replacing cement with powder lime. This increase may be due to the ball bearing effect of powder lime. The particles of powder lime assist other particles in moving which shows increase in workability.

Result shows that mixing brick dust and powder lime simultaneously in concrete shows dominant increase in workability. Reasons of improvement in workability may be due to the fineness of brick dust and powder lime particles. Also, these are fine particles with spherical shape so it can move easily with ball bearing effect during mixing.

4.2 Compressive strength

Results of samples made with brick dust reveals that increasing brick dust content decreases strength at all replacements. The strength is lower than control mix. The reduction in strength may be due to the enhancement of workability which provides flowability to spherical particle of sand comparatively less production of C-S-H (Calcium silicate hydrates).

Concrete samples made with powder lime shows that incorporating 5% and 10% powder lime decreases strength compared to control mix where increases compressive strength at 15% replacement. Beyond this, the strength decreases.

Incorporating 5% powder lime and 15% brick dust shows compressive strength lower than control mix where using 15% powder lime and 5% brick dust shows maximum compressive strength. This may be due to the extra C-S-H produced in presence of water.

4.3 Spilt tensile strength

By using brick dust as replacement of cement reduce the tensile strength of concrete in a continuous manner as shown in figure 4. The trend in reduction of tensile strength is continuous; as brick dust percentage is increased the tensile strength is decreased. While for incorporating powder lime separately in concrete by said amount gives an increasing trend in tensile strength. But for 20% or more of cement replaced with powder lime shows a slight reduction tensile strength, although it is still more than the strength of control mix.

5. Cost Analysis

As brick dust is a cheap and easily available material, thus it affects the cost of project. Below is the comparison between the cost of and cost of brick dust and lime.

Table 2. Cost analysis

S. No	Particulars	Cost(PKR)
1	Cost of cement per bag	500
2	Cost of cement per kg	11
3	Cost of brick dust per kg	2
4	Cost of powder lime per kg	12
5	Total cement used in this project	98
6	Cost of total cement used in this project	1078
	If 20 % cement is replaced with 10% brick dust and 10% powder lime	
7	Cost of brick dust	19.2
8	Cost of powder lime	117
9	Total cost of this mix	999.6
10	Percent reduction in cost	7.3%
11	Reduction in cost when 15 % of Brick dust and 5% of Lime powder is used	12.4%

6. Conclusions

The following conclusions can be made from this study.

- Incorporating brick dust enhances the workability of mixes at all replacements. The maximum workability was observed at 15% replacement. At 20% the slump tends to decrease.
- The same pattern was observed for lime powder and simultaneously lime powder and brick dust. The maximum slump was observed at 15%
- Results shows that incorporating 15% lime powder gives maximum compressive strength and tensile strength compared to control sample.
- Incorporating 15% powder lime and 5% brick dust shows maximum strength beyond this ratio strength tends to decrease.
- Using powder lime and brick dust can save cost of cement in range of 7% - 12.5%, which eventually save cost of the project.

References

[1] Nations U, Programme E. UNEP promotes environmentally sound practices globally and in its own activities . Available at: https://www.uncclearn.org/sites/default/files/inventory/unep06.pdf

[2] Zhu P, Mao X, Qu W, Li Z, John Z 2016 Investigation of using recycled powder from waste of clay bricks and cement solids in reactive powder concrete. *Constr Build Mater.* **113**:246–54.

[3] Aliabdo AA, Abd-Elmoaty AEM, Hassan HH 2014 Utilization of crushed clay brick in concrete industry. *Alexandria Eng J.* **53(1)**:151–68.

[4] Aliabdo AA, Hassan HH 2014 Utilization of crushed clay brick in cellular concrete production. *Alexandria Eng J.* **53(1)**:119–30.

[5] Dang J, Zhao J, Hu W, Du Z, Gao D 2018 Properties of mortar with waste clay bricks as fine aggregate. *Constr Build Mater.* **166**:898–907.

[6] Burroughs JF, Shannon J, Rushing TS, Yi K, Gutierrez QB, Harrelson DW 2017 Potential of finely ground limestone powder to benefit ultra-high performance concrete mixtures. *Constr Build Mater.* **141**:335–42.

[7] Git YI, Urkel UKT 2009 The effect of limestone powder , fly ash and silica fume on the properties of self-compacting repair mortars. **34**(April):331–43.

[8] Kumar A, Oey T, Puerta G, Henkensiefken R, Neithalath N, Sant G 2013 A comparison of intergrinding and blending limestone on reaction and strength evolution in cementitious materials. *Constr Build Mater* **43**:428–35.

[9] Liu J, Song S, Sun Y, Wang L 2011 Influence of Ultrafine Limestone Powder on the Performance of High Volume Mineral Admixture Reactive Powder Concrete.**153**:1583–6.

[10] Kanellopoulos A, Nicolaides D, Petrou MF 2014 Mechanical and durability properties of concretes containing recycled lime powder and recycled aggregates. *Constr Build Mater.* **53**:253–9.

IOP Conf. Series: Materials Science and Engineering **414** (2018) 012009 doi:10.1088/1757-899X/414/1/012009

Development of Textile Based Strain Sensor from Polypyrrole

Abdul Malik Rehan Abbasi[1], Zamir Ahmed Abro [1,3] and Syed Kamran Sami[2]

[1]Department of Textile Engineering, BUITEMS, Quetta 87300, Pakistan
[2]Department of Chemical Engineering, BUITEMS, Quetta 87300, Pakistan
[3]College of Textiles, Donghua University, Shangai, China

Email. rehan_abbaci@hotmail.com

Abstract. The conducting polymers are polyconjugated, which possess electronic properties of metals, while retaining the mechanical properties and processability of conventional polymers. Conductive polymers can only withstand limited strain before breaking and cannot perform well in evaluating large strains. The aim of this study was to develop a low cost, small to large strain sensor using Polypyrrole and Latex/Polyamide 6 yarn. The stretchable yarn was chosen as the substrate due to its excellent resilience and elasticity. Polypyrrole was coated as thin film onto the substrate by means of vapour deposition technique. The response of resistance of the samples on 2% deformation and relaxation during 40 cycles was analysed. The sensitivity or the change in resistance per unit deformation was used as a tool to figure out the suitability of strain sensor. The high resistive sample gave better sensitivity as well as uniformity as compared to low resistive sample which made it suitable to use as a strain sensor.
Keywords. Strain sensor, polymers, polypyrrole

1. Introduction

Electrically conducting polymer films can only withstand limited strain before breaking and cannot perform well in evaluating large strains [1]. To overcome this problem, substrates were employed to provide the necessary support and the surface for conducting polymer film deposition. Generally, fabricating a strain sensor using this approach means that the mechanical properties are highly attributed to the substrate while the conducting polymer introduces the electrical conductivity. This practice is commonly found in the research field of smart textiles which are conductive fabrics produced by coating conducting polymers onto commercial fabrics such as nylon, polyester and Lycra [2-5]. Although excellent results have been demonstrated with smart textiles using conducting polymers, the intended applications are mainly aimed at enhancing the usability of fabric beyond its current use as a protective layer. As a general-purpose strain sensor, the substrate requires having some degree of rigidity and fabrics are not an ideal material due to its soft structure. Furthermore, repetitive strain can cause permanent elongation on individual fibres where the strain may not be distributed equally. This can lead to individual fibres having different mechanical properties that will affect the strain sensing performance. The proposed solution is to replace fabric with Latex/PA6 stretchable yarn, which has good combination of rigidity and elasticity. One of the studies has succeeded in fabricating a strain sensor using PPy and natural rubber substrate where PPy powder is embedded into the structure of the rubber directly [6]. Compared to the coating methods, that approach requires knowledge of rubber manufacture as well as an access to the equipment to produce rubber with consistent mechanical properties.

This work is aimed at developing a low cost, small to large strain sensor using PPy and Elastodien/PA6. This stretchable yarn was chosen as the substrate due to its excellent resilience and elasticity. Commercial Elastodien/PA6 strip was purchased and used to produce the strain sensor. PPy as thin film was coated onto this substrate by means of vapour phase polymerisation technique that provides a good adhesion between the two components of the strain sensing element.

2. Materials

2.1 Elastodiene /polyamide-6

For developing the strain sensor, PA6 wrapped Elastodien yarn manufactured by PEGA® Czech Republic, was selected as a substrate for the deposition of PPy. The microscopic image is shown in figure 1 and the specifications of this yarn are described in table 1.

Table 1. The particulars of PA6 wrapped Elastodien yarn

Properties		**Index**
Diameter of Elastodien [mm]		0.6
Effective diameter of whole thread* [mm]		0.68
Linear density of Elastodien [tex]		36.8
Fineness of PA6 [tex]		22x3
No. of turns of PA6 around Elastodien [cm^{-1}]		12
Composition [%]	PA6	42
	Elastodiene	58

*Yarn packing density was assumed as $\mu = 0.525$ [7]

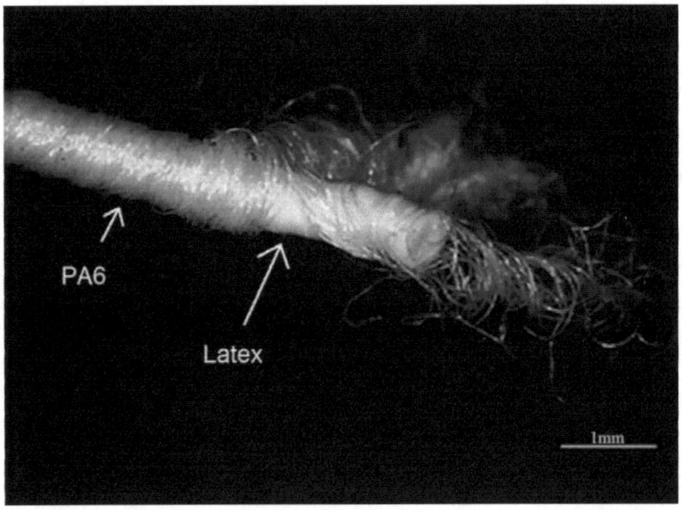

Figure 1 The microscopic image of Elastodien wrapped with PA6

2.2 Chemicals

Pyrrole is available in the market with different synonyms such as 1H-Pyrrole, Divinylenimine, Imidole, Azole, Monopyrrole, 1-Aza-2,4-cyclopentadiene. For this study Pyrrole 98% was received from Alfa Aesar® and was used after distillation to remove colouring impurities.

Anhydrous Iron(III) Chloride 97% was ordered from Sigma-Aldrich® and was used as received. It has molar mass 162.2 $g.mol^{-1}$ and 2.898 $g.cm^{-3}$ density. Tetraethylammonium p-toluenesulfonate also called tosylate (TsO^-) was obtained from Aldrich and was used as received.

3. Methodology

IOP Conf. Series: Materials Science and Engineering **414** (2018) 012009 doi:10.1088/1757-899X/414/1/012009

3.1 Sample preparation

An aqueous liquor bath was prepared using 2:1 molar ratio of $FeCl_3$ and TsO^-. This molar ratio between oxidant and dopant was optimized by the series of experiments in our previous work [8]. As $FeCl_3$ follows exothermic reaction when comes in contact with water, ice was used to reduce the temperature of the solution around 20°C. The yarn sample was immersed in bath for 1 min followed by squeezing at 70% pickup (maximum pickup by yarn) with the help of pneumatic squeezing rollers. Sample was then placed with the help of holders in a glass desiccator filled with pyrrole vapours immediately after padding and squeezing. The temperature of desiccator was kept at 20±2°C and sample was remained into it for 6 h under ambient atmospheric pressure. Sample was then taken out and washed with ethanol in order to stop polymerization and subsequently with plenty of distilled water for several times to remove biproducts and unreacted chemicals. Finally, sample was dried in air at 20±2°C for 24 h. Three different samples were prepared in exactly the same way by varying the concentration of $FeCl_3$ and TsO^- as mentioned in the table 2.

Table 2. Recipes for the preparation of PPy coated Elastodien/PA6 yarn samples.

Sample label Elastodien/PA6	Iron Chloride [mol/L]	Tosylate [mol/L]
SY1	2.0	1.0
SY2	0.6	0.3
SY3	0.1	0.05

3.2 Measurement of electrical resistivity of PPy coated Elastodien/PA6 yarn samples

With the intention of measuring electrical resistance R of PPy coated Elastodien/PA6 samples, stainless steel clamps were used to hold the sample together with connecting wires and a digital multimeter. The electrical resistivity ρ of the yarn sample was calculated from the equation (1), by knowing average cross-sectional area of the yarn sample "a" and the length of the sample "L" between two measuring electrodes.

$$\rho = R\frac{a}{L} \qquad (1)$$

With the objective of determining the dependence of electrical resistance on length of a sample, the resistance was measured by varying the distance from 5 cm to 90 cm between measuring electrodes.

3.3 Measurement of sensitivity of PPy coated Elastodien/PA6 yarn samples

With the objective of characterizing sensitivity of PPy coated Elastodien/PA6 yarn samples against deformation, all the samples from Table 2 were subjected to cyclic loading by using tensile testing machine Labtest from LaborTech® Ltd Czech Republic. The cycle was set for 2% strain at the jaw moving speed of 100mm/min with a rest of 1 second at relaxed or initial position to synchronize normal human breathing rhythm.

One set of measurement includes 40 cycles, therefore in this way 5 sets of readings were recorded and sensitivity (dR/dε) of the yarn sample was monitored. Here dR is the average change in resistance (R_l-R_o) of the specimen calculated from 5 sets of 40 cycles during extension of 2% and dε is the change in length [mm]. R_l is the resistance of specimen at 2mm extension whereas R_o is the resistance at relaxation or initial point.

4. Results and discussion

IOP Conf. Series: Materials Science and Engineering **414** (2018) 012009 doi:10.1088/1757-899X/414/1/012009

4.1 SEM micrographs

The SEM micrographs were taken from TESCAN® VEGA and corresponding to sample SY3 is depicted in figure 2. It can be perceived from the figure that by vapour deposition technique, each individual fibre is covered by PPy thoroughly, even though the structure of Elastodien/PA6 stretchable yarn is very compact as shown in figure 1.

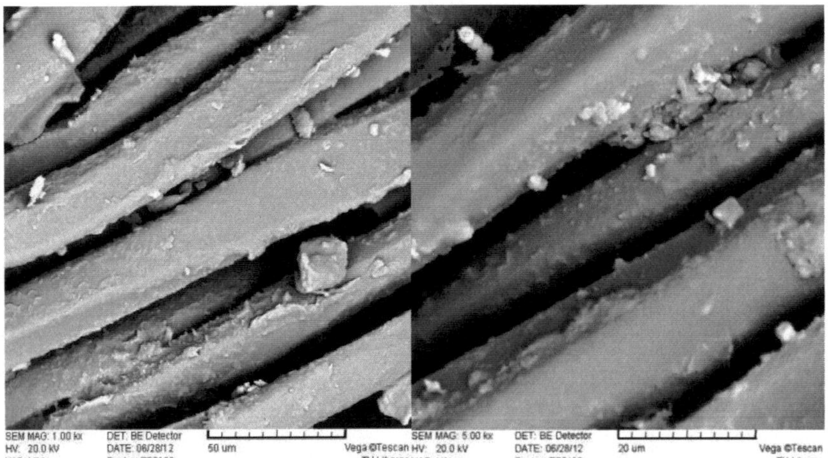

(a) (b)

Figure 2. SEM micrographs of PPy coated Elastodien/PA6 yarn sample SY3 (a) low magnification 1kx (b) high magnification 5kx.

4.2 Linear dependence of resistance on length

In order to determine the dependence of resistance on the length of PPy coated Elastodien/PA6 yarn, sample was clamped between two stainless steel clamp type electrode and resistance was measured by varying the distance between them as described earlier. It was found that resistance of PPy coated Elastodien/PA6 stretchable yarn is a linearly increasing function of the distance between the electrodes holding the yarn as shown in Figure 3. By the least squares regression for this model this relationship can be expressed as:

$$R=ZD+r \tag{2}$$

Where, R is the resistance [kΩ] of PPy coated Elastodien/PA6 yarn sample, D the distance between measuring electrodes and r the resistance of the electrodes found to be 0.0001kΩ approx. in this work. However, the slope Z is the directly proportional to the resistivity ρ in [kΩ.cm] of the PPy coated Elastodien/PA6 yarn samples for this experiment and can be calculated as:

$$Z=275.63\cdot\rho-0.0769 \tag{3}$$

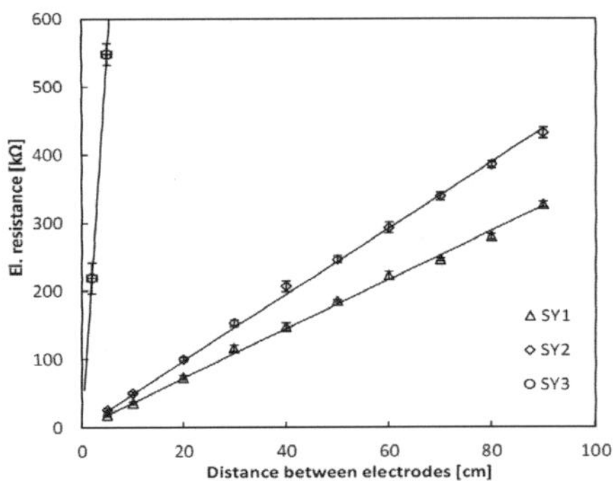

Figure 3. Dependence of electrical resistance of PPy coated Elastodien/PA6 yarn samples on length.

4.3 Sensitivity of strain sensor against cyclic loading

The Elastodien/PA6 stretchable yarn samples were coated by PPy through vapour deposition technique and each PPy coated sample was subjected to cyclic loading for 40 cycles. The response of resistance of the samples on 2% deformation and relaxation during 40 cycles are plotted in figure 4.

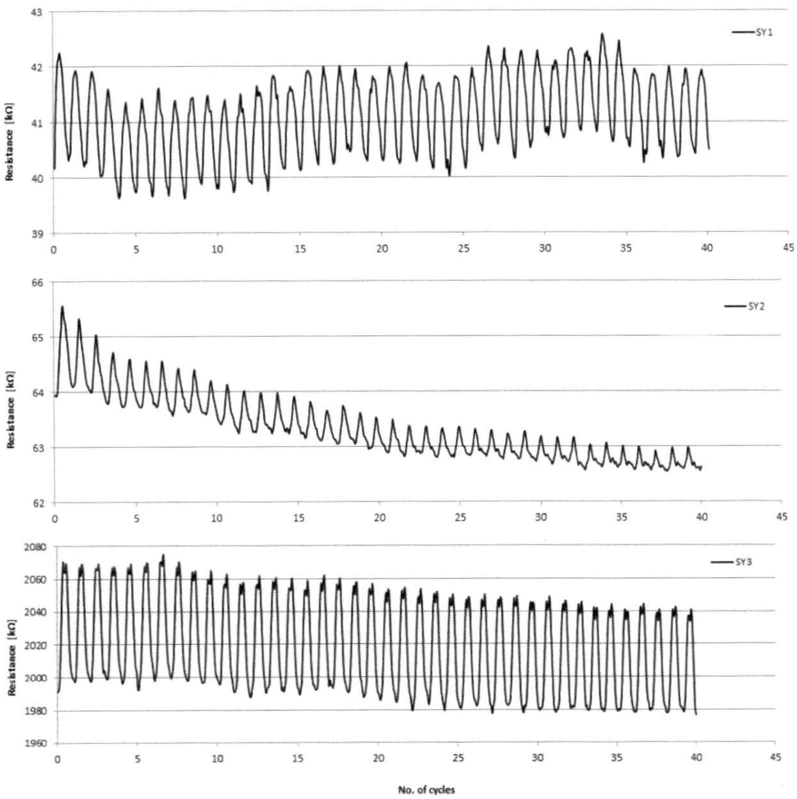

Figure 4. Response of PPy coated Elastodien/PA6 samples at 2% deformation for 40 cycles

From the figure 4 it can be observed that SY1 gives almost equal response dR against deformation in terms of magnitude but this response is not consistent with the number of cycles. The SY2 gives neither an equal response against deformation (decreases with number of cycles) nor the consistency of the response. Whereas SY3 is the best among all the samples and it gives not only an equal response upon deformation but also the level of consistency of the response after each cycle is outstanding. The average response in terms of change in resistance against deformation has been calculated from 40 cycles and named as sensitivity (dR/dε) of each sample. The sensitivity levels of all PPy coated Elastodien/PA6 samples are shown in Figure 5. Although SY3 has the highest resistivity among all three samples under study, however it outperformed SY1 and SY2 in terms of response against small extension. The sample SY2 has been found as the worst in terms of sensitivity and its deviation in the results. The standard deviations of the specimens were calculated as 0.182, 0.315 and 4.49 [kΩ./mm] for SY1, SY2 and SY3 respectively.

Although SY3 has the highest resistivity among all three samples under study, however it outperformed SY1 and SY2 in terms of response against small extension. The sample SY2 has been found as the worst in terms of sensitivity and its deviation in the results. The standard deviations of the specimens were calculated as 0.182, 0.315 and 4.49 [kΩ/mm] for SY1, SY2 and SY3 respectively.

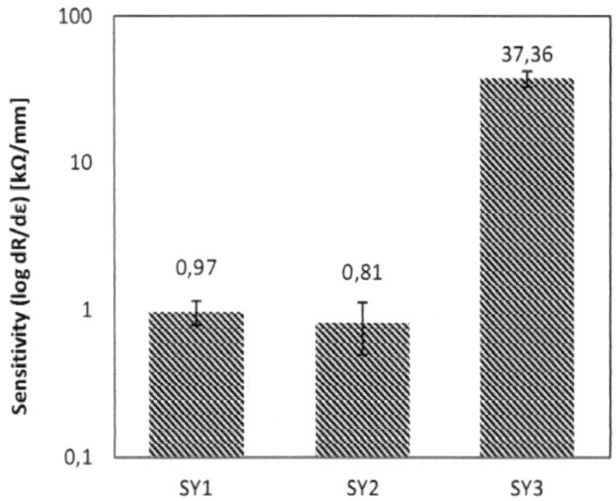

Figure 5. Dependence of sensitivity of the strain sensor on longitudinal deformation.

5. Conclusion

PPy was coated on PA6 wrapped Elastodien yarn samples through vapour phase polymerization by varying concentration of oxidant and dopant thus samples were varied in resistivity levels. It was found that resistance of all the samples follow exactly a linear function of length and hence follow Ohm's law.

The sensitivity or the change in resistance per unit deformation is a useful tool to figure the suitability of the strain sensor out. The high resistive sample gives best sensitivity as well as uniformity or the consistency of the response against small cyclic deformation (2%) whereas low resistive samples fail to give the equal and uniform response. Therefore, it is suggested that this product can be used to monitor human body movements even for very small deformations such as breathing.

References

[1] Murray P, Spinks G, Wallace G, and Burford R 1998 *Synthetic metals*, **97** 117
[2] Li Y, Cheng X, Leung M, Tsang J, Tao X, and Yuen M 2005 *Synthetic metals*, **155** 89
[3] Xue P, Tao X, and Tsang H 2007 *Applied surface science*, **253** 3387
[4] Wu J, Zhou D, Too C O, and Wallace G G 2005 *Synthetic Metals*, **155** 698
[5] Molina J, Del Río A, Bonastre J, and Cases F 2008 *European Polymer Journal*, **44** 2087
[6] Bunsomsit K, Magaraphan R, Edgar A, and Grady B P 2002 *Colloid and Polymer Science*, **280** 509
[7] Militký J, Vik M, Viková M, and Křemenáková D 2004 *Proceedings of 2nd SIENTEX Conference "International Symposium of Textile Engineering* (Natal: Associação Brasileira de Técnicos TEXTEIS)
[8] Abdul Malik R, Ramadan M A, Wiener J, Baheti V, and Militky J 2013 *Journal of Textile Engineering*, **59** 93

Features of Convective heat transfer on MHD peristaltic movement of Williamson fluid with the presence of Joule heating

Naheeda Iftikhar[1], Abdul Rehman[2] and Muhammad Najam Khan[3]

[1]Department of Mathematics, BUITEMS, Quetta, Pakistan
[2]Department of Mathematics, University of Balochistan, Quetta, Pakistan
[3]Department of Chemical Engineering, BUITEMS, Quetta, Pakistan

Email: naheeda_iftikhar@yahoo.com

Abstract. This article reflects the features of convective heat transfer with radially magnetic field of an incompressible Williamson fluid with peristaltic motion and Joule heating in a curved channel. Mathematical study has been carried out with low Reynolds number under consideration of long wavelength approximations. The system of nonlinear equations with small values of Weissenberg number has been obtained and solution of these equations has been constructed. The spectacular characteristics of flow quantities are carried out with special focus on velocity, temperature, heat transfer and trapping. It is examined that pressure gradient heightens for large values of Weissenberg number and magnetic field parameter. The velocity and temperature shows low behaviour for magnetic field parameter and moreover Weissenberg and Biot numbers has opposite effect on temperature distribution. Streamlines pattern are also represented with different parameters.

Keywords: Williamson fluid, peristaltic motion, Joule heating, convective boundary conditions, HPM Solution.

1. Introduction

Involuntary movements of the tubular circular and longitudinal muscles is the peristaltic motion, principally in the digestive tract but once in a while in other vacuous tubes of the body has made grievous attraction to the recent researchers for the expanse of this field in fluid dynamics. This development has found in Physiology; Peristalsis is an inherent property of many syncytial smooth hairs Myotubes, Simulation of some point can cause a contractile ring in the body Sphincter of the intestine, and this ring then spreads along the tube. In this way, Peristalsis occurs in gastrointestinal tract, bile ducts and other gland ducts throughout the body, the uterus and many other smooth muscle tubes of the body. One main field of application of this principle is the construction of finger pumps, rollers pumps that are used in the conveyance of liquids without the contact with the pumping machine. The very first who made the interesting attempt of research on the peristaltic flow are Latham [1] and Shapiro et al [2]. Later on a very substantial study has been reported by the theoretical and experimental approach. One can inquire elite studies [3-8] and many references there in. Besides this interest also exist in the transportation of fluid of

peristaltic MHD flow in a channel linked with some problems of conductive physiological movement of fluids e.g. motility disorders in gastrointestinal tract, Urinary tract treatment problems, Theoretical and experimental investigation on the procedure of peristaltic MHD compressor. Study also exist on Peristaltic motion of (MHD) flow of viscous and non-Newtonian fluid. Temperature and flow of heat are concerned with heat transfer mechanism [9-13]. Temperature is basically about the property of thermodynamic. It depicts the use of amount of thermal energy, while the transfer of thermal energy from one to another place is flow of heat. The major categories of heat transfer are (i) conduction (ii) convection (iii) radiation. Moreover, in the procedure of oxygenation and haemodialysis the significance of thermal effects of blood made learning about the performance of heat transfer in peristalsis.
By literature review and considering heat transfer study for the peristaltic motion of viscous and non-Newtonian fluids [14-20]. Besides the transportation of fluid in an uncoiled channel there also exist curved channels which are used for fluid flows in industry and physiological applications.
Flexibility of the walls may cause of the peristaltic flow in the channel / tube. In the present literature, useful consideration of channel is taken for the peristaltic flows. Mittra and Prasad [21] considered peristaltic motion and studied wall properties effects. Davies [22] examined flow stability with complaint walls of the plane channel. Srivastav [23] studied the peristaltic flow of the mixture of particles and fluids with wall flexibility. Haroun [24] considered an asymmetric channel for peristaltic flow and studied the effects of complaint wall. Radhakrishnamacharya [25] analysed the peristaltic transport of heat transfer with effects of flexible walls. Muthu et al. [26] examined wall properties of the micropolar fluid with peristaltic motion in a circular cylindrical tube. Another study by Elnaby [27], where he studied wall properties effects on a viscous fluid with peristaltic motion.
Hayat et al. [28-29] considered the Johnson-Segalman and Jeffery fluids with peristaltic flow and examined the influence of complaint walls respectively. The same phenomenon considered by Ali et al. [30] with Maxwell fluid in a channel. Kothandapani [31] also examined the effect of wall properties by considering MHD peristaltic flow of viscous fluid but in a porous channel. The more work has been carried out by Srinivas et al. [32] in the existence of slip effects. Srinivas [33] also considered porous space and examined the effect of complaint walls on MHD peristaltic flow. Hayat and Hina [34] analyzed the effect of complaint walls by considering Maxwell fluid.
In view of literature cited above, it is cleared that insufficient studies have been done with curved channel on the peristaltic mechanism. Sato et al. [35] discussed peristaltic motion in a curved channel due to transverse rebounds of the walls. [36] Ali et al carried out analytical study of long wavelength approximation in a curved channel and peristaltic motion. Ali et al. [37-38] extended the investigation of ref. [34] for heat transfer characteristics of third grade fluid. Hayat et al. [39] studied Newtonian fluid with peristaltic motion with effect 0f complaint walls in a curved channel. This study is extended by himself [40] and Hina et al. [41] for third-grade and Johnson-Segalman fluids. The joint effect of heat and mass transfer with peristaltic flow of Johnson-Segalman fluid was studied by Hina et al. [42]. Author also discussed effect of wall properties with peristaltic motion of pseudoplastic fluid in a curved channel [43]. Another study [44] the slip effects of third grade fluid with the peristaltic transport. And Finally, [45] carried out numerical study of nanofluid with peristaltic motion in a curved channel. Few studies associated with our discussion are [46-49].

The present investigation based on the study of Williamson fluid with peristaltic flow and effects of radially magnetic field with analysis of convective heat transfer. Williamson fluid is the generalization of non-Newtonian fluid model. For the better understanding of the investigation, four main sections are developed. First section comprises of problem formulation under consideration of non-dimensional governing equations with geometry of the problem and convective boundary conditions in wave frame. Second section gives the solution of considered problem with large number of constants. Section three shows the graphical

IOP Conf. Series: Materials Science and Engineering **414** (2018) 012010 doi:10.1088/1757-899X/414/1/012010

results of emerging parameters on velocity, temperature, pressure gradient and coefficient of heat transfer and trapping phenomenon has also visualized in this section. And finally conclusion has summarized in last fourth section.

2. Formulation of the problem

The Peristaltic wave of an electrically conducting non-Newtonian incompressible substance in a curved channel of L^* with unvarying broadness $2\,a_0$ coiled in a circle with centre at o' is considered. By choosing a curvilinear coordinate system $(\widehat{G},\widehat{L})$ with unmoving frame and the circulization of wave along the $\widehat{G}$ coordinate and $\widehat{L}-$coordinate is in normal direction of the flow. In laboratory frame of reference, the velocity components are $\widehat{W}_1$ and $\widehat{W}_2$ in radial $(\widehat{L})$ and axial $(\widehat{G})$ direction respectively. The radial magnetic field of force $\widetilde{B}_0$ is subjected to the fluid and heat transfer is directed by the convective conditions at the walls of the channel.
Along the channel walls the motion of fluid is derived by sinusoidal wave circulating with uniform speed C_0. Mathematically we will have

$$\widehat{H}(\widehat{G},\widehat{t}) = \overset{+}{-}\left[a_{00} + a_{11}\,\mathrm{Si}\,n(\frac{2\pi'}{c_0}(\widehat{G} - \lambda c_0\widehat{t}))\right] \tag{1}$$

In this equation c_0 refers the speed, a_{11} is amplitude, λc_0 is wave length and $\widetilde{t}$ is time of the wave in unmoving frame. The velocity

$$\widehat{\mathbf{W}} = (\widehat{W}_1(\widehat{G},\widehat{L},\widehat{t}),\widehat{W}_2(\widehat{G},\widehat{L},\widehat{t}),0) \tag{2}$$

$$\widehat{\boldsymbol{\tau}} = -[\eta'_\infty + (\eta'_0 - \eta'_\infty)(1+\Gamma'|\dot{\boldsymbol{\gamma}}'|)]\dot{\boldsymbol{\gamma}}' \tag{3}$$

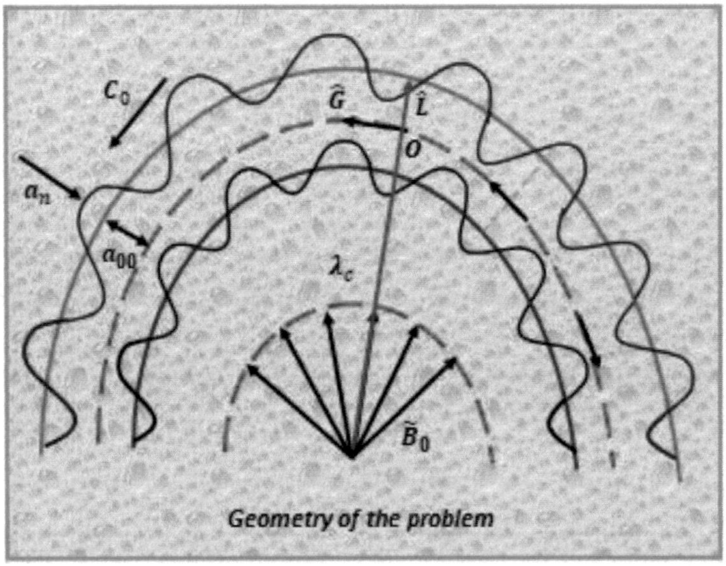

Figure 1. Geometry of the flow

$$\dot{\gamma}' = \sqrt{\frac{1}{2}\sum_j\sum_m \dot{\gamma}'_{jm}\dot{\gamma}'_{mj}} = \sqrt{\frac{\pi'}{2}}. \tag{4}$$

Where π' refers the second unvarying tensor. The constitutional equation for the case of $\eta'_\infty = 0$ bring down

$$\hat{\tau} = -\eta'_0[+(\eta'_0 - \eta'_\infty)(1+\Gamma'|\dot{\gamma}'|)]\dot{\gamma}' \tag{5}$$

In the laboratory frame, the governing equation for the motion of fluid are given by

$$\frac{\partial}{\partial \hat{L}}\{(L^{**}+\hat{L})\hat{W}_1\} + L^{**}\left(\partial \hat{W}_2 \middle/ \partial \hat{G}\right) = 0, \tag{6}$$

$$\rho'\left[\partial \hat{W}_1 \middle/ \partial \hat{t} + \hat{W}_1\left(\partial \hat{W}_1 \middle/ \partial \hat{L}\right) + L^{**}\left(\hat{W}_2 \middle/ L^{**}+\hat{L})\right)\frac{\partial \hat{W}_1}{\partial \hat{G}} - \left(\hat{W}_2^2 \middle/ L^{**}+\hat{L}\right)\right] = -\partial \hat{P} \middle/ \partial \hat{L} + 1 \middle/ (L^{**}+\hat{L})\frac{\partial}{\partial \hat{L}}\{(L^{**}+\hat{L})\hat{\tau}_{\hat{L}\hat{L}}\} +$$
$$\left(\frac{L^{**}}{L^{**}+\hat{L}}\right)\frac{\partial \hat{\tau}_{\hat{L}\hat{G}}}{\partial \hat{G}} - \hat{\tau}_{\hat{G}\hat{G}} \middle/ (L^{**}+\hat{L}) \tag{7}$$

$$\rho'\left[\partial \hat{W}_2 \middle/ \partial \hat{t} + \hat{W}_1\left(\partial \hat{W}_2 \middle/ \partial \hat{L}\right) + L^{**}\left(\hat{W}_2 \middle/ L^{**}+\hat{L}\right)\left(\partial \hat{W}_2 \middle/ \partial \hat{G}\right) - \hat{W}_1\left(\hat{W}_2 \middle/ L^{**}+\hat{L}\right)\right] = \left(L^{**} \middle/ L^{**}+\hat{L}\right)\frac{\partial \hat{P}}{\partial \hat{G}} +$$

$$\left(1 \Big/ (L^{**}+\hat{L})^2 \right) \frac{\partial}{\partial \hat{L}} \{((L^{**}+\hat{L})^2)\hat{\tau}_{\hat{L}\hat{G}}\} + \left(\frac{L^{**}}{L^{**}+\hat{L}} \right) \partial \hat{\tau}_{\hat{G}\hat{G}} \Big/ \partial \hat{G} - \sigma B_0^2 \hat{W}_2 \Big/ (L^{**}+\hat{L})^2 \tag{8}$$

$$\rho' c'_p \left[\partial \hat{T} \Big/ \partial \hat{t} + \hat{W}_1 \left(\partial \hat{T} \Big/ \partial \hat{L} \right) + \hat{W}_2 \frac{L^{**}}{L^{**}+\hat{L}} \partial \hat{T} \Big/ \partial \hat{G} \right] = k' \left[\left(\partial^2 \hat{T} \Big/ \partial \hat{L}^2 \right) + \frac{1}{L^{**}+\hat{L}} \left(\partial \hat{T} \Big/ \partial \hat{L} \right) + \frac{L^{**2}}{(L^{**}+\hat{L})^2} \left(\partial^2 \hat{T} \Big/ \partial \hat{G}^2 \right) \right]$$

$$+ \mu_0 \left[\left(\partial \hat{W}_2 \Big/ \partial \hat{L} - \hat{W}_2 \Big/ (L^{**}+\hat{L}) + \frac{L^{**}}{(L^{**}+\hat{L})} \left(\partial \hat{W}_1 \Big/ \partial \hat{G} \right) \right) \hat{\tau}_{\hat{L}\hat{G}} + \left(\partial \hat{W}_2 \Big/ \partial \hat{L} \right) \left(\hat{\tau}_{\hat{L}\hat{L}} - \hat{\tau}_{\hat{G}\hat{G}} \right) \right] \tag{9}$$

With

$$\hat{\tau}_{\hat{G}\hat{G}} = -2\mu_0 \eta_0 \left[1 + \Gamma' |\dot{\gamma}'| \right] \left(\frac{L^{**}}{(L^{**}+\hat{L})} \left(\partial \hat{W}_2 \Big/ \partial \hat{G} \right) + \hat{W}_1 \Big/ (L^{**}+\hat{L}) \right), \tag{10}$$

$$\hat{\tau}_{\hat{L}\hat{G}} = \mu_0 \eta'_0 \left[1 + \Gamma' |\dot{\gamma}'| \right] \left(\partial \hat{W}_2 \Big/ \partial \hat{L} + \frac{L^{**}}{(L^{**}+\hat{L})} \left(\partial \hat{W}_1 \Big/ \partial \hat{G} \right) - \hat{W}_2 \Big/ (L^{**}+\hat{L}) \right), \tag{11}$$

$$\hat{\tau}_{\hat{L}\hat{L}} = -2\mu_0 \eta'_0 \left[1 + \Gamma' |\dot{\gamma}'| \right] \left(\partial \hat{W}_1 \Big/ \partial \hat{L} \right). \tag{12}$$

Here $\hat{P}$ refers the pressure, ρ' is fluid density, c'_p is specific heat, $\hat{t}$ is time, $\hat{k}$ is thermal conductivity and $\hat{T}$ is the temperature of fluid. Unsteady flow is reasoned in unmoving frame $(\hat{G}, \hat{L})$ whereas in moving frame $(\hat{g}, \hat{l})$ the flow in the channel can be reasoned as steady flow.

The heat transfers because of convective cooling at the boundaries proceeds

$$\begin{aligned} \hat{k} \frac{\partial \hat{T}}{\partial \hat{L}} &= -\hat{\eta}_1 (\hat{T} - \hat{T}_0) \; at \; \hat{L} = \hat{\eta} \\ \hat{k} \frac{\partial \hat{T}}{\partial \hat{L}} &= -\hat{\eta}_2 (\hat{T} - \hat{T}_0) \; at \; \hat{L} = -\hat{\eta} \end{aligned} \tag{13}$$

By Preceding a wave frame $(\hat{g}, \hat{l})$ with wave moving with speed c_0 departed from the fixed frame $(\hat{G}, \hat{L})$ by the transformations

$$\begin{aligned} &\hat{g} = \hat{G} - c_0 \hat{t}, \; \hat{l} = \hat{L}, \; \hat{w}_1(\hat{g}, \hat{l}) = \hat{W}_1(\hat{G}, \hat{L}, \hat{t}), \; \hat{w}_2(\hat{g}, \hat{l}) = \hat{W}_2(\hat{G}, \hat{L}, \hat{t}) - c_0, \\ &\hat{T}(\hat{g}, \hat{l}) = \hat{T}(\hat{G}, \hat{L}, \hat{t}), \; \hat{p}(\hat{g}, \hat{l}) = \hat{P}(\hat{G}, \hat{L}, \hat{t}) \end{aligned} \tag{14}$$

And equations (6) -(12) yields

$$\frac{\partial}{\partial \hat{l}} \{ (L^{**}+\hat{l}) \hat{w}_1 \} + L^{**} \left(\partial \hat{w}_2 \Big/ \partial \hat{g} \right) = 0, \tag{15}$$

$$\rho' \left[\hat{w}_1 \left(\partial \hat{w}_1 \Big/ \partial \hat{l} \right) + \frac{L^{**}(\hat{w}_2 + c_0)}{(L^{**}+\hat{l})} \left(\partial \hat{w}_1 \Big/ \partial \hat{g} \right) - \frac{(\hat{w}_2 + c_0)^2}{(L^{**}+\hat{l})} \right] = -\frac{\partial P'}{\partial \hat{l}} + \frac{1}{(L^{**}+\hat{l})} \frac{\partial}{\partial \hat{l}} \{ (L^{**}+\hat{l}) \hat{\tau}_{\hat{l}\hat{l}} \}$$

$$+\left(\frac{L^{**}}{L^{**}+\widehat{l}}\right)\frac{\partial\widehat{\tau}_{\widehat{l}\widehat{g}}}{\partial\widehat{g}}-\frac{\widehat{\tau}_{\widehat{g}\widehat{g}}}{(L^{**}+\widehat{l})} \quad (16)$$

$$\rho'\left[-c_0\left(\partial\widehat{w}_2\big/\partial\widehat{g}\right)+\widehat{w}_1\left(\partial\widehat{w}_2\big/\partial\widehat{l}\right)+\frac{L^{**}(\widehat{w}_2+c_0)}{(L^{**}+\widehat{l})}\left(\partial\widehat{w}_2\big/\partial\widehat{g}\right)-\frac{\widehat{w}_1\widehat{w}_2}{(L^{**}+\widehat{l})}\right]=-\left(L^{**}\big/L^{**}+\widehat{l}\right)\frac{\partial P'}{\partial\widehat{g}}+\frac{1}{(L^{**}+\widehat{l})^2}\frac{\partial}{\partial\widehat{l}}\left\{(L^{**}+\widehat{l})^2\widehat{\tau}_{\widehat{l}\widehat{g}}\right.$$

$$+\left(L^{**}\big/L^{**}+\widehat{l}\right)\frac{\partial\widehat{\tau}_{\widehat{g}\widehat{g}}}{\partial\widehat{g}}-\frac{\sigma B_0^2(\widehat{w}_2+c_0)}{(L^{**}+\widehat{l})}$$

(17)

$$\rho'c_p'\left[\partial\widehat{T}\big/\partial\widehat{g}+\widehat{w}_1\left(\partial\widehat{T}\big/\partial\widehat{l}\right)+\frac{L^{**}(\widehat{w}_2+c_0)}{L^{**}+\widehat{l}}\left(\partial\widehat{T}\big/\partial\widehat{g}\right)\right]=\widehat{k}\left[\partial^2\widehat{T}\big/\partial\widehat{l}^2+\frac{1}{L^{**}+\widehat{l}}\left(\partial\widehat{T}\big/\partial\widehat{l}\right)+\left(\frac{L^{**}}{L^{**}+\widehat{l}}\right)\partial^2\widehat{T}\big/\partial\widehat{g}^2\right]+$$

$$\mu_0\left[\left(\frac{\partial\widehat{w}_2}{\partial\widehat{l}}-\frac{(\widehat{w}_2+c_0)}{(L^{**}+\widehat{l})}+\frac{L^{**}}{(L^{**}+\widehat{l})}\frac{\partial\widehat{w}_1}{\partial\widehat{g}}\right)\widehat{\tau}_{\widehat{l}\widehat{g}}+\frac{\partial\widehat{w}_2}{\partial\widehat{l}}(\widehat{\tau}_{\widehat{l}\widehat{l}}-\widehat{\tau}_{\widehat{g}\widehat{g}})\right]$$

(18)

$$\widehat{\tau}_{\widehat{g}\widehat{g}}=-2\mu_0\eta_0'\left[1+\Gamma'|\dot{\gamma}'|\right]\left(\frac{L^{**}}{(L^{**}+\widehat{l})}\frac{\partial(\widehat{w}_2+c_0)}{\partial\widehat{g}}+\frac{\widehat{w}_1}{(L^{**}+\widehat{l})}\right) \quad (19)$$

$$\widehat{\tau}_{\widehat{l}\widehat{g}}=\mu_0\eta_0'\left[1+\Gamma'|\dot{\gamma}'|\right]\left(\frac{\partial\widehat{w}_2}{\partial\widehat{l}}+\frac{L^{**}}{(L^{**}+\widehat{l})}\frac{\partial\widehat{w}_1}{\partial\widehat{g}}-\frac{\widehat{w}_2}{(L^{**}+\widehat{l})}\right) \quad (20)$$

$$\widehat{\tau}_{\widehat{l}\widehat{l}}=-2\mu_0\eta_0'\left[1+\Gamma'|\dot{\gamma}'|\right]\left(\frac{\partial\widehat{w}_1}{\partial\widehat{l}}\right) \quad (21)$$

By defining

$$g'=\frac{2\pi\widehat{g}}{\lambda_{c_0}},\ l'=\frac{\widehat{l}}{a_{00}},\ \hat{w}_1=\frac{\widehat{w}_1}{c_0},\hat{w}_2=\frac{\widehat{w}_2}{c_0},\ \hat{\delta}=\frac{2\pi' a_{00}}{\lambda_{c_0}},\ \widehat{\eta}=\pm\frac{\widehat{\mathbf{H}}}{a_{00}},\ P'=\frac{2\pi a_{00}^2\widehat{p}}{c_0\mu_0\lambda_{c_0}},\ W_e'=\frac{\Gamma' c_0}{a_{00}},$$

$$M'=\left(\frac{\sigma'}{\mu_0}\right)^{1/2}\widetilde{B}_0a_{00},\ \mathrm{Re}=\frac{\rho' c_0a_{00}}{\mu_0},\ \hat{k}=\frac{\overset{**}{L}}{a_{00}},\ P_r=\frac{\mu_0c_p'}{\hat{k}},\ \hat{t}=\frac{c_0\widehat{t}}{\lambda_{c_0}},\ \hat{\theta}=\frac{\widehat{T}-\widehat{T}_0}{\widehat{T}_0},$$

$$E_c'=\frac{c_0^2}{T_0'c_p},\ \hat{B}_r=P_rE_c' \quad (22)$$

$$\hat{w}_1=\hat{\delta}\frac{\widehat{k}}{l'+\hat{k}}\frac{\partial\Omega}{\partial g'},\ \hat{w}_2=-\frac{\partial\Omega}{\partial l'}$$

(23)

Equation (10) fulfilled and former expressions under wavelength and low Reynolds number come down to

$$\frac{dP'}{dl'}=0, \quad (24)$$

$$\frac{\hat{k}}{l'+\hat{k}}\frac{dP'}{dg'}=\frac{1}{(l'+\hat{k})^2}\frac{\partial}{\partial l'}\left\{(l'+\hat{k})^2\hat{\tau}_{l'\mathrm{g}}\right\}-\frac{M'}{(l'+\hat{k})^2}(1-\frac{\partial\Omega}{\partial l'}),$$

(25)

$$\frac{\partial^2\hat{\theta}}{\partial l'^2}+\frac{1}{l'+\hat{k}}\frac{\partial\hat{\theta}}{\partial l'}+B_r'\left[(\frac{\partial^2\Omega}{\partial l'^2}-\frac{1-\frac{\partial\Omega}{\partial l'}}{l'+\hat{k}})\hat{\tau}_{l'g'}\right]=0.$$

(26)

$$\hat{\tau}_{l'g'}=\left[1+W_e'(-\frac{\partial^2\Omega}{\partial l'^2}-\frac{1-\frac{\partial\Omega}{\partial l'}}{l'+\hat{k}})\right]\left(-\frac{\partial^2\Omega}{\partial l'^2}-\frac{1-\frac{\partial\Omega}{\partial l'}}{l'+\hat{k}}\right), \quad (27)$$

Where dimensionless boundary conditions are

$$\Omega=-\frac{\hat{F}}{2},\ \Omega_{l'}=1,\ \hat{\theta}l'+\hat{\beta}_1\hat{\theta}=0 \qquad at\ l'=+\eta',$$
$$\Omega=-\frac{\hat{F}}{2},\ \Omega_{l'}=1,\ \hat{\theta}l'+\hat{\beta}_1\hat{\theta}=0 \qquad at\ l'=-\eta'. \quad (28)$$

$$\eta(\hat{g}) = [1+\varepsilon' \sin(2\Pi\,\hat{g})] \quad (29)$$

Whereas $\hat{\beta}_1=\frac{\eta_1' d}{\hat{k}}$ and $\hat{\beta}_2=\frac{\eta_2' d}{\hat{k}}$ refers the Biot numbers at the upper and lower walls.
Coefficient of heat transfer is defined as

$$Z'=\eta_{g'}'\hat{\theta}l'(\eta'). \quad (30)$$

3. Solution Methodology

By letting

$$\Omega=\Omega_0+\hat{W}_e\Omega_1...,$$
$$\hat{\tau}_{l'\mathrm{g}}=\hat{\tau}_{0l'g'}+\hat{W}_e\hat{\tau}_{1l'\mathrm{g}}...,$$
$$\frac{dP'}{dg'}=\frac{dP_0'}{dg'}+\hat{W}_e\frac{dP_1'}{dg'}..., \quad (31)$$
$$\hat{\theta}=\hat{\theta}_0+\hat{W}_e\hat{\theta}_1...,$$
$$\hat{F}=\hat{F}_0+\hat{W}_e\hat{F}_1..$$

For the solution system of zeroth and first order are

Zeroth order system

$$\frac{\hat{k}}{l'+\hat{k}}\frac{dP_0'}{dg'}=\frac{1}{(l'+\hat{k})^2}\frac{\partial}{\partial l'}\left\{(l'+\hat{k})^2\hat{\tau}_{0l'g'}\right\}-\frac{M'}{(l'+\hat{k})^2}(1-\frac{\partial\Omega_0}{\partial l'}), \quad (32)$$

$$\frac{\partial^2\hat{\theta}_0}{\partial l'^2}+\frac{1}{l'+\hat{k}}\frac{\partial\hat{\theta}_0}{\partial l'}+\hat{B}_r\left[(\frac{\partial^2\Omega_0}{\partial l'^2}-\frac{1-\frac{\partial\Omega_0}{\partial l'}}{l'+\hat{k}})\hat{\tau}_{0l'g'}\right]=0$$

(33)

$$\widehat{\tau}_{01'g'} = \left[1 + W_e'(-\frac{\partial^2\Omega_0}{\partial l'^2} - \frac{1-\frac{\partial\Omega_0}{\partial l'}}{l'+\hat{k}})\right]\left(-\frac{\partial^2\Omega_0}{\partial l'^2} - \frac{1-\frac{\partial\Omega_0}{\partial l'}}{l'+\hat{k}}\right) \tag{34}$$

$$\Omega_0 = -\frac{\widehat{F}_0}{2},\ \Omega_{0l'} = 1,\ \theta_{0l'} + \widehat{\beta}_1\hat{\theta}_0 = 0 \quad at\ l' = +\eta',$$
$$\Omega_0 = -\frac{\widehat{F}_0}{2},\ \Omega_{0l'} = 1,\ \theta_{0l'} + \widehat{\beta}_1\hat{\theta}_0 = 0 \quad at\ l' = -\eta' \tag{35}$$

First order system

$$\frac{\hat{k}}{l'+\hat{k}}\frac{dP_1'}{dg'} = \frac{1}{(l'+\hat{k})^2}\frac{\partial}{\partial l'}\left\{(l'+\hat{k})^2\hat{\tau}_{11'g'}\right\} - \frac{M'}{(l'+\hat{k})^2}(1-\frac{\partial\Omega_1}{\partial l'}), \tag{36}$$

$$\frac{\partial^2\hat{\theta}_1}{\partial l'^2} + \frac{1}{l'+\hat{k}}\frac{\partial\hat{\theta}_1}{\partial l'} + B_r'\left[(\frac{\partial^2\Omega_1}{\partial l'^2} - \frac{1-\frac{\partial\Omega_1}{\partial l'}}{l'+\hat{k}})\hat{\tau}_{11'g'}\right] = 0 \tag{37}$$

$$\hat{\tau}_{11'g'} = -\left[\left(\frac{\partial^2\Omega_1}{\partial l'^2} - \frac{1-\frac{\partial\Omega_1}{\partial l'}}{l'+\hat{k}}\right) + \left(-\frac{\partial^2\Omega_0}{\partial l'^2} - \frac{1-\frac{\partial\Omega_0}{\partial l'}}{l'+\hat{k}}\right)^2\right] \tag{38}$$

$$\Omega_1 = -\frac{\hat{F}_1}{2},\ \Omega_{1l'} = 0,\ \hat{\theta}_{1l'} + \hat{\beta}_1\hat{\theta}_1 = 0 \quad at\ l' = +\eta', \tag{39}$$

$$\Omega_1 = -\frac{\hat{F}_1}{2},\ \Omega_{1l'} = 0,\ \hat{\theta}_{1l'} - \beta_1'\hat{\theta}_1 = 0 \quad at\ l' = +\eta'. \tag{40}$$

Solution obtained for the zeroth and first order systems are

$$\Omega_0 = C_1(l'+\hat{k})^{M'} + C_2(l'+\hat{k})^{-M'} - \frac{(l'+\hat{k})^2\ \frac{dP_0'}{dg'})}{2(4-M'^2)} - \frac{\hat{k}^3\ \frac{dP_0'}{dg'}}{M'^2} + l', \tag{41}$$

$$\Omega_1 = a_{11} + \left(\frac{dP_0'}{dg'}\right)^2 a_{20} + a_{13} + a_{14} + \frac{dP_0'}{dg'}\left[a_{19} + a_{21} + a_{22}\right] - a_{15} - \frac{2(l'+\hat{k})}{1-M'^2} + a_{16} - a_{17} + a_{18} - \frac{4}{M'^2} - \frac{dP_1'}{dg'}a_{23},$$
(42)

$$\theta_0 = C_5 + C_6\ln(l'+\hat{k}) - B_r'\left[\frac{dP_0'}{dg'}b_{22} - b_{11} + b_{12} + b_{13}\frac{dP_0'}{dg'}(b_{20}+b_{21}) + b_{14} - b_{15} + b_{16} - b_{17} - b_{18} + \frac{2(\frac{dP_0'}{dg'})^2(l'+\hat{k})^3}{9(4-\hat{M}^2)^2} - b_{19}\right]$$
(43)

$$\theta_1 = f_1 + f_2\ln(l'+\hat{k}) + B_r'\left[\begin{array}{l}\frac{dP_0'}{dg}(c_{11}+c_{12}+c_{16}) + (\frac{dP_0'}{dg})^2(c_{13}+c_{14}-c_{17}) + (\frac{dP_0'}{dg})^3 c_{15} + \frac{dP_1'}{dg}c_{18} + c_{19} \\ + \frac{dP_0'}{dg'}\frac{dP_1'}{dg'}\left(-\frac{5(l'+\hat{k})^2}{32} + \frac{(l'+\hat{k})^3}{16}\right) - 2C_1(\ln(l'+\hat{k}))^2 + \frac{5C_2(\ln(l'+\hat{k}))^2}{4}\end{array}\right] \tag{44}$$

$$C_1 = \frac{-(\hat{k}+\eta')^{-M'}\left(\frac{\hat{F}_0}{2} + \frac{\hat{k}^3 \frac{dP_0}{dg'}}{M'^2} + \frac{\frac{dP_0'}{dg}(\hat{k}-\eta')^2}{2(4-M'^2)} + \eta'\right) + (\hat{k}-\eta')^{-M'}\left(\frac{-\hat{F}_0'}{2} + \frac{\hat{k}^3 \frac{dP_0'}{dg'}}{M'^2} + \frac{\frac{dP_0'}{dg'}(\hat{k}+\eta')^2}{2(4-M'^2)} - \eta'\right)}{(\hat{k}-\eta')^{M'}(\hat{k}+\eta')^{-M'} - (\hat{k}-\eta')^{-M'}(\hat{k}+\eta')^{M'}}$$

$$C_2 = \frac{(\hat{k}+\eta')^{M'}\left(\frac{-\hat{F}_0'}{2} + \frac{\hat{k}^3 \frac{dP_0'}{dg'}}{M'^2} + \frac{\frac{dP_0'}{dg'}(\hat{k}+\eta')^2}{2(4-M'^2)} - \eta'\right) + (\hat{k}+\eta')^{2M'}\left(\begin{array}{c} -(\hat{k}+\eta')^{-M'}\left(\frac{\hat{F}_0}{2} + \frac{\hat{k}^3 \frac{dP_0'}{dg'}}{M'^2} + \frac{\frac{dP_0'}{dg'}(\hat{k}-\eta')^2}{2(4-M'^2)} + \eta'\right) + \\ (\hat{k}-\eta')^{-M'}\left(\frac{-\hat{F}_0}{2} + \frac{\hat{k}^3 \frac{dP_0'}{dg'}}{M'^2} + \frac{\frac{dP_0'}{dg'}(\hat{k}+\eta')^2}{2(4-M'^2)} - \eta'\right) \end{array}\right)}{(\hat{k}-\eta')^{M'}(\hat{k}+\eta')^{-M'} - (\hat{k}-\eta')^{-M'}(\hat{k}+\eta')^{M'}} \tag{45}$$

4. RESULTS AND DISCUSSION

This segment has developed to attain and show the phenomenon of distinct physical deeply involved parameters on velocity, temperature, pressure gradient, coefficient of flow of heat and stream function respectively. For excellence output of the existing study there is a sectional distribution of this discussion which has clear picture of graphical results.

(i) Velocity distribution

In this segment the conduct of velocity dispersion in curved channel for assorted parameters is depicted over figures. It is visible that velocity dispersion has opposite manner, at the walls of the channel for different valued of W_e' (Weissenberg number) and M' (Magnetic field). Observations clearly showed that velocity is less for viscous substance in comparison with Williamson fluid at lower wall, whereas it heightens at upward wall of the channel. Figure 2. (a) exhibit that there does not exist symmetry of the velocity dispersion through curvature. Effects of magnitude field (radial) M' is pictured in Figure 2. (b). It is resulted that magnetic field (radial) has opposing behavior for velocity. In the case, when magnetic field increases in magnitude the resulting situation is the gradually decrease in the flow velocity due to prejudicial effect of the electromagnetic body force. In addition, for higher values of M' Substantial reduction has been seen in the flow velocity closest to the centre of the curved channel inspite the near of the walls.

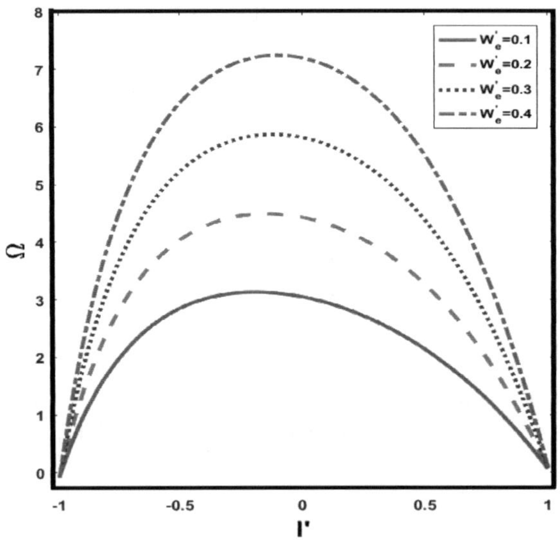

Figure 2. (a) Behavior of velocity distribution for W_e' and M' =4.0, $\hat{k}$ =2.5, ε' =0.01, g' =0.0

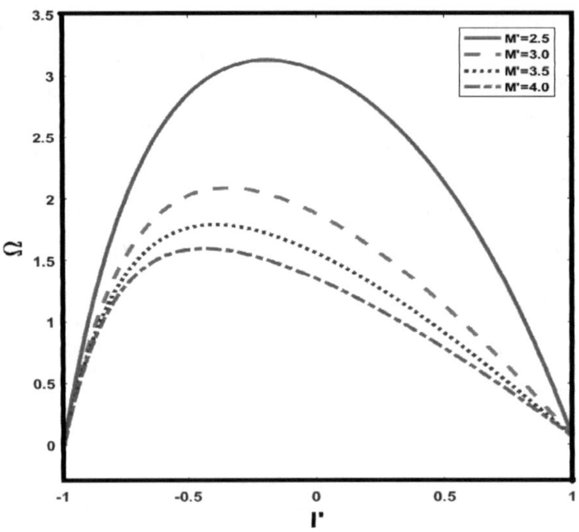

Figure 2. (b) Behavior of velocity distribution for M' and W_e' =0.1, $\hat{k}$ =2.5, ε' =0.01, g' =0.0

(ii) Temperature distribution

Visibility of temperature attributes for W_e' (Weissenberg number) and M' (Magnetic parameter) are drawn and talk about them in the Figures 3. (a) - Figure 3. (f), Figure 3. (a) represents that temperature rises for Williamson fluid (when $W_e' \neq 0$). One another thing for noticing that for Williamson fluid there is less resistance in temperature dispersion near the lower wall when compared to viscous material. Figure 3. (b) and Figure 3. (c) shows the behaviour that temperature is increasing due to viscous dissipation effects while increasing Brashof number and Magnetic field parameter. Actually viscosity assists kinetic energy of the motion of the fluid particles and change into internal energy which raise the temperature distributions.
In Figure 3. (d) it is cleared that for the higher values of $\hat{k}$ (curvature parameter) the temperature dispersion reduces. Another observation resulted that magnetic parameter M' (radial) has resistivity impact on temperature and it decreases with increasing M'. In early discussions it is observed that temperature is anti-symmetric for large values of $\hat{k}$ in curved channel whereas it is symmetric for small values of $\hat{k}$ in unbent channel. Figure 3. (b) illustrates that temperature distribution shows increase while increasing Brashof number. And Figure 3. (e) and Figure 3. (f) is prominently showing that β_1' and β_2' gives opposite behaviour on temperature profile in comparison on other parameters.

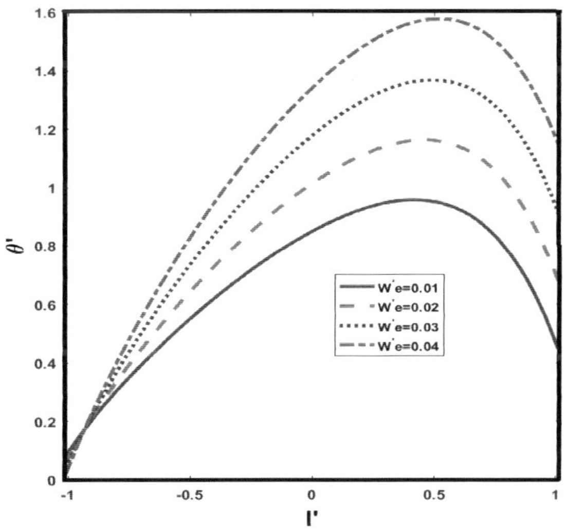

Figure 3. (a) Behavior of Temperature distribution for W_e' and M' =3.5, ε' =0.1, B_r' =0.1, β_1' =10.0, β_2' =10.0

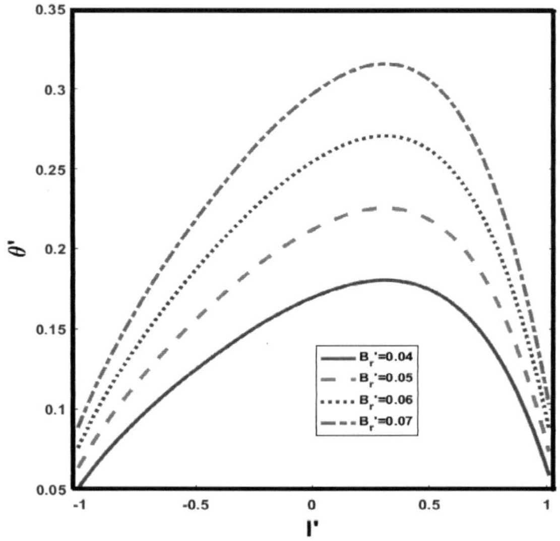

Figure 3. (b) Behavio r of Temperature distribution for B_r' and M' =3.5, $\hat{k}$ =2.0, ε' =0.1, W_e' =0.01, β_1' =10.0, β_2' =10.0

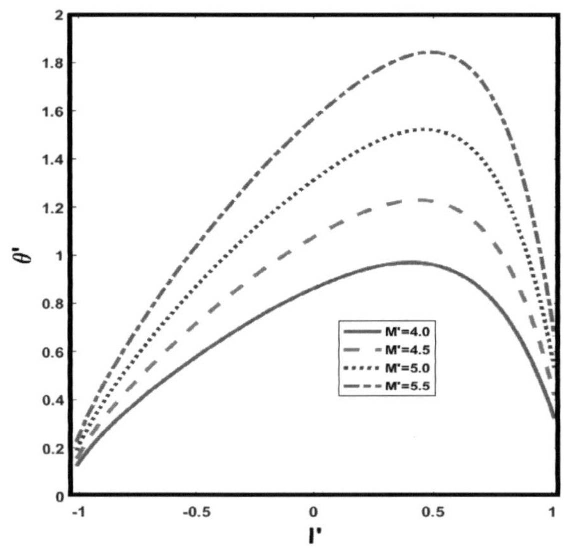

Figure 3. (c) Behavior of Temperature distribution for M' and $\hat{k}$ =2.0, ε' =0.1, W_e' =0.01, B_r' =0.1, β_1' =10.0, β_2' =10.0

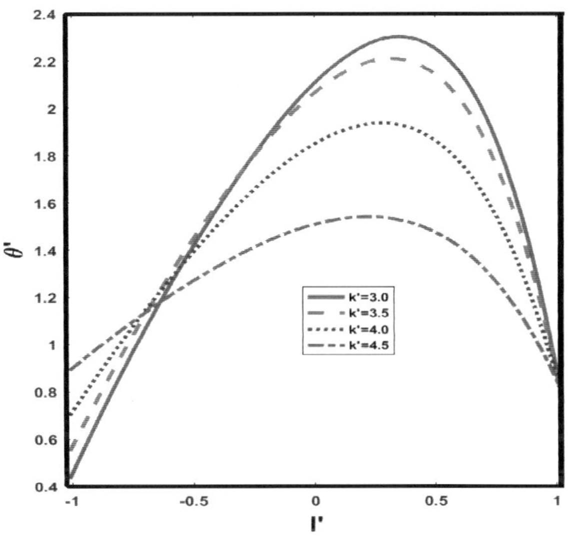

Figure 3. (d) Behavior of Temperature distribution for $\hat{k}$ and M' =3.5, B_r'=0.1, ε'=0.1, W_e'=0.01, β_1'=10.0, β_2'=10.0

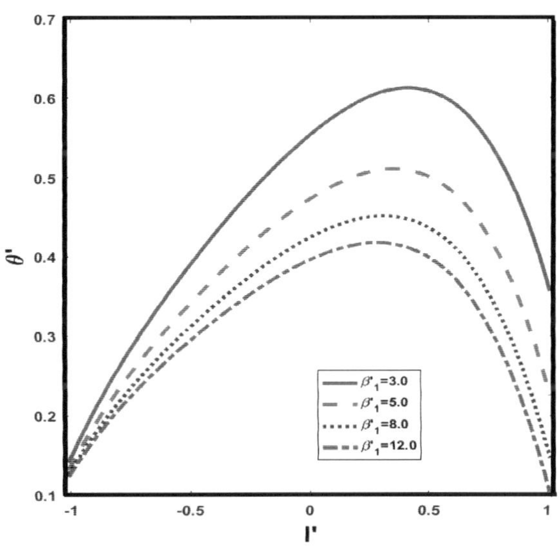

Figure 3. (e) Behavior of Temperature distribution for β_1'=10.0 and β_2'=10.0, M' =3.5, $\hat{k}$ =2.0, ε' =0.1, B_r'=0.1, W_e'=0.01, β_2'=7.0

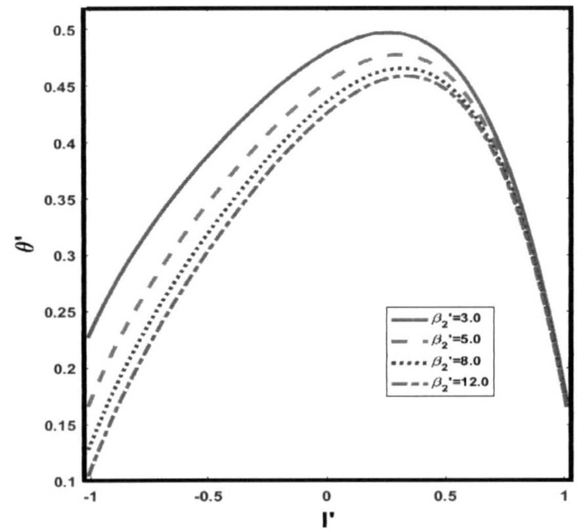

Figure 3. (f) Behavior of Temperature distribution for β_2' and M'=3.5, $\hat{k}$=2.0, ε'=0.1, B_r'=0.1, W_e'=0.01, β_1'=7.0

(iii) Pumping characteristics

Influence of $\widehat{W}_e$, M' and ε' on pressure gradient is discussed and plotted in Fig 4(a)-4(c).Pressure gradient is maximum for small value of W_e'. One can observe enhancement in pressure gradient for large M' magnetic field (radial) in Fig 4(b). Similarly effect of ε' on pressure gradient can also be illustrated in Fig 4(c) by observing that pressure gradient is increasing while increasing ε' and has same behaviour as M' and finally it is resulted that electromagnetic forces are useful in the enhancement of pressure gradient.

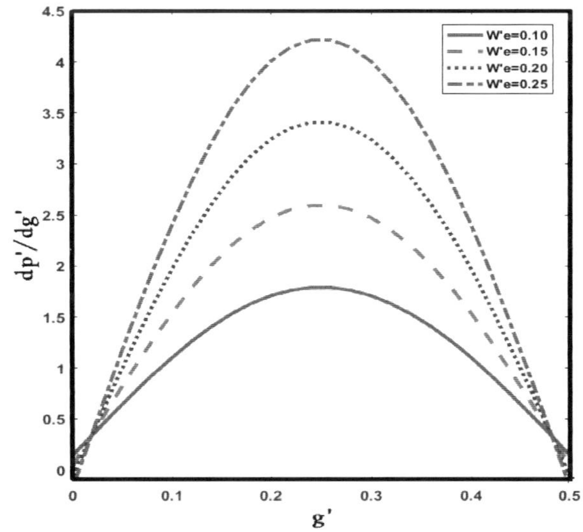

Figure 4. (a) Behavior of Pressure gradient for W_e' and M' =3.5, $\hat{k}$ =4.0, ε' =0.1, Q=0.01

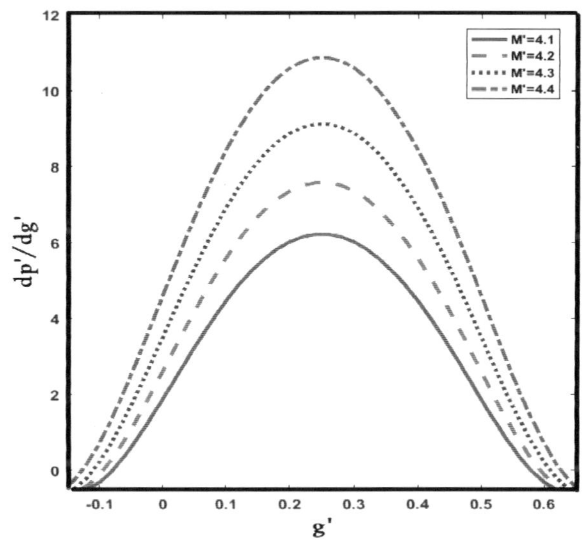

Figure 4. (b) Behavior of Pressure gradient for M' and W_e' =0.1, $\hat{k}$ =4.0, ε' =0.1, Q=0.01

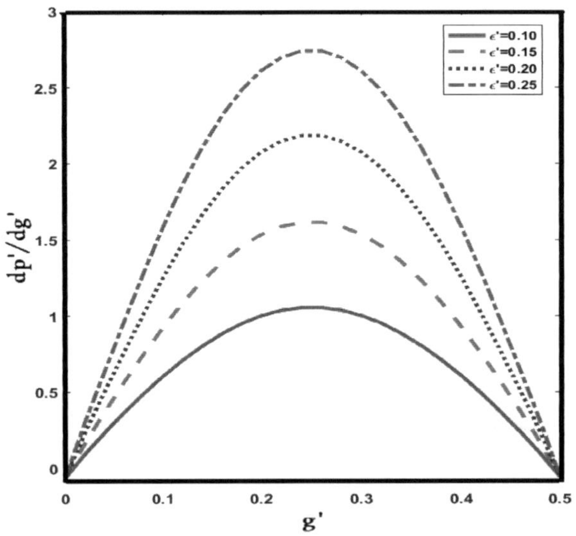

Figure 4. (c) Behavior of Pressure gradient for ε' and W_e' =0.1, $\hat{k}$ =4.0, Q=0.01

Figure 5. (a) - Figure 5. (d) is built to demonstrate the behavior of flow of heat or heat discharge for different material parameters at the upper wall of the channel. Coefficient of heat transfer or flow of heat is computed with the expression ($Z' = \eta'_{g'}\hat{\theta}(\eta')$). It is clearly observed in Figure 5. (a) that heat transfer coefficient Z' is increased by increasing value of W_e' . Flow of heat rate Z' behaves in a same way for B_r' in Figure 5. (b) as for W_e'. Figure 5. (c) and Figure 5. (d) shows the demonstration of the effects of Biot numbers on rate of flow of heat. It makes us visible that β_1' gives rise in the behaviour of Z' and falling attitude of β_2' respectively.

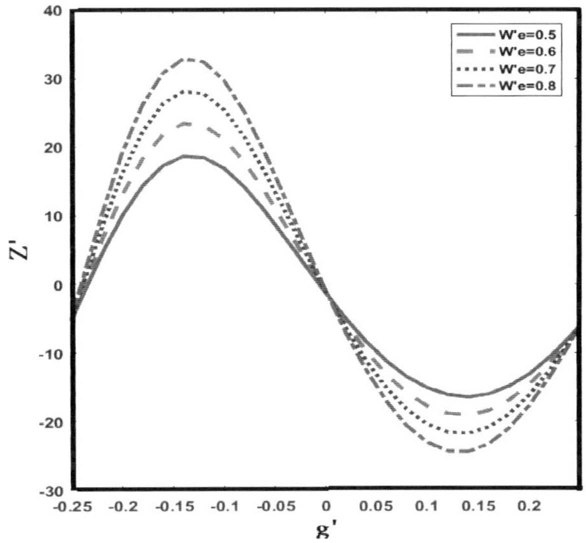

Figure 5. (a) Behavior of heat transfer for W_e' and $M'=5, \hat{k}=3.0, \beta_1'=0.1, B_r'=0.2$

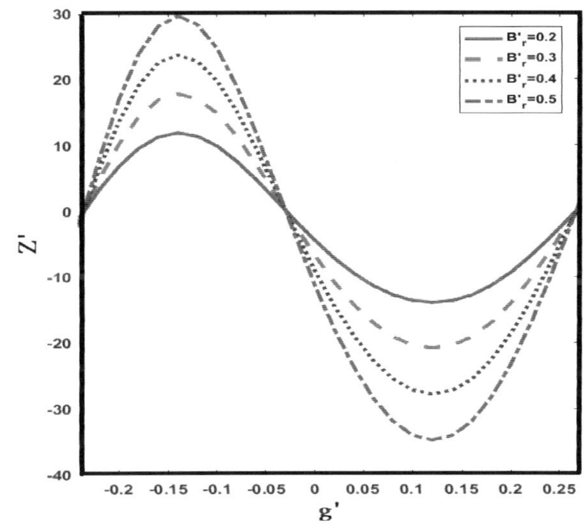

Figure 5. (b) Behavior of heat transfer for B_r' and $M'=5,\ W_e'=0.9,\ \hat{k}=3.0, \beta_1'=0.1, \beta_2'=0.1$

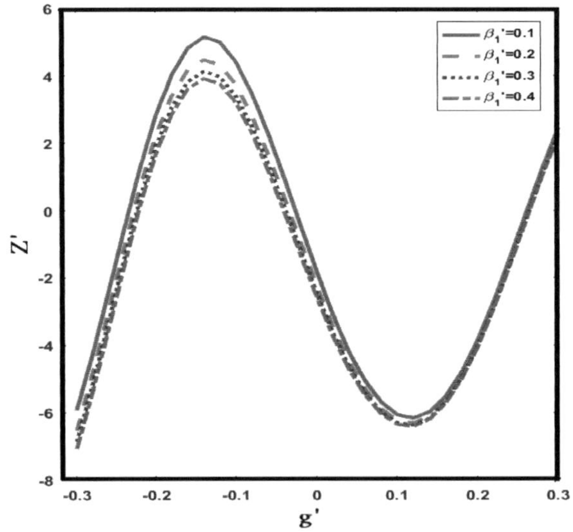

Figure 5. (c) Behavior of heat transfer for β_1' and M' =4, W_e' =0.6, $\hat{k}$ =3, β_2' =0.1, B_r' =0.1

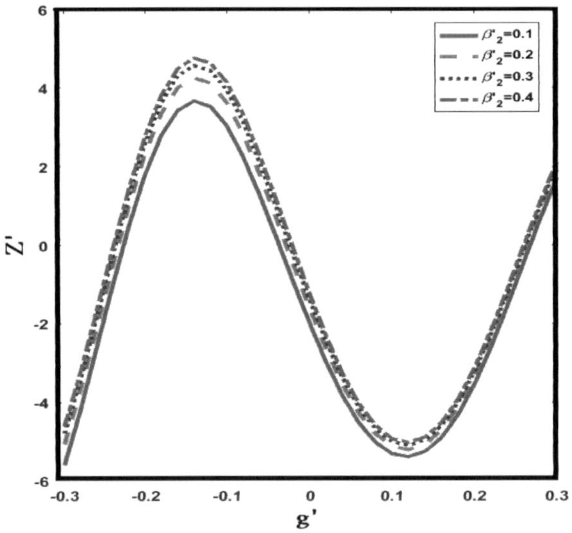

Figure 5. (d) Behavior of heat transfer for β_2' and M' =4, W_e' =0.6, β_1' =0.1, B_r' =0.2

(iv) Trapping

Another an interesting process of fluid motion is the trapping that is the shaping and formation of an interior circulation bolus of fluid which moves along the wave. There are some examples which illustrates this phenomenon are the food bolus movement in the gastrointestinal tract and movement of blood clot. Figs Figure 6. (a) - Figure 6. (f) are constructed to notice the deportment of trapped bolus. Figure 6. (a) and Figure 6. (a) shows the deportment of bolus for small values of W_e' and it is clearly observe that bolus size

IOP Conf. Series: Materials Science and Engineering **414** (2018) 012010 doi:10.1088/1757-899X/414/1/012010

is expanded with small value of W_e' Figure 6. (c) and Figure 6. (d) shows the size of trapped bolus for M' and it is also notice that bolus expanded with large values of $M.'$ Whereas Figure 6. (e) and Figure 6. (f) shows the behaviour of bolus with the change of $\hat{k}$ (curvature parameter) and size of the trapping bolus increases with the increase of $\hat{k}$.

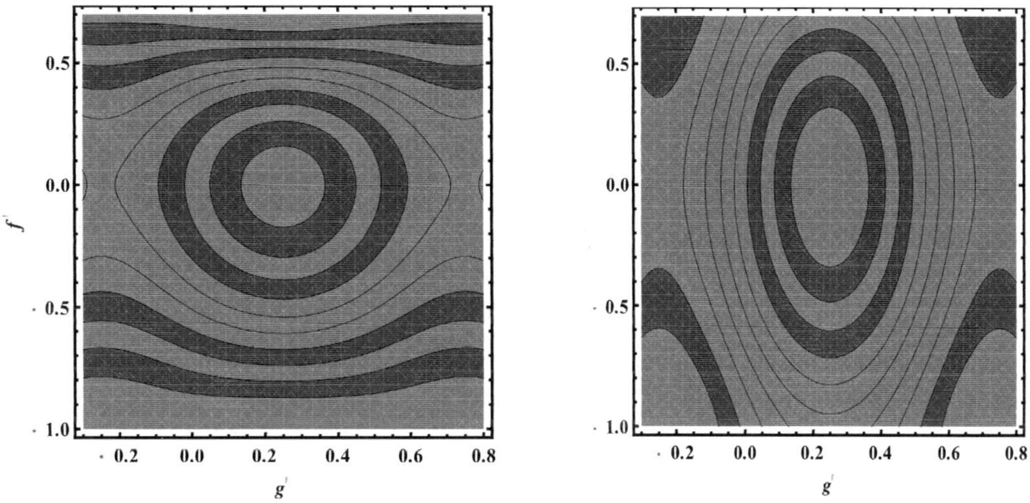

Figure 6. (a) - 6. (b) Trapped bolus for different values of W_e' with M' =5, $\hat{k}$ =3, ε' =0.1.

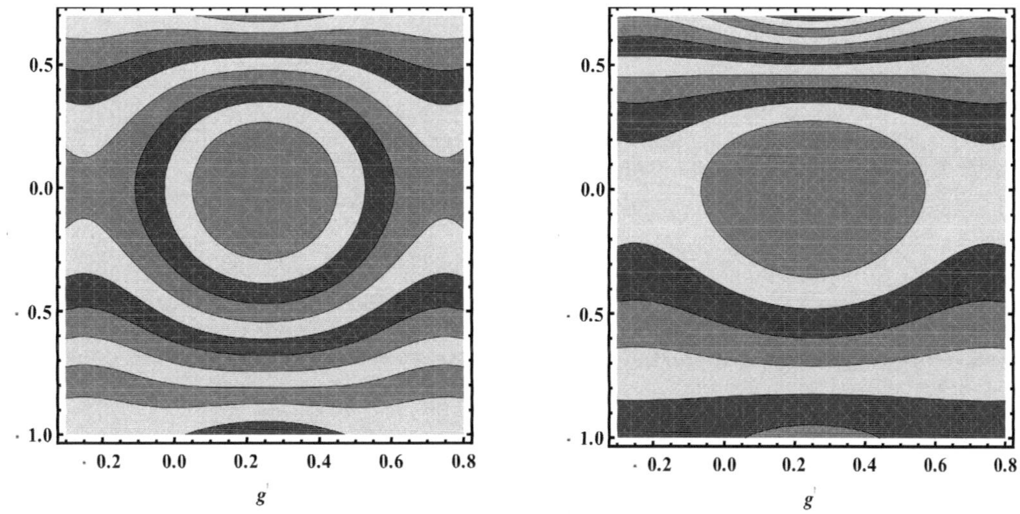

Figure 6. (c) - 6. (d) Trapped bolus for different values of M' with W_e' =0.01, $\hat{k}$ =3, ε' =0.1.

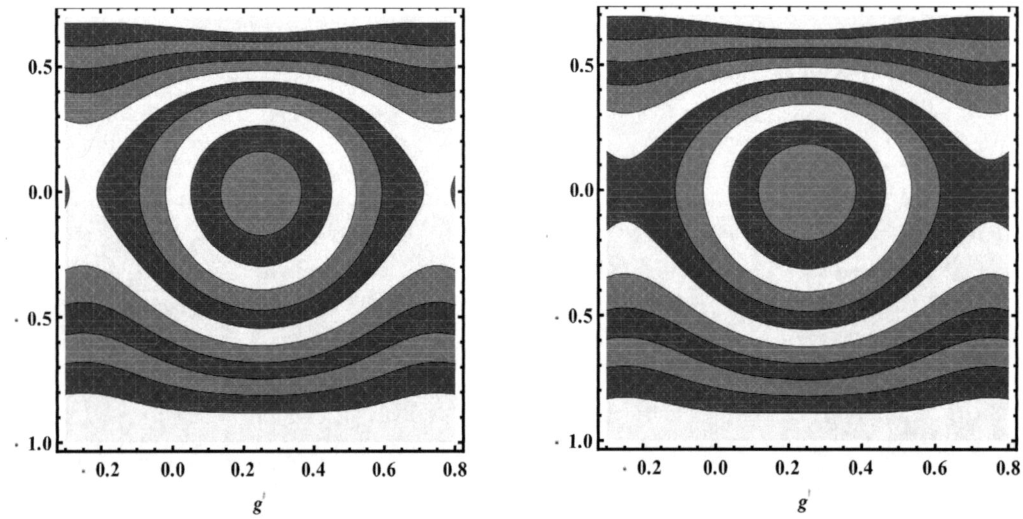

Figure 6. (e) - 6. (f) Trapped bolus for different values of $\hat{k}$ with M' =5, W_e' =0.01, ε' =0.1.

(v) Conclusion

This research study presents the peristaltic transportation of Williamson fluid with MHD and Joule heating effects with convective heating in curved channel. The striking attention of the model on this regard are given as

1. Pressure gradient increases with increasing W_e', M' and ε'. While it decreases by increasing $\hat{k}$ respectively.
2. Weissenberg number has same behavior for velocity and temperature profile
3. M' has opposite behaviour for velocity and temperature profile.
4. Curvature parameter $\hat{k}$ also has opposite deportment for velocity and temperature distribution.
5. Oscillatory behaviour of Heat transfer coefficient Z' for all parameters.
6. Biot numbers β_1' and β_2' has decreasing behaviour on pressure gradient.

Appendix

$$a_{11} = \frac{M'^2(M'-1)^2C_1^2(l'+\hat{k})^{2M'+2}}{(M'+2)(3M'+2)} + \frac{M'^2C_2^2(M'-1)^2(l+\hat{k})^{-(2M'+2)}}{(3M'+2)(M'+2)}$$

$$a_{13} = \frac{2M'(M'^2-1)C_1C_2(l'+\hat{k})^{-2}}{(4-M'^2)} + \frac{2M'^2C_1^2(l'+\hat{k})^{2M'-1}}{(M'-1)(3M'-1)} - \frac{2M'^2C_1C_2(l'+\hat{k})^{-1}}{(1-M'^2)}$$

$$a_{14} = C_1(l'+\hat{k})^{M'}\ln(l'+\hat{k})$$

$$a_{15} = \frac{2M'^2C_1C_2(l'+\hat{k})^{-1}}{(1-M'^2)} - \frac{C_2(l'+\hat{k})^{-M'}\ln(l'+\hat{k})}{(4-M'^2)} - \frac{2M'^2C_2^2(l'+\hat{k})^{-2M'-1}}{(3M'+1)(M'+1)}$$

$$a_{16} = \frac{M'^2C_1^2(l+\hat{k})^{2(M'-1)}}{(M'-2)(3M'-2)} - \frac{M'C_2(l'+\hat{k})^{-(M'+1)}}{(2M'+1)}\frac{M'C_1(l'+\hat{k})^{M'-1}}{2M'+1} - \frac{M'C_2(l'+\hat{k})^{-(M'+1)}}{2M'+1}$$

$$a_{17} = \frac{2M'^2C_1C_2(l'+\hat{k})^{-2}}{(4-M'^2)} - \frac{2M'C_2(l'+\hat{k})^{-(M'+1)}}{2M'+1} - \frac{2(l'+\hat{k})}{1-M'^2} - \frac{2M'C_1(l'+\hat{k})^{(M'-1)}}{2M'+1}$$

$$a_{18} = \frac{M'^2C_2^2(l'+\hat{k})^{-2}}{(4-M'^2)} + \frac{M'^2C_2^2(l'+\hat{k})^{-2(M'+1)}}{(3M'+2)(M'+2)} - \frac{M'C_1(l'+\hat{k})^{M'-1}}{2M'-1}$$

$$a_{19} = \frac{4M'C_1(l'+\hat{k})^{\hat{M}+1}}{(2M'+1)} + \frac{2M'C_2(l'+\hat{k})^{1-M'}}{(1-2M')} + \frac{2(l'+\hat{k})^2}{(4-M'^2)^2}$$

$$a_{20} = \frac{2(l'+\hat{k})^2}{(4-M')^3} + \frac{(l'+\hat{k})^3}{(4-M'^2)^2(1-M'^2)}$$

$$a_{21} = \frac{(M'-1)C_2(l'+\hat{k})^{-M'}\ln(l'+\hat{k})}{(4-M'^2)} - \frac{(M'-1)C_1(l'+\hat{k})^{M'}\ln(l'+\hat{k})}{(4-M'^2)}$$

$$a_{22} = \frac{C_2(l'+\hat{k})^{-M'}\ln(l'+\hat{k})}{2(4-M'^2)} - \frac{C_1(l'+\hat{k})^{M'}\ln(l'+\hat{k})}{2(4-M'^2)} - \frac{C_2(l'+\hat{k})^{-M'}\ln(l'+\hat{k})}{(4-M'^2)} + \frac{C_1(l'+\hat{k})^{M'}\ln(l'+\hat{k})}{2}$$

$$a_{23} = \frac{(l'+\hat{k})^2}{2(4-M'^2)} - \frac{\hat{k}^3}{2M'^2}$$

$$b_{11} = \frac{2C_1C_2M'^2(l'+\hat{k})^{-2}}{4}$$

$$b_{12} = \frac{2C_1C_2M'^4(l'+\hat{k})^{-2}}{4}$$

$$b_{13} = \frac{C_2^2M'^2(l'+\hat{k})^{-2(M'+1)}}{(2+2M')^2}$$

$$b_{14} = \frac{C_1^2M'^2(l'+\hat{k})^{2M'-2}}{(2M'-2)^2}$$

$$b_{15} = \frac{2C_1^2M'^3(l'+\hat{k})^{2M'-2}}{(2M'-2)^2}$$

$$b_{16} = \frac{C_1^2 M'^4 (l'+\hat{k})^{2M'-2}}{(2M'-2)^2}$$

$$b_{17} = \frac{2C_2(1+M')(l'+\hat{k})^{-M'}}{M'(4-M'^2)}$$

$$b_{18} = \frac{2C_1C_2(1+M')M'^2(l'+\hat{k})^{-2}}{4}$$

$$b_{19} = \frac{2C_1C_2M'^2(M'-1)(l'+\hat{k})^{-2}}{4}$$

$$b_{20} = -\frac{2C_2(l'+\hat{k})^{-M'}}{(4-M'^2)} + \frac{2C_1(l'+\hat{k})^{M'}}{M'(4-M'^2)} - \frac{2C_1(l'+\hat{k})^{M'}}{(4-M'^2)}$$

$$b_{21} = \frac{2C_2(l'+\hat{k})^{-1}}{M'(4-M'^2)} - \frac{2C_1(l'+\hat{k})^{\hat{M}}}{M'(4-M'^2)} - \frac{2C_1(M'-1)(l'+\hat{k})}{M'(4-M'^2)}$$

$$b_{22} = \frac{(l'+\hat{k})^2}{4(4-M'^2)^2} - \frac{2C_2\frac{dP_0}{dg'}(l'+\hat{k})^{-\hat{M}}}{\hat{M}(4-M'^2)}$$

$$c_{11} = -\frac{1}{4} + C_1 + \frac{5C_2}{8} + \frac{25C_1(l'+\hat{k})^{-1}}{8} - \frac{7C_2(l'+\hat{k})^{-1}}{4} + \frac{72(l'+\hat{k})^3}{36} - \frac{3C_2}{8}$$

$$c_{12} = \frac{6(l'+\hat{k})}{4} - \frac{C_1(l'+\hat{k})}{2} - \frac{5C_2(l'+\hat{k})}{4} - \frac{5C_1(l'+\hat{k})^2}{32} - \frac{5C_1(l'+\hat{k})^{-1}}{8} + \frac{3C_2(l'+\hat{k})^{-1}}{8} + \frac{5C_1}{8}$$

$$c_{13} = -\frac{1}{8} + \frac{(l'+\hat{k})}{34} + \frac{5C_1}{16} - \frac{(l'+\hat{k})}{8} - \frac{3C_2}{16} - \frac{51(l'+\hat{k})^4}{(16)^2} + \frac{9(l'+\hat{k})}{8} - \frac{3(l+\hat{k})^4}{128} - \frac{39(l'+\hat{k})^2}{192}$$

$$c_{14} = \frac{7(l'+\hat{k})^2}{64} + \frac{3C_2(l'+\hat{k})}{32}$$

$$c_{15} = \frac{(l'+\hat{k})^2}{240} + \frac{(l'+\hat{k})^4}{1024} - \frac{(l'+\hat{k})^3}{64} + \frac{3(l'+\hat{k})^5}{1600} + \frac{3(l'+\hat{k})^2}{256} - \frac{(l'+\hat{k})^2}{128} + \frac{(l'+\hat{k})^2}{34}$$

$$c_{16} = -\frac{3C_2(\ln(l'+\hat{k}))^2}{8} + \frac{12(\ln(l'+\hat{k}))^2}{4} + C_1(\ln(l'+\hat{k}))^2 - \frac{5C_1(\ln(l'+\hat{k}))^2}{8}$$

$$c_{17} = \frac{5C_1(\ln(l'+\hat{k}))^2}{16} - \frac{3C_2(\ln(l'+\hat{k}))^2}{16}$$

$$c_{18} = \frac{(l'+\hat{k})}{2} - \frac{(l'+\hat{k})^2}{4} + \frac{(\ln(l'+\hat{k}))^2}{4}$$

$$c_{19} = -3(l'+\hat{k})^2 - \frac{3(l'+\hat{k})^2}{4} - 6(l'+\hat{k}) + 4C_1(l'+\hat{k})^{-1} + \frac{10C_1C2}{4} - 4C_1(l'+\hat{k})^{-1} + \frac{5C_2(l'+\hat{k})^{-1}}{2}$$

References

[1] Latham T (1966) Fluid motion in a Peristaltic Pump. MS thesis, MIT Cambridge MA.

[2] Shapiro A, Jaffrin M, and Weinberg H (1969) Peristaltic pumping with long wavelength at low Reynolds number. *J. Fluid Mech.* **35** 799-825.

[3] Hayat T, Farooq S, Ahmed B and Alsaedi A (2016) Characteristics of Convective heat transfer in the MHD peristalsis of Carreau fluid with Joule heating. *AIP Advances.* **6** 045302.

[4] Akbar N, Nadeem S, and Khan Z (2014) Numerical simulation of peristaltic flow of a Carreau nanofluid in a asymmetric channel. *Alexandria Eng. J.* **53** 191-197.

[5] Mustafa M, Abbasbandy S, Hina S and Hayat T (2014) Numerical investigation on mixed convective peristaltic flow of fourth grade fluid with Dufour and Soret effects. *J. Taiwan Inst. Chem. Eng.* **45** 308-316.

[6] Mekheimer K, Abd Elmabound A and Abdellateef A (2013) particulate suspension flow induced by sinusoidal peristaltic waves through eccentric cylinders: thread annular. *Int. j. biomath.* **06** 1350026.

[7] Hina S, Shit G, Ranjit N (2015) Peristaltic transport of MHD Flow and heat transfer in an asymmetrical channel: effects of variable viscosity, velocity-slip and temperature jump. *Alexandria Eng. J.* **54** 691-704.

[8] Gad N (2014) Effects of Hall currents on peristaltic transport with complaint walls. *Appl. Math. Comput.* **235** 546-554.

[9] Vafai K, Khan A, Sajjad S, and Ellahi R (2015) The study of peristaltic motion of third grade fluid under the effects of Hall current and heat transfer. *Z Naturforsch A.* **70** 281-293.

[10] Rehman M, Akbar N, Haider A and Azam (2015) H Effects of heat sink/source on peristaltic flow of Jeffery fluid through a symmetrical channel. *Alexandria Eng. J.* **54** 733-743.

[11] Hayat T, Abbasi F, Alsaedi A and Alsaedi F (2014) Hall and ohmic heating effects on the peristaltic transport of Carreau- Yasuda fluid in an asymmetric channel. *Z. Naturforsch A.* **69a** 43-51.

[12] Awais M, Farooq S, Yasmin H, Hayat T and Alsaedi A (2014) Convective heat transfer analysis for MHD peristaltic flow of Jeffery fluid in an asymmetric channel. *Int. J. Biomath.* 7 145002

[13] Kothandapani M and Prakash J (2015) Effects of thermal radiation parameter and magnetic field on the peristaltic motion of Williamson nanofluids in a tapered asymmetric channel. *Int. J. Heat Mass Tranf.* **81** 234-245.

[14] Hameed M, Khan A, Ellahi R and Raza M (2015) Study of magnetic and heat transfer on the peristaltic transport of a fractional second grade fluid in a vertical tube. *Eng. Sci. Tech. Int. J.* **18** 496-502.

[15] Saleem M and Haider A (2014) Heat and mass transfer on the peristaltic transport of non-Newtonian fluid with creeping flow. *Int. J. Heat Mass Trans.* **68** 514-526.

[16] Hayat T, Tanveer A, Yasmin H and Alsaedi A (2014) Homogeneous-Heterogeneous reactions in peristaltic flow with convective conditions. *PLOS One*. (**9**) e113851.

[17] Hina S (2016) MHD peristaltic transport of Eyring-Powell fluid with/mass transfer wall properties and slip conditions. *J. Mag. Mag. Mater.* **404** 148-158.

[18] Akbar N, Raza M and Ellahi R (2015) Influence of induced magnetic field and heat flux with the suspension of carbon nanotubes for the peristaltic flow in a permeable channel. *J. Mag. Mag. Mater.* **381** 405-415.

[19] Hayat T, Yasmin H and Alsaedi A (2015) Convective heat transfer analysis for peristaltic flow of power-law fluid in a channel. *J. Braz. Soci. Mech. Sci and Eng.* **37** 463-477.

[20] Sato H, Kawai T, Fujita T and (2000) Okabe M Two dimensional peristaltic flow in curved channels. *Trans. Jpn. Soc. Mech. Eng. B.* **66** 679-685.

[21] Mitra T, Prasad N (1973) On the influence of wall properties and Poiseuille flow in peristalsis. *J Biomech* **6** 68 1–693.

[22] Davies C, Carpenter W (1997) Instabilities in a plane channel flow between compliant walls. *J Fluid Mech.* **352** 205–243.

[23] Srivastava V, Srivastava L (1997) Influence of wall elasticity and poiseuille flow on peristaltic induced flow of a particle-fluid mixture. *Int J Eng Sci.* **35** 1359–1386.

[24] Haroun M (2006) Effect of wall compliance on peristaltic transport of a Newtonian fluid in an asymmetric channel. *Math Prob Eng*. DOI:10.1155/MPE/2006/61475.

[25] Radhakrishnamacharya G, Srinivasulu C (2007) Influence of wall properties on peristaltic transport with heat transfer. *Compt Rend Mecan.* **335** 369–373.

[26] Muthu P, Kumar B, Chandra P (2008) Peristaltic motion of micropolar fluid in circular cylindrical tubes: Effect of wall properties. *Appl Math Model.* **32** 2019–2033.

[27] Elnaby M, Haroun M (2008) A new model for study the effect of wall properties on peristaltic transport of a viscous fluid. *Commun Nonlinear Sci Numer Simul* **13** 752–762.

[28] Hayat T, Javed M, Asghar S (2008) MHD peristaltic motion of Johnson–Segalman fluid in a channel with compliant walls. *Phys Lett A.* **372** 5026–5036.

[29] Hayat T, Javed M, Ali N (2008) MHD peristaltic transport of Jeffery fluid in a channel with compliant walls and porous space. *Transp Porous Med.* **74** 259–274.

[30] Ali N, Hayat T, Asghar S (2009) Peristaltic flow of a Maxwell fluid in a channel with compliant walls. *Chaos Solitons Fractals.* **39** 407–416.

[31] Kothandapani M, Srinivas S (2008) On the influence of wall properties in the MHD peristaltic transport with heat transfer and porous medium. *Phys Lett A.* **372** 4586–4591.

[32] Srinivas S, Gayathri R and Kothandapani M (2009) The influence of slip conditions, wall properties and heat transfer on MHD peristaltic transport. *Comput Phys Commun.* **180** 2115–2122.

[33] Srinivas S, Kothandapani M (2009) The influence of heat and mass transfer on MHD peristaltic flow through a porous space with compliant walls. *Appl Math Comput.* **213** 197–208.

[34] Hayat T, Hina S (2010) The influence of wall properties on the MHD peristaltic flow of a Maxwell fluid with heat and mass transfer. *Nonlinear Anal: Real World Appl.* **11** 3155–3169.

[35] Sato H, Kawai T, Fujita T and Okabe M (2000) Two dimensional peristaltic flow in curved channels. Trans *The Japan Soc Mech Eng B.* **66** 679–685.

[36] Ali N, Sajid M and Hayat T (2010) Long wavelength flow analysis in a curved channel. *Z Naturforsch A.* **65** 191–196.

[37] Ali N, Sajid M and Javed T (2000) Heat transfer analysis of peristaltic flow in a curved channel. *Int J Heat Mass Transf.* **53** 3319–3325.

[38] Ali N, Sajid M, Abbas Z and Javed T (2010) Non-Newtonian fluid flow induced by peristaltic waves in a curved channel. *Europ J Mech- B/Fluids* 29:387–394.

[39] Hayat T, Javed M, Hendi A (2011) Peristaltic transport of viscous fluid in a curved channel with compliant walls. *Int J Heat Mass Transf.* (2011) **54** 1615–1621.

[40] Hayat T, Hina S, Hendi A and Asghar S (2011) Effect of wall properties on the peristaltic flow of a third grade fluid in a curved channel with heat and mass transfer. *Int J Heat Mass Trans.* **54**:5126–5136.

[41] Hina S, Hayat T, Asghar S (2012) Peristaltic transport of Johnson-Segalman fluid in a curved channel with wall properties. *Nonlinear Anal: Model Control.* **17** 297–311.

[42] Hina S, Hayat T, Alsaedi A (2012) Heat and mass transfer effects on the peristaltic flow of Johnson–Segalman fluid in a curved channel with compliant walls. *Int J Heat Mass Transf.* **55** 3511–3521.

[43] Hina S, Mustafa M, Hayat T and Alsaedi A (2013) Peristaltic flow of pseudoplastic fluid in a curved channel with wall properties. *J Appl Mech Trans ASME.* **80** 024501 (doi:10.1115/1.4007433).

[44] Hina S, Mustafa M, Hayat T and Alsaedi A (2014) Peristaltic motion of third grade fluid in curved channel. *Appl Math Mech-Engl* Ed. **35** 73–84.

[45] Hina S, Mustafa M, Abbasbandy S, Hayat T and Alsaedi A (2014) Peristaltic motion of nanofluid in a curved channel. *J Heat Transf -Trans ASME.* **136** 052001 (doi:10.1115/1.4026168).

[46] Khan M, Hinai M, Al-Hinai A and Dutta J (2014) Visible light photocatalysis of mixed phase zinc stannate/zinc oxide nanostructures precipitated at room temperature in aqueous media *Ceramics International.*

[47] Baruah S, Khan M, J Dutta (2015) Pollutants in Buildings, *Water and Living Organisms.*

[48] Baruah S, M Khan and Dutta J (2016) Perspectives and applications of nanotechnology in water treatment - *Environmental chemistry letters.*

[49] Iftikhar N, Rehman A, Sadaf H and M Khan (2018) Impact of wall properties on the peristaltic flow of Cu-water nano fluid in a non-uniform inclined tube. *Int J Heat Mass Transf.* **125** 772-779

IOP Conf. Series: Materials Science and Engineering **414** (2018) 012011 doi:10.1088/1757-899X/414/1/012011

SWOT analysis of institutional framework for engineering diplomas in technical and vocational education and training system in Pakistan

Aziz Ahmed[1] and A H Khan[2]

[1]Department of Economics, Baluchistan University of Information Technology Engineering and Management Sciences, Quetta, Pakistan
[2]Department of Economics, School of Economics, Quaid-i-Azam University, Islamabad, Pakistan

Email: barshore2006@yahoo.com

Abstract. The theoretical underpinnings of SWOT analysis for institutional framework of engineering diplomas is used to highlight its components of strength, weaknesses, opportunities, and threats. The technical and vocational education and training (TVET) system disseminates approximately 29 associate engineering diplomas in one, three, and four years of duration in Pakistan. Strength includes the establishment of provincial TEVTAs as apex TVET authorities, TVET system, technical education and vocational training in private and public sectors, trade testing boards, and signatory of ILO and UNESCO. Weaknesses capture TVET deficiency, administrative and legal issues, highly skewed technical education, inadequate number of technical colleges, traditional engineering technologies, trade quality compromises, and non-collaboration among TVET institutions and authorities. Opportunities lie in demand is driven market diplomas, local and international markets for earnings and employment, national skill policies, and economic fortunes of mega projects. The threats are present in low skill formation, technical skill deficiency for females and rural areas, international in-competitiveness in labor markets, Chinese skilled immigrants, expansion of TVET trades to modern and emerging fields, imbalanced technical education in TVET set-ups, and low socio-economic standards of common Pakistanis for the development of TVET system in Pakistan. Policy recommendations are devised to develop TVET system for human resource development in Pakistan.
Keywords. SWOT, TVET, associate engineering diploma, Pakistan

1. Introduction

Pakistan is one of the emerging countries on economic fronts. Its economic performance and new initiations of an economic venture like China-Pakistan Economic Corridor (CPEC) is a multitask project of US$ 60 plus billion. CPEC would probably open avenues of economic fortunes of economic prosperity for the mass population of 21 million Pakistani people to eradicate poverty, conventionality, terrorism, and unemployment in years to come. The skilled labor force of Pakistan is yet 6% skilled in technical and vocational trades that is not enough supply to meet the existing market demands and provide services to upcoming CPEC projects and its demand in Pakistan [1]. The technologically lower labor force of Pakistan may face less likelihood to win economically on competitive productivity and efficiency levels in most of the economic fortunes with the mention of human capital formation of 6%. The number of technical colleges and technical training centers is far less in numbers, enrollment capacity, pass outs, and human capital formation, as compared to general schooling in Pakistan [2, 3].

The comparative advantage with only 6% of technically skilled manpower may possess potential threats for the economy of Pakistan. It will result in lower rewards in trade, industry, services inflows and outflows, and upcoming CPEC projects with Chinese labor force inflow in Pakistan. A complete analysis is needed for Pakistani TVET system especially; it's technical education and training, in all the provinces to assess its strengths, weaknesses, opportunities, and threats (SWOT), which has not been assessed in literature concerned. The SWOT analysis may provide a clear picture of existing and deficient TVET system. It will help in devising productive policy options and recommendations for the development of human capital formation in technical and vocational fields of manpower development.

2. Literature review

The system of existing TVET provision for engineering skills consists upon one year, three years, and four years in the shape of associate engineering diplomas in Pakistan. It is disseminated by both the public & private sectors in all the provinces of Pakistan. Both the sectors are working for the sole purpose of achieving the objectives set for skilling Pakistan and formation of human capitalism of technical skills in the labor force of Pakistan [1, 4-7].

National TVET Policy [1] documents have eight objectives for skill development in case of Pakistan. Among these eight objectives, skill development, particularly to the youth of Pakistan, is also aimed at targeting a milestone of 20 million TVET skilled labor force by 2020. The policy focuses upon up-skilling and re-skilling, export of skilled labor, and achieving labor market demand for skilled workers. The policy mentions that skilled labor force will ultimately achieve increased productivity, sustained economic growth, better job opportunities, and earnings to the technical labor force [4-6].

Similarly, all the provincial TVET authorities (TEVTAs) also work for TVET dissemination and skill provisions to the unskilled population in Pakistan. All the TEVTAs are established for the sole purpose of skillful Pakistanis in technology proliferation, skill development, human resource development, and maintaining labor market information systems. Skill development in associate engineering diplomas will yield in job opportunities, reducing unemployment, and prepare skilled youth for mega projects and economic corridors in Pakistan. The provincial TEVTAs are also responsible for managing and promoting TVET systems, collaborating TVET systems, making linkages to labor markets, and many more purposes and functions related to TVET system strengthening for the provinces in Pakistan [1, 4-7].

Private sector TVET institution is also working for a limited number of engineering technologies in all the provinces of Pakistan. Similarly, different NGOs, NPOs, & other organizations registered with NAVTTC and/or other public-sector departments are also functional in Pakistan. These TVET institutions are working in urban and metropolitan areas of big cities to work for those areas of TVET dissemination subject to income earnings and fund hunting from national and international donor agencies. Their programs are mostly irregular and programs functionality are based on funds availability and external financing schemes, which is like the case faced by many developing countries of the world [8-10].

Relevant literature about TVET system for the economy of Pakistan did not mention any kind of comprehensive analysis to highlight SWOT analysis for understanding the existing TVET system and its linkages with the human capital formation and socio-economic development [2, 8, 9, 11-14].

3. Theoretical framework

The theoretical framework of SWOT analysis is used for this study. Acronym "SWOT" is composed of S = strengths, W = weaknesses, O = opportunities, and T = threats. The first two alphabets show internal mechanism in a system and the latter two alphabets show external environment to a system. SWOT analysis gives in-depth analytical underpinnings for strategic exploration, highlighting, modification, formulation, devising, designing, and implementation of a system and series of activities at individual and collective policy levels alike. This approach is commonly used for studies to highlight the strengths, weaknesses, opportunities, and threats to any system or program.

The purpose of SWOT analysis is described to efficiently analyze the existing mechanism for better understanding, formulating, and making policy formulations for system developments. This framework is also used for effective policy options in better development in the areas like education, businesses, organizational development, science and technology, and technological progress. The theoretical framework of SWOT analysis is used for institutional framework of engineering diplomas to highlight its components of strength, weaknesses, opportunities, and threats in TVET system, following the research conventions mentioned by [15-19], for the case of Pakistan. The theoretical framework of SWOT analysis is shown in the figure below.

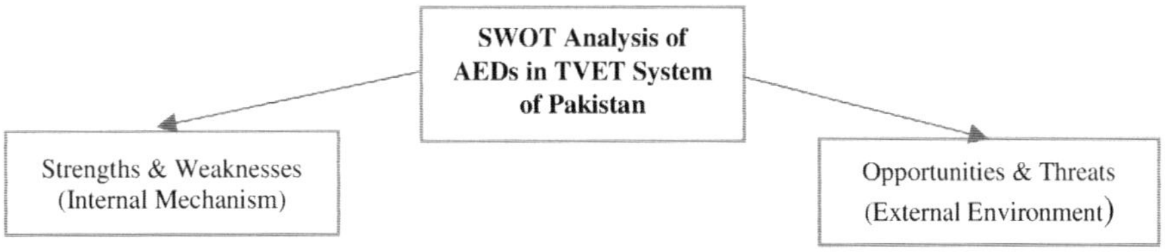

Figure 1. SWOT Analysis of Associate Engineering Diplomas in Pakistan (Designed for theoretical framework about engineering diplomas in TVET the system of Pakistan by following the studies of [16-18]).

4. Research methodology

Research methodology consists upon baseline survey including field work, desktop survey, and content analysis approaches in this study. Fieldwork is undertaken in multiple personal visits to TVET departments, provincial TVETAs, NAVTTC, brainstorming discussions with subject matter experts of TVET, and contacts with associate diploma holders in the labor markets. The desktop survey consists upon browsing internet sources, web portals, and virtual information related to TVET sector and TVET affiliated institutions in Pakistan. Content analysis is undertaken for official TVET documents, TVET acts and rules and regulations, and TVET policy papers/documents to bring forth the SWOT Analysis, in its true spirit, of engineering technologies and associate engineering diplomas, in overall TVET system of Pakistan [17, 19-23].

5. Research Analysis

The SWOT analysis of institutional framework for engineering diplomas in TVET System for Pakistan was conducted. The detailed analysis of strengths, weaknesses, opportunities, and threats were analyzed and discussed separately in the following paragraphs.

5.1. Strengths

There are legislation and implementation of relevant Rules of Business [9] for technical education and training and its provisions for TVET system in Pakistan. The engineering associate diplomas are disseminated through the provincial TEVTAs, allied technical boards, and technical skill departments in all the provinces of Pakistan. The engineering technologies and a proper mechanism for its propagation are undertaken though government technical colleges in Pakistan. There are mono and polytechnical colleges working at all provincial levels. Private sector and NGOs provision of technical skills are also well established for the provision of technical education and training in Pakistan [1, 3].

The complementary part of technical education is vocational training and the system of the vocational system is also established and properly functional for all the provinces of Pakistan. Both the technical and vocational systems collectively constitute the whole TVET system in Pakistan. National levels of regulatory bodies like, N-TEVTA, NAVTTC, NAVTEC, Skill Development Councils, and other TVET regulatory and apex bodies are present in Pakistan. Donors funded programs for skill development are discontinuously functional in Pakistan. CPEC skill programs are started in the province of Punjab, and these programs are yet to be started in other provinces as well. The accessibility of technological education in Pakistan is a quota based on district level in all the provinces of Pakistan, that is why a wide range of accessibility to almost all the districts of all the provinces are the strength of national TVET system [1, 24, 25].

Technical boards, trade testing boards, and technical education board are established for quality maintenance, examination system, diplomas and certification systems in each province of Pakistan. Establishment of NAVTTC regional directorates for skill development initiatives of the prime minister programs for skill development and skilling Pakistan in all the provinces are present. Pakistan is a signatory of ILO, UNEVOC, GIZ, and UNESCO TVET Strategy [20], and other international bodies *"for technological development, achieving SDG4 and inclusive skill development for greening economies"*. It is integrated and a complementary mechanism established for the infrastructure of large number NGOs and private sector TVET institutions across the whole country for TVET system and TVET trades proliferation in Pakistan [3, 25].

There are strong linkages between TVET systems and labor market outcomes of employment generation, livelihood earnings, business start-ups, entrepreneurship, and other economic and labor market fortunes to all the diploma and certificate holders of technical education and training in Pakistan. Almost all the technological based economic sectors of Pakistan's economy like industry, services, Nanotechnology, computer science and IT, dot.com technology, e-business, banking, building and roads, civil works and construction, petroleum-chemical industry, agriculture and poultry, and many more sub-sectors accommodate and demands for technically expert labor force in Pakistan [1, 3, 25-27].

5.2. Weaknesses

Pakistan is a developing economy in the world of leading economies based on technological progress and development. Low levels of education and other socio-economic indicators present the case of Pakistan not suitable for technological progress and high ventures of technological knowledge. There are very limited numbers of technologies in engineering education and training present in Pakistan. According to NAVTTC that the skill dissemination of technical knowledge is confined to 12 technologies, the highest number, in Punjab province and only six technologies in Baluchistan province [1].

The engineering diplomas are confined to traditional technologies of civil engineering, electronics, mechanical, and computer and IT only in all the provinces of Pakistan. No high technology based engineering diplomas like Nanotechnology and microbiology related programs are started in Pakistan. There is not a unified system of quality assurance in terms of quality measurement of associate diplomas of engineering and their examination systems of theory and workshop assessments in Pakistan. Provincial boards of technical educations are responsible without having a unified and common curriculum and course and practice contents for quality assurance and diploma assessment. National qualifying verification framework (NQVF) for quality assurance of associate diplomas in engineering technologies is not fully implemented in Pakistan. Most of the technical colleges do not possess well-established workshops and laboratory rooms for practical exposure and training for their trainees in almost all the provinces. The conditions of physical infrastructure and scientific equipment are miserable in polytechnic colleges in Pakistan. One of the reports of Asian Development Bank mentions that the targets for technical education promotion in whole Baluchistan and parts of other provinces are not met for its TVET strengthening programs in Pakistan [8, 11, 28].

No memorandum of understandings (MoUs) is signed among provincial TEVTAs for TVET collaborations as mentioned in National Skill Strategy [29]. There is a lack of coordination at provincial levels for TVET programs and terms of references for TVET allied departments in all the provinces of Pakistan. There are strong traces of trust deficit, legal, and administrative issues among TVET departments & provincial TEVTAs, especially for the case of B-TEVTA and allied TVET departments, in the provinces. No coordination and collaboration among the TEVTAs, NAVTTC, NAVTEC, Skill Development Councils, and other bodies for technical education in Pakistan. The are many technical, legal, administrative, and coordination problems and constraints among the TVET ministries and departments and among public and private sector TVET organizations for establishing one of the coordinated and well established technical education systems in Pakistan. The technical education system has less capacity to meet the demands of technical skills for domestic and international labor markets. The response of markets for conventional and technical engineering diplomas is lower wages, unemployment, lower female participation, and no regular jobs in Pakistan. No collaboration between provincial TEVTAs and NAVTTC regional directorates for the promotion of technical education in the provinces. Skill deficiency in Baluchistan is reported around 80% in case of technical training and 20% in terms of technical education as compared to national levels of technical training and engineering diplomas. The legal and administrative status of B-TEVTA to work and function as a provincial authority for technical education and training is not fully recognized as it is for the functional status of P-TEVTA in Punjab [1, 5, 6, 29].

Private sector and NGOs have very limited scope for the dissemination of technical education and training in Pakistan. They mostly provide computer and IT related certificates and diplomas in Pakistan. Since these TVET sectors mostly depend upon donors and funds from NAVTTC, that is why their TVET programs are seasonal, discontinuous, and often for funds hunting purposes. So, investment in TVET from the private sector is very limited and confined to big cities and market-based in Pakistan. No skill mapping has been done for technical education and training for the provinces, labor market needs, CPEC projects, mega and mono projects, and demand-driven skills in Pakistan. Very limited number of studies highlights the technical and vocational aspects of skill development, human capital formation, human resource development, linking TVET with economy and development, and exploring labor market outcomes for TVET systems in Pakistan. TVET sector in Pakistan is highly skewed in favor of vocational training as compared to technical education and training. Out of 189 trades, only less than 50 skill trades are technical training, and less than 15 skill trades are offered as associate engineering diplomas in Pakistan [1, 26].

5.3. Opportunities

TVET system is present with officially and legally recognized TVET departments, provincial TEVTAs, and NGOs/private sector in all the provinces for the establishment of its effective system in Pakistan. National Skill Policy [1] is designed for quality skill development and skilling Pakistani youth to have internationally recognized and affiliated standards of technical education in Pakistan. In Pakistan, approximately, *"6% of the whole population acquires skills (Technical Vocational Education & Training). There are 1,647 public & private TVET Institutes in Pakistan"*. The province and ownership-type wise TVET institute are highest in the province of Punjab and the least in Baluchistan. There are approximately one hundred plus (103 to 119) TVET institutions working for associate engineering technologies in Pakistan. The ratio of technical colleges is highest for Punjab and the lowest for Baluchistan, where only five out of total eight technical colleges are functional. NAVTTC has established its regional directorates in all the provincial headquarters of Pakistan. Prime minister programs of technical and vocational for all the four to five phases are successfully completed due to TVET system and its coordinating efforts exerted and initiated by NAVTTC in Pakistan [1, 29].

Private sector colleges of technologies, NGOs, and some international bodies like UNESCO, ILO, UNICEF, GIZ, and UNHCR are also working in technical education and training in very purposive and selected areas of Pakistan. Establishment of all the provincial TEVTAs is no less than a blessing for the regularization of TVET system for the development of skill formation in Pakistan. It may coordinate all the disparities and differences in almost all the TVET study segments of course contents, workshops, diploma quality assurance, institutional development, teachers' training, technological levels, and other types of innovation for all the engineering technologies in Pakistan. Vibrant labor markets for almost all kinds of engineering diplomas in most of the sub-sectors of industries, agriculture, and services sectors of the economy of Pakistan [7, 26, 27].

Associate engineering diplomas are offered for both genders of male and female in rural and urban areas of all the districts of Pakistan. Labor market opportunities for Pakistani skilled labor force present in Arab & Gulf countries, Central Asian countries and European Union for the export of TVET qualified & skilled labor force to earn handsome remittances and capital inflow to their families in

Pakistan. Similarly, CPEC projects, upcoming FIFA World Cup, Dubai Expo, and Gwadar Deep Sea Port, Energy infrastructure, and three economic corridors of rail, roads, buildings, and civil inside Pakistan are big and prospective opportunities for all the engineering diploma holders. There would be high demand for associate engineering diploma holders in these projects and their technologies will be required and absorbed in these national and international economic projects. The futuristic need of human resource development in engineering technologies is likely to be dynamic and prospective for earnings and employment generating for the technologically equipped Pakistani youth in the contexts of mega projects like, CPEC projects, provincial economic activities, demographic dynamism, and labor market needs that will get flourishing in Pakistan [26, 30].

5.4. Threats

Expert and skilled Chinese labor force and its arrival in Pakistan are one of the potential threats for unskilled and semi-skilled Pakistani labor force to compete with them for skills, expertise, productivity, earnings, and employment opportunities. Less coordination between national and provincial TVET institutions also pose threats to the legal and administrative status of their present premises of working and functional authorities in Pakistan. The differentiation of working and funding status of different TEVTAs in different provinces also pose threats to these provincial authorities in Pakistan. Similarly, there are less coordination and lack of uniform policy designed, devised, implement, and execute for engineering skill development that comes through diverse sources of NAVTTC, Skill Councils, provincial TEVTAs, and technical departments of the provincial governments. Weak educational system and low literacy rate do not support enrollment in technical education and training in Pakistan. Currently, only 6% of Pakistani labor force possesses both technical and vocational skills, of which only 30 to 35% consists upon technical education and training, a very serious threat to technical development for Pakistan [1, 5, 6, 24].

Technical diplomas and technical training for the female of the population are very limited. Only limited and very conventional technical skills are allowed for females to take part due to social, gender, tribal, and low mobility constraints and issues in Pakistan. Provincial authorities are not given full authority to work independently and focus labor markets and conduct independent collaborations with international markets and overseas foundations for technical skill development, institutional collaboration, funds rising, financing higher costs, construction of technical workshops and technical laboratories, teachers' training, and designing skill course for local trainees. The provincial authorities also face constraints in building industry linkages, increasing graduates' enrollments, and recruitments of TVET qualified instructors for associate engineering diplomas and linkages building with national and international markets. Challenges of interdepartmental legal issues, administrative tussles, and red tape regulations are yet to be overcome by the TVET sector in Pakistan. The threats are also faced in TVET areas like improper designing and implementation of TVET policies, fragmented TVET programs, lack of coordination between private and public sector TVET sectors, and many more general and issues to any kind in any area of Pakistan. The threat of imbalanced TVET programs for male and females, inclination of vocational trades instead of technical ones, imbalanced rural and urban distribution of TVET institutions and its enrollment practices, and regional disparity exist in Pakistani TVET system. Modern, demand driven, and high-tech engineering skills and associate diplomas are needed in Pakistan. The extension of conventional to modern and demand-driven associate engineering trades is one of the big threats to TVET system for Pakistan. The present number of associate diplomas in approximately 30 in different technologies that seem not enough for fulfilling the economic and national demands for all the emerging and technology-based economic sectors of Pakistan. The threat of demand and supply mismatch for the provision of employment to engineering diploma holders and supplying emerging engineering skills to the market demand is not easy challenges for developing the economy of Pakistan [4-7, 24].

The establishment of a labor market information system (LMIS) for engineering associates is one of the potential challenges for all the provincial TEVTAs in the country. The challenges of skill mapping and transparent and purposeful web-portal for the information of skilled labor force are yet to be overcome for the local industry, labor markets, domestic and international markets, and all other stakeholders of TVET system in Pakistan. The challenges and threats of TVET sector are mentioned by GIZ [27] and ADB Report [11] that identify TVET institutional weaknesses as a big threat for the development and institutional strengthening of TVET system in Pakistan [11, 24, 27].

6. Conclusion and policy recommendations

TVET system includes technical training and education in all the provinces of Pakistan. Provincial levels TEVTAs are established under their legal acts to work as apex TVET authorities for technical and vocational skills dissemination. The authorities also work as regulatory bodies and are responsible for all the concerned TVET aspects and collaboration with other stakeholders in the provinces. SWOT analysis of TVET system, generally, and that of technical education and training, particularly, has shown an opaque picture of overall TVET system in Pakistan. To eliminate the weaknesses and threats associated with TVET system; the following policy recommendations may help to strengthen the technical and vocational education and training system in Pakistan.

- A TVET system at national level collaborated and unified is recommended for all the provinces of Pakistan is required.
- The legal and financial strengthening of all the provincial TEVTAs are recommended for administering their given authorities to proliferate skill development in associate engineering diplomas in Pakistan.
- Skill quality criteria must be ensured in line with an international framework for quality verifications recommended by ILO and UNEVOC standardization for associate diplomas in engineering, technical certificates, and engineering B-Tech degrees.
- The recommendations of the ADB Report [8], should be addressed properly for technical skill development, especially in the case of B-TEVTA in the province of Baluchistan.
- Introduction of market-driven technical education and engineering technologies should be started with national and international collaborators in line with the ILO and UNESCO list of technical trades.
- Technical technologies and trades for upcoming CPEC projects must be launched on a priority basis for Pakistani labor force to competitively win comparative advantage(s) it works in Pakistan.
- Technical colleges for poly or mono technologies, whichever is suitable, should be established for equitable access to labor force for human capital formation in technical skills in each district of each province of Pakistan.
- Technical education and training must be enlarged and expanded for women segment of the population and rural areas from a number of extensions all over the country.
- Private sector and NGOs should be encouraged through funds, subsidies, public-private partnerships, and other incentives for larger coverage of technical skill acquisitions across rural, marginalized, and deprived segments of the population in all the deprived areas of Pakistan.
- Technical education and training must be enlarged in scope and purpose of trainees' enrollment, technical skill acquisitions, labor force skilling, manpower development, human capital formation, managing the human resource, and provision of skill development at mass levels for the labor force of Pakistan.
- Labor market information system (LMIS) for technical technologies and associate engineering diplomas are recommended for all the provincial TEVTAs in Pakistan.
- A separate section is recommended to be reserved for associate engineering technologies in public policies that aimed at skill development and human capital formation in Pakistan.

Acknowledgment
The authors are thankful to B-TEVTA, NAVTTC, KP-TEVTA, P-TEVTA, S-TEVTA, Government of Baluchistan, and its Allied TVET Departments, BUITEMS, School of Economics QAU, and government TVET officials for their support. Authors are much thankful to ORIC BUITEMS for the financial support to this study.

References

[1] NAVTTC 2018. National Vocational & Technical Training Commission, (Government of Pakistan Islamabad) [Internet].; [cited 25th February 2018]. Available from: http://www.navttc.org/TVET_Pakistan.aspx.
[2] Pakistan Bureau of Statistics 2015. PSLM (2013/14). Statistics Division, Government of Pakistan, Islamabad. [Internet].; [cited 24th February 2018]. Available from: http://www.pbs.gov.pk/sites/default/files//pslm/publiçations/pslm2013_14/complete_report_pslm_2013_14.pdf.
[3] Labor Force Survey 2014-15. Pakistan Bureau of Statistics, Statistics Division, (Government of Pakistan, Islamabad) [Internet].; [cited 20th January 2018]. Available from: http://www.pbs.gov.pk/content/labour-force-survey-2014-15-annual-report.
[4] Government of Balochistan 2011. B-TEVTA Act no.II of 2011. (S & GA Department Quetta).
[5] KP-TEVTA 2018. The Khyber Pakhtunkhwa Technical Education and Vocational Training Authority. (Government of The Khyber Pakhtunkhwa, Peshawar).
[6] P-TEVTA 2018. Punjab Technical Education & Vocational Training Authority. (Government of Punjab, Lahore).
[7] S-TEVTA 2018. Sindh Technical Education & Vocational Training Authority. (Government of Sindh, Karachi).
[8] Asian Development Bank. 2009. Education and skills: Strategies for accelerated development in Asia and the Pacific. (ADB Manila, Philippines).
[9] Social Welfare Department 2017 Rules of Business: 2012. [G o B Social Welfare Department Quetta (eds)].
[10] Dargentas, M. and Le, Roux D. 2005. Potentials and limits of secondary analysis in a specific applied context: The case of EDF—Verbatim. In: *Forum Qualitative Sozialforschung/Forum: Qualitative Social Research (**6** No5)*
[11] Asian Development Bank 2002. Technical Assistance to Pakistan. *Restructuring of the Technical Education and Vocational Training System Project (Balochistan).* (ADB Manila, Philippines).
[12] Government of Pakistan 2015. National Skill Policy. Skills for Growth & Development: *A Technical and Vocational Education and Training (TVET) Policy for Pakistan*. (Ministry of Federal Education and Professional Training, Islamabad) [Internet].; [cited 25th February 2018]. Available from: http://www.moent.gov.pk/frmDetails.aspx.
[13] Kazmi, S. W., 2007. Vocational education and skills development: A case of Pakistan *SAARC Journal of Human Resource Development* **3** 105-17.
[14] Khilji B. A., Kakar, Z. K. and Subhan, S., 2012. Impact of vocational training and skill development on economic growth in Pakistan *World Appl Sci J* **17** 1298-302.
[15] Helms, M. M. and Nixon, J., 2010. Exploring SWOT analysis–where are we now? A review of academic research from the last decade *Journal of strategy and management* **3** 215-51.
[16] Houben, G., Lenie, K., and Vanhoof, K., 1999. A knowledge-based SWOT-analysis system as an instrument for strategic planning in small and medium sized enterprises *Decision support systems* **26** 125-35.
[17] Grundy, T., 2006. Rethinking and reinventing Michael Porter's five forces model *Strategic Change* **15** 213-29.
[18] Thompson, A. A. Strickland, A. J. and Gamble, J. E., 2007. *Crafting and Executing Strategy-Concepts and Cases* (USA: McGraw-Hill/Irwin).

[19] Halcomb, E. J. and Davidson, P. M., 2006. Is verbatim transcription of interview data always necessary? *Applied Nursing Research* **19** 38-42.
[20] ILO 2003. Education and Training: Skills for Employability Including the Challenge of Youth Unemployment. [Global Employment Forum. (ILO Geneva)].
[21] Hsieh, H-F. and Shannon, S. E. 2005. Three approaches to qualitative content analysis *Qualitative health research* **15** 1277-88
[22] Schulman, P. Castellon, C. and Seligman, M. E., 1989. Assessing explanatory style: The content analysis of verbatim explanations and the Attributional Style Questionnaire *Behaviour research and therapy* **27** 505-9.
[23] Tuckett, A. G., 2005. Applying thematic analysis theory to practice: a researcher's experience *Contemporary nurse* **19** 75-87.
[24] Khan, A. H. and Ahmed, A. 2018. Identification of Relevant Technical & Vocational Skills for Job Growth & Human Resource Development in the Context of CPEC Projects for Balochistan. (CoE-CPEC, PIDE, Islamabad, Pakistan)
[25] NAVTTC 2016. Skills Trend in Textile Industry Pakistan. *A case study of Faisalabad district.* Government of Pakistan Islamabad.
[26] Planning Commission 2016. *CPEC Related Press Release*. Planning Commission. Federal Ministry of Planning & Development, Islamabad
[27] German Federal Ministry for Economic Cooperation and Development 2017-2019. Supporting technical and vocational education and training (TVET) reform in Pakistan.
[28] Khan, M. Akhtar, S. Mehmood, H. Z. and Muhmood, K., 2013. Analysing skills, education and wages in Faisalabad: Implications for labour market *Procedia Economics and Finance* **5** 423-32.
[29] NAVTTC 2016. National Skill Information System. (Government of Pakistan, Islamabad). Internet].; [cited 10th January 2018]. Available from: http://www.navttc.org/TVET_Pakistan.aspx.
[30] Planning Commission 2015. Pakistan Vision-2025. Planning Commission [Federal Ministry of Planning & Development, (Islamabad)].

Investigation of raw materials for cement industry of Upper Hunza, Gilgit-Baltistan, Pakistan.

Zahid Hussain[1], Naseer Muhammad Khan[1], Syed Mohammad Ali Shah[1], Faheem Nawaz[1], Muhammad Najam Khan[1], Sajjad Hussain[2] and Ishaq Ahmad[2].

[1]Balochistan University of Information Technology, Engineering and Management Sciences, Quetta, Pakistan.
[2]Department of Mining Engineering, University of Engineering & Technology, Peshawar, Pakistan.

Email: engrnaseer1@gmail.com

Abstract: Limestone is the main constituent as raw material used in manufacturing of cement. In this study the limestone deposits of Khyber Karimabad, Passu and Morkhun-Gircha Upper Hunza Valley, Gilgitbaltistan were evaluated for its suitability to be used raw material for cement industry through chemically and mineralogical techniques. The representative limestone's samples from different geological beds of the proposed area were collected. These samples were investigated petrographic study, X-ray diffraction (XRD), Scanning electron microscopy coupled with X-ray diffraction (SEM-EDX), X-ray fluorescence (XRF) and reserve estimation of the proposed deposit was carried out using Geological Information System (GIS). Geological review shows that there is a good potential for industrial-grade limestone, the composition of this limestone can be generally expressed in terms of CaO, MgO, Al_2O_3, and SiO_2, that may be used in place of similar commodities of good quality cement raw materials. After analysis of results, it is concluded that limestone and other raw materials of Khyber Karimabad (KKLS) are is of good quality and can fulfills the international standards of cement and while limestone and other raw materials of Passu and Morkhun-Gircha is not suitable to be used as raw material for cement industry.
Keywords: Limestone, dolomite, mineralogical, elemental, Hunza Valley.

1. Introduction

In Pakistan, the cement production has been decrease from 3498 thousand tonnes in month March 2017 to 3327 thousand tonnes in month April 2017. On an average the production of cement has been increased to 3498 tonnes from year 2003 until 2017 [1]. Cement sector in Pakistan is divided into two zones: North and South, Major players in the south zone include Lucky Cement and Attock Cement. North zone comprises of several players including DG Khan Cement, Maple leaf cement, Bestway cement and Cherat cement etc. Lucky Cement has the largest market share equating to 16.2%, followed by Best Way Cement 12.3%, DG Khan Cement 9.2% and Fauji Cement 7.5% etc. Pakistan cements manufacturing association [2]. Raw materials of cement are abundantly distributing at various stratigraphic levels throughout the Pakistan and particularly in Gilgit-Baltistan (GB) at different localities. The materials meet the specifications of British and Pakistan Standard. They should use economically for the manufacturing of cement in the near future. Area of Gilgit-Baltistan is rich in deposits of cement raw materials with high percentage. Therefore it is very important to evaluate the mineralogical and chemical composition of raw material [3]. The contour map of the study area is shown in figure 1.

In cement manufacturing about thirty different raw materials are used but limestone and clay are the primary raw materials [4]. In addition to this, secondary raw materials include clay, shale, laterite, gypsum etc. Cement plant can be established at any region where limestone and clays are sufficient for forty years [5]. To evaluate the suitability of raw materials for cement plant it is very important to observe their chemical and mineralogical properties [6]. In this research study raw material were investigated chemically and mineralogical for cement manufacturing in Gilgit-Baltistan, Pakistan.

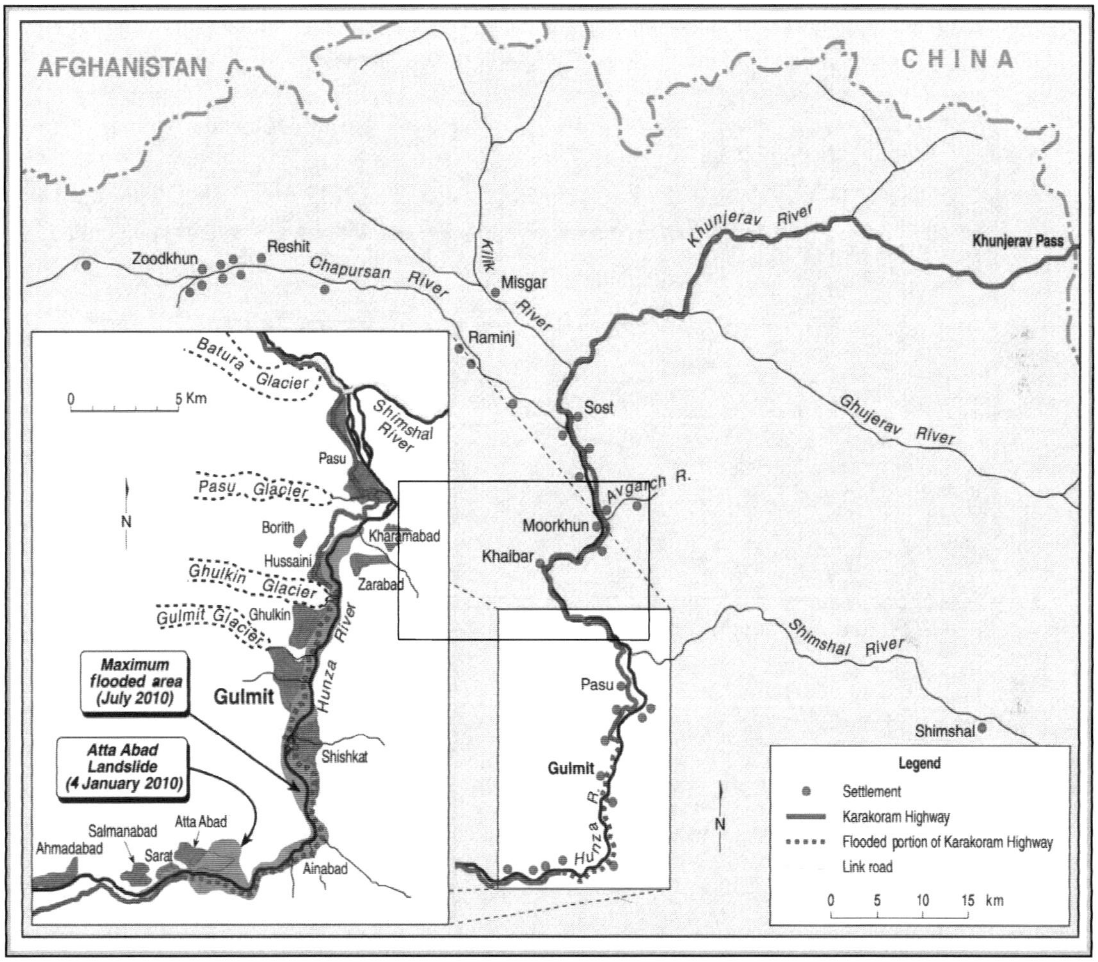

Figure 1. Location map of the study area

2. Geology of study area

Upper Hunza Valley (UHV) is located along the northern most extremity of western Gilgit-Baltistan. Hunza connected with Gilgit through Karakoram Highway (KKH) [3, 7]. It is border with China in north, with Ghizer in south and southwest and with Gilgit in south and southeast [7]. It is about 150 km from Gilgit, a capital city of Gilgit-Baltistan and constitutes southern margin of Eurasian plate [8]. These rock formations range from Permian (Gircha fm) to early Tertiary (Shahnoz Cong). Khyber Karimabad limestone deposits fall in Gojal formation [9]. This formation is mainly composed of extensive carbonate deposit. The age of this deposit is Triassic but later on scientists proved that the age of this unit starts from Permian [10]. The detail information of area are describe in geological map of the area are shown in figure 2.

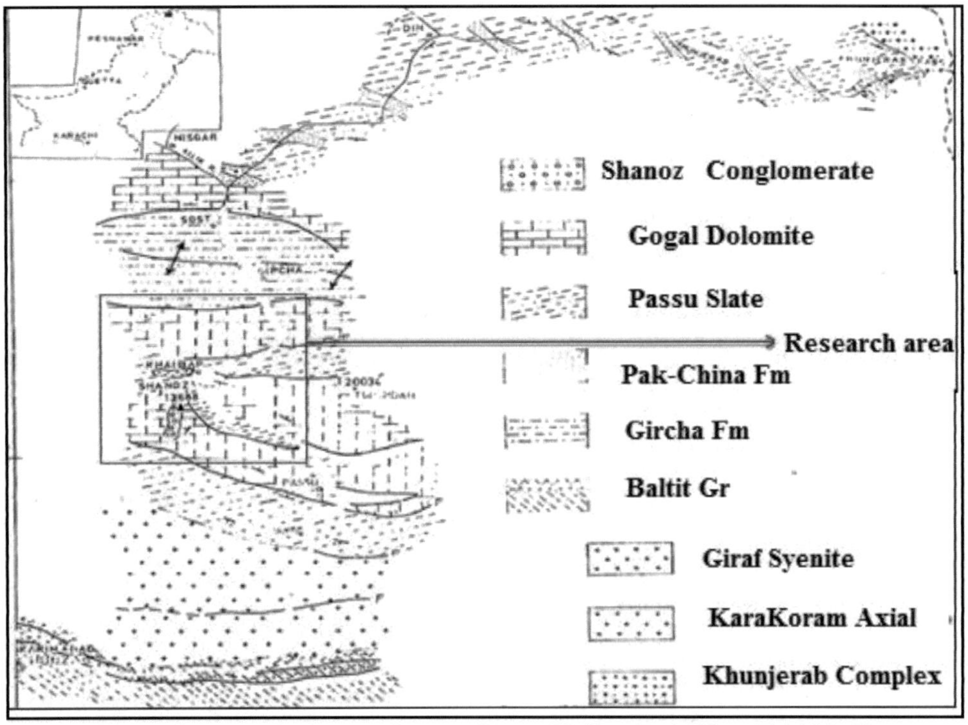

Figure 2. Geological map of Upper Hunza Valley (R.A Khan Tahirkheli)

3. Materials and methods

3.1. Field and lab works

Six representative Samples were collected in bags with noting coordinates from Upper Hunza Valley, Gilgit-Baltistan by channel sampling methods. The collected samples were prepared using crusher and grinder JXC of model LYN to produced mesh size of 120 micrometer. The ground sample was prepared for investigation purpose using conning and quartering by John Riffle box.

3.2. Cement ratios

Cement manufacturing need preparation of suitable raw mix balancing of many oxides are very important. These oxides are CaO, SiO_2, $Al2O_3$ and Fe_2O_3. These oxides are very important for proper burning and for the required qualities of final product. To achieve the balance of these oxides many indices are required. These indices are silica ratio (SR), lime saturation factor (LSF) and alumina ratio (AM) [11].

These ratios are given by

$$SR = \frac{SiO_2}{Al_2O_3+Fe_2O_3} \tag{1}$$

$$LSF = \frac{CaO}{2.80SiO_2+1.18Al_2O3+0.65Fe_2O_5} \tag{2}$$

IOP Conf. Series: Materials Science and Engineering **414** (2018) 012012 doi:10.1088/1757-899X/414/1/012012

The standard range of SR (2.0 -3.0) and AM (1.0-4.0) [12].The calculated cement ratios during the research are, silica ratio is 2.375, and alumina mean is 2.45. These ratios are very close to standard ratios there these raw materials can be used for cement preparation.

3.3. Material characterizations
X-ray diffraction (XRD) analysis was used to find the mineralogical composition of raw materials. The chemical, mineralogical and thin sectioning have been carried out in Geological survey Pakistan laboratory (GSP) Islamabad. Minerals were identified by X-ray diffraction machine (XRD,) X' PERT-PRO MPD) with Cu Kα (λ = 0.15406 nm) irradiation. Chemical analysis was carried out by X-ray fluorescence machine (XRF) model Philips (type PW, 1404). Polarizing light microscope of model LEICA DM 750P was used to find minerals and their arrangement of particles based on optical properties of minerals in thin section analysis. Energy dispersive x-rays (EDX) model (Oxford INCA x-act) were used for elemental analysis of raw materials.

3.4. Evaluation of suitability of raw materials
In order to evaluate the suitability of raw materials for cement manufacturing their chemical and mineralogical results were compared with international standard especially with British standards (BS) because in Pakistan cement is on the basis of British standard [13].

4. Results and discussions

4.1. Chemical analysis
The chemical analyses were carried by X-ray fluorescence (XRF) to investigate percentages of main oxides. They were then compared with international values to evaluate their suitability as raw materials.

4.1.1. Chemical composition of Passu deposit
Table 1 showed that the composition of MgO is 35.135%. The maximum concentration of MgO for cement should be less than 5% according to International cement standards. Because of MgO concentration, the Passu limestone is discarded and concluded that this deposit is not suitable for cement manufacturing.

4.1.2. Chemical composition of Morkhun-Gircha deposit: The chemical composition of this deposit shows that MgO is 35.337% that is unsuitable for cement manufacturing. Hence, the deposit of this region is dolomitic in nature. For manufacturing of cement, the percentage of MgO in limestone must be less than 5%. If it is greater than 5% then the limestone will be unsuitable for cement manufacturing.

Table 1. Chemical composition of Passu and Morkhun-Gircha deposit

Passu Deposit		**Morkhun-Gircha Deposit**	
Chemical compounds	Percentage	Chemical compounds	Percentage
CaO	63.8817	CaO	59.9691
MgO	35.1375	MgO	35.3375
SiO_2	0.3017	SiO_2	0.6217
Fe_2O_3	0.2272	Fe_2O_3	1.4076
Al_2O_3	0.1625	Al_2O_3	0.1898
CuO	0.1398	CuO	1.6617
SO_3	0.0936	SO_3	0.7547
SrO	0.0377	SrO	0.0182
P_2O_5	0.0183	P_2O_5	0.0311

4.1.3. Composition of Khyber Karimabad limestone deposit.

Table 2 shows that the percentage of MgO is within the standard limit and meets the international standards. Moreover, the percentage of CaO also fulfills the international standards. Therefore, it has concluded that the li Khyber Karimabad limestone (KKLS) can be used for manufacturing of indigenous cement.

Table 2. Compostion of Khyber Karimabad limestone deposits

Name	KKLS1	KKLS2	KKLS3	KKLS4	KKLS5	KKLS6	Ave.value	Std.values
SiO_2	3.14	2.12	2.61	4.98	3.18	2.24	3.04	7.41
TiO_2	0.1	0.01	0.09	1.36	0.00	0.01	0.26	1.3
Al_2O_3	0.02	2.15	1.27	1.93	0.04	0.05	0.91	1.55
Fe_2O_3	0.40	0.27	0.34	0.60	0.25	0.37	0.37	1.90
MnO	0.16	0.22	0.20	0.17	0.01	0.02	0.13	0.5
MgO	2.01	1.44	0.09	1.60	0.31	16.44	3.64	3.0-3.5
CaO	50.92	47.35	55.60	47.20	55.59	36.35	48.83	44-45
Na_2O	0.30	0.11	0.21	0.40	0.13	0.16	0.218	0.15
K_2O	0.03	1.33	0.06	0.02	0.01	0.01	0.39	0.15
P_2O_5	0.01	0.50	0.03	0.22	0.00	0.01	0.12	0.16
LOI	45.48	43.34	42.01	44.19	43.48	46.34	44.14	42.22-44

4.1.4. Composition of Khyber Karimabad clay deposit

Table 3 shows that the silica content of in Khyber Karimabad clay (KKCL) was higher than standard values hence this clay is unsuitable for cement manufacturing. Due to this reason, the alternative of clay was very important to investigate at the research area. The alternative to the clay is slate. After chemical analysis of clay, chemical analysis of slate was carried out to check its suitability to be used for cement production.

Table 3. Composition of Khyber Karimabad clay deposit

Mineral	**KKCL**	**Normal Clay**
SiO_2	60.55	50.33
Al_2O_3	17.3	19.17
Fe_2O_3	7.1	6.50
MgO	5.34	3.77
CaO	3.55	1.43
Na_2O	0.00	0.81
K_2O	6.66	2.32

4.1.5. Alternate to clay:

Since the chemical analysis of clay showed that it could not be used for cement manufacturing due high silica content, therefore alternatives to clay called slates were investigated from the field region. There is huge quantity of slates at the region of Khyber Karimabad region. This slate can be used in place of clay to acquire the required standard. For this reason, the chemical composition of slate was compared with standard values and to check their suitability to be used as cement raw material. Table 4, shows the chemical composition of slates. From the chemical composition, it was conformed that the slate can be used as raw material for cement because it fulfills the international standards. This shows that the slate can be used for cement manufacturing in place of clay. Hence, there is no any problem of raw materials at the research area for cement manufacturing. Hence, the raw materials can be used for manufacturing of indigenous cement to fulfill the local requirements.

Table 4. Composition of Passu slate deposit

Mineral	PS 1	PS 2	Normal shale
SiO_2	66.71	68.18	58.10
Al_2O_3	14.57	12.93	15.40
Fe_2O_3	6.54	5.60	4.02
MgO	3.09	2.69	4.44
CaO	1.60	2.20	3.11
Na_2O	1.91	2.00	4.54
TiO_2	0.69	0.66	-*
MnO	0.20	0.16	-*
P_2O_5	0.23	0.22	-*
K_2O	2.68	2.1	2.12
LOI	2.78	3.19	-*

Note: *show traces of TiO_2, MnO, P_2O_5 and loss of ignition less than 0.1%.

4.2 Mineralogical composition of raw materials

4.2.1. Mineralogical composition of limestone

X-ray Diffraction (XRD), model (X' PERT-PRO MPD) with Cu Kα (λ = 0.15406 nm) irradiation, was used for mineral identifications. The identified minerals are present in the following graphs by XRD software X'pert high score. The achieved pattern identified the phases of calcite, dolomite, ankerite and wollastonite, which show intense peaks at 30°, 33.4°, 47.2° and 49°. The XRD pattern of KKSL-5 and KKSL-6 are very sharp which indicates the crystalline nature of raw materials. Figure 1(a)-(b) is the XRD result of KKSL-5, which shows that Khyber Karimabad contains major minerals like calcite, ankerite, dolomite, wollastonite and many other minerals in minor amount. In all samples, calcite is present as major mineral. It is very important to note that the limestone must not contain greater amount of quartz. Otherwise, this will cause the grinding and crushing of limes difficult.

The XRD result of KKSL-5 shows that Khyber Karimabad contains major minerals like Dolomite, Moscovite, Biotite and many other minerals in minor amount. In this sample Dolomite is present in greater percentage and will cause problem but this can be solved by blending with other samples of limestone.

4.2.2. Mineralogical composition of slate:

Figure 1 shows the different peak of different mineral in different samples of Khyber Karimabad limestone (KKSL-5, KKSL-6) and Passu slate (PS-1, PS-2). Figure 1(c)-(d) shows that calcite is the major mineral present in Passu slates with biotite, muscovite and other minerals as minor minerals. The XRD results show that the slates can be used for cement manufacturing as raw materials alternative to Khyber Karimabad clay. This result also shows that the mineralogy of slates fulfills the international standards of slate and shale that are used in cement manufacturing.

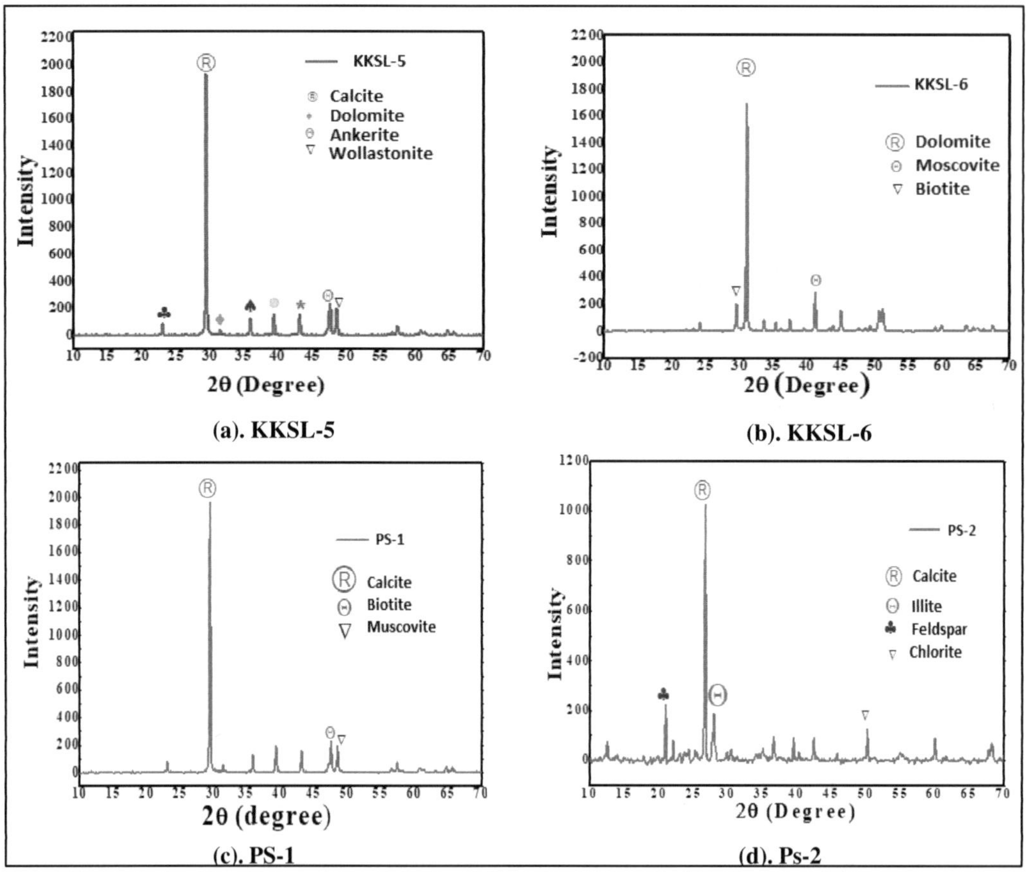

Figure 3. X- Ray diffraction analysis of Khyber Karimabad limestone and Passu slate

4.3 Petrography

The petrographic study of limestone samples from Khyber Karimabad area was carried out. The lime stone is a medium to fine and clastic texture. In thin section analysis different types of minerals present in rocks can be find out on the basis of their optical properties under polarizing microscope. In this study four samples were analyzed to find the minerals present in limestone. This figure 4(a) shows that KKSL-1 contains quartz and volcanic clasts, figure 4(b)-(c) contains calcite crystal while figure 4(d) contain micritic limestone. However, quartz is present in small quantity therefore it will not create grinding problem during cement manufacturing.

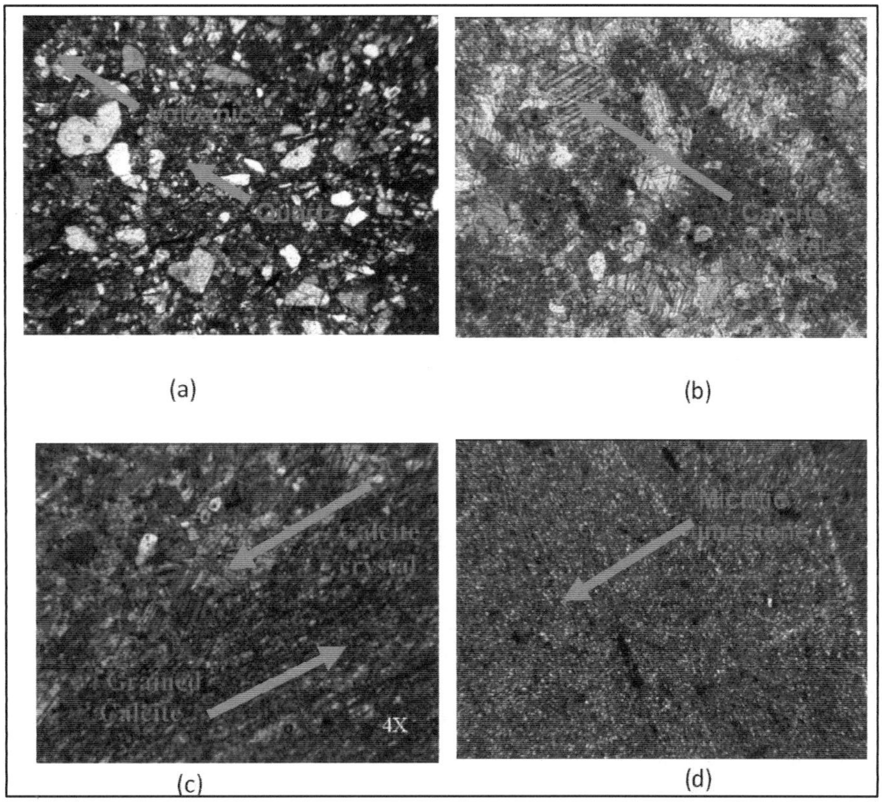

Figure 4. Texture mineralogy of (a) shows different volcanic clast, (b) shows calcite crystal, (c) shows show grain of calcite and calcite crystal and (d) shows micritic limestone.

5. Conclusions

The detailed field and laboratory investigations of limestone and other raw materials were carried out of Khyber Karimabad (KKLS), Passu and Morkhun-Gircha deposits. After analysis of results it concluded that Khyber Karimabad (KKLS) deposit is of good quality which can fulfill the international standards of cement. Therefore, it can be used for manufacturing of cement. Moreover, the reserve of this region is very large where a cement plant can easily be established for production of indigenous cement. However, deposits of Passu, Morkhun-Gircha are not feasible for cement manufacturing because these deposits contain low percentage of CaO and high percentage of MgO. This showed that the limestone of these regions is dolomitic in nature.

6. Acknowledgments

The author greatly appreciates Karakoram International University for scholarship during study leave, Balochistan University of Information Technology and management sciences for cooperation during studying period and Geological Survey of Pakistan, Quetta for cooperation during lab work.

References

[1] Cement PA 2015, Annual report *Flying cement company limited* pp10.

[2] Jaffar A, Ali N, Anwer M, Naeem MA, Riaz MR, Jamal MH, Hussain A and Ming X 2014. The impact of Cement Manufacturing on Economic Development of Pakistan: The Analysis of Cement Industry *International Journal of Research* **2** pp 205-220.

[3] Ullah TQ, Hussain M 2016 Complete Feasibility study of cement raw materials in Gilgit-Baltistan *Geo band work management* pp 55.

[4] Reyes RSA, Alujas AD, Betancourt SR, Alberto CLR and Fernando J M H, Rodríguez SB, Rodríguez CAL, Hernández JFM 2017 Assessment of Cuban Kaolinitic Clays as Source of Supplementary Cementitious Materials to Production of Cement Based on Clinker–Calcined Clay–Limestone *Proc. 2nd Int. Conf. on Calcined Clays for Sustainable Concrete* Springer pp 21-28.

[5] Shah M T, Fayez A, & Ali L 2007 Chemical study of the raw material in Gandghar range district Haripur NWFP for portland cement manufacturing *Journal of the Chemical Society of Pakistan* **29** pp 103-110.

[6] Alp I, Deveci H, Süngün Y, Yilmaz A, Kesimal A and Yilmaz E 2009 Pozzolanic characteristics of a natural raw material for use in blended cements *Iranian Journal of Science and Technology* **33** pp 291.

[7] Desio A and Martina E 1972 A review of the geology and tectonics of the Kohistan island arc north Pakistan *Bollettino della Societa Geologica Italiana* **91** pp 283-314.

[8] Kanwal M S, Shaik F and Price G 2012 Pakistan: Regional Rivalries Local Impacts DIIS Reports *Danish Institute for International Studies* **12** pp 47.

[9] Gaetani M, Angiolini L, Garzanti E, Jadoul F, Leven EY, Nicora A and Sciunnach D 1995 Permian stratigraphy in the northern Karakorum (Pakistan) *Rivista Italiana di Paleontologia i Stratigrafia* **101** pp 107-152.

[10] Özgüner AM 2014 Prospection of Portland cement raw material: A case study in the Marmara region of Turkey *Journal of African Earth Sciences* **97** pp 230-243.

[11] Marzouki A, Lecomte A, Beddey A, Diliberto C and Ouezdou M B 2013 The effects of grinding on the properties of Portland-limestone cement *Construction and Building Materials* **48** pp 1145-1155.

[12] Taylor S, Rankin G and Cleland D 2001. Arching action in high-strength concrete slabs *Proceedings of the Institution of Civil Engineers-Structures and Buildings* **146** pp 353-362.

[13] Bleininger AV 1904. The manufacture of hydraulic cements *Ohio Geol. Survey Bull* **3** pp 391.

Scarcity of water resources in rural area of Quetta District; challenges and preparedness

Zainuddin Kakar[1], Syed Munawar Shah[2], Maqsood Ahmad Khan[3]

1. Faculty of Management Sciences, Balochistan University of Information Technology, Engineering & Management Sciences (BUITEMS), Quetta, Pakistan
2. Department of Economics, Balochistan University of Information Technology, Engineering & Management Sciences(BUITEMS) Quetta, Pakistan
3. Balochistan University of Information Technology, Engineering & Management Sciences (BUITEMS), Quetta, Pakistan

Email: maqsood@buitms.edu.pk

Abstract: This research was motivated by the statement appearing in the media and in the government reports that Quetta is going to run out of water in the near future. The study aims to explore the causes and impact of water scarcity from a socio-economic perspective. The study is based on both the primary and secondary data. The primary data is collected in the rural area of Quetta district from farmers through a well structure questionnaire, whereas secondary data is collected from the online accessible sources. The results indicate that water level has reached to an alarming level in the selected area, primarily due to a subsidy on electric tube wells, indiscriminate installation of tubewells, violation of tubewells spacing norms, growing of high delta crops, population growth, and poor drought management. This affects the people of the area in terms of decline in Kareze irrigation system, depletion of groundwater level, deteriorating health conditions, and uncalled migration from the area. Based on the results, the study recommends effective strategies for putting a restriction on the installation of new tube wells in the study area. Moreover, there is a need of instantaneous campaigns for educating farmers on the efficient use of water and awareness regarding conservation of water resources.
Keywords: water scarcity, enormity, perception, subsidy, efficient use of water.

1. Introduction

Water is one of the most indispensable of all natural resources, it is essential for a human being, economic development and biodiversity (Walter., *et al* 2011). Water is foundation of life and livelihoods, and is key to sustainable development (Guppy and Anderson, 2017). Water is becoming scarce globally, and water trade is taking an important place in water policy discussion (Lenzen., *et al* 2013). Water shortage is one of the most dangerous risks for society, changing hydro-climate and socio-economic conditions have aggravated water scarcity problem throughout the world (Veldkamp., *et al* 2015). Many countries have to face the challenge of rapidly growing of water demand, driven by an increasing population and economic growth, linked to urbanization, industrialization, and mechanization (King, 2004). Water scarcity is one of the most pervasive natural resource allocating problems faced by development planners. One-third of the developing world will face sever water shortage in the twenty-first century (Keller., *et al* 2000). The emergence of water scarcity recognizes that socio-economic, environmental and political factors can influence the occurrence shortage in the availability of ware relative to demand (Damkjaer and Taylor, 2017). Hence, water resources management seems to have become one of the important political, social, and economic issues of the present century and economists face new challenges of growing water demand and changing laws and institution (Louw, 2002). Shatanawi and Naber (2011) pointed out that demographic change like migration and urbanization have increased the demand for water. Social changes such as improvement in lifestyle and rise in standards of living have influenced the perception of people and their attitude toward water. Changing in the global economy and the growth of international trade of goods and services has increased water scarcity. Similarly, Yin *et al* (2017) pointed out that increasing population growth and socio-economic development have put greater pressure on the water resources, and anticipated that climate and socio-economic change may further increase water scarcity in some countries of the world.

Pakistan has moved from water-stressed to water scarce country. Pakistan is facing unprecedented water shortage; owing to climate change and poor water management that has resulted in exploitation of water at a rate faster than it is replenished, directly contributing to the growing water scarcity crises (Khair., *et al* 2011). Balochistan province of Pakistan is the most water scarce province of the country. Land in Balochistan means nothing, whereas water is considered everything as it is a resource at a premium and it adds to the value of land from almost no value to a very high priced commodity in areas where ample groundwater or surface waters are available (Ahmad, 2006). Inefficient water use, wastage of surface water and indiscriminate abstraction of groundwater resources coupled with water scarcity have aggravated the current situation making management of water resources a really complex and a difficult task in the Balochistan province of Pakistan (GoB, 2006). Similarly, Khair *et al* (2014) pointed that in Balochistan the groundwater policies and governance arrangements have resulted in a massive decline in the groundwater table, the uncontrolled installation of tubewells, electric subsidy policy on agriculture tubewells and ineffective governance are key causes of groundwater decline in Balochistan.

Quetta, the provincial headquarter of Balochistan is located at an average elevation of 1,680 m above mean sea level. It is surrounded by Chiltan, Murdar and Zarghoon Mountains. Quetta is facing many socio-economic problems due to its ever increasing population growth and the influx of Afghan refugees. Among other problems, scarcity of water in Quetta has always remained a problem and water supply position becomes acute during the period of low rainfall.

In Quetta district, the groundwater is the only source to fulfill the requirement for domestic, industrial and agriculture purpose. Forty years back the water supply was through karezes which have been dried up. The tubewells then installed for the supply of water and due to high population growth, the numbers of tubewells have been increasing each year. The indiscriminate installations of tubewells have abstracted a huge quantity of water round the clock. Due to these huge numbers of tubewells the rural area of Quetta district is facing a serious problem of water scarcity. The situation become further deteriorated to the mismanagement of water system and lack of coordination among the department supplying water to the study area of Quetta district.

Thus, this paper initially aims to examine the factors that are responsible for water scarcity in the rural areas of Quetta district. The study then explores the impact of water scarcity in the rural areas of Quetta District. Finally, the study suggests measures and techniques for controlling of water scarcity.

2. Material and methods

This section presents the research design to achieve the objectives of the research; it presents the conceptual framework of the study and provides information about the source of data, data collection instruments, sampling techniques, and methodology.

3. Questionnaire Designed

The questionnaire was developed and distributed among exports for their comments and suggestion. Following their feedback, the questionnaire was revised. Keeping in view the objectives of the study all the relevant information included in the questionnaire.

IOP Conf. Series: Materials Science and Engineering **414** (2018) 012013 doi:10.1088/1757-899X/414/1/012013

4. Data Collection

Primary and secondary data were used for this study. Primary data has been obtained through questionnaire, interviews and group discussion from the respondent (i.e. farmers). The farmers' community of the study area of Quetta District is taken as a target population. A sample of 55 farmers selected randomly from all the rural area of Quetta District. The sample size for the above-mentioned area is based on stratified random sampling method. Moreover, secondary data were collected from the relevant government departments which included Department of Agriculture Research and Extension, Irrigation and Power Department, Planning and Development Department, NGOs and other relevant departments.

Conceptual Framework of the study

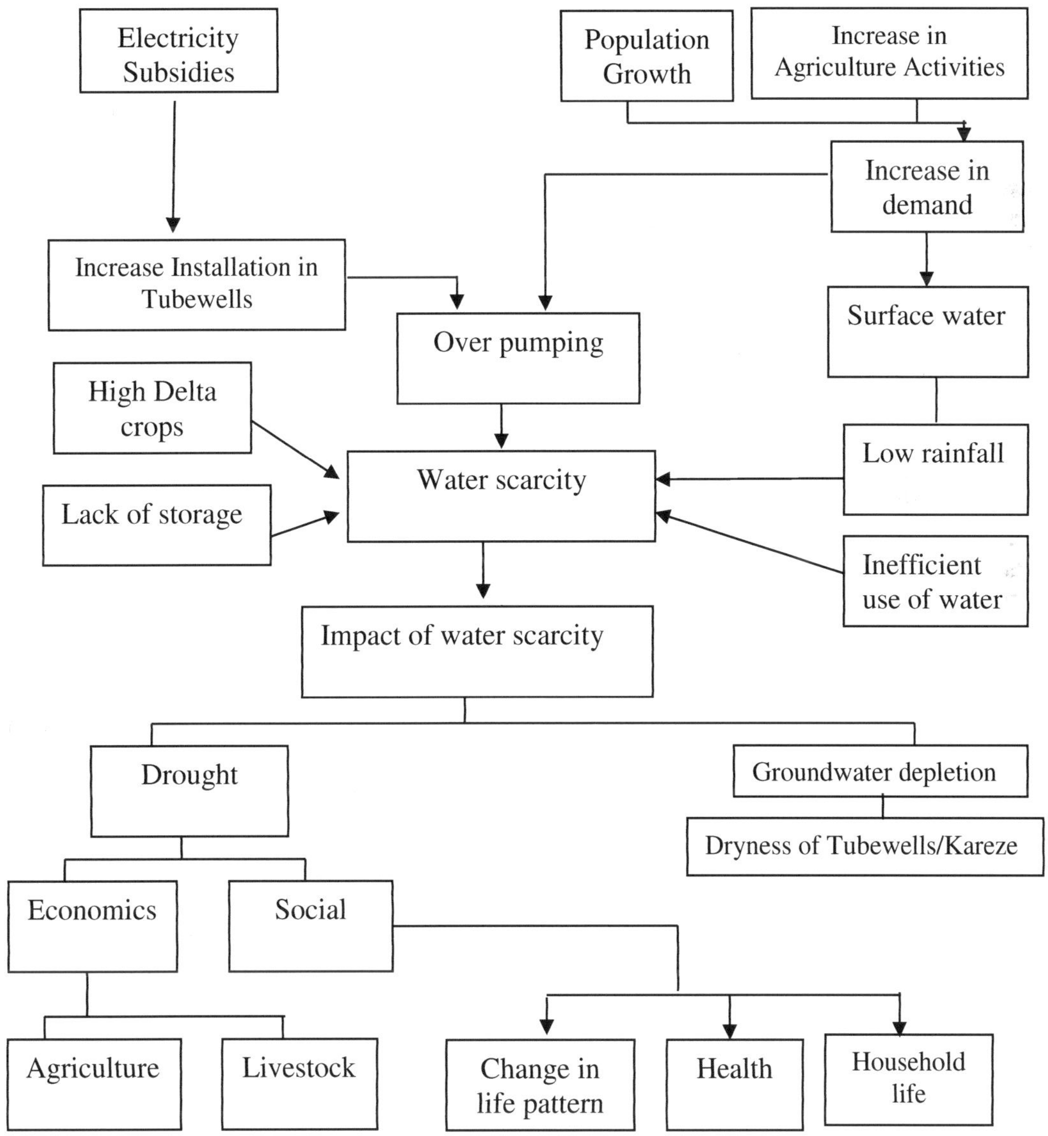

IOP Conf. Series: Materials Science and Engineering **414** (2018) 012013 doi:10.1088/1757-899X/414/1/012013

5. Result and discussion

This section presents the survey results. The purpose of this section is to present the farmers' perception about the present water situation, factors causing of water scarcity and impact of water scarcity through graphs and charts in the study area of Quetta district. The details are given in the following sections.

5.1 Major Factors Causing of Water Scarcity in the Rural Area of Quetta District

Shortage and mismanagement of water have caused great problems to the community in the rural area of Quetta District. The following factors have emerged as of possible importance in explaining the water shortage in the study area of Quetta District.

5.1.1. Subsidy on electricity

The agriculture sector in Balochistan is getting a subsidy on electricity. It was decided that the farmer will pay a flat rate of Rs. 6,000 per month irrespective of consumption and the remaining amount will be shared by the Federal Government, Provincial Government and Water and Power Development Authority (WAPDA) with a ratio of 40, 30 and 30% respectively. The subsidy helped the farmers in the installation of deep tubewells or converting their deep diesel-operated tubewells to electricity. However, subsidy also resulted in lowering of water table and mining of the scarce groundwater resources. The indiscriminate and unplanned development of deep tubewells in the last two decades, with almost no control over their operation, has resulted in a rapid declining of groundwater table in the study area of Quetta District.

According to Ahmad (2005) subsidized electricity to agriculture tubewells encouraged farmers to use groundwater indiscriminately for increasing agriculture production. The tube well subsidy is perhaps the greatest impediment to efficient utilization of water resources and provides a huge disincentive to farmers to adopt more efficient methods of irrigation. During research study when the farmers asked about other reasons for groundwater depletion, the majority of farmer's responses were over-abstraction and increase in the number of tubewells in the area.

5.1.2. Uncontrolled Installation of Tube wells

The government-led electrification of many of the rural area of Quetta District has improved communication networks and generated promising returns for fruit and vegetables, and the tube well irrigation has gained remarkable growth. During research study, it was observed that farmers are installing additional tubewells to meet the requirement of rural agriculture sub-sectors. Anyone anywhere can install tubewells on his lands and any kind of company or individual can be hired for this purpose. This has resulted in adopting any kind of standards in site selection or technique of drilling, tubewells materials, design and testing. In a study, Ahmad (2006) pointed out that the unchecked installation of tubewells in large numbers, indiscriminate pumping of water and highly subsidized electricity were the main reasons for water scarcity in Balochistan.

During survey, it was fined that majority of the farmers having more than one tubewells in the study area of Quetta District. When the farmers were asked a question about the number of tubewells they are using for their irrigation purpose. The following responses were received from the farmers.

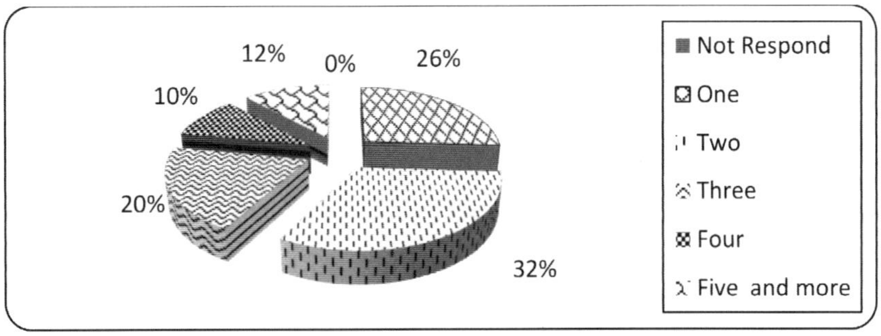

Figure 1. Percentage of Farmers Having Number of Tube wells

Figure 1 showed the responses of the farmers about the number of electric tubewells they are having. About 12% of the respondents were of the view that they have five and more tubewells, 10% mentioned four, and 20% said three, 32% pointed out two, and 26% replied only one. These results clearly show that majority of the farmers having more than one tubewells. This indiscrimination of tubewells is the major cause of over exploitation of groundwater in the rural area of Quetta District. During survey another important thing which was noted that in spite of large number installing of tubewells, the farmers using high Horse Power (HP) motors for running their tubewells. When the farmers were asked about the type of tubewells they were using for their irrigation purpose. The following were the response of the farmers:

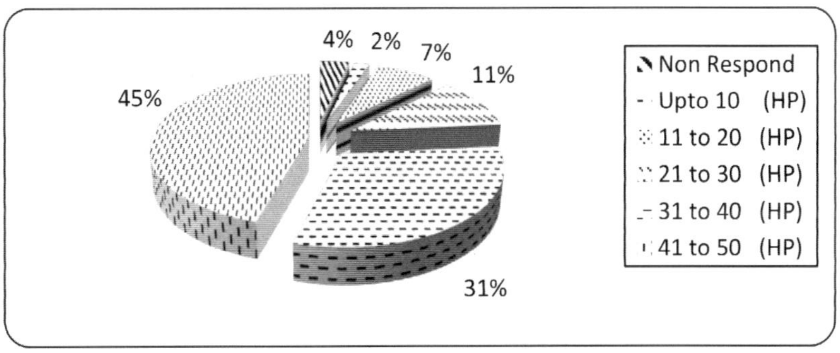

Figure 2. Percentages of Farmers Having Prime Mover Type of T/W

Figure 2 showed a majority of the farmers using high prime mover type of tubewells for irrigation of their fields. 45% of the farmers pointed that they using 41 to 50 HP prime mover types of tubewells. 31% mentioned that 31 to 40 % HP prime mover types of tubewells, 11% were having 21 to 30 HP. These results showed that 87% of the farmers were using the prime mover amongst 31 to 50 HP turned to be an important factor causing water scarcity in Quetta District.

5.1.3 Violation of Tubewells Spacing Norms

There exists a clause that guides regulators about the tubewell to tubewell spacing norms under the Balochistan Groundwater Administration Ordinance 1978 (Halcrow, 2007). Violations of these tubewell spacing norms were often observed during the survey, most of the farmers do not follow the minimum distance regulation between tubewells, which is 750 feet; therefore it also happens to be one of the major contributing factors to the alarming depletion of water table in the study area. During survey when the farmers were asked about the number of tubewells per 1000 square feet the following responses as per figure 3 received.

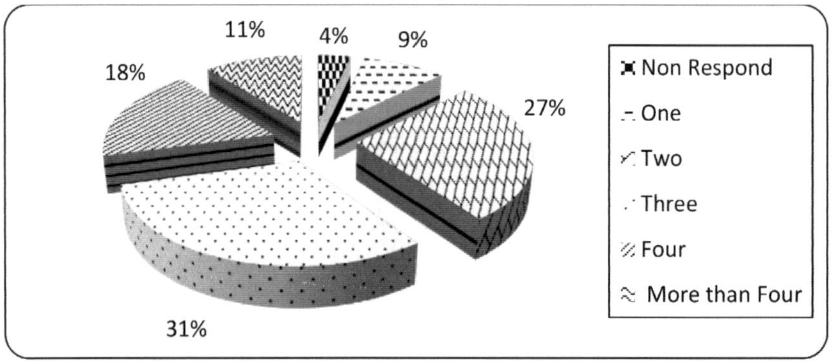

Figure 3. Percentage of T/W Installation within 1000 Square Feet

Figure 3 shows that 11% farmers replied that there are more than 4 tubewells within every 1000 square feet, 18% pointed out that 4, 31% respondent that 3, 27% mentioned that 2. This result clearly shows that the number of installation of the tubewells per 1000 square feet is very high. It has also been noted that no permission is required for the installation of tubewells, because of the lack of proper regularly system both from the federal and provincial government.

5.1.4 Growing of high Delta Crops

After the provision of subsidized electricity to the farmers, the area under high-value crops like apple cultivation steadily increased. It had also become the cash crop so the farmers had cultivated it more than other crops, and the income from apple orchard was more than the other crops. The water requirement of apple was more than other fruits like grapes, pomegranates, peaches, almonds, and plums, but climatically growing apple was not suitable in the area such as Quetta District that has altitude less than 1600m. it may be summed up that apple replaced the traditional crops like pomegranates, grapes plum, peaches and almond which are climatically suited and adapted to that area like Quetta District. Asharaf and Majeed (2006) conducted a study to analyze the water situation in Balochistan. The main finding of the study was that the water resources were scarce both for the domestic and agriculture purposes almost in all parts of the province. The reason for this water scarcity was the growth of high delta crops and lack of crops knowledge among the farmers which led to over irrigation of crops. Similarly, Choudhary *et al* (2008) conducted a comparative study of the cost of production and decision making analysis in case of onion and sunflower crops in Quetta district. This study concluded that water scarcity was the main problem in Balochistan, which caused high delta crops in the study area. During survey when the farmers were asked about the cropping pattern in the study area, the following response as figure 4 was received.

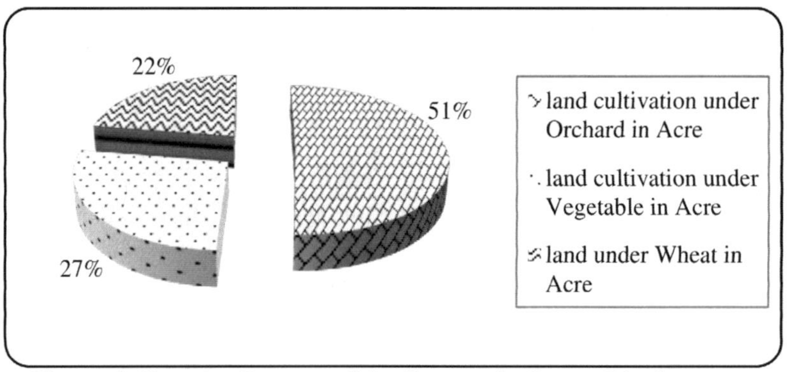

Figure 4. Cropping Patterns of the Farmers in Quetta District

Figure 4 shows the cropping pattern of the farmers in the area. It has been found that 51% of the area is used for orchards, 27% of the area is used for vegetables, and 22% area for wheat. The above figure shows that apples are the major crop in the study area of Quetta District. It is consuming more water than the other crops.

5.1.5 Influx and high Growth Rate of Population

The overall water supply and demand projection for Quetta district show that the water problem will be acute in the future as a result of a rapid increase of population growth. Shah *et al* (2002) pointed that although agriculture sector is highly depended on irrigation water; however overexploitation and overpopulation growth facing sever water shortage in upland of Balochistan. Quetta District will require more water for its population, which is increasing at a rate of 3 to 6 percent per year. The water supply capacity is limited and cannot meet the total demand. Poverty and lack of economic opportunities have pushed labor from the rural areas because of the lack of job opportunities, education, and healthcare facilities. During summer season this problem became acuter with the increase in population due to the influx of seasonal visitors from other parts of provinces of the country due to the harsher climatic condition.

5.1.6 Inefficient use of Irrigation Water

The irrigation system in the study area of Quetta District is inefficient due to poor maintenance of watercourses and channels and traditional methods of irrigation (mainly flooding). During the survey, it has been observed that most of the respondent did not have any information and education about the efficient use of water. In response to the adoption of high-efficiency irrigation system such as sprinkler and trickle, most of the farmers have no information about such irrigation system. Qureshi and Mujeeb (2004) observed that water was scarce in Balochistan because farmers failed to make efficient use of water and that they at times over irrigated crops due to their ignorance. In response to a question asked from farmers about information of high efficiency irrigation system, the following response as per figure 5 received

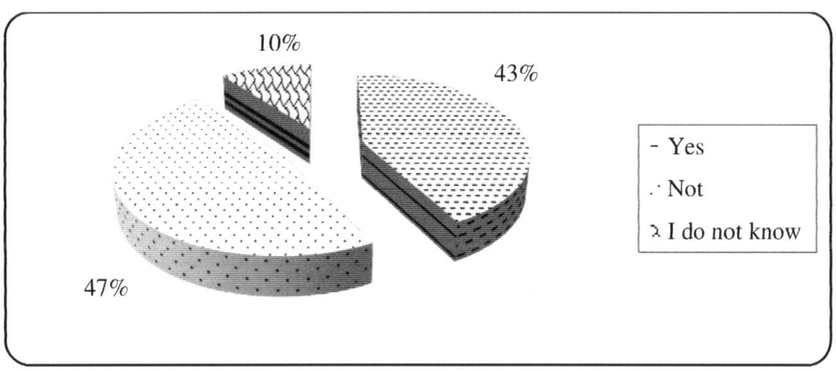

Figure 5. Farmers Perception about High-Efficiency Irrigation System.

Figure 5 shows the information about the farmers' perception of the use of the high efficiency irrigation system. 47% farmers responded that they think high-efficiency irrigation system put no impact on their on their output production. Whereas, 43% of the farmers stated that they have information about the usage of high-efficiency irrigation system and it can result in the increase of their output production. Hence the lack of awareness about the high-efficiency irrigation system was also one the major factors that caused water depletion in the area.

5.1.7. Poor Drought Management

There have been many droughts in the history of Balochistan resulting from erratic rainfall and diminishing precipitation in successive years. The drought of 1998-2006 exhibited a substantial loss of rainfall to a very low level. The average rainfall 50-60% below the average annual rainfall caused reduction of river flows as the recharge declined and the rate of decline of water table further accelerated; further worsened the situation. The impacts of drought were many folds; first it prompted the more abstraction from groundwater in the form of tube well installation as drought mitigation measures, secondly, the annual recharge to groundwater recharge almost touched zero levels. Halcrow, (2008) mentioned in a study Supporting Public Resource Management in Balochistan that in an alluvial aquifer of Quetta District; the decline of water table was around 18-24 meters per year. The production of fruits, crops, and vegetables fell by about one-third of total production.

5.1.8. Impact of Water Scarcity

From the discussion of the respondents, it was revealed that scarcity of the water has a great impact on the inhabitant of Quetta District. Followings are the impact of water scarcity on the people of the rural area of Quetta District.

5.1.9. Falling Water tables

Water scarcity affected the groundwater situation and water tables are falling at an alarming rate in the study area of Quetta District. According to Halcrow (2008), the decline of water table was around 18-24 feet per year in the alluvial aquifer of Quetta Valley. Similarly, Ahmad (2006) studied that the water table in the three main basins-Pishin Lora Basin (PLB), Nari River Basin (NRB), and Zhob River Basin (ZRB) of Balochistan are declining at the rate of more than four to five meters annually. The overall water table decline from 50-100 feet to 200-700 feet in a period of around 30 years is an excellent example of water mining. The falling water tables are not only increasing the cost of pumping but also causing the groundwater quality to deteriorate, and this is a major issue in the groundwater dependent area of Balochistan (Qureshi *et al.*, 2008). During survey when the farmers were asked a question to know the rate of annual decline in the water table in the study area of Quetta District, the following responses have been received from all the farmers.

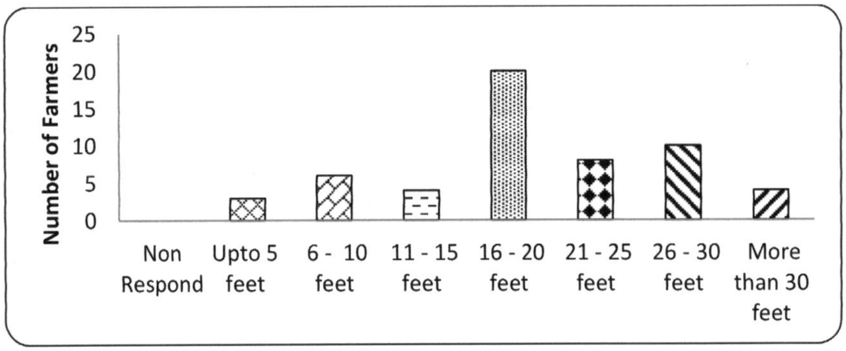

Figure 6. Farmers Perception about Depletion of Ground Water in Quetta District

Figure 6 shows the farmers response about the annual depletion of water table in the study area. It can be seen that majority of the farmers i.e., 20 out of 55 farmers mentioned that that water is declining at an annual rate of 16 to 20 feet per year. Similarly, 10 farmers pointed out that, water is declining at a rate of 26 to 30 feet annually, 8 farmers mentioned that 21 to 25 feet per year and 4 farmers stated that the rate of depletion is even greater than 30 feet annually. 6 farmers mentioned that the rate is 6 to 10 feet, 4 showed 11 to 15 feet and 3 stated 5 feet per year. Overall, out of 55 farmers

42 stated that the groundwater is depleting 25% annually which is a very high rate especially in an arid zone like Quetta.

5.1.10. Decline of Kareze Irrigation System

Kareze irrigation had been the ancient irrigation system prevailing in many districts of Balochistan for many centuries for drinking and irrigation purposes. It had been the main source of irrigation in Quetta, Pishin, Killa Abdullah, Mastung, Kalat and Ziarat districts of Balochistan. About 60% of the area used to be irrigated by kareze/springs in the Quetta and Pishin districts in 1904 with some381 karezes in the Mastung and Kalat districts, 123 in Zhob, 115 in Lorelai, 127 in Panjgure and 23 in Chaghi (Balochistan Gazetteer, 1906 quoted by IUCN, 2000). A survey conducted by the Irrigation and Power department during 1998-2000 revealed that the total number of karezes in the province were some 1,146 in the different districts of Balochistan. Karezes, despite having been the practice for centuries, have fallen prey to electrification and the subsequent mechanized groundwater development, and its decline is concomitant with the introduction of Persian wheels, dug wheels and later on deep tube wells. Mustafa and Qazi (2007) argued that the ecologically sustainable kareze system, which had been the locus of community life, has been under strain due to the excessive groundwater drawdown from electric and diesel tube wells.

Now the century's old system of kareze irrigation is so seriously endangered that it is near to complete collapse and there are only a small number of karezes that are still running. During survey when the farmers asked a question about the number of karezes/tubewells dried in the study area of Quetta District, the following response received from the farmers as per figure.

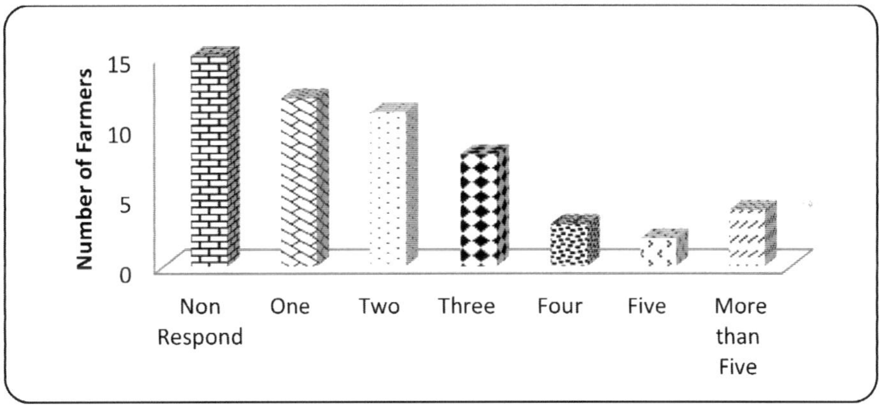

Figure 7. Farmers Perception about the Number of Tubewells/Karezes Dried.

Out of 55 farmers, 40 farmers replied that there tubewells/ karezes have been dried due to water scarcity in the area. 12 pointed one, 11 stated two, 8 showed three, 3 said four, 2 indicated five, and 4 farmers mentioned more than five tube wells/karezes have been dried out due to water scarcity. These results clearly show that water scarcity resulted in the drying of tube wells/ karezes in the study area of Quetta District.

5.1.11. Socio-economic Impact of Water Scarcity

Water scarcity had adversely affected the social and economic condition of the people of study area of Quetta District. Followings are the socio-economic impact of water scarcity, which are discussing below;

5.1.12. Migration of people

According to Ahmad *et al* (2004) that water scarcity has caused widespread migration of people. This has converted a large population of settle villagers into migrants termed as refugees. Similarly,

Nasurullah *et al* (2011) pointed out that water scarcity has a serious socio-economic effect on the population of Tehsil Karezat District Pishin. Majority of people are migrated from the tehsil to the urban centers to get job especially in government services and business in some households few members are left their homes in search of job, the farmers having their own land now working as daily wages in other farmers' fields. During survey as per information received from farmers that, around 30% population migrated from their villages to other area where water and agricultural satisfactory. The competition for local work added to the pressure on the job market and the relation between the local population and the migrants has led the situation of conflict in the area.

5.1.13. Impact on Health Situation

Water scarcity adversely affected human health in the study area. According to Ahmad et al (2006) that water scarcity causes for the poor state of human health. The most common health complains due to water scarcity related to diarrhea, vomiting, and fever among children. During survey information revealed from the farmers in group discussion indicates that spending on health has decreased approximately 50% as compared to the situation before water scarcity. As a guess, the farmers mentioned that they used to spend 20% of their income on health which declined to around 10% at present. The scarcity of water has resulted in the increase in communicable disease, especially waterborne disease and also in nutritional disorder in the study area.

5.1.14. Rise in Conflict

The water scarcity has caused a rise in conflicts among the farmers in the rural area of Quetta District. As everyone abstract groundwater they blame each other for the decreasing discharge capacity of their tube wells. Where ever anyone has to drill a new well it always causes a conflict and badly affected even the relationship of the near relatives there is also a stress and agony which affected the normal life of the society. Due to the over-exploitation of water and down the water table every year a considerable amount of money is being spent on digging deeper their tube wells each or every ultimate year. According to Ahmad (2006) that with the current rate of groundwater depletion, water shortage may cause serious inter-generational issues and inhabitant of different basins may have serious conflict among themselves due to water scarcity.

6. Conclusion and recommendation

The main finding of the study is that the water scarcity issues have been proved as recognized by the respondents as well as by the direct observation by the researcher. The reason for water scarcity is rising population growth, the growth of high delta crops, low rainfall and over-exploitation of groundwater. Similarly, it has also been noted that installation of new tubewells, violation of tubewells spacing norms and subsidy on electric tube wells were also found to be the key factors led to the water shortage of water in the study area of Quetta district. Moreover, the results also showed that most of the farmers were not aware of the modern high-efficiency irrigation system led to the mismanagement of water resources in the study area. Regarding the impact of water scarcity on the dwellers of the study area, this water shortage led to adversely affect the social and economic condition of farmers in the study area. Moreover, this water shortage also led to the migration of farmers from their native place, adversely affected the health of farmers and conflict among farmers rises due to water shortage. On the basis of above finding the followings suggestions are recommended for the solution of water scarcity problem in the rural area of Quetta District.

- The Government should ban on installation of new tubewells in the valley.
- The subsidized flat rate for electricity encourages wastage, charging electric supply for all kind of tube wells should be metered and the flat rate should be linked with high-efficiency irrigation system.

- Incentives should be given to the farmer by subsidizing mechanism, introducing of new high yield crop varieties and improving the irrigation efficiency coordination among the agencies governing the water resources
- The mass awareness campaign is made through electronic and print media for controlling waste of water in households is must and should be undertaken on an urgent basis.

References

[1] Ahmed I 2006 Electric Tariff and Tubewells Subsidy in Balochistan-Building Technical Argument for Capping the Subsidy. Consultant Study No. TA 4560 (PAK) - #23, Irrigation and Power Department Quetta.

[2] Ahmad S 2005 Restructuring and Strengthening of Water Resources Planning, Development and Monitoring Directorate of Irrigation and Power Department, Balochistan. Vol. 4 No.1 TA-4560 (PAK) Quetta.

[3] Ahmad S 2006 Balochistan Economic Report, Background Paper Balochistan's Water Sector: Issues and Opportunities. *Joint Mission of World Bank and Asian Development Bank.*

[4] Ahmad S, Hussain Z, Qureshi A S, Majeed R and Saleem M 2004 Drought Mitigation in Pakistan: Current Status and Option for Future Strategies, International Water Management Institute, *Working Paper, 85.*

[5] Ashraf M and Majeed A 2006 Water Requirement of Major Crops for Different Agro-Climatic Zone of Balochistan. IUCN-The World *Conservation*: 39.

[6] Choudhary M A, Lodhi A S and Ahmad M 2008 A Comparative Study of Cost of Production and Decision Making Analysis in Case of Onion and Sunflower Crops in Quetta District. *Sarhad Journal of Agriculture*. vol. 24, No.3 pp 469-478.

[7] Damkjaer S and Taylor R 2017 The measurement of water scarcity: Defining a meaningful indicator. *Ambio*, Vol. 46 pp 513-531.

[8] GoB 2006 Integrated Water Resources Management Policy Balochistan, Irrigation and Power Department, Government of Balochistan Quetta.

[9] Guppy L and Anderson K 2017 Global water crises: The Facts. Water Crises Reports, United Nations University, *Institute for Water, Environment and Health*

[10] Halcrow 2008 Supporting Public Resource Management in Balochistan, Pishin Lora Basin Management Plan: Final Report. Irrigation and Power department, Government of Balochistan, Royal Netherland Government, *Halcrow Pakistan (PVT) in association with Cameos.*

[11] Halcrow 2008 Supporting Public Resource Management in Balochistan. Identification of Recharge Potential Zone in three Over-drawn Basins (PLB, Nari and Zhob, final report). Irrigation and Power Department, Government of Balochistan, Royal Netherland Government. *Halcrow Pakistan (Pvt) in association with Cameos.*

[12] Halcrow 2007 Supporting Public Resource Management in Balochistan. Bain-wise Water Resources Availability and Use: Final Report. Irrigation and Power Department, Government of Balochistan, Royal Netherland Government. *Halcrow Pakistan (Pvt) in association with Cameos.*

[13] IUCN 2000 Water Background Paper of the Balochistan Conservation Strategy. *Government of Balochistan and the Royal Government of Netherland.*

[14] Khair S M, Mushtaq S and Smith K R 2014 Ground Governance in a Water-Starved Country: Public policy, farmers' Perceptions, and Drivers of Tubewells Adoption in Balochistan Pakistan.

[15] Khair S M, Mushtaq S, Culas R J and Hafeez M 2011 Groundwater Market under the Water Scarcity: The Upland Balochistan Region of Pakistan. Paper Presented at the 40th Australian Conference of Economics July, 11-13.

[16] Keller A, Sakhivadivel R and Seckler D 2000 Water Scarcity and the Role of Storage in Development. International Water Management Institute, Research Report, No. **39**.
[17] King N 2004 The Economic Value of Water in South Africa. In: Blignaut, J. and de Witt, M., (eds): Sustainable Option: Development Lesson from Applied Environmental Economics: 207-238. *University of Cape Town Press*, Cape Town.
[18] Lenzen M, Moran D, Bhaduri A, Kenemoto K, Bekchanov M, Geschke A and Foran B 2013 International Trade of Water. Ecological Economics, vol. 94 pp 78-85.
[19] Louw 2002 The Development of a Methodology to Determine the True Value of Water and the Impact of a Potential Water Market on the Efficient Utilization of Water in the Berg River Basin. Water Research Commission (WRC), Pretoria.
[20] Mustafa D and Qazi M U 2007 Transition from kareze to tubewells irrigation: development, modernization and social capital in Balochistan Pakistan. *World Development*, vol. 35 No.10 pp 1796-1813.
[21] Nasurullah M A, Ahmad M, Malghani M A K and Kakar E 2011 Socio-Economic Effect of Water Scarcity in Tehsil Karezat District Pishin Balochistan. *Journal of Applied & Emerging Science,* vol. 2, No. 2 pp 116-123.
[22] Qureshi A S and Majeed A 2004 Analysis of Drought Coping Strategies in Balochistan and Sindh Province of Pakistan. International Water Management Institute, Working Paper 86.
[23] Qureshi A S, Gill M A and Sarwar A 2008 Sustainable groundwater management in Pakistan: Challenges and opportunities. Irrigation and Drainage, DOI: 10.1002/ird.
[24] Shah T C, Scott A, Kishore A and Sharma A 2002 Energy-Irrigation Nexus in South Asia: Approach to agrarian prosperity with viable power industry. International Water Management Institute (IWMI), India.
[25] Shatanawi M and Naber S 2011 Valuing Water from Social, Economic and Environmental Perspective, *CIHEAM*: 109-117.
[26] Veldkamp T I, Wada Y, De Moel H, Kummu M, Eisner S, Aerts J C and Ward P J 2015 Changing mechanism of global water scarcity events: Impact of socio-economic changes and inter annual hydro-climate variability. *Global Environmental Change*, vol. 32 pp 18-29.
[27] Walter T, Kloos J and Tsegai D 2011 Improving Water use Efficiency under Worsening Scarcity: Evidence from the Middle Olifants Sub-basin in South Africa. Water SA, vol. 37, No. 3 pp 357-370.
[28] Yin Y, Tang Q, Liu X and Zhang X 2017 Water scarcity under various socio-economic pathways and its potential effects on food production in the Yellow River basin. Hydrol. Earth Syst. Sci. Vol. 21 pp 791-804

Air Pollution Deterministic Index Modeling: Application in Quetta, Pakistan

Akhtar[1] F, Shahkhan[2] S
[1]Department of Environment Management and Policy, Balochistan University of Information Technology, Engineering & Management Sciences, Takatu Campus, Quetta, Pakistan.
[2]Department of Physics, Gen. Muhammad Musa Postgraduate College, Quetta, Pakistan.

E-mail: faizaakhtar42@yahoo.com

Abstract: Mathematical model are applied to predict the sensitivity of climate to changes produced by natural phenomena and human activities. In this paper Air Pollution Deterministic Index Modeling (APDIM) for Pakistan is developed with the practical implication in Quetta City. The modeling is based on deterministic model and Pollution Indices to monitor the Ambient Air condition in Quetta City. The novelty of the model is adding a coefficient in the basic deterministic formula. The values of the coefficient balance the theoretical and experimental validations for Ambient Air. The tool gives the alerts about the weather conditions in simple indices or colour coding displays. These indices indicate the air pollution situation in the city by a single number or colour code. The application of the applied model is done in Quetta city, six criteria pollutants (CO, SOx, NOx, O_3, TSP and PM_{10}) are chosen for modeling according to the WHO criteria which have harmful health effect on the communities. The results of the modeling indicate that the gases (CO, SOx, NOx, and O_3) are touching the boundary of satisfactory to un-satisfactory zone. The cause of concern is TSP and PM_{10}, which lies in the hazardous zone. APDIM is an important tool of decision making, which determine the risk assessment for communities. The aim of the tool is to inform the general public of the local area about the severity of ambient air pollution, and the possible health hazard it causes, particularly on vulnerable groups or people with existing cardiac, lungs and breathing diseases. Further the tool helps the environmentalists and policy makers to modify the policy and strategies according to the provided air data.
Keywords. Air Pollution Deterministic Index Modeling, Ambient Air Pollution Indices, Deterministic modeling.

1. Introduction:

Human being in the conquest for food and better living conditions is modifying the environment. In this craze for progress and development, he added a lot of dust, smoke, poisonous gases, hazardous waste and harmful radiation on the face of the earth [1]. Through environmental modifications either man made or artificial, not only the physical, geological, geographical and topographic surfaces but also the biological configuration of the planet is liable to constant change. Such alterations involve not only the gradual disappearance of the developed ecosystems and the reappearance of new ones but also the processes may continue long after the human species have been extinct [2].

The major constituents of the ecosystems and all the ecological complex units of the earth-water-atmosphere systems are being somehow interconnected and intimately, but naturally linked together. Any interaction of one will trigger repercussions on the others and vice versa culminating to a general state of imbalance, disruption of equilibrium and deteriorations [3]. Degradations and modifications of the physical and the biological specifications of man's environment are potentially capable of inducing responses and repercussions in the human mind, human reflexes, human metabolism and human activities, etc. [4]. The extent of such responsive attitudes could only be revealed long after the phenomenal changes have taken place and ultimately culminating in physical, mental and genetic modifications of the human species [5, 6].

Many of the urban centers in the developing countries are beset with the environmental problem especially air pollution due to increasing population, industrialization growing transportation. Exposure to air pollution has become an unavoidable part of metropolitan life and millions of people faces health and environmental problems due to air pollution besides owing to poor shelter and lack of safe water, sanitation, drainage and other environmental contamination. These problems are compounded by dangerous releases caused mainly by fossil fuels, automobiles emission and smoke from industries / domestic combustion. The proportion at which these metropolises are developing with the general absence of pollution control methods in many of them urban air quality will continue to deteriorate.

2. Mathematical Modeling in Environmental Management

A model is defined as the creation of an idealized representation of reality in order to demonstrate its most important properties. The model building is a complementary way of thinking about the world. Environmental mathematical models (Empirical & Simulation) facilitate, to a certain extent, the study, analysis and investigation of certain basic results to be expected regarding the environmental disruption and the mean of prediction [7, 8]. The desired prediction of environmental trends and developmental disruption phenomena are preceded by a series of concrete factual data properly analyzed and processed [9]. Such results will enable scientists to project a mathematical model representation culminating to the desired prediction and preventive objective [9].

3. Development of Air Pollution Deterministic Index Modeling

APDIM is based on the deterministic model which is used to develop the Pollution Indices for the Air quality. Air Pollution Deterministic Index Modeling is typically based on a function α, where α is a number indicating conditions of air quality. The model takes the weighted values of individual pollutants parameters measured at spatial points and then compares this to the single number in the air quality standards. The attraction of this model is that the number α is a non-dimensional number (not in ppm or mg/m^3) [10]. The Pollution indices have fixed values to represent air quality. Each value is presented by a Colour code that's level is understandable by the public.

The applied methodology used for the preparation of model depends upon on the concentrations from ambient air monitoring data. The data is presented by mean and standard deviation. The pollutants consider for the modeling are CO, SOx, NOx, O_3, TSP and PM_{10}. A modification is made in the deterministic model by introducing a coefficient. The theoretical and experimental values are different for Air data that why a coefficient balance these values. The calculated value of the range for the coefficient for different Pollutants lies in-between 0.1 to 0.3; other researchers also use these types of constants with different situations [11, 12, and 13]. The non-dimensional α is calculated by the following formula;

$$\alpha = \alpha_i + \frac{\alpha_{i+1} - \alpha_i}{C_{i+1} - C_i}(C - C_i) \times \text{K (coefficient)}$$

Where

α = Pollution Standard Index
α_i = break point from one quantity to another
C = Ambient pollution concentration
C_i = Corresponding concentration in ppm
K = coefficient- Variation Range 0.1 to 0.3

In APDIM only comparison is made with the Pollution Indices. Pollution Indices have fixed values ranges and is used only to indicate the quality of air i.e. good, satisfactory, unhealthy and hazardous. Therefore, these levels are divided into seven categorizing zones and are also presented by a colour code i.e.

1. Good Quality
2. Satisfactory
3. Unhealthy
4. Hazardous – Alert
5. Hazardous – warning
6. Hazardous – emergency
7. Hazardous – serious harm

The break points for Breakpoints Pollution Indices Category Showing Pollution Indices, Categorizing Range, Color coding and Concentration time, are given in table-1:

Table 1. Pollution Indices, Categorizing and Concentration Range Source: Compiled

Category Range	Pollution Indices (PI)	Concentration (ppm)	Color Code
Good	<50	≤0.07	Green
Satisfactory	50- 99	0.14	Yellow
unhealthy	100- 199	0.3	Orange
Hazardous – Alert	200- 299	0.35	Blue
Hazardous – warning	300- 399	0.6	Purple
Hazardous – emergency	400- 499	0.8	Maroon
Hazardous – critically harm	500- above	1.0	Red

The health outcome information used with the Pollution Indices is pollutant-specific. Associated likely health impacts for different PI categories and pollutants have also been suggested, shown in table 2.

Table 2. Health effect information for Air Pollutant-Specific Indicator

Pollution Indices (PI)	Category Range	Associated Health Impacts
<50	Good	The pollutants have insignificant health effects
50- 99	Satisfactory	Problems occur for those having respiratory problems - Minor breathing discomfort to sensitive people.
100- 199	Unhealthy	Respiratory discomfort when peoples are exposed to longer time Breathing, and uneasiness to people those have hearth problem.
200- 299	Hazardous – Alert	Asthma and other respiratory attacks, lung problems and heart problems may occur especially in children and elders.
300- 399	Hazardous – Warning	If the exposure prolongs then causes lung heart and respiratory illness.
400- 499	Hazardous – Emergency	causes lung heart and respiratory illness respiratory impact even on healthy people,
500- above	Hazardous – Critically Harm	Serious health impacts on people with lung/heart disease. The health impacts may be experienced even during light physical activity.

4. Application of Air Pollution Deterministic Index Modeling (APDIM) at Quetta City

Quetta is a funnel shape valley and the capital city of Balochistan, Pakistan and lies between $66^0 15^{/} 00^{//}$ to $67^0 15^{/} 00^{//}$ Easting and $30^0 00^{/} 00^{//}$ to $30^0 25^{/} 00^{//}$ Northing in the survey of Pakistan Topo-sheets 34 N/3 & 4 and 34 J/15 & 16. The main valley of Quetta or Shal proper (comprising an area of 1720 sq. km), which unites the two sides at the toe of the horse-shoe, is a parallelogram about sixteen miles by eight [14]. The ambient Air data is collected at three sites of Quetta city. These sites are selected as per guidelines provided by UNEPA. The monitoring sites are shown in figure 1. The data of the Ambient Air for Quetta city is given in the table-3.

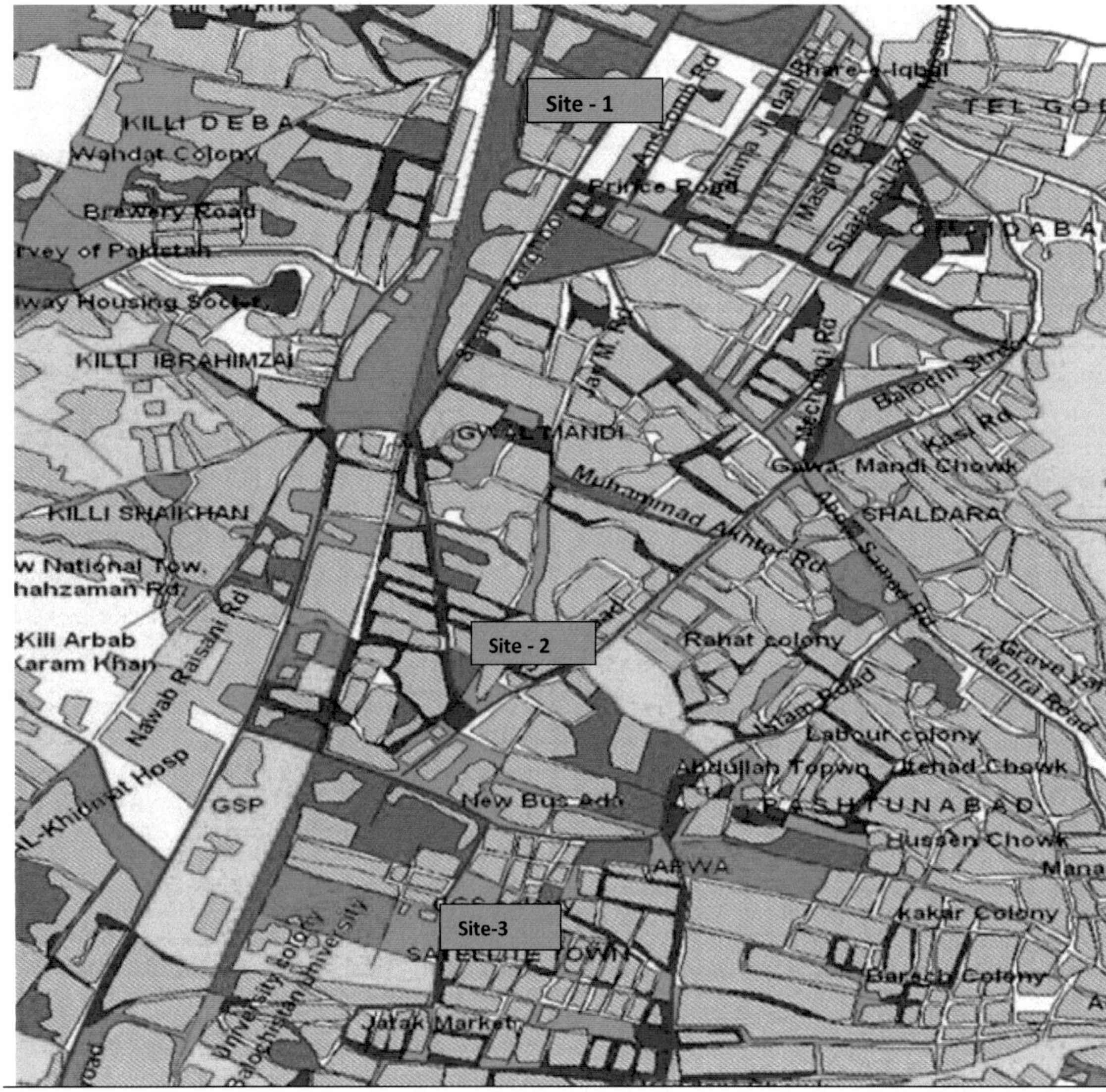

Figure 1. Map showing the monitoring sites in Quetta city Source: Complied

Table 3. Ambient Air for Quetta city: source: compiled

S. No	Criteria Pollutants	Concentration of Ambient Air in Quetta City (Annual Average)
1	Carbon monoxide	9.6 ppm
2	Sulphur dioxide	0.029 ppm
3	Nitrogen dioxide	0.037 ppm
4	Ozone	0.02761 ppm
5	Hydro Carbon	0.841 ppm
6	Total Suspended Particle	544 µg/m^3
7	PM_{10}	304.83 µg/m^3

The APDIM is applied to calculate the situation of air in Quetta city. The seven Pollution Indices categories in APDIM are Good, Satisfactory, Unhealthy Quality and Hazardous situation. The range of hazardous level are further divided into three stages- Hazardous – Alert, Hazardous – Warning, Hazardous – Emergency and Hazardous – Critically Harm. These indices' scale starts from very low (0) and to very high (7) that measures the relativity sensitivity of the health effects due to air pollution. The six pollutants consider for modeling are: CO, NOx, SOx, O_3, TSP and PM_{10}. And are considered for modeling are (NOx, SOx, CO, O_3, TSP and PM_{10}). The recorded time for was 24-hourly average. Results of Air Pollution Deterministic Index Modeling in Quetta City is presented in table-4.

Table 4. Air Pollution Deterministic Index Modeling for Quetta City; source compiled

Criteria Pollutants	Concentration of Ambient Air (Annual Avg. C)	Calculated values for (α)	Break Points ($α_i$)	Pollution Indices	WHO Standards	Colour Code
Carbon monoxide	9.6 ppm	50	1.1-2.0	50 -99	10 ppm	Satisfactory
Sulphur dioxide	0.029 ppm	50	0–40	≥ 50	0.0437 ppm	Good
Nitrogen dioxide	0.037 ppm	172	0–40	100-199	0.080 ppm	Unhealthy
Ozone	0.02761 ppm	50	0.07–0.14	≥ 50	0.055_{ave} ppm	Good
Total Suspended Particle	544 µg/m^3	510	500-above	500-above	190_{ave} g/m^3	Hazardous–serious harm
PM_{10}	304.83 µg/m^3	411	300-399	421 – 570	70 µg/m^3	Hazardous–serious harm

5. Discussion and Conclusions

Air pollution and its public health impacts are drawing increase concern from the environment point of view. Air quality has a direct influence on a public's health status and plays a significant role in quality of natural life, years of vigorous life lived, and health discrepancies. The quality of air, indoor and outdoor is closely related to morbidity to mortality from respiratory infection to cardiovascular diseases especially from particulate matters [15]. Exposure to air pollution is associated with premature death, significant number of hospital admission, exacerbation of symptoms, chronic diseases or allergies, bronchitis, asthma, cancer, heart disease, and respiratory disease.

Recent research indicates that air pollution harms vulnerable groups more due to their age. Therefore, children and older peoples are more effected than the younger groups but depend on multiple factors. That includes age, genetic, social conditions, health conditions, personal characteristics, substantial portion of their time outside home and depends upon individual's daily activities. Researchers notices that cities with improved air quality have longer life expectancy. A reduction of 10 μg/m^3 in ambient air pollution is associated with an estimated increase in average life expectancy of 0.61 years [16].

Environmental data and facts are the best authentic prediction tools that predict the health effects, economic effects and ecological effects that are generated by the environmental degradation. Based upon this study environmental modeling can lead to a new aspect of future planning and appropriate remedial measures. Therefore, mathematical modeling which is based on documentary data and information-collected from various independent sources and processed by modern computerized techniques provide base line value for the contemplated predictions and guide the environmentalist in their planning and development program.

Generally, standards are difficult to understand by the general public so a single digit standard which shows that how much a particular pollutant is dangerous to human health. APDIM is a simple, color-coded, unit-less index that is an effective way to communicate air pollution concentrations to the general public. The APDIM provides an indication of the quality of the air and its health effects. The aim of this paper is to better understand the environmental problems and health hazardous by a single number or condition to the public. The communities can use the APDIM to benefit them to take decisions on alfresco events; for example, different organizations such as schools and sports can check the latest info on air conditions and can decide whether outdoor sporting event should be conducted or not. The newest way to report the Air Reporting Index Modeling System is via an Internet or different computerized sign boards that can be positioned at the prominent places of the city. The impetus System APDIM is to grab the local population and government officials to take pollution control measures with in their jurisdictions and work for the improvement of air quality.

Acknowledgement
The authors are thankful to Pakistan Space and Upper Atmosphere Research Commission (SUPARCO), Pakistan Meteorological Department and Geological Survey of Pakistan to support the study.

References

[1] Park R J, Jaccob D J, Chin M, and Martin RV 2003.Sources of Carbonaceous aerosols over the United States and implication for natural visibility, *Journal of Geophysics Res.* **108**, pp-4355-67.

[2] Frank J K and Julia C F 2015. Air pollution and public health: emerging hazards and improved understanding of risk, *Environ Geochem Health.* 2015; **374**, 631–649. Published online 2015 Jun. 4. doi: 10.1007/s10653-015-9720-1

[3] Yamamoto SS, Phalkey R and Malik A 2014. A systematic review of air pollution as a risk factor for cardiovascular disease in South Asia: Limited evidence from India and Pakistan. *In.t J. Hyg. Environ Health.* **217**:133–44.

[4] World Commission on Environmental Development "1987. *Our Common Future*, Oxford University Press, New York, NY.

[5] Kan H, Chen B, Zhao N, London SJ, Song G, Chen G 2010. Part 1. A time-series study of ambient air pollution and daily mortality in Shanghai, China. *Res Rep Health Eff Inst.* **154**:17–78.

[6] Mellouki A, George C, Chai F, Mu Y, Chen J, Li H 2016. Sources, chemistry, impacts and regulations of complex air pollution: Preface. *J Environ Sci* (China), **40:1**–2.

[7] Mueller S F, Boylan J W, Russell A G, Odman M T and Wilkinson J G 2001. "*Performance of an air quality model across a range of environmental conditions*,"10th Int. Symposium on Transport & Air Pollution, Colorado.

[8] Laigna KJ, Potashnik E L1990. Mathematical Model for Predicting the Atmospheric Pollution State in the Areas of Local and Regional Scale, Estonian Scientific *Research Institute of Scientific-Technical Information and Technical-Economic Research, Tallinn*, p -342.

[9] Kaasik M, Kimmel V and Kaasik 2000. Air quality modeling system for a medium-sized town: a case study in Estonia. *Int. Jour. of Environment & Pollution* **16;** 519-527.

[10] Garanrd K1996, *Environmental Engineering*, McGraw- Hill international Editions, pp-867, 936-939.

[11] Masters GM 1991. *Introduction to Environmental Engineering and Science*, Prentice-Hall, Englewood Cliffs, New Jersey.

[12] Ball, DJ, Hamilton RS and Harrison R M 1991. The Influence of Highway Related Pollutants on Environmental Quality in Highway Pollution, *Studies in Environmental Science*. **44**, 1-47.

[13] Bernarie M M 1980. *Urban Air Modeling*, Mcmillan Press Ltd, New York.

[14] Shahid N k, Muhammad Y, Mohsin A K and Syed A H. 1986.Environmental geology, mineral, raw material and aggregate resources of Quetta valley, Balochistan, Pakistan P- 1-2 GSP Rec. **Vol. LXXVI-** *Environmental Geology*.

[15] Xi Anglue, Luke p, Nacher, A Review of traffic related pollution exposure assessment studies in the developing world. *Environmental International,* **32**-pp106-120, 2006.

[16] Marchwinska- E,Wyrwal G, Dziubanek I, Hajok M, Rusin K.,Oleksiuk and Kubasiak 2011.*The Impact of Air Pollution on Health, Economy, Environment and Agricultural Sources"*, book edited by Mohamed K. Khallaf, ISBN 978-953-307-528-0, Published: under CC BY-NC-SA 3.0 license.

Empirical SPT-CPT correlation for soils from Lahore, Pakistan

Mehtab Alam[1], Muhammad Aaqib[2], Shamsher Sadiq[2], Saeedullah Jan Mandokhail[3], Muhammad Bilal Adeel[2], Maqsood-ur-Rehman[4], Nawaz Ali Kakar[3]

[1] Graduate Student, Institute of Mountain Hazards and Environment, Chinese Academy of Sciences, Chengdu, P.R China
[2] Graduate Student, Department of Civil and Environment Engineering, Hanyang University, Seoul, South Korea
[3] Assistant Professor, Department of Civil Engineering, BUITEMS, Quetta, Pakistan
[4] Senior Engineer, National Engineering Services of Pakistan (NESPAK)

Email: ershrajput@hotmail.com

Abstract. Standard Penetration Test (SPT) and cone penetration test (CPT) are the most widely used in situ tests to depict the soil stratigraphy and determine the geotechnical engineering of the subsurface soils. The CPT leaps out because of its capability to trace the resistance continuously and due to its accuracy, it is considered to be reliable than the SPT. The outcome of these tests is very important for the design procedures to be implemented for various geotechnical purposes. To effectively utilize all the available data, there is a need of updating the correlation between these two widely used in situ tests. This study implements the statistical linear regression model using a 107 SPT and 49 CPT measurements across the city of Lahore, Pakistan to develop correlations between the the uncorrected SPT blow counts (N) and cone resistance (qc) for various soils. The developed correlation is compared in terms of qc/N ratios with the previous published studies.
Keywords. Empirical SPT-CPT, correlation, Lahore Pakistan

1. Introduction

Standard Penetration Test (SPT) is the widely used in-situ test in Pakistan and around the world for geotechnical investigations. It is also common practice in Pakistan to design most of the foundations based on SPT-N values and soil physical properties. The SPT-N values is empirically correlated with most of engineering properties of the soil. Despite of the most common and economic mean of subsurface investigation SPT has disadvantage such as potential inconsistency of measured resistances depending on operator inconsistency and possibility of missing delicate changes of soil properties due to the unavoidable discrete record.

On the other hand, Cone Penetration Test (CPT) also leaps out because of its capability to trace the resistance continuously. Currently CPT is also in practice to be used as in situ test for geotechnical investigations and design [1]. Most of the geotechnical engineers are well versed in foundation design using soil parameters correlated to SPT-N value.

It is common at many of the construction projects to use Standard Penetration test for estimation of soil parameters at preliminary design stage, whereas static cone penetration test results are used for final design and construction quality control, or vice versa. For effective utilization of the available database of field performances and property correlated with SPT, it is valuable to correlate static cone tip resistance-qc to SPT N-value. Therefore, in those cases where only SPT results are available, engineers, who are more familiar with CPT interpretations, will translate the SPT blow counts (N-values) into CPT cone resistances (qc-values).

A considerable amount of studies has been taken place over the years to correlate SPT N-value and CPT cone resistance qc. Most of the correlations are based on a constant value of qc/N while some

IOP Conf. Series: Materials Science and Engineering **414** (2018) 012015 doi:10.1088/1757-899X/414/1/012015

other defined n=qc+fs /N for different soil types. Some researchers suggest n=qc/N as a function of mean grain size [2-5] or the fines content [3, 6-8].

Now the SPT-CPT database is gradually increasing in Pakistan, however, SPT-CPT correlation for the local soil conditions are not published. In this study, empirical SPT-CPT correlations for various local soil conditions are proposed.

2. Data Collection

In this study in-situ tests (SPT & CPT) data of Orange Line Metro Train Project, Lahore has been used. Orange Line Metro Train is a rapid transit line built as part of Lahore Metro, Pakistan. This orange metro line is 27.1 km in length, out of which 25.4 km is elevated, while 1.72 km is constructed bellow the ground surface, and transition zone between elevated and underground sections is 0.7 km. The location plan of the area under study is given in Figure 1.

To delineate the subsurface soil a total of 109 SPT and 49 CPT tests were conducted along the project route. The SPT was performed at one-meter interval ranging up to a maximum depth of 50 m and CPT soundings were recorded at each 10 cm interval up to 20 m depth below NSL. Sieve analysis tests were performed on the samples collected in SPT sampler while the CPT data were interpreted in term of Soil Behavior Type (SBT) as proposed by Lunne et al. and Robertson et al. [9, 10]. Sieve Analysis and SBT shows that four types of soil exist along the project route, which are Lean Clay, silty Clay, Clayey Silt and Silty Sand.

In CPT reading is collected for every .01m while performing SPT at each meter only one SPT – N value was collected for 0.5m. Therefore, CPT results have been averaged for 0.5m at every meter depth to make it compatible to SPT results. While choosing the level for CPT tip resistance, the first thing considered was the depth at which SPT was performed. The correlation between SPT N-value and CPT tip resistance was then established for each 0.5m up to the maximum depth of CPT sounding (20 m)

Five hundred and thirty-eight (538) data pairs (N and qc) were available to establish correlation from forty-nine (49) CPT soundings and SPT points.

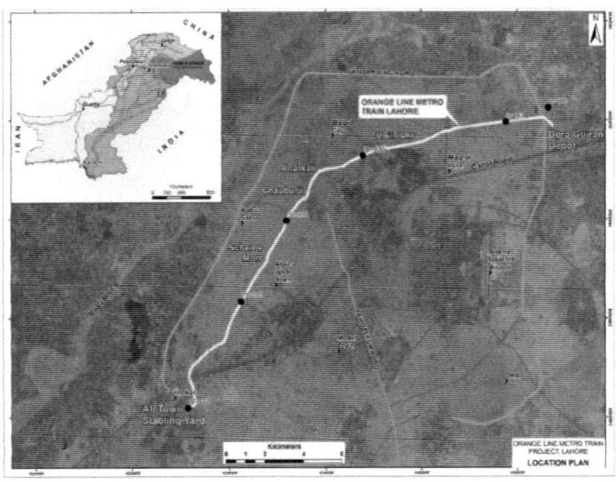

Figure 1. Location Plan of Orange Line Metro Train Project, Lahore, Pakistan [11]

3. Methodology Adopted for Data Processing and proposed correlation

Each type of soil identified along the project route has been separated and correlation was developed. Four soil types have been identified along the project route as Silty Sand, Sandy Silt, Silty Clay and Lean Clay.

Other soil types were regarded infrequent, and if present, they consisted of thin layer. Techniques adopted for calculation of n (qc/N) value is discussed as below:

3.1. Calculation of n ratio

For each soil type as identified along the project route the available data (q_c and N) were combined and n (q_c/N) was calculated, Table 1 summarizes the range of CPT (q_c) and SPT (N) value observed at the project site. The method of arithmetic average is then applied for final calculation of n-value (qc/N) and summary of the result is given in Table 2.

Table 1. Range of SPT-N Value and CPT-q_c

Soil Type	SPT-N Value		CPT q_c Value (MPa)	
	Min	Max.	Min	Max.
Silty Sand	4	50	2.258	20
Sandy Silt	5	37	0.97	17.40
Silty Clay	4	45	0.06	16.95
Lean Clay	5	33	0.164	11

Table 2. Arithmetic Average Method Results

Soil Type	Total No. of Pairs (No. of n_f)	$\sum n_1$ ($n_1=q_c/N$)	$\sum n_1$/(No.of n_f)	n value
Silty Sand	385	159.7	159.70/385	0.41
Sandy Silt	33	11.37	11.37/33	0.34
Silty Clay	85	27.00	27/85	0.32
Lean Clay	33	9.934	9.934/40	0.30

3.2. Proposed correlations

Linear regression is performed to establish a correlation between SPT and CPT. For analysis a line q_c = nN with zero intercept has been adopted using least squares method to establish a linear relation. Correlations were established for all the selected soil cases. the correlations are shown in **Error! Reference source not found.** to Figure 5. The relations along with the correlation coefficients are tabulated in Table 3. Proposed correlations are also compared with the literature and the q_c/N values comparison can be seen in

Table *4*.

Table 3. Proposed correlations

Soil Type	Correlation Equation	Correlation Coefficient (R^2)
Silty Sand	$q_c = 0.427N$	0.303
Sandy Silt	$q_c = 0.337N$	0.438
Silty Clay	$q_c = 0.319N$	0.604
Lean Clay	$q_c = 0.291N$	0.602

Table 4. Comparison of proposed qc/N values with the literature

Authors	Soil Type	Proposed	This study
De Alencar Velloso (1959) [12]	Clay and silty clay	0.35	-
	Sandy clay and silty sand	0.2	0.43
	Sandy silt	0.35	0.34
Franki Piles (1960) [13]	Silty sand	0.5	0.43
	Sandy clay	0.4	0.34
	Silty clay	0.3	-
	Clays	0.2	0.29
Schmertmann (1970) [14]	Silt, sandy silt and silt-sand mix.	0.2	0.34
	Fine to medium sand, silty sand	0.3-0.4	0.43
	Coarse sand, sand with gravel	0.5-0.6	-
Akca (2003) [13]	Silty sand	0.702	0.43
	Sandy silt	0.58	0.34
Robertson (1986) [15]	Silty sand	0.3-0.4	0.43
Danziger and De Valleso (1995) [16]	Fine to medium sand, Silty sand	0.3-0.42	0.43

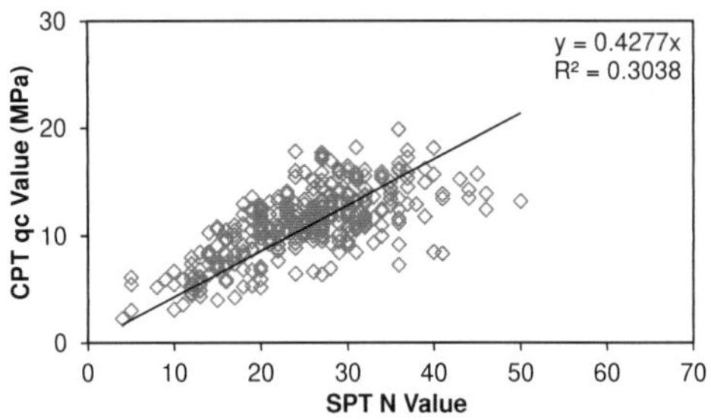

Figure 2. q_c and N relationship for sity sand

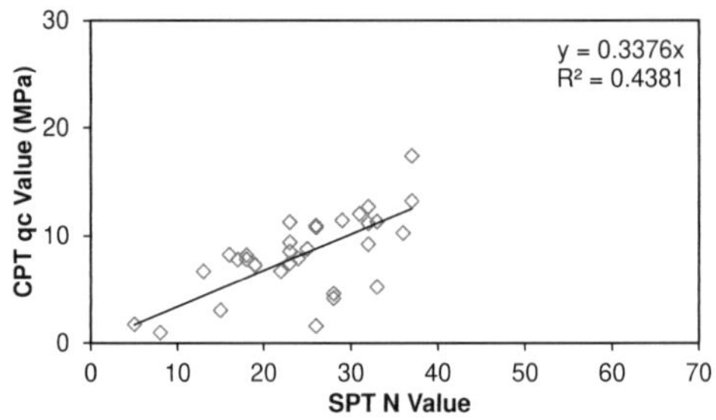

Figure 3. q_c and N relationship for sandy silt

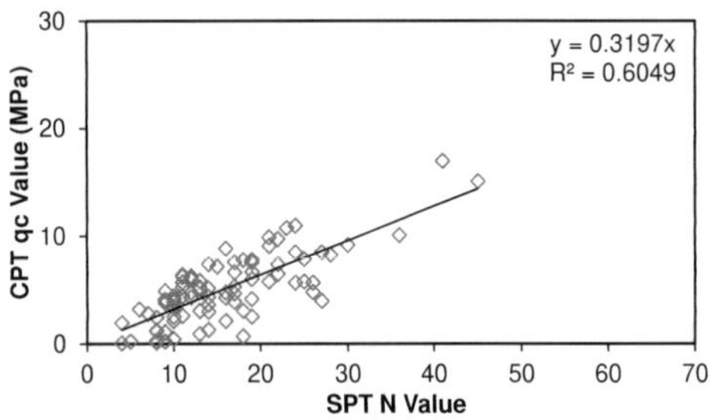

Figure 4. q_c and N relationship for silty clay

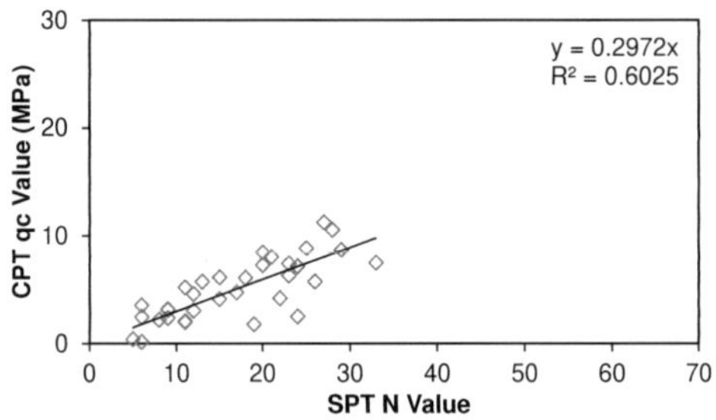

Figure 5. q_c and N relationship for Lean clays

3.3. Discussions

It can be seen in the comparison Table 4 that the results of the current study are within the range of the available correlations in the literature. In fact, the qc/N ratios matches well in most of the cases except those of Akca [13], which are larger comparing with this study. This can be described by the fact that the values proposed by Akca [13] are for the soils which have cemented layers and most of the times showing bands of gypsum and gravel. Presence of stiff bands in those soils could be the reason of difference between the values proposed in current study and those proposed by Acka [13].

4. Conclusions

This paper presents the correlation for different soil types from Lahore Pakistan using arithmetic average method and linear regression. In the studied area, the soil types found are Silty Sand, Sandy Silt, Silty Clay and Lean Clay. The qc/N ratios from the developed correlations have been compared with the previously proposed correlations available in literature. The comparison show that the correlations developed in this research are in good agreement for fine grained soils while it show a little variation for coarse grained soils.

References

[1] Suzuki, Y. Correlation between SPT and seismic CPT. in Proc. of the First International Conference on Site Characterization. 1998.

[2] Seed, H.B. and P. De Alba. Use of SPT and CPT tests for evaluating the liquefaction resistance of sands. in Use of in situ tests in geotechnical engineering. 1986. ASCE.

[3] Kulhawy, F.H. and P.W. Mayne, Manual on estimating soil properties for foundation design. 1990, Electric Power Research Inst., Palo Alto, CA (USA); Cornell Univ., Ithaca, NY (USA). Geotechnical Engineering Group.

[4] Stark, T.D. and S.M. Olson, Liquefaction resistance using CPT and field case histories. Journal of geotechnical engineering, 1995. 121(12): p. 856-869.

[5] Mitchell, J.K., et al., Performance of Improved Ground and Earth Structures. Earthquake Spectra, 2000. 16(S1): p. 191-225.
[6] JAMIOLKOWSKI, M. New developments in field and laboratory testing or soils. in Proc. 11th. Int. Conf. on SMFE., San Francisco, CA, 1985. 1985.
[7] Kasim, A.G., M.-Y. Chu, and C.N. Jensen, Field correlation of cone and standard penetration tests. Journal of geotechnical engineering, 1986. 112(3): p. 368-372.
[8] Jefferies, M.G. and M.P. Davies, Use of CPTU to estimate equivalent SPT N 60. Geotechnical Testing Journal, 1993. 16(4): p. 458-468.
[9] Lunne, T., P. Robertson, and J. Powell, Cone penetration testing. Geotechnical Practice, 1997. 20: p. 23-35.
[10] Robertson, P., CPT-DMT Correlations. Journal of Geotechnical and Geoenvironmental Engineering, 2009. 135(11): p. 1762-1771.
[11] Sadiq Shamsher , M.A., Maarijh Shaikh,Saeed Ullah Jan Mandokhail,Muhammad Bilal Adeel, Numerical Evaluation of Pile Axial Capacity. in Proceedings, International Conference on Sustainable Development in Civil Engineering, 2017.
[12] De Alencar Velloso, D., O ensaio de diepsondeering ea determinacao da capacidade de cargo do solo. Rodovia, 1959: p. 29.
[13] Akca, N., Correlation of SPT–CPT data from the United Arab Emirates. Engineering Geology, 2003. 67(3-4): p. 219-231.
[14] Schmertmann, J.H., Static cone to compute static settlement over sand. Journal of Soil Mechanics & Foundations Div, 1970.
[15] Robertson, P.K., et al. Use of piezometer cone data. in Use of in situ tests in geotechnical engineering. 1986. ASCE.
[16] Danziger, B. and D. Velloso. Correlations between the CPT and the SPT for some Brazilian soils. in Proc. of the International Symposium on Cone Penetration Testing. 1995.

IOP Conf. Series: Materials Science and Engineering **414** (2018) 012016 doi:10.1088/1757-899X/414/1/012016

Process optimization of microwave assisted co-pyrolysis of coal and oil palm shell blend with carbon surfaces

F Mushtaq[1], S K Sami[1], R Mat[2] and F N Ani[3]

[1]Faculty of Engineering and Architecture, Department of Chemical Engineering, Balochistan University of Information Technology, Engineering and Management Sciences (BUITEMS), Airport Road, Baleli, Quetta, Balochistan, Pakistan.
[2]Faculty of Chemical and Energy Engineering, Universiti Teknologi Malaysia (UTM), Malaysia.
[3]Faculty of Mechanical Engineering, Department of Thermo-fluids, Universiti Teknologi Malaysia (UTM), Malaysia.

Email. faisal.mushtaq@buitms.edu.pk

Abstract. In this study, Oil Palm Shell (OPS) waste biomass and coal were subjected to Microwave (MW) co-pyrolysis conditions. The effects of process parameters on the efficiency of co-pyrolysis blended fuel were tested to identify the optimal processing conditions. The results of this study suggest that co-pyrolysis blended fuel over 35-55-75wt% carbon loading with 3-Layer method at 600W and 4LPM of N_2 flow rate suggests that increasing the carbon loading increases the oil yield. This suggests that experimental design space for carbon loading can be narrowed down. For that reason, 55wt% carbon loading was selected as the minimum and 75wt% carbon loading as the maximum range. The levels of MW selected power were 600W and 720W, where the highest is based on the maximum rated capacity of multimode MW oven (800W). The response surface methodology was used to establish model yields. The process temperature of blended fuel solids with 3-Layer method were recorded by online thermocouples and discussed. The co-pyrolysis oil obtained under various conditions were tested for chemical composition and reported.
Keywords. Oil palm shell, carbon surfaces, microwave co-pyrolysis, blend, optimization.

1. Introduction

The non-renewable crude oil, Natural Gas (NG) and coal resources still dominated as the major energy sources to power our world. These natural and finite energy resources are exploited to drive our modern societies and are at the verge of getting extinct. In addition, the dramatic energy demands in transportation, aviation and industrial power sectors of many countries have ensured increasing production from these fossil fuel resources. More importantly, fossil fuel utilization through combustion augmented CO_2, sulfur and nitrogen oxides emissions. To address the growing energy insecurity and emissions motivated scientists and researchers to explore renewable and cleaner fuel resources. Coal still remains the largest and cheapest source of solid fuel across the globe [1, 2], and is mainly used to produce electricity in Coal Power Plants (CPPs) [3]. Coal combustion offers an inexpensive energy but at the expense of some serious environmental concerns, such as CO_2, SOx, NOx, particulate matter and mercury emissions [4]. The search for cleaner energy production from coal is driving many countries to partially replace renewable waste biomass in CPPs. The motivation to use waste biomass in CPPs is growing but with some serious technical challenges, such as ignition and combustion problems (flame stability), fouling and slagging, and high temperature corrosion etc.

[5-9]. The second interesting option is co-pyrolyze coal with waste biomass to recover valuable fuel and energy products.

Microwave (MW) assisted pyrolysis has been demonstrated as a promising alternative to conventional pyrolysis [10-13]. The integration of MW technology to pyrolysis provides a challenging environment to treat various materials, which can improve liquid and gaseous fuel products. Several authors demonstrated that MW assisted pyrolysis offers low temperature and energy efficient route to convert biomass solids to valuable fuels [14-16]. More interestingly, MW assisted pyrolysis of biomass solids with carbon based Microwave absorber (MWA) demonstrated, improved bio-oil selectivity in terms of phenol yield [17-20], high content syngas [21, 22] and high quality bio-char [23]. Previous studies on MW assisted pyrolysis of waste biomass indicated that the process factors influencing bio-oil yield are: MWA loading [24, 25] and MW power [26]. In addition, factors that are assumed to have minimal effects on bio-oil yield can be eliminated, as well as the factors that are categorical [27]. Most MW assisted pyrolysis techniques developed previously mainly focused towards intimately mixing of MWA with pyrolysis solids [17, 20, 25, 28] with much less attention towards uniform distribution method. The use of non-uniform distribution of MWA with pyrolysis solids or intimately mixing of heterogeneous solids under MW irradiation conditions initiate hotspots phenomena, a phenomena when the MWA absorbs much higher MW energy than the surrounded pyrolysis solids and results in intense heating of MWA (commonly observed a spark). These hotspots are mainly responsible of non-uniform process heating and deteriorate pyrolysis reaction mechanism. Further, difficulty in online temperature were noticed when thermocouple tip in contact with hotspots resulted in thermocouple meltdown. In these concerns, the distribution of MWA with pyrolysis material was paid attention to design an energy efficient system to minimize and control hotspots. We proposed and tested novel method of controlling the hotspots by surface methods, the detail of the surface method and preliminary results are discussed [29-34]. In addition, till to date, few studies were carried out to optimize fuel products from MW assisted pyrolysis of biomass solids [20, 35-39].

To the best of our knowledge, no study has been carried out to optimize co-pyrolysis products of coal and biomass blend under MW assisted pyrolysis environment in the presence of MWA and with surface methods. The primary aim of this study was to investigate the effects of uniformly distributed Coconut Activated Carbon (CAC) absorber over coal and Oil Palm Shell (OPS) blend at various levels of CAC loading, MW power and N_2 flow rate on heating profile and pyrolysis product yield. The response surface methodology was used to establish model to predict oil, blend-char and gas yields.

2. Materials and Methods

For this study, co-pyrolysis of Oil Palm Shell (OPS) and coal blended fuel arrangement was selected for the optimization study. This is based on the preliminary results and following three reasons; (i) ease of operation, (ii) better heating performance, (iii) higher oil yield compared to segregated fuel arrangement. More interesting, the preliminary results of co-pyrolysis blend over 35-55-75wt% carbon loading with 3-Layer method at 600W and 4LPM of N_2 flow rate suggested that increasing the carbon loading increases the oil yield. The oil yield was measured 24.23wt%, 28.20wt% and 31.72wt% using 35wt%, 55wt% and 75wt% carbon loading, respectively. This suggests that experimental design space for carbon loading can be narrowed down. Therefore, 55wt% carbon loading was selected as the minimum and 75wt% carbon loading as the maximum range. The levels of MW selected power were 600-720W, where the highest is based on the maximum rated capacity of MMW oven (i.e. 800W). In addition, the effect of N_2 flow rate on MW pyrolysis performance of isolated fuels suggests that the test levels for the co-pyrolysis blend is 4-8 Liters per minute (LPM) [30, 31].

With three factors, full factorial CCD experimental design with three blocks plus star was studied to investigate the significant effects on oil, blend-char and gas yield. The effects of CAC loading (X_1), MW power (X_2) and N_2 gas flow rate (X_3) was investigated at five different levels i.e. star low, low, medium, high and star high and are coded as $-\alpha$, -1, 0, $+1$, $+\alpha$ respectively. The response surface methodology was used to establish model yields. The MW assisted co-pyrolysis experimental setup is similar to our study and can be seen in [34]. The co-pyrolysis oil obtained under various MW operating conditions was tested in GC-MS for chemical composition. The co-pyrolysis product yields were calculated using equation (1), (2) and (3).

$$\text{Oil yield (wt\%)} = \frac{\text{Weight of oil (g)}}{\text{Weight of blend (g)}} \times 100 \quad (1)$$

$$\text{Blend-char yield (wt\%)} = \frac{\text{Weight of oil (g)}}{\text{Weight of blend (g)}} \times 100 \quad (2)$$

$$\text{Gas yield (wt\%)} = 100 - \text{Oil yield (wt\%)} - \text{Blend-char yield (wt\%)} \quad (3)$$

The process temperature of blended fuel solids with 3-Layer were measured by labeling thermocouples T_{UML}, T_{MML} and T_{BML} representing upper mixed layer, middle mixed layer and bottom mixed layer, respectively. Where M in the subscript stands for mixed (i.e. blend). With three selected factors (i.e. CAC loading, MW power and N_2 flow rate), n = 3, the experimental design is required to study the possible interaction between factors and to develop response surfaces, so 2^n factorial design was eliminated [40]. Moreover, 3^n factorial design can be used to develop the response surfaces with 27 experiments required. However, a CCD (commonly called Box-Wilson Design) is often used instead of 3^n factorials to minimize number of experiments, which can save time and expenses. Therefore, CCD was selected as suitable design for the response surface methodology (RSM) studies. For model analysis, equation (4) was used.

$$Y_i = b_0 + \sum_{i=1}^{3} b_i x_i + \sum_{i=1}^{3} b_{ii} x_i^2 + \sum_{i=1}^{2}\sum_{j=i+1}^{3} b_{ij} x_i x_j \quad (4)$$

Where Y_i= Predicted response, x_i= Independent variables, b_o= Intercept coefficient, b_i= Coefficient for linear effects, b_{ii}= Coefficient for quadratic effects, b_{ij}= Coefficient for interaction effects
This experimental design uses 9 coefficients, given by the equation (5) are used to predict the response surfaces.

$$Yi = b_o + b_1x_1 + b_2x_2 + b_3x_3 + b_{11}x_1^2 + b_{22}x_2^2 + b_{33}x_3^2 + b_{12}x_1x_2 + b_{13}x_1x_3 + b_{23}x_2x_3 \quad (5)$$

Where Y_i= predicted response, i.e. oil, blend-char and gas yields. x_1, x_2, x_3= linear effects of CAC loading, MW power and N_2 flow rate, respectively. x_1^2, x_2^2, x_3^3= quadratic effects of CAC loading, MW power and N_2 flow rate, respectively. x_1x_2, x_1x_3, x_2x_3 = Interaction effects for CAC loading-MW power, CAC loading-N_2 flow rate, MW power-N_2 flow rate, respectively. b_o, b_1, b_2, b_3= coefficients for offset term, MWA loading, MW power and N_2 flow rate respectively. b_{11}, b_{22}, b_{33}= quadratic coefficients for the effects of CAC loading, MW power and N_2 flow rate, respectively. b_{12}, b_{13}, b_{23}= Interaction coefficients for CAC loading-MW power, CAC loading-N_2 flow rate, MW power- N_2 flow rate respectively.

STATISTICA® software was utilized to perform regression modeling for oil, blend-char and gas yield. The observed responses were entered into the program at the prescribed experimental design parameters (i.e. CAC loading, MW power and N_2 flow rate). The standard least square model was selected to run the full model as described in equation (4). The model results was analysed statistically and graphically. Initially, the deviations of actual experiment data were observed from the predicted values called residuals, which ensures that the experimental values are not interrelated to one another in some way. The overall fitness of the predicted models was checked by the R^2 value (coefficient of determination) that gives the variation in percentage. Furthermore, the significance of higher R^2 value was reviewed by F-test method using ANOVA table. A standard ANOVA table was generated using STATISTICA® software. The F_{ANOVA} is the ratio of Mean Square Regression (MSR) to the Mean Square Error (MSE). The MSE is based on degree of freedom and sum of square for regression models. The degree of freedom is the number of estimated model based parameters and observations. The standard ANOVA table is shown in Table 1.

Table 1. Standard analysis of variance table

Source	Degree of Freedom	Sum of squares	Mean square	F_{ANOVA}

Regression (model)	m	SSR	MSR= SSR/ m	MSR/MSE
Residual error	N- m-1	SSE*	MSE=[SSE/(N- m -1)]	
Total	N-1	SST=SSR+SSE*		

*SSE = Sum of Square Error

Where, N and m represents the number of experiments and number of model parameters, respectively. The R^2 can be calculated using the ratio of Sum of Square Regression (SSR) and Sum of Square Total (SST), or taking the square root of MSE called Root Mean Square Error (RMSE). RMSE is an important parameter to evaluate the model, which estimates the standard deviation of residual error. The F_{ANOVA} calculated from ANOVA table was compared with $F_{critical}$ value, which accounts for the degrees of freedom and desired confidence level. For this study, 95% confidence level or p value $\leq$ 0.05 was selected for the model coefficients. For $F_{ANOVA} > F_{critical}$, the model is considered significant at 95% confidence level. Otherwise null hypothesis is valid, which implies that the full regression model is insignificant.

The significant of each coefficient was determined using Pareto chart. A student t-distribution was used to estimate the parameter value in the model equations accounting the standard error at confidence limits, whereas the p-value are used to check the significance of each manipulated variables in the process. Therefore, significance of each variable on process was determined by comparing t-distribution with p-value. For this reason, the Pareto chart was analysed for p-value $\leq$ 0.05 and higher student t-distribution. Moreover, the most and the least significant factor and their effects were predicted on oil, blend-char and gas yield. After statistical model analysis, several graphical representations were generated. A three-dimensional response surfaces were prepared to observe influence of multiple factors on given response by keeping other factors constant. The optimized conditions for the oil, blend-char and gas yield were determined using response surfaces. Additional experiments were conducted at optimized conditions and compared with predicted values. The optimized runs were repeated twice to verify the predicated critical value.

The co-pyrolysis oil chemical composition were analyzed using Agilent Technologies 6890 GC–MS with HP-5MS capillary column of 30 m length and 250lm diameter. The GC oven temperature was raised from initial temperature of 80°C to 200°C at 10°C/min, and then to 300°C by 5°C/min. The final oven temperature of 300°C was held constant for about 10 min. Helium gas flow rate was maintained at 2 mL/min. The GC was connected to Agilent Technologies 5975 series Mass Spectroscopy (MS) equipped with inert Mass Selective Detector (MSD) at scanning acquisition mode. The MS conditions were: mode Electron Ionization (EI), ion source temperature 230°C, emission current 34.6l A, ionization energy 70 eV, full scan range of 50–550 and quantization by Selected Ion Monitoring (SIM) mode. The Agilent Chemstation software was used to indentify chemical compounds and peaks with the help of NIST library.

3. Results and Discussion

3.1 Microwave assisted heating behaviour

The real time heating behaviour of blended solids under various levels of carbon loading (wt%), MW power (W) and N_2 flow rate (LPM) is shown in Figure 1. The final pyrolysis temperature reached in blended solids is shown in Figure 2. The heating profiles indicate that the heating rate and final pyrolysis temperature reached in fuel solids are influenced by the process variables.

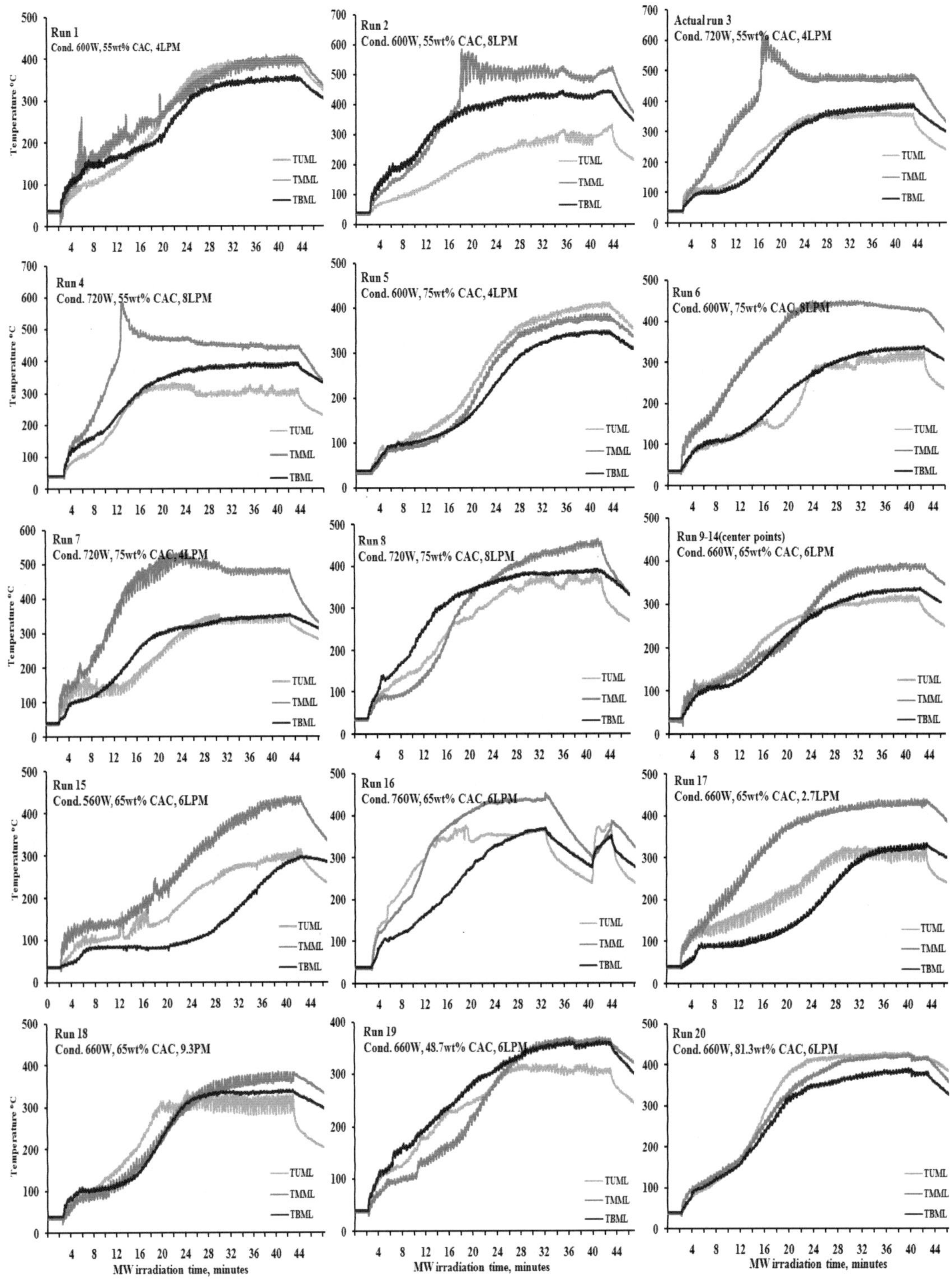

Figure 1. MW heating behaviour of blended solids under various MW pyrolysis operating conditions

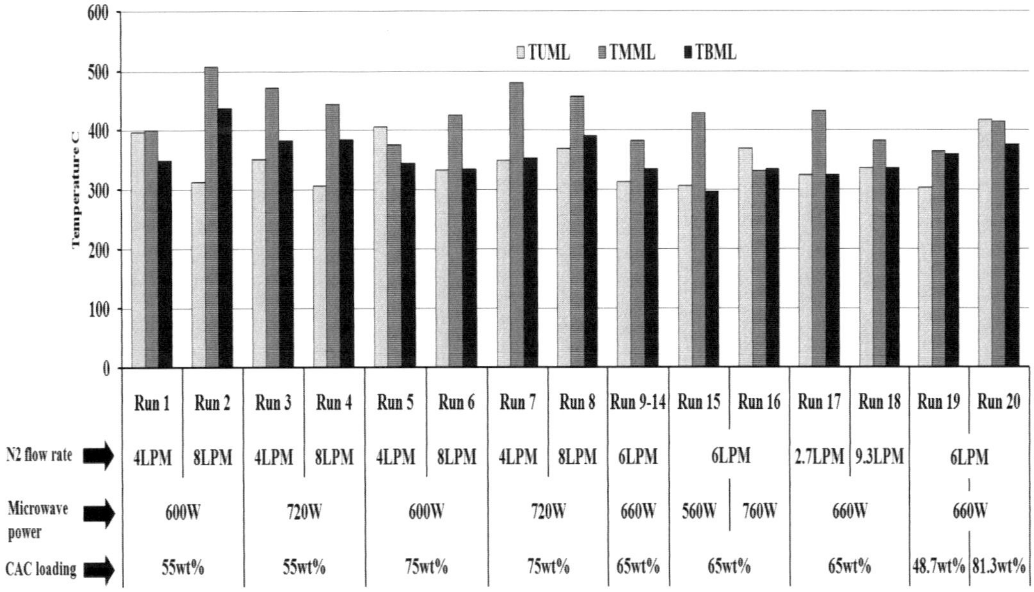

Figure 2. Effects of process variables on maximum final pyrolysis temperature of blended solids

3.1.1 Effects of N_2 flow rate

The heating rate and final pyrolysis temperature reached in fuel solids were found to vary with N_2 flow rate. The upper fuel solids demonstrated significant increase in heating rate and final pyrolysis temperature when pyrolysis was carried out by maintaining N_2 flow rate of 4LPM at 600W and 55wt% carbon loading. However, increasing the N_2 flow to 8LPM decreased the upper fuel solids heating rate and final pyrolysis temperature. On the contrary, the heating rate and final temperature of middle and bottom fuel solids increases significantly with increasing N_2 flow rate (Figure 1 and 2, Run 1-2). The final pyrolysis temperature under these conditions were observed $[T_{UML}=397^oC, T_{MML}=401^oC, T_{BML}=349^oC]_{4LPM}$ and $[T_{UML}=313^oC, T_{MML}=508^oC, T_{BML}=436^oC]_{8LPM}$. With 720W and 55wt% carbon loading produced slight increase in heating rate and final temperature of fuel solids at 8LPM compared to 4LPM (Figure 1 and 2, Run 3-4). Similarly, 600W and 75wt% carbon loading with 4LPM slightly raised the heating rate and final pyrolysis temperature of upper and bottom fuel solids compared to 8LPM. However, the middle fuel solids showed significant increase in heating rate with 4LPM, which then decreases significantly with 8LPM (Figure 1 and 2, Run 5-6). With 720W and 75wt% loading, the upper and bottom fuel solids demonstrated no significant effects on heating rate and final temperature with 4-8LPM N_2 flow rate, but middle fuel solids showed similar heating behaviour to that of 600W, 75wt% carbon loading with 4-8LPM (Figure 1 and 2, Run 7-8).

From the above discussion, it is clear that the heat generated by carbon solids and carried with N_2 flow plays an important role in controlling heat to the supported fuel solids. As carbons are good MWA and respond quickly to MW energy [11]. Therefore, MW reflected by the cavity walls is mainly absorbed by the carbon solids and converted to heat, which can then be transmitted to the supported fuel solids. As a result of high N_2 flow rate, the heat carried with inert gas is higher than the heat generated by the carbon solids, which can reduce the heating rate and final pyrolysis temperature in fuel solids. The heat carrying capacity with N_2 gas was reported to influence MW pyrolysis reaction temperature [41].

More importantly, the use of fairly low N_2 flow rate of 2.7LPM demonstrated non-uniform heating rate and significant difference in final pyrolysis temperature of fuel solids (Figure 1, Run 17), whereas increasing the N_2 flow rate to 6LPM and 9.6LPM controls the heating rate and final pyrolysis temperature of fuel solids by achieving nearly complete uniformity of process temperature (Figure 1, Run 9-14 and 18). This suggests that the heat carried with N_2 gas from carbon and fuel solids, and heat carried with N_2 gas from the MW reaction zone reached equilibrium. This important finding suggests that the uniform process heating of fuel solids in the presence of carbon surfaces can be achieved by controlling N_2 flow rate.

3.1.2 Effects of microwave power

Increasing MW power from 600W to 720W with 55wt% carbon loading and 4LPM indicated no effects on heating rate and final pyrolysis temperature of upper and bottom fuel solids. However, significant increase in heating rate and final temperature of middle fuel solids was observed with 720W (Figure 1 and 2, Run 1 and 3). With 8LPM and 75wt% loading, increasing MW power increases the heating rate of upper and middle fuel solids but remains constant for bottom fuel solids. Moreover, no significant effects on final temperature were noticed under these conditions (Figure 1 and 2, Run 2 and 4). With 75wt% and 4LPM, the heating rate and final temperature of upper and bottom fuel solids remained fairly constant with MW power but the heating rate and final temperature of middle fuel solids significantly increases with 720W (Figure 1 and 2, Run 5 and 7). With 75wt% and 8LPM, increase in MW increases the heating rate of upper and bottom fuel solids, but decreases significantly for middle fuel solids. In addition, the final pyrolysis temperature of fuel solids under these conditions was not affected with MW power (Figure 2, Run 6 and 8). Another noticeable development is the significant increase in fuel solids heating rate with 760W, 55wt% carbon loading and 6LPM (Figure 1, Run 16). Moreover, comparison of heating profiles under similar carbon loading and N_2 flow rate but with 560W and 660W showed much lower heating rate of fuel solids (Figure 1, Run 15 and 9-14). However, no significant difference in final pyrolysis temperature was observed under these conditions. The higher heating rate of fuel solids is mainly the result of increase in electron intensity at higher power and increased heat generation rate from carbon solids. In addition, the synergistic effects of MW coupling with carbon absorber loading and the heat carried with inert gas can encourage fuel solid heating rate.

3.1.3 Effects of carbon loading

Comparison of heating profiles with increase in carbon loading from 55-75wt% showed no significant effects on upper fuel solids heating rate at 600W and 720W with 4LPM of N_2 gas, but 55wt% showed higher heating rate of middle and bottom fuel solids, which then deceases with 75wt% loading. Moreover, the final temperature reached in fuel solids maintains much similar conditions (Figure 1 and 2, Run 1 and 5). Similar heating conditions of fuel solids were observed with 600W and 8LPM, but the middle and bottom fuel solids maintains much higher final temperature at sustained pyrolysis conditions for longer duration with 35wt%, which decreases with 75wt% loading (Figure 1, Run 2 and 6). The final pyrolysis temperature with under these conditions were observed $[T_{UML}=313°C, T_{MML}=508°C, T_{BML}=436°C]_{35wt\%}$, which changes to $[T_{UML}=333°C, T_{MML}=426°C, T_{BML}=335°C]_{75wt\%}$ (Figure 2, Run 2 and 6). The use of 720W and 4LPM showed no effects on heating rate and final pyrolysis temperature of fuel solids (Figure 1 and 2, Run 4 and 8). However, 720W and 8LPM produced much higher heating rate of middle fuel solids with 55wt% for longer sustained pyrolysis duration, which decreases significantly at 75wt%. Moreover, final temperature conditions under these conditions were not affected with carbon loading (Figure 1 and 2, Run 4 and 8). Another noticeable development is the increase in fuel solids heating rate with 660W, 81.3wt% carbon loading and 6LPM (Figure 1, Run 20). Moreover, comparison of heating profiles under similar MW power and N_2 flow rate with 48.7wt% and 55wt% showed slow heating rate of fuel solids (Figure 1, Run 19 and 9-14). In addition, the final pyrolysis temperature was observed higher for sustained longer pyrolysis duration with 81.3wt% compared to 48.3wt% and 55wt% loading (Figure 2, Run 19, 9-14 and 20). The final pyrolysis temperature under these conditions remained higher at $[T_{UML}=417°C, T_{MML}=414°C, T_{BML}=376°C]_{81.3wt\%}$, which decreases to $[T_{UML}=304°C, T_{MML}=364°C, T_{BML}=360°C]_{48.7wt\%}$ and $[T_{UML}=313°C, T_{MML}=382°C, T_{BML}=335°C]_{55wt\%}$. It is also important to note that the use of higher loading of 81.3wt% showed fairly smooth heating of fuel solids by reaching nearly complete uniformity process temperature at higher temperature for longer sustained pyrolysis condition (Figure 1, Run 20) with minimum difference in final temperature of fuel solids (Figure 2, Run 20). This suggests that the heat generated by the carbon solids at higher loading and carried with N_2 gas to fuel solids can maintain sustained pyrolysis conditions for longer duration. This important finding also suggests that the uniform process heating of fuel solids can be achieved by increasing the carbon loading.

3.2 Response surface methodology

Statistical response surface methodology provides reliable method of analyzing process conditions and interaction effects. The design of experiment with three process variables, i.e. CAC loading (wt%), MW power (W) and N_2 flow rate (LPM) at five levels with twenty experiments are shown in Table 2.

Table 2. Central composite design with coded values at various levels and observed responses

DOE	Actual Run	Factors: CAC loading (wt%) x_1	MW power (W) x_2	N_2 flow rate (LPM) x_3		Block	Oil Yield (wt%) Y_1	Blend-char Yield (wt%) Y_2	Gas Yield (wt%) Y_3
1	1	55 (−1)	600 (−1)	4 (−1)	2^n points	1	28.20	47.47	24.33
2	4	55 (−1)	720 (+1)	8 (+1)			30.29	39.51	30.19
3	6	75 (+1)	600 (−1)	8 (+1)			33.04	37.99	28.97
4	7	75 (+1)	720 (+1)	4 (−1)			22.99	38.72	38.29
5	9	65 (0)	660 (0)	6 (0)	n_{F1}	1	31.27	41.04	27.69
6	10	65 (0)	660 (0)	6 (0)			32.38	43.47	24.15
7	2	55 (−1)	600 (−1)	8 (+1)	2^n points	2	24.68	40.98	34.34
8	3	55 (−1)	720 (+1)	4 (−1)			33.09	40.45	26.47
9	5	75 (+1)	600 (−1)	4 (−1)			31.72	45.33	22.95
10	8	75 (+1)	720 (+1)	8 (+1)			31.36	38.77	29.87
11	11	65 (0)	660 (0)	6 (0)	n_{F2}	2	33.13	41.51	25.35
12	12	65 (0)	660 (0)	6 (0)			32.55	42.00	25.45
13	19	48.7 (−α)	660 (0)	6 (0)	2n points	3	26.46	46.18	27.36
14	20	81.3 (+α)	660 (0)	6 (0)			32.64	39.14	28.22
15	15	65 (0)	560 (−α)	6 (0)			31.49	44.49	24.02
16	16	65 (0)	760 (+α)	6 (0)			29.17	38.87	31.97
17	17	65 (0)	660 (0)	2.7 (−α)			34.53	40.27	25.20
18	18	65 (0)	660 (0)	9.3 (+α)			32.61	39.45	27.93
19	13	65 (0)	660 (0)	6 (0)	n_A	3	33.21	41.50	25.29
20	14	65 (0)	660 (0)	6 (0)			33.00	42.23	24.77

CCD design with three factors and 3 blocks: Standard Design Summary: 2^3 cube plus star (central composite design), number of factors: 3, number of blocks: 3, number of runs: 20 [$2^n = 2^3 = 8$ (8 factor points, 4 cube points per factor block), 2n = 6 (6 axial points) and M = 2 (1 center point per factor block) + 1 (1 center point for axial block) + 3 (1 center point added per block)] ,alpha for rotatability: 1.6818, alpha for orthogonality: 1.6733

3.2.1 Model analysis

The observed responses of oil yield (Y_1), blend-char yield (Y_2) and gas yield (Y_3) responses tested for CAC loading (x_1), MW power (x_2) and N_2 flow rate (x_3) is given in Table 2. The regression coefficients of three responses are given in Table 3.

The second order polynomial model for the oil, blend-char and gas yields are given in equations (6), (7) and (8).

$$Y_1 = -295.4052 + 4.3191\,x_1 + 0.6718\,x_2 - 11.8719\,x_3 - 0.0150\,x_1^2 - 0.0003\,x_2^2 + 0.0025\,x_3^2 - 0.0044\,x_1x_2 + 0.1000\,x_1x_3 + 0.0081\,x_2x_3 \quad (6)$$

$$Y_2 = 135.7895 - 0.7993\,x_1 - 0.0923\,x_2 - 7.0710\,x_3 + 0.0022\,x_1^2 - 0.00004\,x_2^2 - 0.2067\,x_3^2 + 0.0006\,x_1x_2 + 0.0007\,x_1x_3 + 0.0135\,x_2x_3 \quad (7)$$

$$Y_3 = 259.6157 - 3.5198\,x_1 - 0.5794\,x_2 + 18.9429\,x_3 + 0.0128\,x_1^2 + 0.0004\,x_2^2 + 0.2042\,x_3^2 + 0.0038\,x_1x_2 - 0.1008\,x_1x_3 - 0.0216\,x_2x_3 \quad (8)$$

The observed and predicted value for the three responses shown in Figure 3 indicates that the experimental and predicted values are close to each other. The coefficient of determination (R^2) for the oil, blend-char and gas yield model equations are 0.84977, 0.8768 and 0.90307 indicating that 84.977%, 87.68% and 90.307% of the data variability are accounted to the model equations. It also indicates that only 15.023%, 12.32% and 9.693% variations in data are not accounted by the model components.

Table 3. Regression coefficients for observed responses

		Regression coefficients		
		Oil yield (wt%)	Blend-char yield (wt%)	Gas yield (wt%)
		Y_1	Y_2	Y_3
Mean/Intercept		–295.4052	135.7895	259.6157
1. CAC loading (wt%)**(L)**	x_1	4.3191	–0.7993	–3.5198
CAC loading (wt%)**(Q)**	x_1^2	–0.0150	0.0022	0.0128
2. MW power (W)**(L)**	x_2	0.6718	–0.0923	–0.5794
MW power (W)**(Q)**	x_2^2	–0.0003	–0.00004	0.0004
3. N_2 flow rate (LPM)**(L)**	x_3	–11.8719	–7.0710	18.9429
N_2 flow rate (LPM)**(Q)**	x_3^2	0.0025	–0.2067	0.2042
1**L** by 2**L**	$x_1 x_2$	–0.0044	0.0006	0.0038
1**L** by 3**L**	$x_1 x_3$	0.1000	0.0007	–0.1008
2**L** by 3**L**	$x_2 x_3$	0.0081	0.0135	–0.0216

L and Q stands for linear and quadratic terms

The analyses of variance (ANOVA) for the model responses are shown Table 4. It shows that the observed models as described in equations (6), (7) and (8) gives good prediction at 95% confidence level. For all models, F_{value} is higher than $F_{0.05}$ rejecting the null hypothesis. The student's t-distribution and p-values of each model factor is shown in Table 5. Similarly, Pareto chart given in Figure 4 shows the significance of each model coefficient by comparing smaller p_{value} of ≤ 0.05 or greater t-value. The largest effects on oil yield is from interaction terms of CAC loading and MW power (x_1x_2), and CAC loading and N_2 flow rate (x_1x_3) with highest t-values (–4.48178, 3.43027) and lowest p-values (0.001176, 0.006436) at 99.999% and 99.994% significance levels, respectively. The quadratic term of CAC loading (x_1^2) and MW power (x_2^2) can also be regarded as significant factors on oil yield at 99.992% and 99.976% significant levels, respectively. The blend-char yield is affected largely by linear terms of MW power (x_2) and CAC loading (x_1) with highest t-values (–4.95249, –4.02487) and lowest p-values (0.000576, 0.002419) with 99.998% and 99.999% significance levels, respectively. The linear and quadratic terms of N_2 flow rate (x_3 and x_3^2) and interaction term of MW power and N_2 flow rate (x_2x_3) with 99.993%, 95.670% and 99.449% significance levels can also be considered as significant factors on blend-char yield. The gas yield are largely influenced by linear term of MW power (x_2) and interaction terms of CAC loading and MW power (x_1x_2), CAC loading and N_2 flow rate (x_1x_3), and MW power and N_2 flow rate (x_2x_3). The t-values of x_2, x_1x_2, x_2x_3 and x_1x_3 were observed highest at 4.52116, 3.91691, –3.45929 and –4.44602 with significance levels of ≥99.387%. The quadratic terms of CAC loading (x_1^2) and MW power (x_2^2) and linear term of N_2 flow rate (x_3) are also regarded as significant factors on gas yield with 98.170%, 98.621% and 97.466% significance levels, respectively.

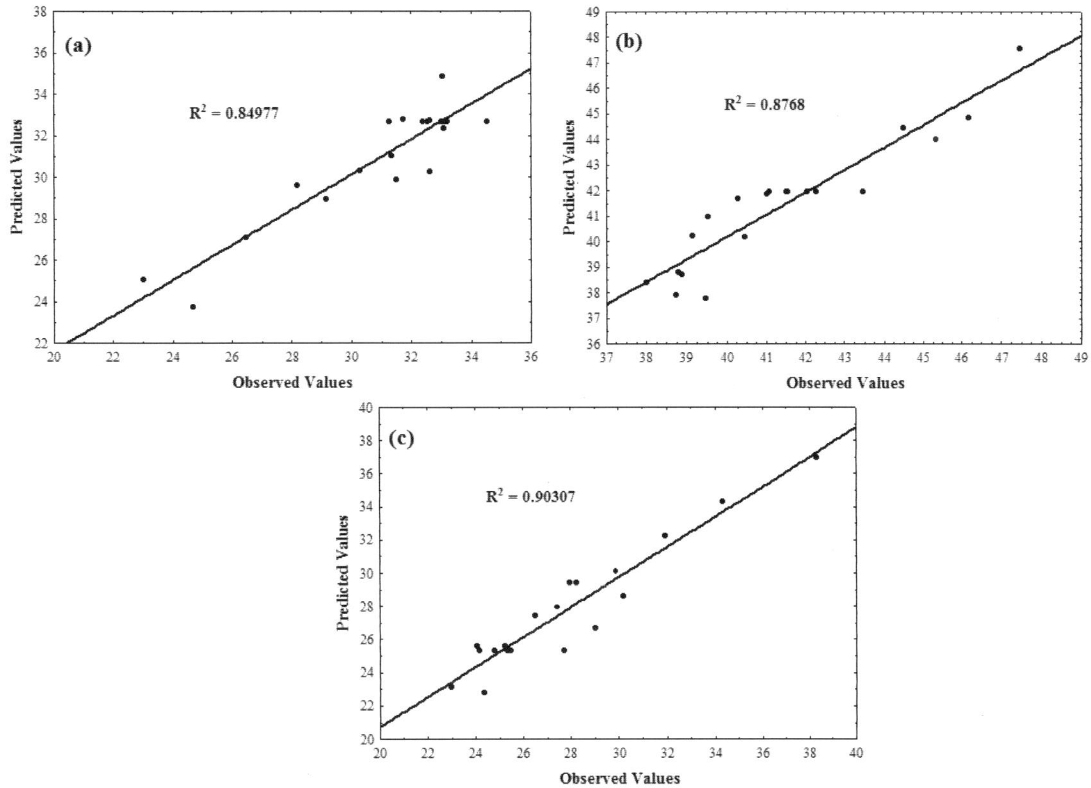

Figure 3. Coefficient of determination and observed versus predicated values, (a) oil yield (wt%), (b) blend-char yield (wt%) and (c) gas yield (wt%)

Table 4. Analysis of variance

Source	Sum of squares	Degree of freedom	Mean square	F_{value}	$F_{0.05}$
Oil yield model					
Regression (SSR)	153.96	9	17.11	6.29	>3.02
Residual error (SSE)	27.22	10	2.72		
Total (SST)	181.18	19			
Blend-char yield model					
Regression (SSR)	120.16	9	13.35	7.91	>3.02
Residual error (SSE)	16.88	10	1.69		
Total (SST)	137.039	19			
Gas yield model					
Regression (SSR)	253.11	9	28.12	10.35	>3.02
Residual error (SSE)	27.17	10	2.72		
Total (SST)	280.28	19			

Table 5. ANOVA for response surface quadratic models

Factor	Sum of squares	Degree of freedom	Mean square	T_{value}	F_{value}	p_{value}	Remarks
Oil yield model parameters							
x_1	12.5555	1	12.55548	2.14778	4.61294	0.057279	

x_1^2	29.6153	1	29.61532	-3.29861	10.88080	0.008032	Significant
x_2	1.0338	1	1.03385	-0.61631	0.37984	0.551460	
x_2^2	19.1759	1	19.17595	-2.65430	7.04533	0.024134	Significant
x_3	0.0049	1	0.00490	0.04242	0.00180	0.967000	
x_3^2	0.0013	1	0.00132	0.02203	0.00049	0.982859	
$x_1 x_2$	54.6709	1	54.67094	-4.48178	20.08635	0.001176	Significant
$x_1 x_3$	32.0267	1	32.02667	3.43027	11.76674	0.006436	Significant
$x_2 x_3$	7.5660	1	7.56605	1.66727	2.77980	0.126425	
Blend-char yield model parameters							
x_1	27.3501	1	27.35007	-4.02487	16.19961	0.002419	Significant
x_1^2	0.6425	1	0.64254	0.61691	0.38058	0.551078	
x_2	41.4097	1	41.40966	-4.95249	24.52719	0.000576	Significant
x_2^2	0.2851	1	0.28508	-0.41092	0.16885	0.689800	
x_3	19.3420	1	19.34199	-3.38473	11.45638	0.006947	Significant
x_3^2	9.0305	1	9.03053	-2.31275	5.34883	0.043304	Significant
$x_1 x_2$	0.8800	1	0.88002	0.72197	0.52124	0.486851	
$x_1 x_3$	0.0018	1	0.00180	0.03265	0.00107	0.974595	
$x_2 x_3$	20.9520	1	20.95202	3.52278	12.41001	0.005513	Significant
Gas yield model parameters							
x_1	2.8438	1	2.84380	1.02314	1.04683	0.330353	
x_1^2	21.5334	1	21.53338	2.81542	7.92660	0.018304	Significant
x_2	55.5296	1	55.52957	4.52116	20.44086	0.001106	Significant
x_2^2	24.1372	1	24.13719	2.98079	8.88508	0.013789	Significant
x_3	18.7313	1	18.73133	2.62586	6.89515	0.025342	Significant
x_3^2	8.8134	1	8.81343	1.80119	3.24429	0.101854	
$x_1 x_2$	41.6785	1	41.67845	3.91691	15.34216	0.002881	Significant
$x_1 x_3$	32.5087	1	32.50867	-3.45929	11.96669	0.006130	Significant
$x_2 x_3$	53.6993	1	53.69934	-4.44602	19.76714	0.001243	Significant

Oil yield model; R^2= 0.84977; adjusted-R^2= 0.71457; MS residual= 2.721796
Blend-char yield model; R^2= 0.8768; adjusted-R^2= 0.76592; MS residual= 1.688316
Gas yield model; R^2= 0.90307; adjusted-R^2= 0.81584; MS residual= 2.716597

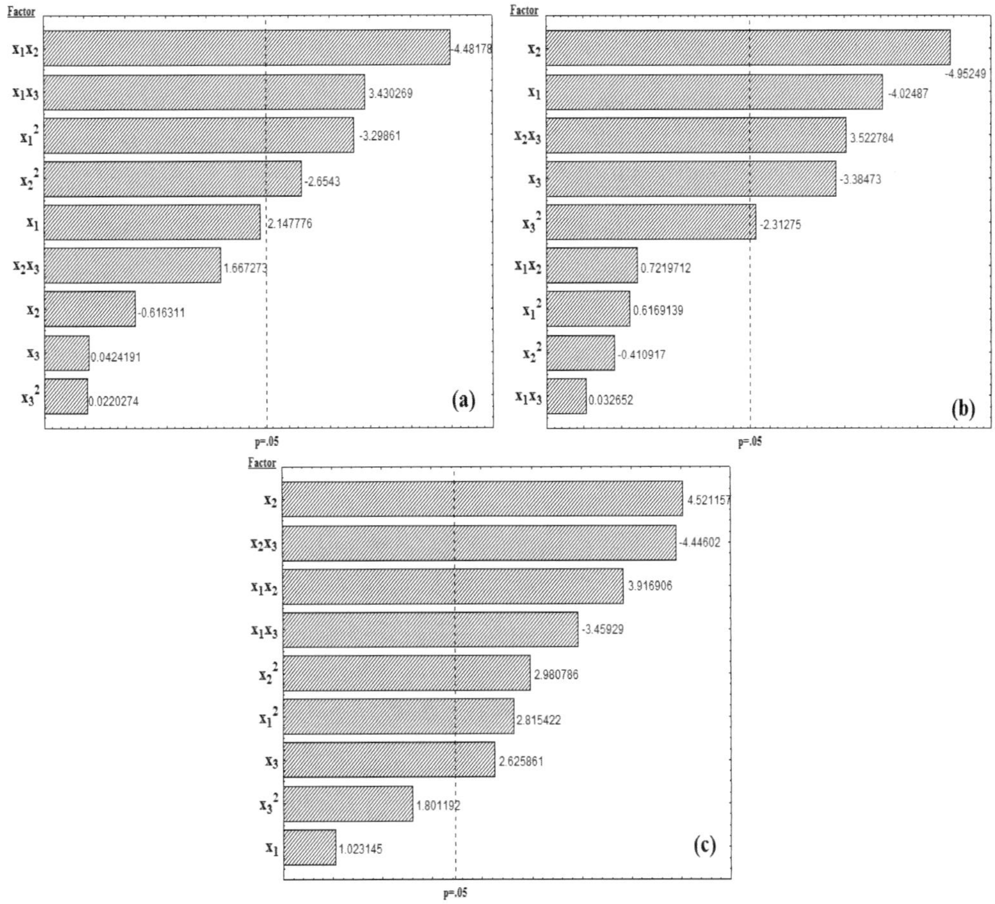

Figure 4. Pareto charts, (a) oil yield, (b) blend-char yield and (c) gas yield

3.2.2 Effects of variables on responses

The 3D response surfaces plotted for three factors on oil, blend-char and gas yield are shown in Figure 5. The oil yield is influenced by quadratic terms of CAC loading and MW power, interaction terms of CAC loading and MW power, and CAC loading and N_2 flow rate. The oil yield increases with increasing CAC loading at low MW power but decreases with increasing CAC loading at high MW power (Figure 5, (a)). This difference in oil production with CAC loading is mainly the result of MW power on heat generated from carbons and heat carried with inert gas to controls the final temperature and pyrolysis duration of fuel solids. At low carbon loading and MW power, decrease in heat generated of carbons and heat carried with N_2 gas can reduce the final temperature and pyrolysis duration resulting in incomplete pyrolysis conditions. On the contrary, increase in carbon loading increases the heat generated of carbons and heat carried with N_2 gas, which can improve final temperature and pyrolysis duration supporting pyrolysis conditions to completion (Figure 1 and 2, Run 19, 9-14 and 20). The decrease in oil yield with increasing CAC loading at high MW power can be explained by higher MW absorption capacity of carbon solids and fast heat transfer to the fuel solids by maintaining quick pyrolysis conditions and higher final pyrolysis temperature, which can promote secondary phase vapor cracking. The interaction effects of CAC loading and N_2 flow rate shown in Figure 5, (b) reveals that the oil yield decreases with increasing N_2 flow rate at low carbon loading but increases with increasing N_2 flow rate at high carbon loading. The decrease in oil yield with higher N_2 flow at low carbon loading suggests that heat carried with inert gas is higher than the heat generated by carbon solids, which can reduce final pyrolysis temperature impeding pyrolysis conditions. The higher oil production with increasing N_2 flow rate at high carbon loading could be the result of increased heat carrying capacity of inert gas by improving final temperature and assisting pyrolysis conditions (Figure 1 and 2, Run 17, 9-14 and 18). The interaction effects of N_2 flow rate and MW power shown in Figure 5, (c) shows that increasing the MW power from 540W to 660W at low N_2 flow rate increases the oil yield, beyond which the oil yield deceases significantly. With higher N_2 flow, increasing MW power increases the oil yield. With low MW power, increasing the N_2 flow slightly decreases the oil yield but increases with high MW power. This difference in oil production can be attributed to the synergetic effects of MW power and heat generated by carbon solids, which can affect the heat carry capacity of inert gas and simultaneous heat transfer to the fuel solids.

The blend-char yield is affected by linear terms of carbon loading, MW power and N_2 flow rate, and interaction of MW power and N_2 flow rate as shown in Figure 5, (d), (e) and (f). The blend-char yield linearly increases with decreasing MW power both at low and high carbon loading (Figure 5, (d)). This is based on the fact that lower MW power generates less heat of carbons, which is then carried with inert gas to impede solid conversion, thereby increasing blend-char yield. However, higher MW power generates more heat of carbons and higher heat carried with inert gas and simultaneous heat transfer to the fuel solids can result in pyrolysis process to completion (Figure 1 and 2, Run 15, 9-14 and 16). The blend-char yield increases with decreasing N_2 flow rate at low and high carbon loading (Figure 5, (e)). The higher blend-char yield can be associated to decrease in heat carrying capacity of inert gas and simultaneous heat transfer to fuel solids, whereas increasing the N_2 flow decreases the blend-char yield, which is mainly of increased heat carry capacity of inert gas and simultaneous heat transfer to fuel solids. The interaction effects of N_2 flow rate and MW power on blend-char yield shown in Figure 5, (f) indicate that blend-char yield decreases with increasing MW power at low N_2 flow rate but increases with increasing MW power at high N_2 flow rate. Moreover, increasing N_2 flow decreases blend-char yield at low MW power but increases at high MW power. The gas yield is influenced mainly by linear term of MW power and interaction terms of CAC loading and MW power, CAC loading and N_2 flow rate, and MW power and N_2 flow rate.

Referring to Figure 5 (g), the gas yield decreases with increasing MW power at low CAC loading but increases with increasing MW power at high CAC loading. Similar behaviour of gas production was noticed by analyzing the interaction effects of MW power and N_2 flow rate as shown in Figure 5, (i). The interaction effects of CAC loading and N_2 flow rate shown in Figure 5, (h) shows that the gas yield decrease with decreasing N_2 flow rate at low carbon loading but increases significantly with high carbon loading. Moreover, the gas yield decreases with increasing CAC loading at high N_2 flow rate but increases significantly at low N_2 flow rate. The higher gas yield can be correlated to non-uniform heating behaviour of fuel solids and significant temperature difference for longer pyrolysis duration. Referring to Figure 1, Run 7, the middle fuel solids quickly maintained pyrolysis condition at much higher heating rate and final pyrolysis temperature for longer period compared to upper and bottom fuel solids, which resulted in highest gas yield of 38.29wt% and lowest oil yield of 22.99wt% (Table 2, Actual Run 7). Similar increase in gas yield was observed with Run 2 shown in Figure 1, where the middle and bottom fuel solids maintained quick pyrolysis conditions at much higher heating rate and final temperature compared to upper fuel solids. More importantly, lowest gas yield of 22.95wt% with 31.72 wt% oil yield occurred when the fuel solids maintained fairly uniform process heating rate with minimum difference in final pyrolysis temperature (Figure 1, Run 5). At this stage, it can be suggested that differential heating nature of fuel solids in the presence of carbon absorber for extended pyrolysis duration at elevated temperature can result in increased gas production by promoting secondary phase vapor cracking.

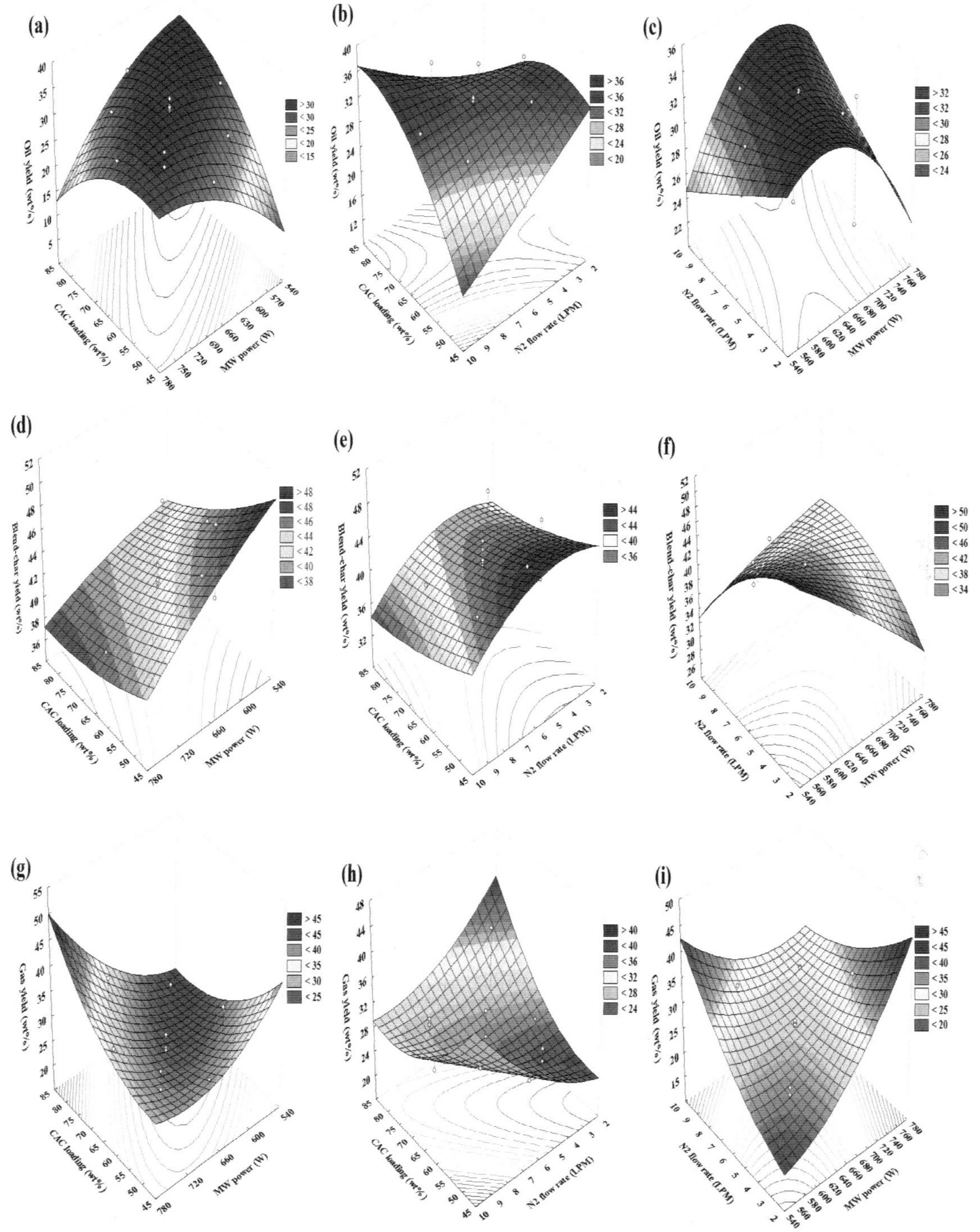

Figure 5. 3D response surfaces; (a), (b) and (c) effects on oil yield; (d), (e) and (f) effects on blend-char yield; (g), (h) and (i) effects on gas yield

3.2.3 Optimization by response surface modeling

The developed mathematical models for oil, blend-char and gas yields were investigated to predict critical values. The quadratic mathematical model predicted that the critical process factors for oil yield are 582W, 71.38wt% CAC loading, 3.52LPM of N_2 flow rate, which gives 33.17wt% oil. The observed oil yield was 31.33 wt% indicating 5.52% error between the observed and predicted values.

The critical process factors for blend-char and gas yield quadratic mathematical models are shown in Table 6 with observed and predicted values. The heating profiles under optimized process conditions are shown in Figure 6. Several experiments were performed at the critical values to verify the accuracy of the proposed models.

Table 6. Optimization conditions and observed values

Factors	Observed minimum (wt%)	Critical values (wt%)	Observed maximum (wt%)	Predicted critical value (wt%)	Observed value (wt%)	% Error
Oil yield model						
CAC loading (wt%)	48.67	71.38	81.33	33.17	31.33	5.52
MW power (W)	562	582	758			
N_2 flow rate (LPM)	2.7	3.5	9.3			
Blend-char yield model						
CAC loading (wt%)	48.67	83.76	81.33	38.44	36.75	4.41
MW power (W)	562	767	758			
N_2 flow rate (LPM)	2.7	8.1	9.3			
Gas yield model						
CAC loading (wt%)	48.67	74.48	81.33	30.09	32.43	7.79
MW power (W)	562	790	758			
N_2 flow rate (LPM)	2.7	13.8	9.3			

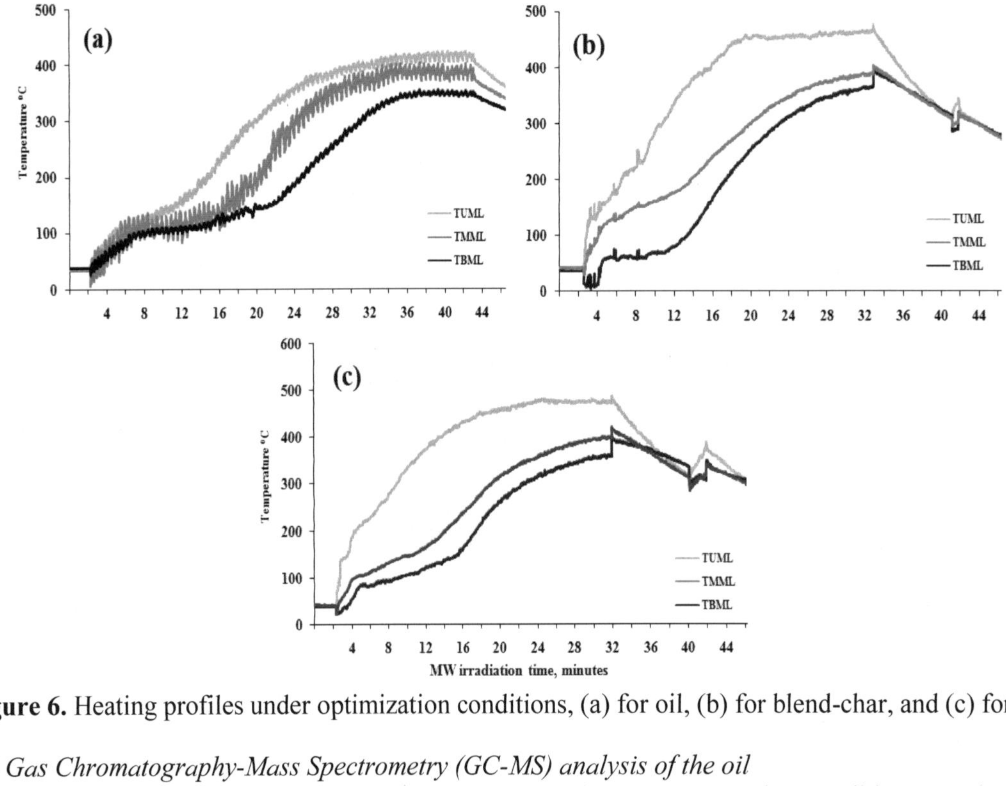

Figure 6. Heating profiles under optimization conditions, (a) for oil, (b) for blend-char, and (c) for gas

3.3 Gas Chromatography-Mass Spectrometry (GC-MS) analysis of the oil

The chemical composition of the oil obtained under various MW operating conditions consisted of several groups of chemical compounds. The chemical species were classified into twelve major groups; phenol, phenolics, guaiacols, naphthalenes, benzenes, xylene, aldehydes/ketones, alkanes, alkenes, acids, alcohols and others. The main components of the oil are phenol, phenolics, guaiacols, naphthalenes and benzenes, whereas the minor includes, xylene, alkanes, alkenes, aldehydes/ketones, acids, alcohols. The composition of the oil obtained under various MW co-pyrolysis conditions by chemical groups and by chemical classification is shown in Figure 7, (a) and (b).

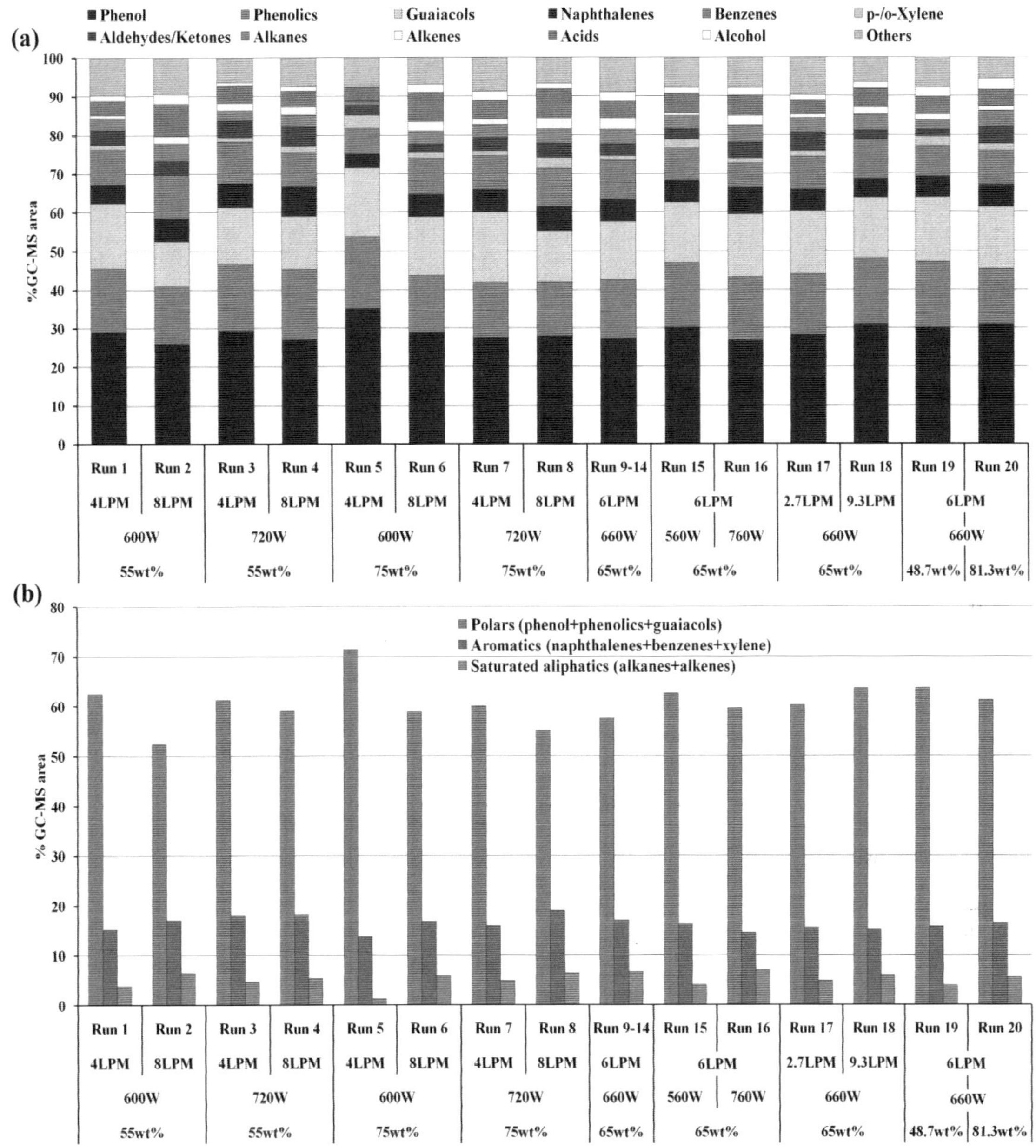

Figure 7. Chemical composition of the oil under various MW co-pyrolysis heating conditions, (a) by chemical groups, (b) by chemical classification

The co-pyrolysis oil was dominated with phenol, phenolics and guaiacols chemical species consisting of 26.00-35.10%, 14.14-18.61% and 11.68-18.19% GC-MS area, respectively. The phenolics were mainly composed of 2/3-methyl phenol, 2,4/3,5-dimethyl phenol and 2/3-ethyl phenol. The guaiacols consisted mainly of 2-methoxy-phenol, 2-methoxy-4-methyl-phenol, 4-ethyl-2-methoxy-phenol, 2,6-dimethoxy-phenol, 2-methoxy-4-(1-propenyl)-phenol, 2,6-dimethoxy-4-(2-propenyl)-phenol. The naphthalenes and benzenes detected remained 3.63-7.59% and 6.30-11.07% GC-MS area, respectively. The naphthalenes were mainly composed of 1/2-methyl naphthalene and 2,6/2,7-dimethyl napthalene. The alkanes and alkenes in the oil remained fairly low of 1.10-4.44% and 0.25-2.88% GC-MS area, respectively. It can be observed that the chemical composition of the oil under various conditions remained fairly unchanged. However, higher phenol content of 35.10% GC-MS was detected when the pyrolysis was carried out at 600W, 75wt% CAC loading and 4LPM (Figure 7, (a), Run 5). This higher phenol production can be associated to the slow heating rate and fairly uniform process heating conditions of fuel solids (Figure 1, Run 5). More importantly, the maximum final temperature recorded under these conditions of T_{UML}=405°C, T_{MML}=376°C, T_{BML}=344°C were observed for fairly shorter pyrolysis duration, which might have reduced secondary phase vapor cracking since the pyrolysis vapor generated from upper fuel solids flows over fairly low temperature environment.

More interesting, the fuel solids showed fairly smooth heating rate and uniform process heating conditions with Run 9-14 and Run 18 as shown in Figure 1. However, phenol content under these conditions is lower than Run 5. This decrease in phenol content is mainly of the higher heating rate and final pyrolysis temperature of middle and bottom fuel solids for sustained pyrolysis conditions for longer duration. Similar uniform heating behaviour of fuel solids was also observed with Run 20 in Figure 1. However, the heating rate and final temperature reached in fuel solids under these conditions was observed higher compared to Run 5. The decrease in phenol content under most other operating conditions can be related to significant difference in heating rate and final temperature reached in fuel solids. More importantly, the higher temperature region formed in the middle and bottom of the reaction chamber as a result of non-uniformity of fuel solids can favors secondary phase vapor cracking. It is therefore suggested that uniform heating behaviour of fuel solids in the presence of carbon surfaces can minimize secondary reactions. However, the final temperature reached in fuel solids needs special attention to optimize oil for higher phenol production.

The chemical classification of the oil shown in Figure 7, (b) shows that the polars (phenol, phenolics and guaiacols) detected in fuel oil remained fairly high at 52.51-71.48% GC-MS area. The aromatics (napthalenes, benzenes and xylene) and saturated aliphatic (alkanes and alkenes) chemical species remained 13.74-19.06% and 1.35-6.99% GC-MS area, respectively. The MW assisted co-pyrolysis of coal and OPS waste biomass with uniform distribution of carbon absorber achieved higher polar yields. This improved polar formation suggests that coal can be utilized effectively with renewable OPS biomass in MW pyrolysis system for the recovery of valuable chemicals.

4. Conclusion

This study investigated the effects of process parameters, carbon loading, MW power and N_2 flow rate on optimization of pyrolysis products from MW assisted co-pyrolysis of coal and OPS waste biomass blend with uniformly distributed coconut activated carbon under various operating conditions. The results showed that N_2 atmosphere control the fuel solids temperature, MW power influenced the fuel solid heating rate and carbon loading control the final pyrolysis temperature. The co-utilization of coal and OPS blend under MW operating conditions with carbon surfaces achieved nearly uniformity of pyrolysis heating conditions. This uniform process heating achieved either by increasing carbon loading or N_2 flow rate improved oil yield. In this study, uniform process heating of fuel solids with absorber medium was found essential to improve fuel conversion and selectivity. The oil was enriched in polar chemical compounds suggesting potential use as an alternative and renewable feedstock for petroleum based resins.

Acknowledgement

Authors gratefully acknowledge the support of Ministry of Higher Education Malaysia and Universiti Teknologi Malaysia for the award of UTM-Research University Grant Q.J.130000.2524.01H03 and Q.J.130000.2544.04H68. First author is grateful to Higher Education Commission (HEC), Islamabad, Pakistan for the financial support under Faculty Development Program (FDP) and Balochistan University of Information Technology, Engineering and Management Sciences (BUITEMS), Quetta, Balochistan, Pakistan for exploring new areas of research in the field of thermo-electromagnetic conversion of coal and waste biomass for fuel production.

References

[1] WCA 2014 Coal Price World Coal Association
[2] BP 2013 Statistical Review of World Energy
[3] WCA 2014 Coal and Electricity World Coal Association

[4] Zheng L 2013 *Cleaner Combustion and Sustainable World,* ed H Qi and B Zhao: Springer Berlin Heidelberg) pp 19-21

[5] Van Loo S and Koppejan J 2008 *The Handbook of Biomass Combustion and Co-firing*: Earthscan

[6] Wang L, Hustad J E, Skreiberg Ø, Skjevrak G and Grønli M 2012 A Critical Review on Additives to Reduce Ash Related Operation Problems in Biomass Combustion Applications *Energy Procedia* **20** 20-9

[7] Gilbe C, Öhman M, Lindström E, Boström D, Backman R, Samuelsson R and Burvall J 2008 Slagging Characteristics during Residential Combustion of Biomass Pellets *Energy & Fuels* **22** 3536-43

[8] Baxter L L 1993 Ash Deposition during Biomass and Coal Combustion: A Mechanistic Approach *Biomass and Bioenergy* **4** 85-102

[9] Tillman, David A, Duong D and Miller B 2009 Chlorine in Solid Fuels Fired in Pulverized Fuel Boilers Sources, Forms, Reactions, and Consequences: a Literature Review *Energy & Fuels* **23** 3379-91

[10] Domínguez A, Menéndez J A, Fernández Y, Pis J J, Nabais J M V, Carrott P J M and Carrott M M L R 2007 Conventional and Microwave Induced Pyrolysis of Coffee Hulls for the Production of a Hydrogen Rich Fuel Gas *Journal of Analytical and Applied Pyrolysis* **79** 128-35

[11] Menéndez J A, Arenillas A, Fidalgo B, Fernández Y, Zubizarreta L, Calvo E G and Bermúdez J M 2010 Microwave Heating Processes involving Carbon Materials *Fuel Processing Technology* **91** 1-8

[12] Yin C 2012 Microwave-Assisted Pyrolysis of Biomass for Liquid Biofuels Production *Bioresource Technology* **120** 273-84

[13] Song Y, Shi J, Fu J, Lan X, Zhang Q and Zhou J 2012 Analysis of Products by Conventional and Microwave Induced Pyrolysis for Low Rank Coal. In: *Advanced Materials Research*: Trans Tech Publ). pp 871-5

[14] Zhang Z, Macquarrie D J, Budarin V L, Hunt A J, Gronnow M J, Fan J, Shuttleworth P S, Clark J H and Matharu A S 2015 Low-temperature Microwave-Aassisted Pyrolysis of Waste Office Paper and the Application of Bio-oil as an Al Adhesive *Green Chemistry* **17** 260-70

[15] Budarin V L, Clark J H, Lanigan B A, Shuttleworth P and Macquarrie D J 2010 Microwave Assisted Decomposition of Cellulose: A New Thermochemical Route for Biomass Exploitation *Bioresource technology* **101** 3776-9

[16] Shuttleworth P, Budarin V, Gronnow M, Clark J H and Luque R 2012 Low Temperature Microwave-Assisted vs Conventional Pyrolysis of various Biomass Feedstocks *Journal of Natural Gas Chemistry* **21** 270-4

[17] Abubakar Z, Salema A A and Ani F N 2013 A New Technique to Pyrolyse Biomass in a Microwave System: Effect of Stirrer Speed *Bioresource technology* **128** 578-85

[18] Bu Q, Lei H, Wang L, Wei Y, Zhu L, Liu Y, Liang J and Tang J 2013 Renewable Phenols Production by Catalytic Microwave Pyrolysis of Douglas Fir Sawdust Pellets with Activated Carbon Catalysts *Bioresource Technology* **142** 546-52

[19] Bu Q, Lei H, Ren S, Wang L, Holladay J, Zhang Q, Tang J and Ruan R 2011 Phenol and Phenolics from Lignocellulosic Biomass by Catalytic Microwave Pyrolysis *Bioresource Technology* **102** 7004-7

[20] Bu Q, Lei H, Wang L, Wei Y, Zhu L, Zhang X, Liu Y, Yadavalli G and Tang J 2014 Bio-based Phenols and Fuel Production from Catalytic Microwave Pyrolysis of Lignin by Activated Carbons *Bioresource technology* **162** 142-7

[21] Domínguez A, Fidalgo B, Fernández Y, Pis J and Menéndez J 2007 Microwave-assisted Catalytic Decomposition of Methane over Activated Carbon for CO2-free Hydrogen Production *International Journal of Hydrogen Energy* **32** 4792-9

[22] Fidalgo B, Fernández Y, Domínguez A, Pis J and Menéndez J 2008 Microwave-assisted Pyrolysis of CH4/N2 Mixtures over Activated Carbon *Journal of Analytical and Applied Pyrolysis* **82** 158-62

[23] Salema A A and Ani F N 2011 Microwave Induced Pyrolysis of Oil Palm Biomass *Bioresource Technology* **102** 3388-95
[24] Bu Q, Lei H, Ren S, Wang L, Zhang Q, Tang J and Ruan R 2012 Production of Phenols and Biofuels by Catalytic Microwave Pyrolysis of Lignocellulosic Biomass *Bioresource Technology* **108** 274-9
[25] Salema A A and Ani F N 2012 Microwave-Assisted Pyrolysis of Oil Palm Shell Biomass Using an Overhead Stirrer *Journal of Analytical and Applied Pyrolysis* **96** 162-72
[26] Wang X, Morrison W, Du Z, Wan Y, Lin X, Chen P and Ruan R 2012 Biomass Temperature Profile Development and its Implications under the Microwave-assisted Pyrolysis Condition *Applied Energy* **99** 386-92
[27] Mushtaq F, Mat R and Ani F N 2014 A review on microwave assisted pyrolysis of coal and biomass for fuel production *Renewable and Sustainable Energy Reviews* **39** 555-74
[28] Bu Q, Lei H, Wang L, Wei Y, Zhu L, Liu Y, Liang J and Tang J 2013 Renewable Phenols Production by Catalytic Microwave Pyrolysis of Douglas Fir Sawdust Pellets with Activated Carbon Catalysts *Bioresource Technology* **142** 546-52
[29] Mushtaq F, Mat R and Ani F N 2014 The Performances of Intimately Mix and Layer Methods in Microwave Assisted Pyrolysis System *Applied Mechanics and Materials* **554** 150-4
[30] Mushtaq F, Mat R and Ani F N 2014 Pyrolysis of Solid Palm Waste Biomass with Microwave Absorber under Microwave Irradiation. In: *Applied Mechanics and Materials*: Trans Tech Publ) pp 73-7
[31] Mushtaq F, Mat R and Ani F N 2016 Fuel production from microwave assisted pyrolysis of coal with carbon surfaces *Energy Conversion and Management* **110** 142-53
[32] Mushtaq F, Malghani M N K, Nasar M S, Mengal A N, Mat R and Ani F N 2017 Pyrolysis Heating Performance of Oil Palm Shell Waste Biomass with Carbon Surfaces *Journal of Applied and Emerging Sciences* **7** pp70-5
[33] Mushtaq F, Channa A S, Mat R and Ani F N 2014 Microwave Assisted Pyrolysis of Waste Biomass Resources for Bio-Oil Production *Applied Mechanics and Materials* **554** 307-11
[34] Mushtaq F, Abdullah T A T, Mat R and Ani F N 2015 Optimization and characterization of bio-oil produced by microwave assisted pyrolysis of oil palm shell waste biomass with microwave absorber *Bioresource technology* **190** 442-50
[35] Robinson J, Dodds C, Stavrinides A, Kingman S, Katrib J, Wu Z, Medrano J and Overend R 2015 Microwave Pyrolysis of Biomass: Control of Process Parameters for High Pyrolysis Oil Yields and Enhanced Oil Quality *Energy & Fuels* **29** 1701-9
[36] Zhang X, Lei H, Wang L, Zhu L, Wei Y, Liu Y, Yadavalli G and Yan D 2015 Renewable Gasoline-Range Aromatics and Hydrogen-enriched Fuel Gas from Biomass via Catalytic Microwave-induced Pyrolysis *Green Chemistry* **17** 4029-36
[37] Bu Q, Lei H, Wang L, Yadavalli G, Wei Y, Zhang X, Zhu L and Liu Y 2015 Biofuel Production from Catalytic Microwave Pyrolysis of Douglas Fir Pellets over Ferrum-modified Activated Carbon Catalyst *Journal of Analytical and Applied Pyrolysis* **112** 74-9
[38] Borges F C, Du Z, Xie Q, Trierweiler J O, Cheng Y, Wan Y, Liu Y, Zhu R, Lin X and Chen P 2014 Fast Microwave Assisted Pyrolysis of Biomass Using Microwave Absorbent *Bioresource Technology* **156** 267-74
[39] Jamaluddin M A, Ismail K, Mohd Ishak M A, Ab Ghani Z, Abdullah M F, Safian M T-u, Idris S S, Tahiruddin S, Mohammed Yunus M F and Mohd Hakimi N I N 2013 Microwave-assisted Pyrolysis of Palm Kernel Shell: Optimization using Response Surface Methodology (RSM) *Renewable Energy* **55** 357-65
[40] Kuehl R O and Kuehl R 2000 Design of Experiments: Statistical Principles of Research Design and Analysis. Duxbury Press
[41] Huang Y-F, Kuan W-H, Chang C-C and Tzou Y-M 2013 Catalytic and Atmospheric Effects on Microwave Pyrolysis of Corn Stover *Bioresource Technology* **131** 274-80

Role of masonry infill wall on the seismic behavior of typical four-storey building in Pakistan

A Ahmed[1], A Ali[1], Hammaad Khalid[1] and M Ahmad[2]

[1]Department of Civil Engineering, Capital University of Science and Technology, Islamabad, Pakistan
[2]Professor, Department of Civil Engineering, Capital University of Science and Technology, Islamabad, Pakistan

Email: m.awaisali8899@gmail.com

Abstract. Masonry infill walls are used as partition in reinforced concrete frames and are considered as a non-engineering structure for design and analysis purposes. The goal of this study is to compare different response parameters such as displacement, storey drift, base shear and ground overturning moment of a multi-storey RC frame structure with and without infill walls. For this purpose, a four-storey building is selected, which is supposed to be situated on stiff soil type (S_D) and seismic zone 2B according to the Building Codes of Pakistan. Openings are neglected in the building. Non-linear static pushover analysis method is used to check the behavior of building during earthquake using SAP-2000. To find the width of compression struts, FEMA-356 is used. The results are compared and shown in the form of graphs. It is concluded that the masonry infill walls may have significant effects on the seismic response of the reinforced concrete frame structures, therefore, infill walls need to be considered during the design and analysis of the building.
Keywords: Seismic, Pakistan, reinforced concrete, SAP-2000

1. Introduction

Masonry infill walls are used all over the world and are a common practice in developing countries for partition purposes due to their better functionality, accessibility and less cost. They are used to enclose the frame structure or divide the structure. Infill walls are made of bricks, concrete blocks, hollow concrete blocks or any solid material. Design standards for infill walls cannot be found because infill walls are considered as non-structural. Many researchers have termed infilled frame structures as "Earthquake Risk" structures. Masonry infill walls may alter the behavior of frame structure during earthquakes and respond to characteristics like energy dissipation and stiffness. Masonry infill walls in RC frame structure might also decrease the displacement of the structure. The effect of masonry infill walls on seismic response of RC frame buildings and concluded that the infill walls contribute significantly to lateral stiffness, strength and overall ductility, [1, 2]. The earthquake analysis of mutli-storey building with and without infill walls and concluded that the presence of infill walls reduces the displacement and increases the base shear, [3] The response of infilled panels and found that it is necessary to include masonry infill's in analysis of Reinforced Concrete (RC) moment resisting frame structure. The seismic assessment of RC structures with infill panels and concluded that relative storey displacement is significantly affected by earthquake [4]. The seismic behavior of RC frame structure with masonry infills, [5]. He concluded that the presence of infill wall significantly affects the base shear. The Waleed (2012) [6] conducted the parametric study on masonry infill walls and concluded that masonry infill walls can increase the base shear capacity and reduce the displacement capacity of masonry reinforced frame buildings. The Ioannis (2011) [7] performed numerical modeling of RC frames with infill walls and concluded that finite element method shows that the masonry infill walls significantly affect the resistance to seismicity of RC frame structure.

Multi-storey buildings with infill walls subjected to dead and live load do not pose too much threat. The problem arises when earthquake generate lateral force on the RC frame structures with infill walls. The buildings are normally designed with the assumption that the total lateral load is absorbed by bare frame during earthquake without considering the effect of infill walls in RC frames. If masonry infill walls are modeled and analyzed with frame only, then they may add beneficial effects on the earthquake response behavior of the RC frame building.

2. Modeling and Analysis of Reinforced Cement Concrete Building

2.1 General description

The building is a reinforced concrete frame building structure. The dimensions of the building are 110'x56'. The height of each storey is 12' and the height of total building is 64' including mumty and base storey. The building was designed for zone 2B and stiff soil type S_D. The building was modeled and analyzed using SAP2000 according to building code of Pakistan and UBC-97. The plan of the building is as shown in Figure-1 and the data of the building is shown in Table-1 and Table-2. The Time Period of the modeled building without infill wall is 1.22 second and for infilled frame is 0.52 seconds. The peak ground acceleration for the respective zone 2B is 0.20g. Push over load cases were defined by introducing target displacement. The formula to calculate the target displacement is mentioned below which is taken from FEMA-356.

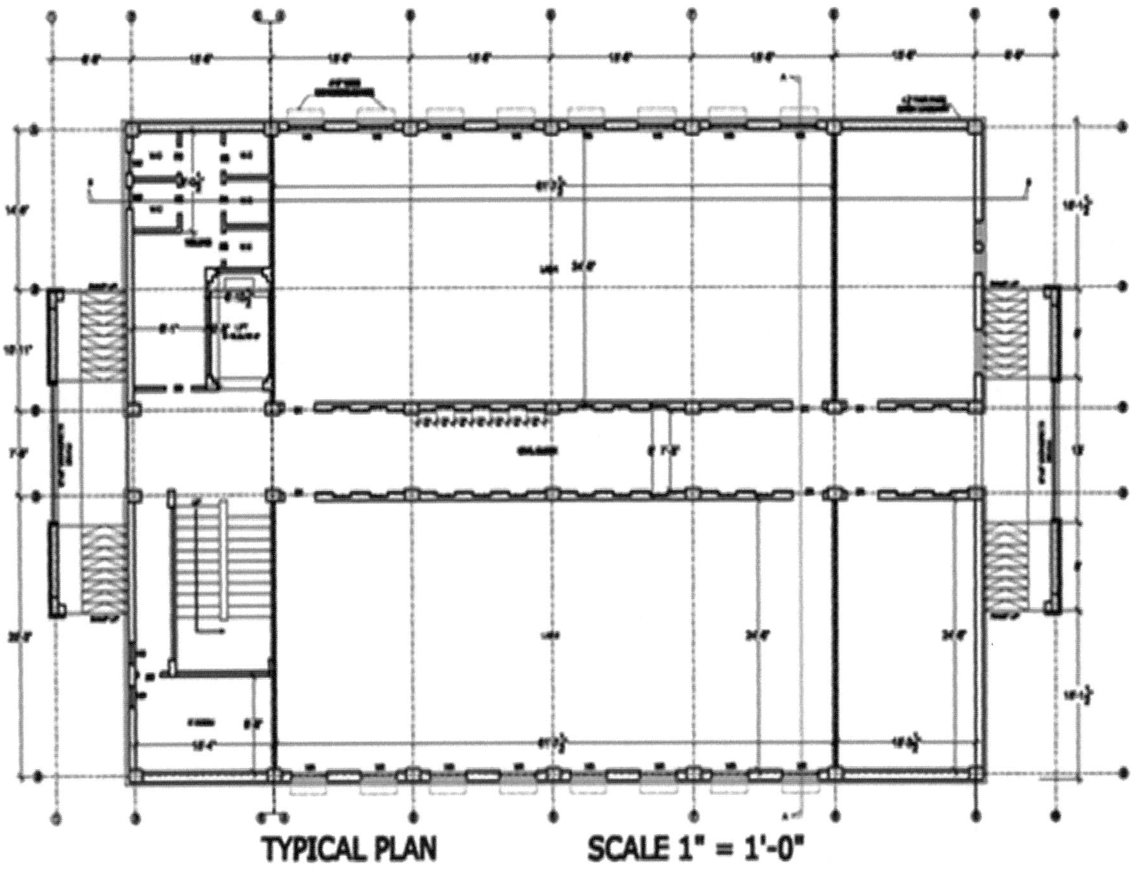

Figure 1. Plan of Building

Table 1. Geometrical parameters of frame members

Structural element	Transverse Section (in)	Transverse section area (in$^{2)}$	Stiffness modification factor
Longitudinal beams	15"x18"	270	0.35
Transverse beams	18"x21"	378	0.35
Column	21"x24"	504	0.7
Slab	6" thick	-	0.35

Table 2. Properties of the material

Frame element	Compressive strength (ksi)	Modulus of elasticity	Poison ratio
Longitudinal beam	3	3122	0.2
Transverse beam	3	3122	0.2
Column	3	3122	0.2
Masonry	1.7	598.28	0.25
Slab	3	3122	0.2

2.2 Calculation for Compression Strut

Various equivalent diagonal strut formulas and methods are available to find the width of compression struts. Diagonal struts were connected to beam column joint with hinge moment free connection, so they could take only compressive forces. In this study, FEMA-356 is used to find the width of compression struts.

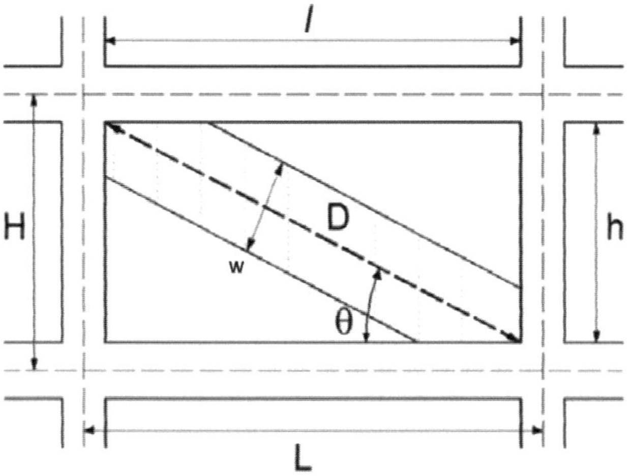

Where:
w= width of equivalent strut
λ_h=width of diagonal strut
H=height of column
d_{inf}=diagonal length of infill panel
$t_{inf.}$=thickness of infill panel
E_c=elastic modulus of frame
I_c=moment of inertia of column
h=height of infill panel
θ=angle of diagonal strut
L=center to center distance of frame
l=inner length of frame

Figure 2. Diagonal Strut Parameters

$$w = 0.172(\lambda_h . H)^{-0.4} d_{inf}$$

$$\lambda_h = \left[\frac{E_{inf} t_{inf.} \sin 2\theta}{4 E_c I_c h}\right]^{0.24}$$

From the above given formula, the width of compression struts were calculated, the value of which is given below in the Table-3.

Table 3. Strut details

S No.	Width (in)	Thickness(in)
1	17.76"	9"
2	20.35"	9"
3	40.85"	9"
4	36.00"	9"

2.3 Non-Linear Modeling

Mathematical model was developed in SAP2000. Hinges were assign using Auto hinge FEMA-356 option. After which Pushover load cases were generated. The modal was analyzed, without pushover load case. Then modals were analyzed again using pushover load cases, following concrete check. After the results were extracted (pushover curve, displacement, storey drift, shear and over turning moment). Above procedure is adopted for infilled frame, compression struts as link elements are assigned. Stress-strain masonry curve (1:3 mortar ratio) is converted into force-displacement curve respectively which is than further assigned to link elements.

$$F = \delta . A$$

$$\Delta = \varepsilon . l$$

Where,

F = Force, A = Cross-Sectional Area of Strut, δ = Stress

Δ = Displacement, ε = Strain, l = Diagonal Length of infill

2.4 Target Displacement

The target displacement of the bare and infill frame was calculated using the following formula from FEMA 356. The following parameters were substituted; the target displacement for bare frame is 0.41' and 0.14' with infill walls which shows that the stiffness of structure is increased. The reinforcement of the building remained unchanged due to existing structure.

$$\delta_t = C_0 C_1 C_2 C_3 S_a \frac{T_e^2}{4\pi^2} g$$

Where, δ_t= target displacement
C_0=$C_1 = C_2 = C_3$ = modification factors
S_a = response spectrum acceleration
Te= time period
g= gravitational acceleration

3. Results and Discussion

Following results are extracted from both infill and without infill models. The results are shown for pushover curve, displacement, storey drift, storey shear and storey turning moment in Fig-3 to 7 for these structural responses.

IOP Conf. Series: Materials Science and Engineering **414** (2018) 012017 doi:10.1088/1757-899X/414/1/012017

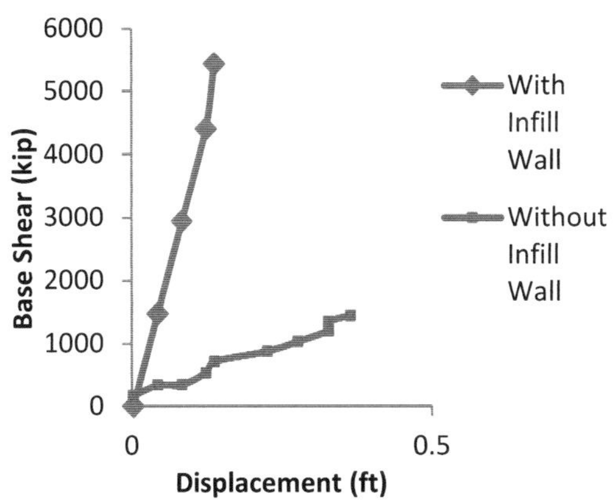

Figure 3. Displacement vs. Base Shear

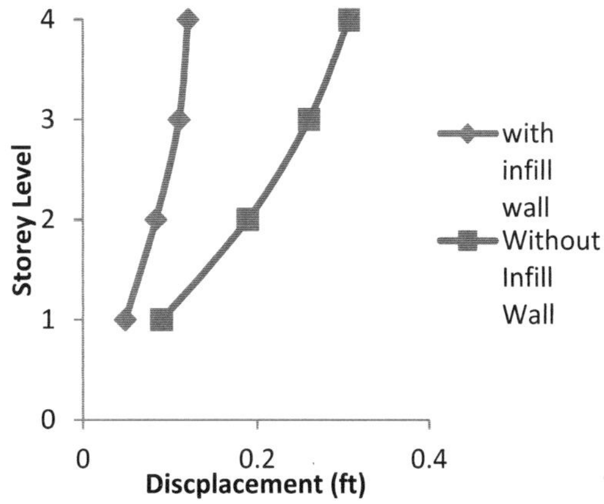

Figure 4. Displacement vs. Storey Level

Figure 3 shows the pushover curve. The base shear of infilled frame is 72% more than that of bare frame. Figure 4 shows the relationship between displacements versus storey level. The difference between displacements of bottom and top storey of bare and infilled frame is 22% and 66% respectively. Figure 5 represents the trends between storey drift and storey level. Bare frame storey drift is 48% more than that of infilled frame. The graph between storey shear and storey level is shown in figure 6. The storey shear of infilled is 83% more than that of bare frame.

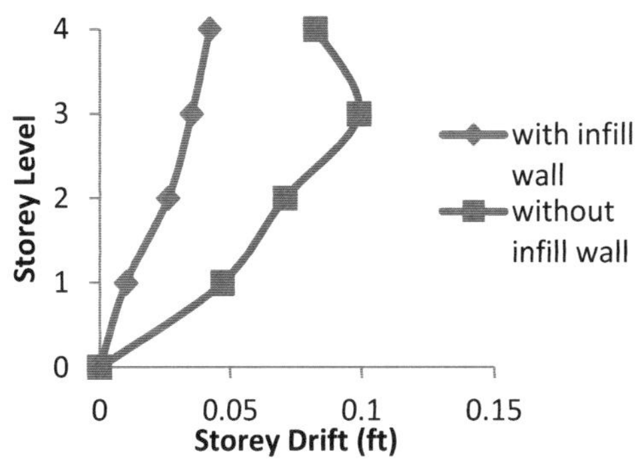

Figure 5. Storey Drift vs. Storey Level

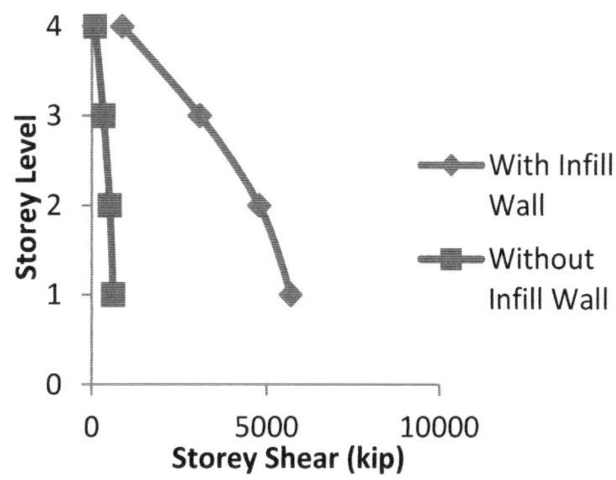

Figure 6. Storey Shear vs. Storey Level

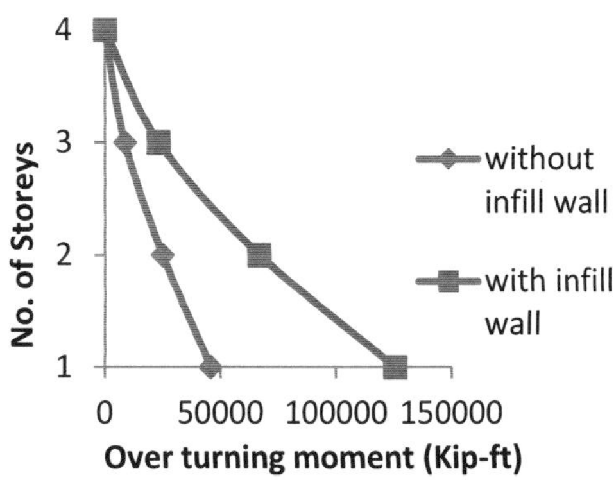

Figure 7. Storey Moment vs. Storey Level

IOP Conf. Series: Materials Science and Engineering **414** (2018) 012017 doi:10.1088/1757-899X/414/1/012017

Figure 7 shows the relationship between over-turning moment and number of storeys. Over-turning moment is maximum on ground storey and minimum on the top storey. There is an increase of 77% of over-turning moment of fully infill frame.

Figure 8 and figure 9 represent the area of steel of bare and fully in-filled frame and it can be seen that there is significant difference in the area of steel. Difference of 22% was observed between the bare frame and in-filled frame due to the braced frame action.

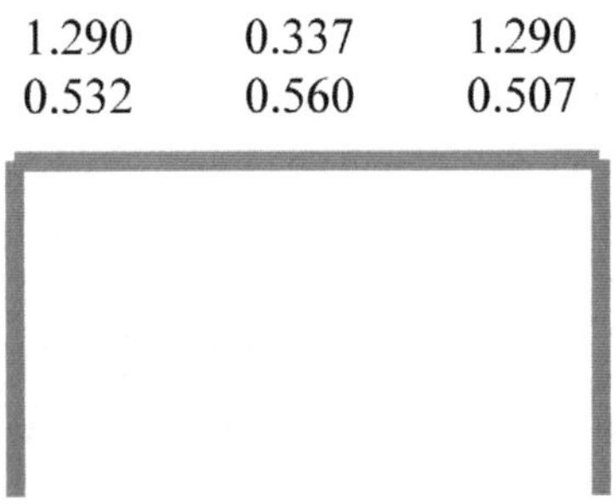

Figure 8. Area of steel in bare frame.

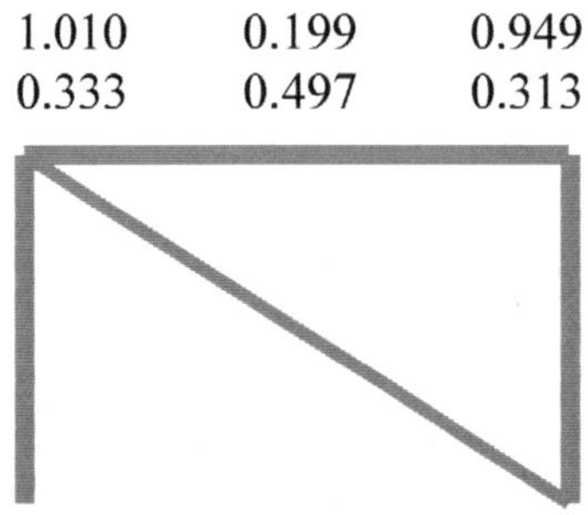

Figure 9. Area of steel in Fully In-filled Frame.

4. Conclusion and Recommendation

The important conclusions of this study are elaborated as: The base shears of fully in-filled frame is 2.5 times more than that of bare frame. This is due to the braced frame action which obviously is stiffer than the bare frame. The displacement of bare frame is more than that of fully in-filled panels. The result shows a difference of 66% on the highest floor, respectively.

Ground over turning moment of fully in-filled frame has higher values than that of bare frame. There is an increase of 77% of over-turning moment of fully infill frame. Storey shear decreases of each storey as the number of floors are increased. Relative storey displacement is the least in fully in-filled frame and for bare frame have higher values of drift. The area of steel in beams in bare frame is more than that of the in-filled frame structure a difference of 22% was observed, it's due to the braced frame action. The structure is more-stiff. As stated in the above-mentioned results, there are significant differences in seismic response of frames with and without in-fill walls. Building analysis and design shows that in-fill walls can also economize the cost of structures as the in-fill walls contribute towards lateral stiffness due to braced frame action. The results also show that area of steel can be reduced. Moreover, due to availability of advanced analysis and design tools it is recommended to include the effect of in-fill walls for determining the realistic behavior as well as to achieve economy.

References

[1] Murty C V R and Jain S K 2000 Beneficial influence of masonry infill walls on seismic performance of RC frame buildings. In *12th W Confr. on EQ Eng*

[2] Wakchaure M R and Ped S P 2012 Earthquake analysis of high rise building with and without in filled walls. *J. EIT* Vol **2**

[3] Girgin K and Darılmaz K 2002 Seismic response of infilled framed buildings using pushover analysis. *Deptt. of Civil Engineering, Istanbul Technical University, 34469.*

[4] Korkmaz K A, Demir F and Sivri M 2007 Earthquake assessment of R/C structures with masonry infill walls. *IJST* **2** 155-64

[5] Mulgund G V and Kulkarni A B 2011 Seismic assessment of RC frame buildings with brick masonry infills. IJAEST **2** 140-47

[6] Mohamed W A E 2012 Parametric study on the effect of masonry infill walls on the seismic resistance of RC buildings. *J. ES* **3** 701-21

[7] Koutromanos I, Stavridis A, Shing P B and Willam K 2011 Numerical modeling of masonry-infilled RC frames subjected to seismic loads. *CS* **11** 1026-37

IOP Conf. Series: Materials Science and Engineering **414** (2018) 012018 doi:10.1088/1757-899X/414/1/012018

Rainwater harvesting, a measure to meet domestic water requirement; a case study Islamabad, Pakistan.

O Rashid[1], F M Awan[1], Z Ullah[1], and I Hassan[1]

[1]Department of Civil Engineering, Capital University of Science and Technology, Islamabad, Pakistan

Email: saami.avon@gmail.com

Abstract. Pakistan is located in the semi-arid region of the world where water scarcity is a major issue. In such areas, rainwater harvesting (RWH) is considered to be the best alternate source of domestic water supply. Rainwater harvesting is being considered as an integral part of the sustainable water management in many parts of the world. Rainfall patterns in arid areas are typically highly variable, both spatially and temporally. Hence, there is always a need to evaluate rainwater harvesting keeping in view the seasonal variations for a specific area. In rainwater harvesting, a mechanism is designed to collect surface runoffs effectively during rainfall events from residential rooftops. In this paper, it has also been analyzed that rainwater has a great potential to be taken as a source of water demands in residential colonies of major cities like Islamabad being our study area. Here, 116 m^2 (five marlas) housing pocket has been considered as a model site for which various water demands have been calculated depending upon residential needs and horticulture requirements of the study pocket. Rainwater from rooftops has been used to reduce freshwater needs of the houses. As a result, residential water demands for fresh water have reduced by almost 22% average.
Keywords. Rainwater harvesting, rooftop collection, residential water demand, Islamabad.

1. Introduction

Pakistan has been facing rapid climate change since last few years and the water resources of the country are at extremely vulnerable state. Water scarcity is the principal issue in most of the developed cities of Pakistan. The Capital Islamabad has been prone to a major migration of people since last decade which incredibly increases the water demand. The city is growing at the rate of 5.7% per year. According to the Capital Development Authority (CDA) [1] the average water demand within the city is 800 Million Liters/day (L/d), whereas average 481 Million L/d water shortage has to be encountered by the city every year. Islamabad relies on Simly dam and Khanpur dam to meet the water demand, tube wells and bores are also a key source of water in the surrounding areas of the capital city. The two major sources contribute 381 Million L/d collectively to meet the water demand. The city has to face acute water shortage in summer and pre-monsoon period. In June 2017 the Simly and Khanpur dams found drying at the rate of 50mm/d, as a result, the water supply was cut down further to 86 Million L/d from 109 Million L/d by the authorities [1]

Inappropriate usage of local resources has also led to water scarcity in the city. Low water prices, little amount of water bills and easy access to water allows the user to misuse water without concerning the vulnerability of water resources. Moreover, unplanned and miserable domestic water supply lines led to the leakages of water. There are less appropriate rules and regulations to monitor the water usage, and less awareness ultimately reduces the supply and increases the demand. According to UN-Habitat report [2] the groundwater level in Islamabad city is depleting at an alarming rate of 1.7 m/year. The groundwater level was observed to drop by 20m in various sectors of the city. This extreme drop links to the amplified extraction of water and reduced percolation of water into the soil because of urbanization and population growth within the city.

IOP Conf. Series: Materials Science and Engineering **414** (2018) 012018 doi:10.1088/1757-899X/414/1/012018

2. Rainwater harvesting

Rainwater Harvesting (RWH), a famous technique that has been practiced for hundreds of years. It is a method for accumulating and saving rainwater from various elements such as rooftops, surface runoff, and other catchments. Most of the time water downpour is so intense that a sufficient amount of water can be saved by simple harvesting techniques for a few months. To ensure a sustainable urban future, society must adapt effectual application behavior towards water usage. RWH has a very contributing role to minimize the risk of scarcity. Furthermore, in an urban environment, RWH is a scheme that carries certain profits and it may help to overcome water scarcity issues, degradation of land and flooding. Harvested water can be used for all domestic purposes like drinking water, cooking and washing etc. Moreover, this harvested water has a potential to meet agricultural and crop's water needs according to the available economy, the storage can be managed either on-surface or sub-surface.

UNESCO [3] reported in 2008 there were one billion people who don't approach protected and safe drinking water. Food and Agriculture Organization (FAO) also demonstrated almost 840 million people still encounters the evil impacts of undernourishment. According to the UN [4] that urban water demand will become double from 2007 to 2050. The climate of Pakistan is uneven and almost two-thirds of the country demonstrate arid type climate. Groundwater table in the Potohar region (Rawalpindi district and its surroundings) has depleted by 116m in the last 30 years as reported by Water and Sanitation Agency (WASA), Rawalpindi [5] Per capita availability of water is running out on a fast pace as it dropped to 850 m^3 in 2013 as compared to 5,300 m^3 in 1951. Pressure on water assets is increasing day by day with developing demand but water resources are getting limited and this expanding demand decreases sources of freshwater [7] Asian Development Bank (ADB) placed Pakistan in red zone declared it as a water-stressed country and is likely to face an acute shortage of water in the next five years [5] In Rawalpindi, precipitation regularly changes in a scope of 900-1000 mm per year but in case of potential evapotranspiration, there is a fluctuation of 1200-1400 mm [8] All these facts and figures emphasize the need to adjust water administration to natural and modern techniques. RWH, among different choices, may assume a focal part for providing freshwater even with increasing water shortage and raising interest.

RWH is a technique in which water is collected from rooftops and then used for flushing, washing and different household purposes. The aim of the study is to apply this technique to meet the daily water demand. The rainwater harvesting has various advantages such as water is free; its direct collection and storage near the place of use eliminates need of distribution system and its maintenance cost; good quality for various applications; it reduces need for fresh potable water for uses other than drinking; it decreases the costs of utility bills and helps to recharge the groundwater [5] RWH decreases the requirement for mains water and rainwater discharge also decreases and henceforth may add to lessening flood dangers and the load on storm drainage and sewer frameworks [9] Burkhard [10] claimed that due to more softening of rainwater, it reduces the usage of washing powders.

"Sant Cugat del Vallès" a town in Spain, RWH from housetop could conceivably meet around 16% of the aggregate residential water demand of that town and in Jordan, RWH saved 0.270 % to 16.6% [11] In the case of south-eastern Brazil, RWH saves 41%, average running from 12% to 80% varies with the study area [12] Hassan [5] suggested [12] that in Rawalpindi, 110 m^2 houses with flat roof have sufficient size to accommodate 2 persons and 4 persons at the most. According to a study on RWH in Amman and Irbid, rainwater meets the WHO standards chemically and physically but when it comes to biological tests, it does not meet with WHO guidelines [13] Chang [14] stated that the quantity of pollutants in roof rainwater depends upon surface, atmosphere and pollutants properties.

According to Alpaslan [15] rainwater quality and its collection efficiency greatly depend upon roof area and construction material and best roof for RWH are tiled and Cement roofs in terms of durability, economy and water quality. Sazakli [7] investigated that water from housetops, by and large, meets the universal rules of drinking water. Helmreich [15] claimed that the quality of rainwater in a rural area is far better than the urban area due to their climate. Kahinda [17] concluded that for RWH, it is not right to expect that the harvested water achieves the drinking water standard.

Fewster [18] stated that there are various choices to achieve rainwater of good quality like the washing of rooftop, avoiding the primary flush from rooftops and chlorine must be used for cleansing of the tank. Slow sand filtration is a modest strategy to enhance the bacteriological nature of water. Wegelin [19] concluded that pasteurization by sun-oriented innovation is otherwise called a cheap disinfection method. This Sunlight technique is good for small families as it only gives us the 100L of water on an area of 1 Square meter of sunlight. According to Kim [20] rapid Sand Filter is also used to remove suspended particles from the water.

Domenech [21] suggested for the single-family system, the tank limit extended from 2 to 20 m^3 and was by and large of 12 m^3. A promising end-use for water which is especially engaging in single-family houses is flushing. The water-powered reenactment display demonstrates that a tank of 17 m^3 can completely take care of flushing demand in a standard single-family house. The most proficient tank for flushing is 11 m^3 which empowers to take care of 97.9% of the demand and can spare 28.4 m^3 of drinking water every year or 25.9 LCD. In the case of a single family, where water utilization per capita is high, water displayed an essential potential to diminish per capita drinking water utilization. If there is a low precipitation in any year, little tanks could take care of 100% of the demand for those utilizations with a steady request, for example, latrine flushing, and laundry requirements. According to Sazakali [7] RWH seems by all accounts, to be a standout amongst the most encouraging choices for providing freshwater even with expanding water shortage and raising interest.

RWH technique is very commonly used in the Northern part of Pakistan where the climate is very humid and often rainy. Residents of Northern cities of Khyber Pakhtunkhwa like Nathiagali, Ayubia, Batgram, some parts of Abbottabad and Mansehra rely on RWH as the main source of portable water in these cities are from perennial streams and seasonal streams. The roofing system of houses in these areas are sloping, made of steel sheets which minimize the absorption and other roof losses. In said areas, the maximum amount of water could easily be harvested without any additional loss. The system is designed as the harvested water enters a primary tank (flash tank) first where sedimentation takes place and clean water enters into the secondary tank from where water is drawn for domestic use. Moreover, in some remote areas of Azad Jammu and Kashmir RWH is the main source for domestic water utilization. In these hilly areas like Khabli, Ghari Dupatta, Peer Hasimaar and Majhoi, there is no proper municipal water supply system and due to hard soil strata, bore-holes are near to impossible. The residents of these areas face difficulty in getting water to meet their daily requirements. Due to the economic status of these areas people cannot afford water tankers from private dealers as the cost is too high for them to afford. In order to meet their water need, they mostly rely on RWH.

The major concern of municipal authorities of Islamabad is the speedy decline of the groundwater table which is an alarming threat to the future of the city. Increasing the growth rate of the city's population and deforestation are setting new environmental challenges for the municipal authorities. This study aims to promote an alternative source of water for the inhabitants so that the dependency on conventional sources could be minimized. RWH system proved to be an efficient system to meet the partial water demand of inhabitants plus it minimizes the reliance and conventional water sources. This study would also suggest the sectors where harvested water could be utilized effectively.

3. Study area

The study area is in the vicinity of Islamabad (Zone V) and it is highlighted in figure 1. This site is accessible only from Islamabad Highway through Japan Road at a distance of 6.1 Km's. The site is surrounded by Gulberg Greens, Japan Road, Naval Anchorage Block F and OPF Housing Society from North, South, West, and East respectively. This site consists of 500 houses. The plot size for each house is 116 m^2 and effective roof size of each house is 94 m^2 as by building bye-laws.

4. Water availability survey and demand

A reconnaissance survey has been conducted in the surrounding of the study area in a village named "Dadoo Maira" which consists of 50-60 houses. The aim of this survey is to know about the condition of water sources. Figure 1 shows the localities which were surveyed to find various facts about water in the surroundings of the study area.

Statistics show a shortage of water in the surrounding areas which ask for development of RWH as an alternative source of water to meet the water demand of the society. Different questions that were asked from the people about the source of their water supply, water depth, water level, change in water level in past few decades, problems about their water demand and quality of water. The water availability survey data is presented in the table 1.

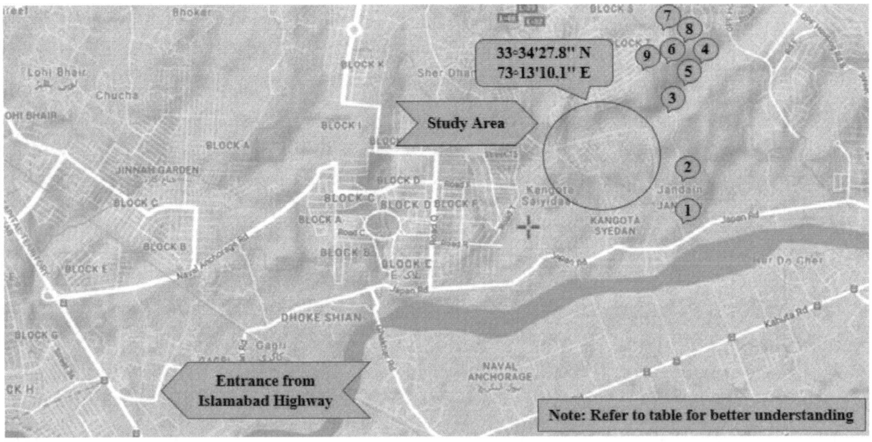

Figure 1. Location map of the study and survey Area.

Table 1. Survey Results.

No	Name of Person	Main Source	Village Daado Maira			Availability issue during	Quality
			Water Level (m)	Depletion in Summer (m)	Depletion in past decades (m)		
1	Shabbir	Bore	52	Dry	12-15	Full Year	Good
2	Abdul Irfan	Bore	49	15-18	14-15	Summer	Good
3	Raja Abrar	Bore	43	6	10-14	Summer	Good
4	Imran	Bore	37	1.5-3	10-12	Summer	Good
5	M. Saeed	Bore	61	Dry	17	Full Year	Good
6	Mujtaba	Bore	55	21-24	13-17	Summer	Good
7	M. Aslam	Bore	46	15	14-17	Summer	Good
8	M. Aslam	Bore	20	00	6-8	Summer	Good
9	Tahir	Bore	33	18-20	9-12	Summer	Good

4.1 Water demand analysis

According to CDA [23] water requirement per capita per day is 227 LPCD. When we apply this figure to our 116 m^2 (5 marlas) house having 4 people living in it, what we got for our monthly water demand for a single house is 27,240 L and 326,880 L for the whole year. Half of this demand is used for the summer season and another half for the winter season as roof absorption losses are different for each season. Table 2 shows the water demand of a single house having 4 persons in it.

Table 2. Water demand for a single house having 4-person accommodation.

Water demand	Quantity (liters)
CDA Standard	227 Liters/person/day
Daily	908
Monthly	27,240
Seasonally (6 Months)	163,440
Yearly	326,880

5. Weather data analysis

In this section we will discuss and analyze the precipitation, and evaporation data. The number of wet days, effective rainfall losses, and roof absorption losses is also discussed in the following paragraphs.

5.1 Precipitation data analysis

Precipitation data was taken from the World Bank Group's official website [22] suggest an average yearly precipitation of 1,063 mm which is sufficient to harvest down water and to use it in many effective ways. Minimum precipitation was observed in the months of October and November whereas maximum was in the months of July and August with an average of 89 mm per month. For a 116 m^2 (5 marlas) house with an effective roof size of 94 m^2, total rainfall generated during a year is 98,700 L which is more than 30% of yearly water demand. Figure 2 shows average monthly precipitation that Islamabad receives per year. This data will be used to carry out a study to find out how much rainwater can be harvested and how much daily water requirements can be fulfilled using that harvested rainwater.

5.2 Calculation of evaporation losses

As per Nestle Pakistan report [8] the average evapotranspiration per day has been picked which was further processed to calculate average evaporation per day. According to Herbst and Lodger [24] determined evaporation loss as 35.65% of evapotranspiration.

Jasechko S [25] worked out evaporation loss as 42.5% in evapotranspiration. This study is based on taking average 40% evaporation losses from the roof. On the basis of these values, it is estimated that yearly evaporation losses are 11,059 L of rainfall. The average evaporation per day for each month of a year is presented graphically in figure 3.

5.3 Calculation of wet days

A number of average wet days in Rawalpindi region are collected from the weather sparks website [26] The minimum wet days are 2 in the month of November and maximum is in the month of August that is 14 with an average of 8 days. These wet days were multiplied with evaporation losses of that particular month then subtracted from the average precipitation of that month. All data about wet days is shown in figure 4.

5.4 Roof absorption losses

Alpaslan N [15] claims that the roof losses during an event of a rainfall are up to 20% of the total downpour. This study carries an average 12.5% (10%-15%) for winter season and 17.5% (15%-20%) for summer season [15] The winter span is from October – March and summer span is from April –

IOP Conf. Series: Materials Science and Engineering **414** (2018) 012018 doi:10.1088/1757-899X/414/1/012018

September in Pakistan. On the basis of facts explained above, roof absorption losses for a year are 15,490 Liters. Average roof losses percentage is shown in figure 5.

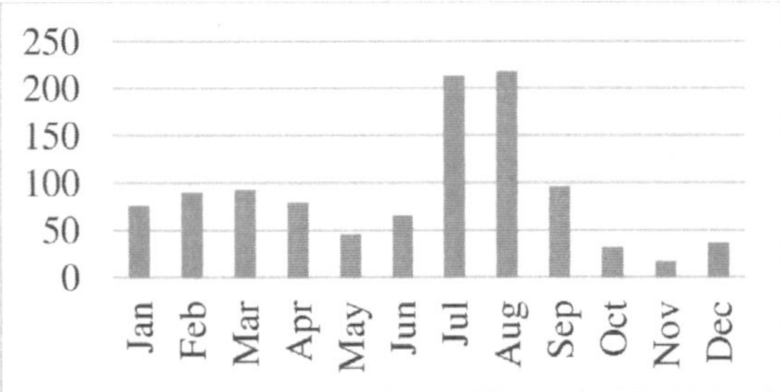

Figure 2. Average rainfall per month (mm).

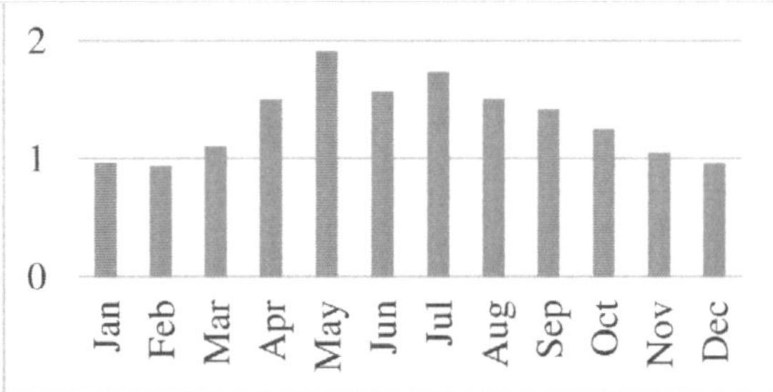

Figure 3. Average Evaporation per day (mm).

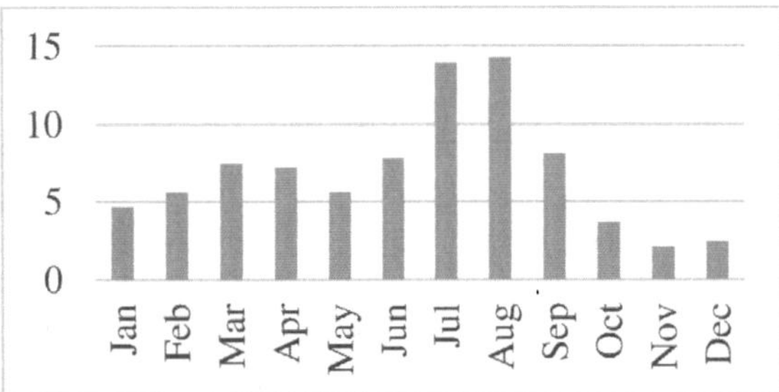

Figure 4. Average wet days in a month.

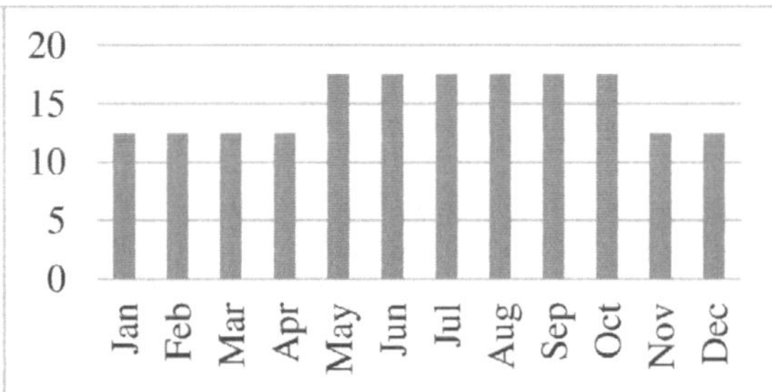

Figure 5. Average roof losses percentage.

5.5 Effective rainfall analysis

Evaporation losses and roof absorption losses are subtracted from the actual amount of precipitation to get the effective precipitation that will generate the water flow. Effective average precipitation per day is shown in the figure 6.

For the plot size of 116 m^2 (5 Marla) with an effective roof size of 94 m^2 with evaporation losses and roof absorption losses, the amount of rainwater that can be harvested for each month is shown in figure 7. The total amount of rainwater that can be harvested is 72,154 Liters.

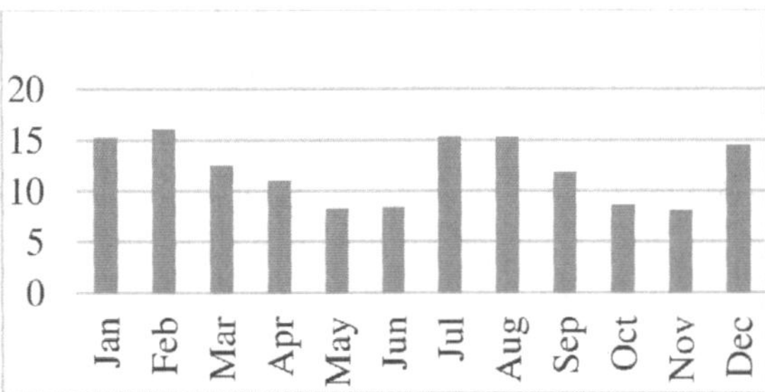

Figure 6. Average effective rainfall per day in a month (mm).

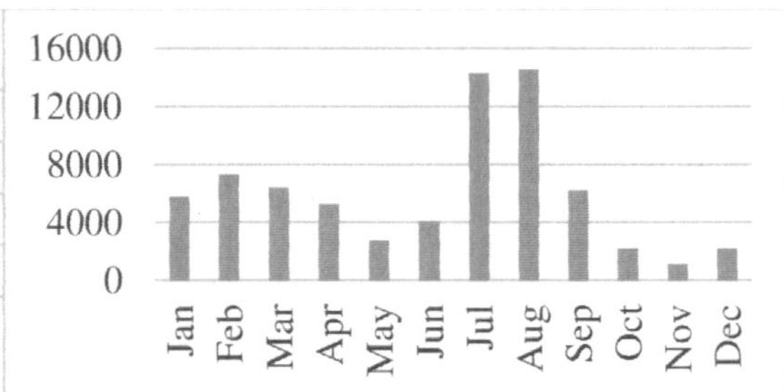

Figure 7. Harvested Rainwater (liter per month).

6. Result and discussion

According to CDA, the yearly water demand for a single house having four people accommodations is 326,880 L as given in table 2, and the maximum amount of water that can be saved using RWH is 72,154 L which is more than 22% of the yearly demand.

By considering all relevant factors like roof absorption losses, evaporation losses, and system inefficiencies, RWH still meets the 22% of the yearly water demand in this specific area. For water bifurcation, the percentage for different uses are shown in figure 8 from which quantity of water for different purposes is calculated. By utilizing water demand bifurcation, yearly water demand and yearly harvested rainwater are bifurcated into four different purposes. The drinking and cooking is 4%, black water is 16% of total water demand. The grey water demand is 36% while demand for all other usage is 44% of the total water demand. The amount of rainwater that can be used for different purposes like cooking and drinking, black water for flushing and the grey water for laundry, and washing is illustrated in the table 3.

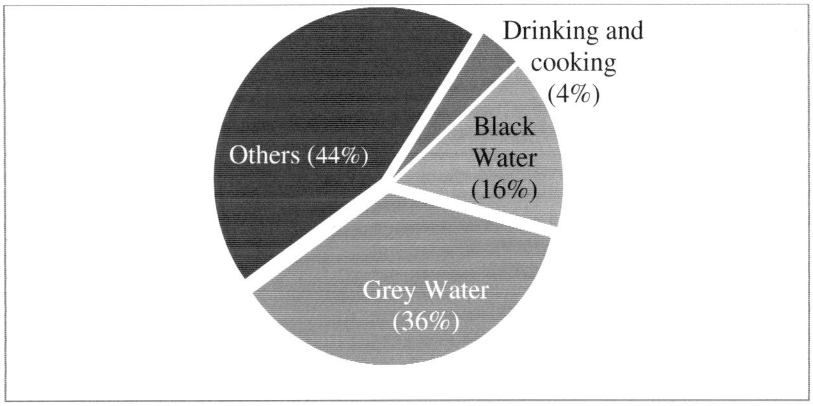

Figure 8. Water demand bifurcation.

For blackwater usage like flushing, rainwater can meet 100% requirement annually and for grey water usage after appropriate treatment, rainwater can meet almost 17% of its requirements annually for laundry, washing floors, in the kitchen etc. For the months of July and August.

Rainwater can meet more than 50% of monthly water demand in which black water and grey water usage can be 100% accommodated at the same time. But in the months of May, June, and October to December, harvested rainwater is not enough even for flushing so we have to save some water from previous two months to meet grey water demand here. The monthly harvested rainwater analysis is presented in table 4.

Table 3. Yearly water usage.

Purpose / Use	Water demand (L)	Harvested Rainwater used (L)	Percentage met by RWH	Remarks
Drinking and Cooking	13,075	0	0%	Not suitable
Black Water	52,300	52,300	100%	Suitable without any treatment
Grey Water	117,677	19,853	17%	Suitable with treatment

Purpose / Use	Water demand (L)	Harvested Rainwater used (L)	Percentage met by RWH	Remarks
Others	143,827	0	0%	-
Total for Single House (4 persons)	326,880	72,154	22%	-

Table 4. Monthly harvested rainwater analysis

Monthly	Jan	Feb	Mar	Apr	May	Jun	Jul	Aug	Sept	Oct	Nov	Dec
Harvested RW (L)	5757	7355	6429	5298	2738	4045	14269	14514	6247	2184	1126	2193
Water demand (L)	27240	27240	27240	27240	27240	27240	27240	27240	27240	27240	27240	27240
Water demand % meet by RWH	21.2	27.0	23.6	19.5	10.1	14.9	52.4	53.3	22.9	8.0	4.1	8.0
Flushing demand % meet by RWH	100	100	100	100	62.8	92.8	100	100	100	50.1	25.9	50.3

7. Conclusion

Rainwater Harvesting system incorporation at the household level is the need of the hour to meet water demand for sustainable development even in the capital city of Islamabad. The same concept should be studies in different cities and necessary amendments in building bye-laws be made to make it compulsory to construct a RWH system.

References

[1] Shahid J and Abbasi K 2017. *Dawn Newspaper.* Karachi, Pakistan
[2] United Nations Human Settlements Program 2018. HS Number: HS/100/14E ISBN Number (Series) 9789211324006 ISBN Number 9789211326383
[3] United Nations Educational Scientific and Cultural Organisation, Official Homepage 2008 *www.unesco.org*
[4] The UN 2009 World water development report 3: water in a changing world. *WWAP Paris/London UNESCO Earthscan*
[5] Hassan I 2016 Rainwater Harvesting-an alternative water supply in the Future for Pakistan. *J. Bio. Envi. Sci.* **8** 213-22
[6] Fletcher T D, Deletic A, Mitchell V G and Hatt B E 2008 Reuse of urban runoff in Australia: a review of recent advances and remaining challenges. *J Environ Qual* **37** 116–27
[7] Sazakli E, Alexopoulos A, and Leotsinidis M 2007 Rainwater harvesting, quality assessment and utilization in Kefalonia Island. *Greece. Wat. Res.* **41** 2039–47
[8] Nestlé Waters Pakistan 2016 Islamabad Water Factory Constructive critical review of hydrogeological reports. *International Division Immeuble Axeo - 29 avenue Aristide Briand - CS 10006 94117 Arcueil cedex FRANCE* **A83640C**. Islamabad, Pakistan.
[9] Environment Agency 2010 Harvesting rainwater for domestic uses: an information guide. *Rio House Waterside Drive.* Bristol BS **32** 4UD
[10] Burkhard R, Deletic A and Craig A 2000 Techniques for water and wastewater management:

a review of techniques and their integration in planning. *Urb.* **2** 197-221
[11] Abdulla F A and Al-Shareef AW 2009. Roof rainwater harvesting system for household water supply in Jordan. *Des.* **243** 195–207
[12] Ghisi E, Bressan L and Martini D M 2007. Rainwater tank capacity and potential for potable water savings by using rainwater in the residential sector of southeastern Brazil. *Build. Environ.* **42** 1654–1666
[13] WHO 1993 Center of environmental health activities guidelines on technology for water supply systems in small communities. *Amman. CEHA* D. No. TLM-**05**
[14] Chang M, McBroom M W and Scott B R 2004 Roofing as a source of nonpoint water pollution. *J. Envir.* **73** 307–15
[15] Alpaslan N, Harmancioglu N B and Singh V P 1992. *J. IAH* **15** 1–13
[16] Helmreich B and Horn H 2008 Opportunities in rainwater harvesting. *J. Desa* ***248*** 118-24
[17] Kahinda J M, Akpofure E and Taigbenu J R B 2007. *Domestic rainwater harvesting to improve water supply in rural South Africa.* Physics and Chemistry of the Earth **32** 1050 – 1057
[18] Fewster E, Mol A and Wiessent-Brandsma C 2004 The longterm sustainability of household bio-sand filtration. *Proceedings of the 30th WEDC International Conference Vientiane Laos PDR.* 1–3
[19] Wegelin M, Canonica S, Mechsner K, Fleischmann T, Pescaro F and Metzler 1994 A solar water disinfection: scope of the process and analysis of radiation experiments, *J. W S R Technol.* **43** 154–69
[20] Kim R H, Lee S and Kim J O 2005 Application of a metal membrane for rainwater utilization: filtration characteristics and membrane fouling. *J. Desa.* **177** 121–32
[21] Domènech L and Saurí D 2011 A comparative appraisal of the use of rainwater harvesting in single and multifamily buildings of the Metropolitan Area of Barcelona (Spain): social experience, drinking water savings and economic costs. *J. Cle. Prod* **19(6-7)** 598-608
[22] The World Bank in Pakistan 2012 Open Data on Pakistan. *www.worldbank.org/en /country/Pakistan*
[23] Capital Development Authority 1993 Modalities and Procedures. Islamabad, Pakistan
[24] Herbst M and Kappen L 1999 The ratio of transpiration versus evaporation in a reed belt as influenced by weather conditions. *Aquatic Botany* **63(2)**:113-25
[25] Jasechko S, Gerrits A M J C, Van R J, Bogaard T A, Erlandsson L W, Hrachowitz M and Savenije H H G 2013 Uncertainties in transpiration estimates. *Nat.* **496** 347–50
[26] Weather Spark 2018 Average weather in Islamabad Pakistan-Year-Round. www.weatherspark.com/y/107761

IOP Conf. Series: Materials Science and Engineering **414** (2018) 012019 doi:10.1088/1757-899X/414/1/012019

Development of nanofibers based neuropathic patch loaded with Lidocaine to deal with nerve pain in burn patients

S Abid[1], T Hussain[1], A Nazir[1], N Khenoussi[2], A Zahir[1] and S Riaz[1]

[1]National Textile Research Center, National Textile University, Pakistan
[2]Université de Haute-Alsace, Laboratoire de Physique et Mecanique Textiles, France
E-mail: m.sharjeel.abid@gmail.com

Abstract. Burns wounds are difficult and different to treat when compared with other wounds. The management of burn wounds is divided into three main categories; pain management, infection management and healing. Various commercial products are available to treat and prevent infection in burn patients but, for the management of pain, intravenous (IV) route is preferred which is associated with different side-effects. The local release of analgesic agent for nerve pain can reduce the IV related side effects and can provide quick and effective nerve pain management in burn patients. In this study, electrospun nanofibers of sodium alginate/PEO were loaded with lidocaine to reduce nerve pain and the effect of parameters were studied to get optimized bead free nanofibers. The drug release was tunable (from minutes to hours) and other properties like liquid absorption were studied against distilled, saline and solution A. The combination of moist environment and strong nerve pain inhibitor could be salient features to be used as contact layer for burn wound as well as the use of antidepressant drugs could be skipped.

1. Introduction

To deal with burn wounds is critical and tricky as compared with other wounds [1]. They are broadly categorized as first, second and third-degree wounds depending on the depth of damage [2]. The major problem is the destruction of the obstacle "skin" which is a protective barrier for pathogens, therefore, the burn patients are much more favorable to different microbes [3]. Thus, major focus of medical practitioners and researchers is to develop different tools that may aid in infection prevention. Whereas, first approach is to reduce the pain associated with burns which in turn depends on the degree of burns. For example, first degree burns are superficial and less painful, while, the second and third-degree burns require proper analgesic and pain killers to reduce the pain due to the damaged nerves ends reside in dermis and subcutis. The changing pattern of pain as well as the patient's response to pain requires adequate monitoring for proper analgesic dosages [4].

Lidocaine, also known as lignocaine or xylocaine, is known as an effective analgesic drug; being used for over half century [5]. It has been used extensively either by IV route or as topical gel [6]. A patch loaded with 5% lidocaine has been reported in literature to provide peripheral analgesic action with improved pain relief as per clinical study [7]. Commercial available lidocaine patch is known as Lidoderm which uses 5% active drug attached with polyester felt as backing materials and polyethylene terephthalate as release liner. On the other hand, nanofibers are getting diverse attention in almost every field of life due to the exceptional properties they offer [8]. They are being used in medical and other industries as an effective tool for various applications [9]. Polyethylene oxide (PEO) is a synthetic hydrophilic but biocompatible polymer approved by FDA to be used for various medical applications including the drug delivery [10]. Sodium alginate (SAlg) is a natural biopolymer known to provide favorable environment for wound healing and a good drug delivery tool [11-13].

Thus, the purpose of present study is to fabricate lidocaine loaded nanocomposite nanofibers of PEO-SAlg that can provide localized analgesic effect, appropriate environment for healing and avoid the other IV related side effects.

2. Materials and Methods

2.1 Materials

Sodium alginate, Triton X-100 and PEO were purchased from Sigma-Aldrich. Lidocaine was kindly donated by Saffron Pharmacy Private Ltd.

2.2 Solution Parameters

Sodium alginate and PEO solutions were made in double distilled water in separate container at 400 rpm and 80 °C for 24 hours to get homogenous solutions. The electrospinning solution was made by adding 20 parts of SAlg (2% w/v) and 80 parts of PEO (5% w/v). To reduce the surface tension of the final solution, 1% Triton X-100 was added and stirred for 4 hours at room temperature. Then, lidocaine (5% OWF) was added in the above solution and stirred for further 5-6 hours at 400 rpm for complete blending of drug. The final viscosity of the solution was about 5200 cP.

2.3 Electrospinning

Electrospinning was done in needle electrospinning machine Fluidnatek LE-100 Bioinica, Spain. Flow rate of 200-300 μl, voltage range from 6 to 12 kV. The syringe was connected with a positive voltage and collector was grounded. The voltage range mentioned earlier was used to get stabilized Taylor cone so that smooth and bead free fibers could be collected. The fibers were collected on an aluminum sheet and separated (peeled) for further testing.

2.4 Surface Morphology of Nanofibers

The resultant nanofibers were sputter coated with gold and then analyzed using SEM-FEI Quanta 250 series machine equipped with EDX. The diameter of the nanofibers was measured by built in scale at various points on fibers and the mean diameter was reported.

2.5 Fourier Transform Infra-Red Spectroscopy and XRD

The drug loaded and without drug nanofibers were characterized by FTIR (Spectrum two-Perkin Elmer, USA) within scanning range of 600-4000 cm^{-1}. Samples were placed in a desiccator prior to analysis to remove moisture from the sample. The nanofibers were characterized using XRD (PANalytical X'pert Pro, Netherland) analysis for the change in crystallinity of the nanofibers after the incorporation of drug.

2.6 Liquid Absorption and Drug Release

The liquid absorption of nanofibers was tested against solution A, saline and distilled water. Drug release of produced nanofibers was analyzed at specific wavelength which was determined earlier by preliminary experiments using UV-VIS-NIR (Perkin Elmer, Lambda 950, UK). Then, fibers were immersed in phosphate buffer solution at 7.4 pH and after specific interval 2ml of solution was taken and analyzed in UV-VIS-NIR.

3. Results and Discussions

3.1 Morphology of produced Nanofibers

The selected blended of polymer produced smooth fibers at all selected electrospinning process parameters. The average size of nanofibers produced without drug was around 271 ± 51 nm.

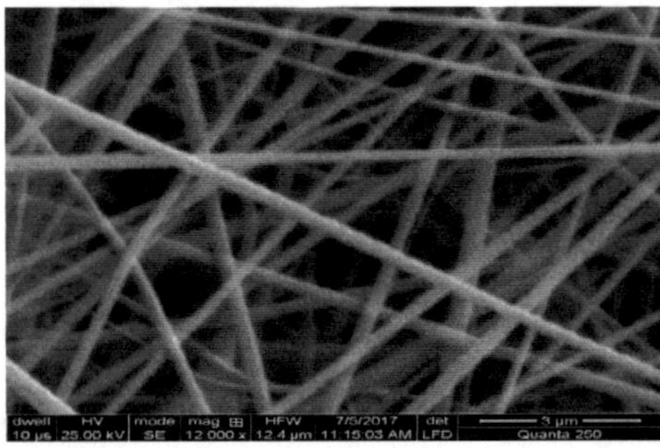

Figure 1: Nanofibers without Drug

Figure 2: Nanofibers with Drug

When the drug was incorporated, the average diameter of resultant nanofibers was 242 ± 55 nm. The SEM images of the produced nanofibers with and without drug loading can be seen in figure 1 and 2, respectively. The reduction in average diameter could be attributed to the changed solution properties after the addition of drug in dope solution. It was observed that the viscosity of the final solution after the addition of drug was reduced slightly. This reduction in viscosity is known to be one of the factors for diameter reduction in nanofibers.

3.2 XRD and FTIR Analysis

X-ray Diffraction is an effective technique to evaluate the crystal size as well as crystallinity of the polymer material. XRD patterns of drug and without drug loaded nanofibers are shown in figure 3. Characteristic peaks at 19.6° and 23.5° exhibits PEO/Alginate within the nanofibers. However, additional sharp peaks at 28.8° and 41° represent Lidocaine drug within the nanoweb [14]. Crystal size of the nanofiber web was determined by Deby-Scherrer formula described in eq. 1. [15].

$$D = \frac{C\lambda}{\beta cos\theta} \quad (1)$$

Where C is the Deby-Scherrer constant, λ is wavelength of the x-ray (K-Alpha = 1.504 Å), β is the full widh of half maximum and θ is the Braggs diffraction angel. With the loading of Lidocaine drug, the crystal size of the Nanofiber web is reduced to 64 Å and 58 Å from 140 Å and 90 Å respectively. Loading of drug within the core of the nanofiber web removes free spaces between the polymeric chains of the PEO /Alginate thus, reduces the crystal size of the nanofiber web and increases its crystallinity [16].

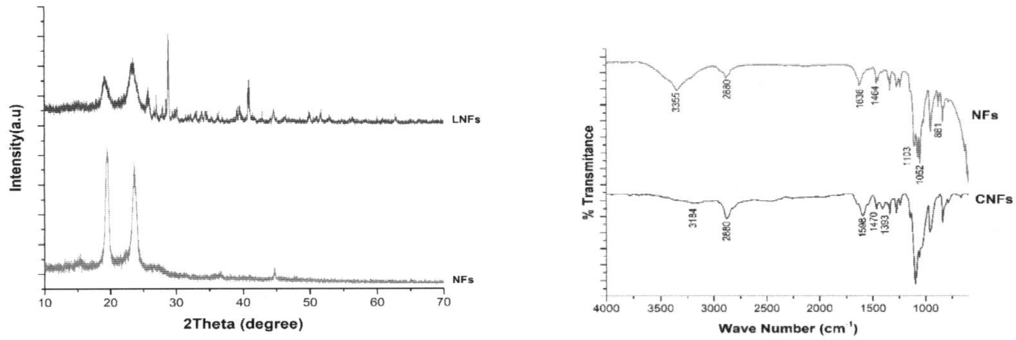

Figure 3: XRD pattern of Nanofibers (NFs) and Lidocaine loaded Nanofibers (LNFs)

IR spectra of PEO/Alginate nanoweb with and without Lidocaine are shown in figure. For PEO/sodium Alginate, broad shallow characteristic peak at 3355 corresponds to -OH, peak at 2880 represents -CH stretching within the pyranose ring. Sharp transmittance peaks at 1636 and 1469 is attributed to asymmetric and symmetric C=O stretching vibrations respectively [16, 17]. Loading of Lidocaine drug within the nanofiber web showed prominent changes in IR spectra as can be seen in figure. Transmittance peak 1470 cm^{-1} is due to joint bend vibration of O-H and C-H, shifting of peak from 1636 to 1598 is due to the C-N stretching.

3.3 Liquid Absorption and Drug Release of Nanofibers

The nanofibers with and without drug were tested against solution A, saline and distilled water and showed changed liquid absorption capacity. Both fibers showed high liquid absorption for saline solution, followed solution A and then the distilled water. Without drug nanofibers showed better liquid absorption capacity than drug loaded nanofibers as shown in figure 4.

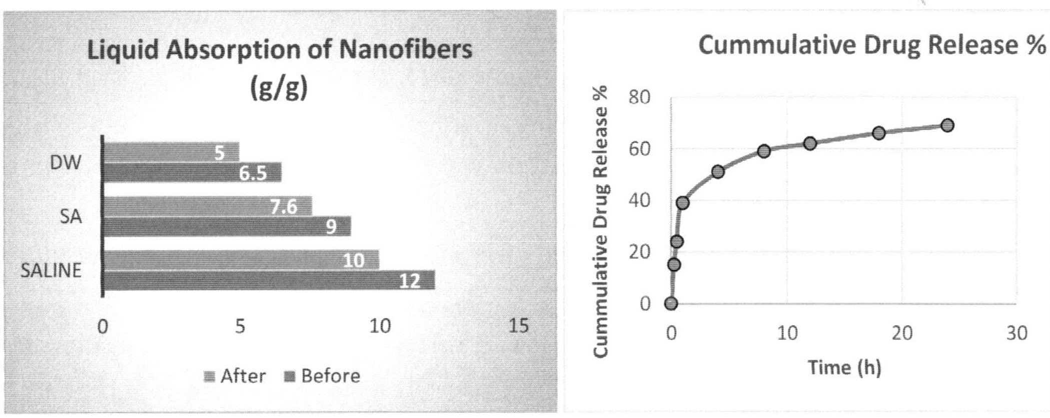

Figure 4: Liquid Absorption and Drug release of Nanofibers

Though, the liquid absorption was lowered a bit, but still, these fibers could hold a high amount of liquid and thus, are suitable to be used for wounds with exudate. The produced nanofibers showed less absorption as compared to our previously reported results in the case of alginate/PVA nanofibers, but, this could be attributed to the composition of co-polymer [18]. The produced nanofibers showed burst release of drug within first hour and this could be beneficial for effective analgesic effect as per requirement for pain management. It was observed that around 60% drug was release within 10 hours which was due to the solubility of PEO in water.

4 Conclusion

Lidocaine loaded defect free nanofibers of about 242 nm were successfully developed by electrospinning technique. The produced fibers showed good liquid absorption as well as desirable drug release for effective swear nerve pain management in burn wounds. FTIR analysis of nanofibers confirmed the incorporation of lidocaine drug in the produced nanofibers. The XRD analysis showed increased crystallinity of drug loaded fibers. The overall results confirmed that the promising results of produced nanofibers can make them a choice to deal with burn wounds for pain relief.

References

1. Tiwari VK 2012 Indian Burn wound: How it differs from other wounds? *J. Plast. Surg.* **45** 364–73

2. Church D, Elsayed S, Reid O, et al 2006 Burn wound infections *Clin. Microbiol. Rev.* **19** 403–34

3. Jr. B A P, McManus A T, Kim S H, Goodwin CW 1998 Burn Wound Infections: Current Status. *World J. Surg.* **22** 135–145

4. Castro R J A de, Leal P C, Sakata R K 2013 Tratamento da dor em queimados. *Rev. Bras. Anestesiol* **63** 154–158

5. Eipe N, Gupta S, Penning J 2016 Intravenous lidocaine for acute pain: an evidence-based clinical update. *B.J.A. Educ.* **16** 292–298

6. Wolf D, Otto J 2015 Efficacy and Safety of a Lidocaine Gel in Patients from 6 Months up to 8 Years with Acute Painful Sites in the Oral Cavity: A Randomized, Placebo-Controlled, Double-Blind, Comparative Study. *Int. J. Pediatr.* **2015** 1–6

7. White W T, Patel N, Drass M, Nalamachu S 2003 Lidocaine patch 5% with systemic analgesics such as gabapentin: a rational polypharmacy approach for the treatment of chronic pain. *Pain Med.* **4** 321–30

8. Vasita R, Katti D S 2006 Nanofibers and their applications in tissue engineering. *Int. J. Nanomedicine* **1** 15–30

9. Kenry, Lim C T 2017 Nanofiber technology: current status and emerging developments. *Prog. Polym. Sci.* **70** 1–17

10. Ma L, Deng L, Chen J 2014 Applications of poly(ethylene oxide) in controlled release tablet systems: a review. *Drug Dev. Ind. Pharm.* **40** 845–851

11. Lee K Y, Mooney D J 2012 Alginate: properties and biomedical applications. *Prog. Polym. Sci.* **37** 106–126

12. Tonnesen H H, Karlsen J 2002 Alginate in Drug Delivery Systems. *Drug. Dev. Ind. Pharm.* **28** 621–630

13. Sosnik A 2014 Alginate Particles as Platform for Drug Delivery by the Oral Route: State-

of-the-Art. *ISRN. Pharm.* **2014** 926157

14. Hu C, Gong R H, Zhou F L 2015 Electrospun Sodium Alginate/Polyethylene Oxide Fibers and Nanocoated Yarns. *Int. J. Polym. Sci.* **2015** 1–12

15. Zahir A, Aslam Z, Kamal M S, et al 2017 Development of novel cross-linked chitosan for the removal of anionic Congo red dye. *J. Mol. Liq.* **244** 211–218

16. Li H, Pu X, Yin J, et al 2016 Effect of Crystallite Size and Crystallinity on Photoluminescence Properties and Energy Transfer of Y_6MoO_{12}:*Eu. J. Am. Ceram. Soc.* **99** 954–961

17. Daemi H, Barikani M 2012 Synthesis and characterization of calcium alginate nanoparticles, sodium homopolymannuronate salt and its calcium nanoparticles. *Sci. Iran.* **19** 2023–2028

18. M. Rafiq, T. Hussain, S. Abid, A. Nazir, and R. Masood 2018 Development of sodium alginate/PVA antibacterial nanofibers by the incorporation of essential oils. *Mater. Res. Express*, **5** 035007

Development of antibacterial fibers and study on effect of guar-gum addition on properties of carboxymethylcellulose (CMC)/alginate fibers

S Riaz*[1], S Malik[1], T Hussain[1], M Ashraf*[1], F Iftikhar[1], A Younus[2], S Abid[1] and A Zahir[1]

[1]Functional Textiles Research Group, National Textile University, Sheikhupura road, Faisalabad (37610), Pakistan
[2] Physics department, University of Agriculture Faisalabad, Jail road, Faisalabad (38000), Pakistan

*Corresponding Authors E-mail: munir.ashraf01@gmail.com, shaguftariaz84@gmail.com

Abstract. The effect of carboxymethylcellulose (CMC) addition on the tensile and absorption properties of alginate fibers had been reported in the previous research. Great increase in absorbency was observed due to addition of CMC to calcium alginate fibers but was not balanced with mechanical strength. By addition of CMC dry and wet strength of composite fibers was reduced, therefore, a balance was needed between the absorbency of the fibers and mechanical strength. The effect to introduce another component i.e. Guar-gum on overall mechanical and absorbency characteristics is reported here. In comparison to CMC/alginate co-spun fibers, it was shown that inclusion of guar-gum improves the tensile strength as well as absorbency of composite fibers. CMC/guar-gum/alginate fibers were extruded through two different coagulation baths. In first step the dope solution was extruded through $CaCl_2$ coagulation bath. The dope of optimized composite fiber formulation with adjusted mechanical property and absorption was extruded, then, through 50:50 $CaCl_2$ and $ZnCl_2$ coagulation bath. The formed CMC/guar-gum/Zn/alginate composite fibers were found excellent for medical applications with their super absorbency along with equivalent mechanical strength and high antibacterial efficacy.

1. Introduction

Seaweeds have been considered a rich source of polysaccharide because of their durability, reliability, availability, and novel applications with high efficiency. Various research studies have been conducted owing to their biocompatibility, biodegradability and eco-friendliness. Alginate from brown seaweed has been used in biomaterials for last many years due to its unique properties. It forms gel by exchanging its calcium ion with sodium ions of the blood. Gel forming property depends upon the quantity of β-D-mannuronic acid and α-L-guluronic acid [3]. Its swelling, hemocompatibility and hemostatic properties make it ideal for wound care materials [1- 4]. In wound management, alginate dressings are widely accepted and most prevalent [5], because, they are highly absorbent due of gel formation [6], hemostatic due to calcium ions exchange with fluid of wound and good in fluid handling permitting better moisture and oxygen transfer. These essential attributes make alginate best choice for highly exuding wounds [3], [4].

Wound occurs due to epithelial cell destruction of skin by some external agent. Skin has inherent ability to reconstruct this tissue discontinuity at wound site described as innate wound healing property that is a complex and vital natural response [7]. The prime objective is to heal the wound in minimum possible time without infecting it that occurs when the number of pathogens at wound site increases than immune system. Usually infected wound is responsible for delayed wound healing process [8]. Mostly, skin pathogens like *Staphylococcus aureus, Escherichia coli* and *Pseudomonas aeruginosa* are responsible for such infection [9]. Absorption, tensile strength, healing, antibacterial activity and protection of wound from toxicity are main functions of wound care material [10].

The properties of alginate dressings can be enhanced by the addition of non-alginate polymers. Quin and gilding [11] made fibers by using alginate and CMC. The concentration of CMC determines some of the properties such as solubility, swelling, chemical and bacterial resistance due to their entrapment in gel. The major problem in the alginate and CMC co-spun fibers was the high absorbency but not balanced with tensile strength. This issue was resolved by addition of guar gum. Guar gum also know by name of cluster bean, exhausted from refined endosperm of cluster bean is a natural nonionic polysaccharide which is soluble in water. It has the capability to change the physical and chemical properties and to form gels [12]. So, it can be used in wound care materials for the absorption of exudate [13] and to increase the strength of composite fibers. The aim of the study was to obtain optimum values of absorbency of composite fibers without compromising the mechanical properties.

2. Material and Method

2.1. Materials

Sodium alginate (protnal LF 60/10, guluronic acid content 65-75%) supplied by FMC biopolymer,Norway. Zinc chloride $ZnCl_2$ (UNI.Chem, lab grade extra pure), Guar gum and CMC of commercial grade were purchased from local market. Calcium chloride dihydrate ($CaCl_2.2H_2O$, RdH Laboratories GmbH & Co, analytical grade), Acetone (99%, Sigma Aldrich), and Sodium Chloride (NaCl, reagent grade UNI.Chem.).

2.2. Methods

2.2.1. Fiber Production. A multifunctional wet extruder ESM 4480 designed in house by NTRC and manufactured by PACE electronics Pvt. Ltd had been used for fibers production. Solution of CMC/guar-gum/alginate was prepared by 3-4 hour stirring of different concentrations of each biopolymer in distilled water to make it uniform at room temperature as shown in table 1.

Table 1. Design of experiment for fiber formation.

Sr #	Na Alginate (%)	CMC (%)	Guar gum (%)
S1	2.5	0.5	1
S2	2.5	0.75	1
S3	2.5	1	1
S4	3	0.5	0.5
S5	3	0.75	0.5
S6	3	1	0.5
S7	4	0.5	0.75
S8	4	0.75	0.75
S9	4	1	0.75

CMC was slowly fed in and stirred in which total solid contents was not more than 5% w/v as shown in Table 1. The solution pH was maintained at 7 ±1 to avoid degradation of polymeric solution at higher pH [1]. The prepared dope was filtered, shifted into the hose and left overnight to release gases at ambient conditions. The dope was pumped through extruder with 12 holes spinneret at the constant rate of 11 rpm, draw ratio of 38 rpm, into coagulation bath with concentration of 2%. The spun filaments were washed with distilled water and dried in aqueous acetone. The whole process of fiber formation is shown in Figure 1.

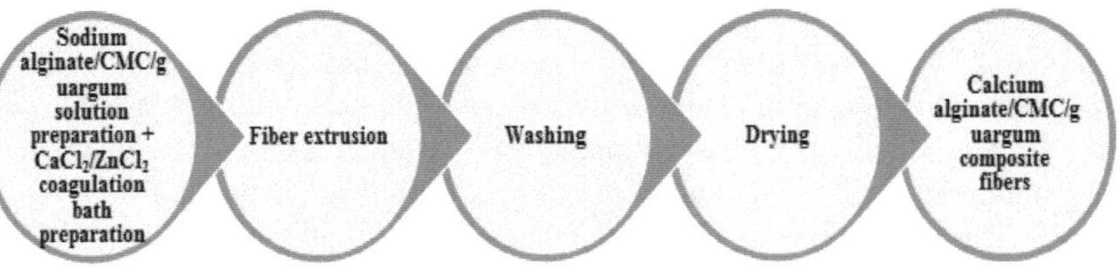

Figure 1. Flow chart for fiber development.

2.2.2. Liquid absorption testing. Liquid absorption of CMC/guar-gum/alginate composite fibers was calculated by placing fibers in three different solutions; Normal saline solution (0.9 % NaCl), distilled water and solution A (0.8298 % NaCl and 0.0368 % $CaCl_2$) by using the formula,

$$\text{Absorbance (gram/gram)} = W_w\text{-}W_d/W_d \quad (1)$$

Where, W_w is wet weight and W_d is oven dry weigh

2.2.3. Tensile strength measurement. Tensile strength, percentage elongation, and tenacity of composite fibers were measured by using single fiber strength tester according to standard EN ISO 5079/ ASTM D 5079 which was based on constant load.

2.2.4. Antibacterial testing. The fibers were investigated for their qualitative antibacterial activity by standard test procedure using AATCC 147-1998. Inhibition zone was observed for antibacterial fibers.

3. Result and discussion

3.1. SEM analysis

It can be seen from SEM micrographs (figure 2) that alginate fiber surface has more cracks without CMC/guar-gum (figure 2a). But, after their addition smoother surface was obtained (Figure 2b).

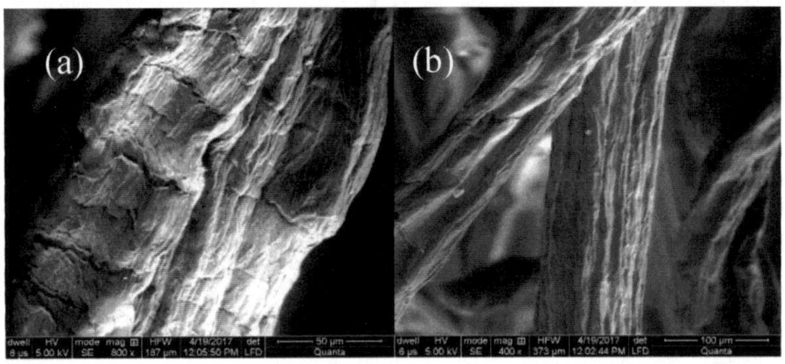

Figure 2. SEM micrograph of fibers; (a) alginate fibers; (b) CMC/guar-gum/alginate fibers

3.2. Liquid absorption testing

Basic chemical structure of all polysaccharide is nearly same and they are highly absorbent biodegradable biopolymers. Different polysaccharides are blended together to develop synergistic mixture with improved gelation characteristics. Absorption characteristics are highly dependent on Ion-exchange mechanism. Insoluble three-dimensional gel network is formed by inter-linkage of two or more molecular chains known as "Junction zones". These "junction zones" are cross-linking of negatively charged polymeric chains with ions typically cations like (Ca^{++}, Zn^{++}). Ions (Ca^{++}, Zn^{++}) in composite fibers were changed by sodium ions in any fluid (wound exodus, Saline solution or Solution A) that come in contact with them and changing these fibers into soluble sodium alginate that absorbs high quantity of water. Thus, ion-exchange mechanism is responsible for swelling and gel formation of fibers [2]. Results are shown by graph (Figure 3). It is clearly visible that absorbency of saline solution by all samples is highest and that of water is least. While absorbency of samples for solution A exists in between these two limits. The main reason behind was the highest amount of sodium ions contained in saline solution that were exchanged by calcium ions in the fibers and made the fibers highly absorbent. Also, by the addition of CMC the performance of fibers can be improved in the same way as by the addition of sodium ions.

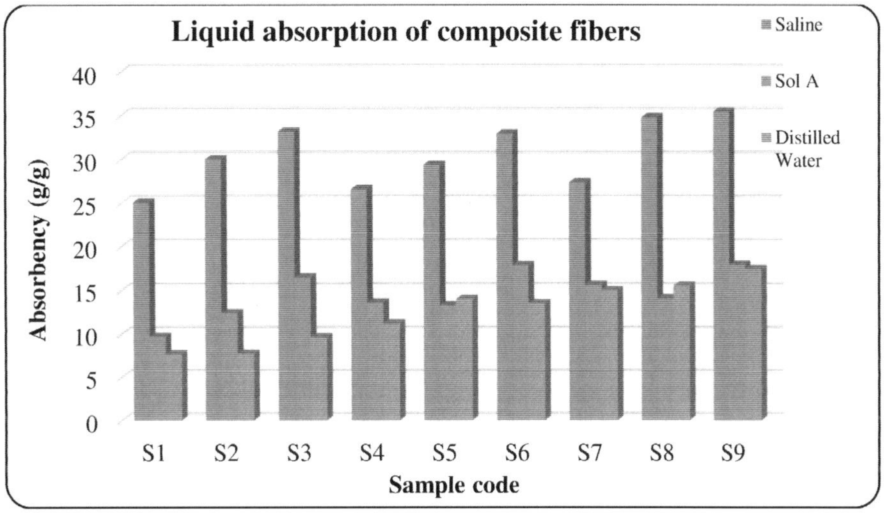

Figure 3. Liquid absorption of composite fibers for saline, solution A and distilled water.

Therefore, due to addition of this polysaccharides into composite fibers, the absorbency was increased too many folds than the simple alginate fibers. On other side, water didn't contain any ion therefore, no ion exchange occurred between fibers and water solution. But, due to addition of CMC absorbency was higher than simple alginate fibers but less than both saline and solution A. As, solution A contained both sodium and calcium ions, but less sodium ions than saline was having medium absorbency and swelling. Guar-gum showed little effect on absorbency of fibers compared to the CMC. By increasing concentration of alginate, CMC and guar-gum, the absorbency was increased. When the effect of each polysaccharide (CMC and guar-gum) for one concentration of alginate fibers was observed, increase in the concentration of CMC had more affect than guar-gum. It can be seen clearly by graph (figure 3). Highest absorbency was observed for sample 9 that contained highest amount of alginate and CMC. Due to addition of CMC regular structure of alginate is disrupted and due to more flexible structure the fibers could be easily swollen [14].

3.3. Mechanical strength

The fibers must have enough strength (6-8 cN/Tex) to be able to bear the stresses of non-woven process during formation of dressing and dressing application on wound. Therefore, balance was made between absorbency and mechanical strength of alginate fibers by addition of CMC and guar-gum.Absorbency and swelling properties can be enhanced by CMC but at the same time dry and wet strength of fibers can be negatively affected most likely due to interference of CMC within the regular polymeric chain structure of alginate fibers. Regular pattern of alginate had been broken and mechanical strength of formed fibers was reduced [14].

Average of ten testing was taken in which the effect of different concentrations of CMC, sodium alginate and guar gum on fiber strength and elongation at break was analyzed. It can be seen from (figure 4) that by increasing concentration of sodium alginate, decrease in mechanical properties was observed. As the concentration increased, the fibers became more harsh and brittle, thus, the fibers lose strength with small elongation at break. By increasing the concentration of CMC % decrease of mechanical properties was observed due to interference in regular structure of alginate. Increase in guar gum % showed increase in mechanical properties. Highest mechanical properties were found for S1, in which concentration of alginate and CMC was minimum but of guar gum was maximum. If we consider the effect of CMC due to its less concentration, alginate structure interference was less, chains were better oriented and can move smoothly [15], due to which the strength and tenacity were higher and fibres elongated more before their breakage.

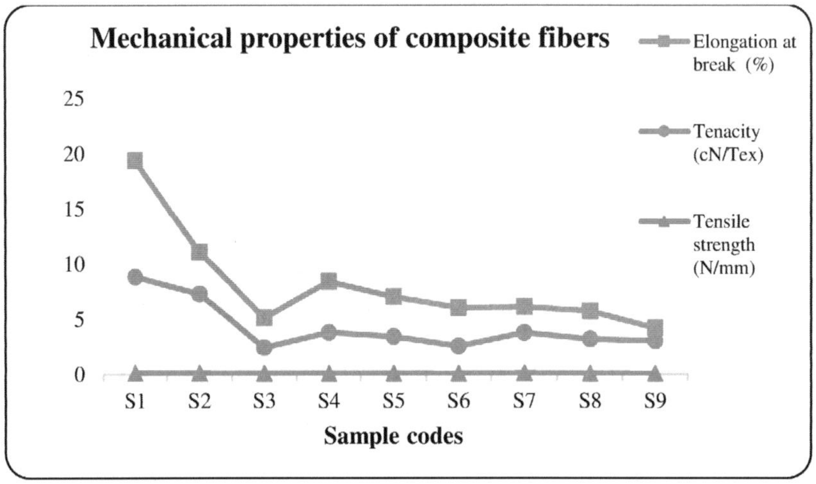

Figure 4: Graphical representation showing effect of sodium alginate, CMC and guar-gum on mechanical properties

Tenacity of CMC/alginate fibers without guar-gum was 2 cN/Tex (2.5% alginate, 0.75% CMC), but on addition of guar-gum (1%) it increased from 2cN/Tex to 7.2 cN/Tex, which is major consideration of wound dressing fibers. Absorbency of alginate fibers (2.5% alginate) was 15g/g for saline solution which on addition of CMC (0.75%) became 27.01 g/g, that further increased upto 30.007 g/g on addition of guar-gum (1%). Though guragum had minimum effect on absorbency but there was great increase in tenacity of fibers. On the basis of above mentioned results sample 2 (S2) was considered as best fibers with maximum strength and absorbency.

3.4. Antibacterial activity

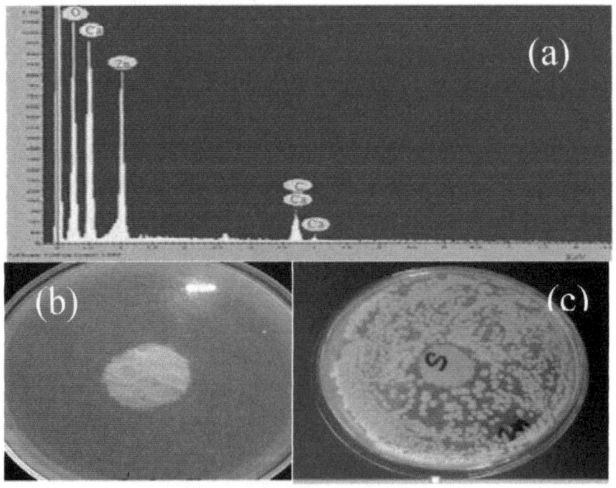

Figure 5. a) EDX analysis of fibers containing zinc ions; (b) Antibacterial property of fibers against *E.coli*; (c) Antibacterial property of fibers against *S.aureus*.

After optimization of mechanical and absorbency characteristics the same dope solution was extruded through coagulation bath containing $CaCl_2$ and $ZnCl_2$ in ratio 50:50. Energy dispersive x-ray (EDX) spectroscopy of the fibers was done for elemental analysis that confirmed the presence of zinc along with calcium ions (figure 5a). These fibers were then investigated for their antibacterial activity against two different bacterial strains. Clear inhibition zones were shown by Gram-negative bacteria (figure 4b) and Gram-positive bacteria (figure 5c). But, zone was bigger for *E.Coli* than zone made by *S.aureus* which has already established in literature [16]. These zones could be due to release of zinc ions from the fibers which were swollen by absorbing water from agar bed thus releasing antibacterial agent that possibly blocked the cell membrane permeability of bacteria causing its death [17].

4. Conclusions

Literature review has shown that increased concentration of sodium alginate increased the water uptake of fiber, but, it decreased tensile strength and elongation at break. This study showed that by increasing the CMC%, 48.2% increase in water take up of fiber occurred, but, at the same time it decreased the wet and dry tensile strength and elongation at break. Guar gum showed little increase in fiber absorbency (9.99%) but mechanical properties increased too many folds (72.2%) by its addition. These composite fibers showed excellent antibacterial efficacy against different bacterial strains and best choice for wounds.

5. References

[1] Qin Y 2008 Alginate fibres: an overview of the production processes and applications in wound management *Polym. Int.* **57** pp 171–80

[2] Qin Y 2004 Absorption characteristics of alginate wound dressings *J. Appl. Polym. Sci.* **91** pp 953–7

[3] Lloyd L L, Kennedy J F, Methacanon P, Paterson M and Knill C J 1998 Carbohydrate polymers as wound management aids *Carbohydr. Polym.* **37** pp 315–22

[4] Blair S D, Backhouse C M, Harper R., Matthews J and McCollum C N 1988 Comparison of absorbable materials for surgical haemostasis *Br. J. Surg.* **75** pp 969–71

[5] Shalumon K T, Anulekha K H, Nair S V, Chennazhi K P and Jayakumar R 2011 Sodium alginate/poly(vinyl alcohol)/nano ZnO composite nanofibers for antibacterial wound dressings *Int. J. Biol. Macromol.* **49** pp 247–54

[6] Qin Y 2004 Gel swelling properties of alginate fibers *J. Appl. Polym. Sci.* **91** pp 1641–5.

[7] Chen M, Przyborowski M and Berthiaume F 2009 Stem cells for skin tissue engineering and wound healing *Crit. Rev. Biomed. Eng.* **37** pp 399–421

[8] Edwards, Ruth, Harding, G Keith 2004 Bacteria and wound healing *Curr. Opin. Infect. Dis.* **17** pp 91–6

[9] Martens M G, Kolrud B L, Faro S, Maccato M and Hammill H 1995 Development of wound infection or separation after cesarean delivery. Prospective evaluation of 2,431 cases *J. Reprod. Med.* **40** pp 171–5

[10] Segal H C, Hunt B J and Gilding K 2016 The effects of alginate and non-Alginate wound dressings on blood coagulation and platelet Activation *J Biomater. Appl.* **12** pp 249–57

[11] Lim S J *et al.* 2010 Effect of sodium carboxymethylcellulose and fucidic acid on the gel characterization of polyvinylalcohol-based wound dressing *Arch. Pharm. Res.* **33** pp 1073–

81

[12] Gupta A P and Verma D K 2015 Preparation of carboxymethyl guar gum/silver nanocomposites film and its optical and antimicrobial properties *Int. J. Adv. Res.* **3** pp 224–9

[13] Carla A and Haynes E L1998 Solid polysaccharide materials for use as wound dressings

[14] Qin, Y 2006 The characterization of alginate wound dressings with different fiber and textile structures *J. Appl. Polym. Sci.* **100** pp 2516–20

[15] Lee KY and Mooney D J 2012 Alginate: Properties and biomedical applications *Prog. Polym. Sci.* **37** pp 106–26

[16] Riaz S, Rehman A, Ashraf M, Hussain T and Hussain M T 2017 Development of functional alginate fibers for medical applications *J. Text. Inst.* **108** pp 2197–204

[17] Qin Y, Zhu C, Chen J, Liang D and Wo G 2007 Absorption and release of zinc and copper ions by chitosan fibers *J. Appl. Polym. Sci.* **105** pp 527–32

Identification of brittle and ductile zones in sandstone reservoir using well log analysis; a case study, Southern Indus Basin, Pakistan

I Khan[1], A Ismail[2], and I Ali[1]

[1] Department of Geological Engineering, Balochistan University of Information Technology, Engineering and Management Sciences, Quetta Pakistan
[2] Department of Petroleum and Gas Engineering, University of Engineering and Technology, Lahore Pakistan

Email: ibraheemgeo12@gmail.com

Abstract. Brittleness and ductility play a very important role in the stimulation design of unconventional reservoirs as it controls the fracture length and shape of hydraulic fractures. The estimation of mechanical properties of the reservoir rock is very important for the successful execution of reservoir drilling, production, and development operations. In this study, the mechanical properties are calculated using well log responses for selected reservoir section after a quality check of the data. Brittleness is estimated from empirical correlations based on mechanical properties estimated from well logs response, uniaxial compression strength and tensile strength. High values of young modulus and corresponding low values of Poison's ratio were indicating the brittle zones in the reservoir. Brittle-ductile zone identification based on mechanical properties were giving consistent results. Brittle-ductile categorization is done for identification of brittle, less brittle, less ductile and ductile reservoir lithology in each studied well. It was found from the computer-assisted approach and graphical presentation that, there was a high brittleness in the lower part of the reservoir which is due to highly compressed shale formation. This part of the reservoir could be a good candidate for hydraulic fracturing if it satisfies all other constraints for a prospective reservoir.
Keywords: Brittleness, ductility, petrophysical properties, Southern Indus Basin, Pakistan

1. Introduction

The failure of the material depends upon the nature of the brittleness which is very effective in terms of the definition of sweet spots for hydraulic fracturing and assessment of the wellbore stability, [1, 2]. If a material breaks without any substantial deformation then it is called as of brittle [3]. This shows that if a material is brittle then there will be high options of fracture potential. Brittle rocks have no characteristics of ductility and plasticity and break straight after the loss of a stress-strain relationship. The transition of ductility to brittleness depends upon the magnitude of the stress applied to the material,[4]. The initiation of the fractures in the rock depends upon the applied stress and strength parameters, [5].

Depending upon stresses, strength attributes and mineralogical properties of the reservoir, numerous researchers have expressed brittleness into different definitions. The calculations of brittleness depend upon the selection of compatible approach according to reservoir conditions [6]. The lack of ductility of the material also shows the brittle nature of it [7, 8]. Brittleness is a very important factor in the identification of sweet spots which require precise indices to quantify [9]. The definition of brittleness on the basis of elastic waves is commonly used [10]. The brittleness of the unconventional reservoirs can be estimated using sonic and density logs. The brittleness index based on wireline log data needs to be correlated with other available indices [11].

The Sawan field lies in the Southern Indus basin extending beside eastern boundary of Pakistan [12]. The present study is confined to the discussion on Cretaceous rocks with special reference to the

IOP Conf. Series: Materials Science and Engineering **414** (2018) 012021 doi:10.1088/1757-899X/414/1/012021

Goru Formation which has the main lithology of sandstone with low to high permeability along the depth section. Cretaceous rocks are widely distributed in different parts of middle Indus Basin. The reservoir is highly heterogeneous in terms of supply of sediment and conditions of the environment. The upper part of the reservoir has very good potential as a candidate of the reservoir in different parts of the Sawan field. In general, the depositional environment of the Goru formation is deep marine while the Lower Goru section is a combination of the barrier to deltaic environments [13, 14].

Figure 1. Base map of the studied area [15]

2. Material and Methods

For the study, three wells; Sawan-7 (Well-A), Sawan-8 (Well-B) and Sawan-9 (Well-C) of a gas producing field from middle Indus basin Pakistan was selected (Figure 1). All well logs suite and routine core analysis data were available for all studied wells (Table 1). The well trajectory of the wells is vertical with completion in sands of the lower Goru formation. All the wells have wireline suite containing gamma ray log, sonic log, density log, resistivity log. Petrophysical information measured from routine core analysis were porosity and permeability. Sonic logs, density log, the gamma-ray log was prominent in the estimation of mechanical properties from wireline log study. The estimated petrophysical properties from wireline log data of the wells were matched against the values of core data for the same reservoir interval.

Table 1. Available Data for Analysis.

Data	GR	DT4P	DT4S	RHOB	DRHO	LLD	Core Data
Well-A	Yes	Yes	Yes	Yes	Yes	Yes	Yes
Well-B	Yes	Yes	Yes	Yes	No	Yes	No
Well-C	Yes	Yes	Yes	Yes	Yes	Yes	Yes

2.1 Quantification of Brittleness

Brittleness of the rock can be used to quantify the reservoir rock. Depending upon the different characteristics and availability of the data, many approaches have been proposed by the researchers [2, 13, 16].

The geomechanical properties of the reservoir; Young's modulus and Poisson's ratio are the properties, which consider both the effects of stress-strain and considered as a better approach to defining the brittleness of the reservoir. If the rock has low Poisson's ratio and high Young's modulus then the rock is considered as a highly brittle rock [17]. Low Poisson's ratio and high Young modulus are considered favorable factors for effective hydraulic fracturing of the brittle rocks [10].

2.2 Unconfined compressive strength (UCS) and Brazilian tensile strength (BTS) approaches

Uniaxial compression strength (UCS) and Brazilian tensile strength (BTS) are geomechanical properties, which can be measured by laboratory experimentation. UCS and BTS can be estimated using empirical equations as well [18, 19]. The ability of the rock to compress is shown by UCS while the force of cohesion between the grains of the rock is expressed by the magnitude of the BTS. The relative difference between UCS and BTS is the brittleness of the rock mass [20].

2.3 Geomechanical Cross plots

The brittleness of the rock can be analyzed using the cross plot of geomechanical properties whose results can be extended to ductile and brittle zones of the reservoir depth. Poisson ratio's and Young's modulus to create a cross plot for the brittleness assessment of the reservoir [21].

3. Research methodology

The research methodology is comprised of six steps. The first step is the identification of horizons and the next two are the estimation of petrophysical and mechanical properties. The last step is the identification of the ductile-brittle zone. The main steps to perform the study are explained in figure 2.

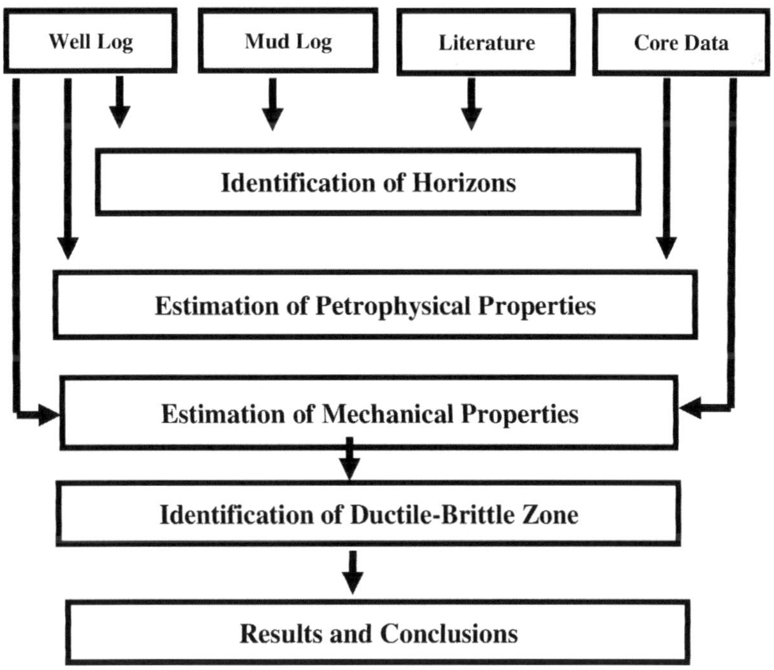

Figure 2. The methodology of workflow.

3.1 Identification of Horizon

The horizon is identified by curve matching technique in which the well log response of particular reservoir is matched with identified well log signature of the same reservoir. The particular signature for the studied reservoir was taken from literature was used for identification for the horizon.

3.2 Young's modulus

Dynamic Young's modulus (D_{ynamic}) and Static Young's modulus (E_{static}) can be estimated using shear wave velocity (Vp) and sonic wave velocity information (Vs) [22, 23].

$$E_{dynamic} = \rho_b V_s^2 \left(\frac{3V_p^2 - 4V_s^2}{V_p^2 - V_s^2} \right) \times 10^{-6}\ Gpa \tag{1}$$

$$E_{static} = 0.96 E_{dynamic} - 2.64\ Gpa \tag{2}$$

3.3 Dynamic Poisson's Ratio

Dynami Poisson's ratio ($\upsilon_{dynamic}$) can be estimated by equation while approach can be used for static Poisson's ratio (υ_{static}) calculation [23, 24].

$$\nu_{dynamic} = \left(\frac{V_p^2 - 2V_s^2}{2V_p^2 - 2V_s^2} \right) \tag{3}$$

$$\nu_{static} = 1.2437\ \nu_{dynamic} \tag{4}$$

3.4 Dynamic Shear Modulus

The dynamic shear modulus ($G_{Dynamic}$) and static shear modulus (G_{Static}) can be estimated as [22]

$$G_{dynamic} = \rho_b V_s^2 \tag{5}$$

$$G_{static} = \frac{E_{static}}{2\ (1 + 2\nu_{static})} \tag{6}$$

3.5 Dynamic Bulk Modulus

Dynamic ($K_{dynamic}$) and static (K_{static}) bulk modulus can be determined by empirical equations as [25].

$$K_{dynamic} = \frac{E_{dynamic}}{3\ (1 - 2\nu_{dynamic})} \tag{7}$$

$$K_{static} = \frac{E_{static}}{3\ (1 - 2\nu_{static})} \tag{8}$$

3.6 Unconfined Compressive Strength

The unconfined compressive strength (UCS) can be determined by using equation [18].

$$UCS = 254\ \ (1 - 2.7\phi)^2 \tag{9}$$

3.7 Brazilian Tensile Strength

The Brazilian Tensile strength can be obtained by using equation [10].

$$BTS = \left(\frac{UCS}{12.38} \right)^{1.0725} \tag{10}$$

3.8 Brittleness index
The brittleness index can be estimated by using equation [10].

$$E_{Brittleness} = \frac{E - E_{\max}}{E_{\max} - E_{\min}} \tag{11}$$

$$v_{Brittleness} = \frac{v - v_{\max}}{v_{\min} - v_{\max}} \tag{12}$$

$$BA = \frac{E_{Brittleness} + v_{Brittleness}}{2} \tag{13}$$

3.9 Identification of Brittle and Ductile Zone
By utilizing a computational approach, the analysis of the estimated mechanical properties is done. Using discriminant approach brittle and ductile zones are identified in two-dimensional space in each well. The brittle and ductile zones are identified and compared with offset wells.

4. Results and Discussions
In all the three wells brittle and ductile zones are identified by using computer analysis and graphical representation of data. Figure 4, shows high values of brittleness in the lower part of the formation as compared to upper section which is due to the presence of highly compressed shale rock whose strength increases than sandstone due to excessive overburden pressure in Well-A. In case, this part of the Zone satisfies the requirements of shale gas prospective, it is suitable depth for execution of hydraulic fracturing [15].

The identified brittle and ductile zones of the Zone were matching with the results concluded from Young's modulus and Poisson's ratio colour coding; the magnitude of higher values of Young's modulus and lower values of Poisson's ratio defines the range of brittleness. The newly drilled wells will have high stability in the lower part of the reservoir as compared to upper section due to the behavior of mechanical properties

From the cross-plot analysis of static Young's modulus and static Poisson's ratio, it is concluded that studied formation has both brittle and ductile zones. In Well-A, most of the data points were falling in the brittle area of the cross plot showing the high capacity of the formation for activities whose success depends upon the brittle behavior of the rock as shown in figure 3.

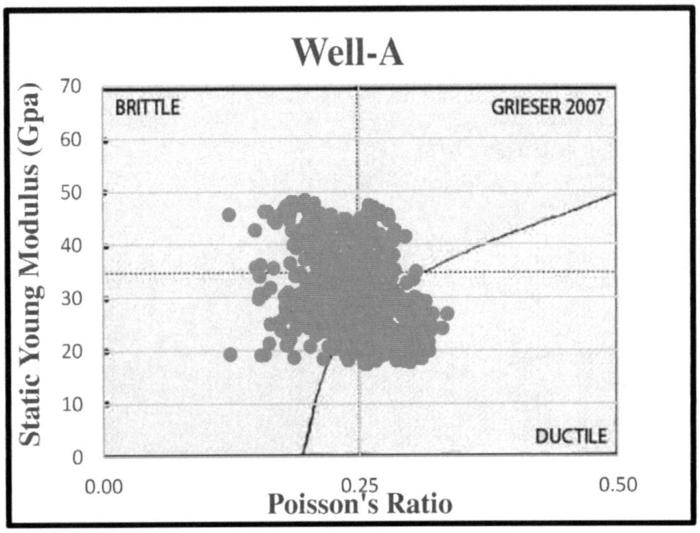

Figure 3. Cross plot of poison's ratio v.s Young's Modulus Well-A; most of the data points were falling in the brittle area of the cross plot in this well.

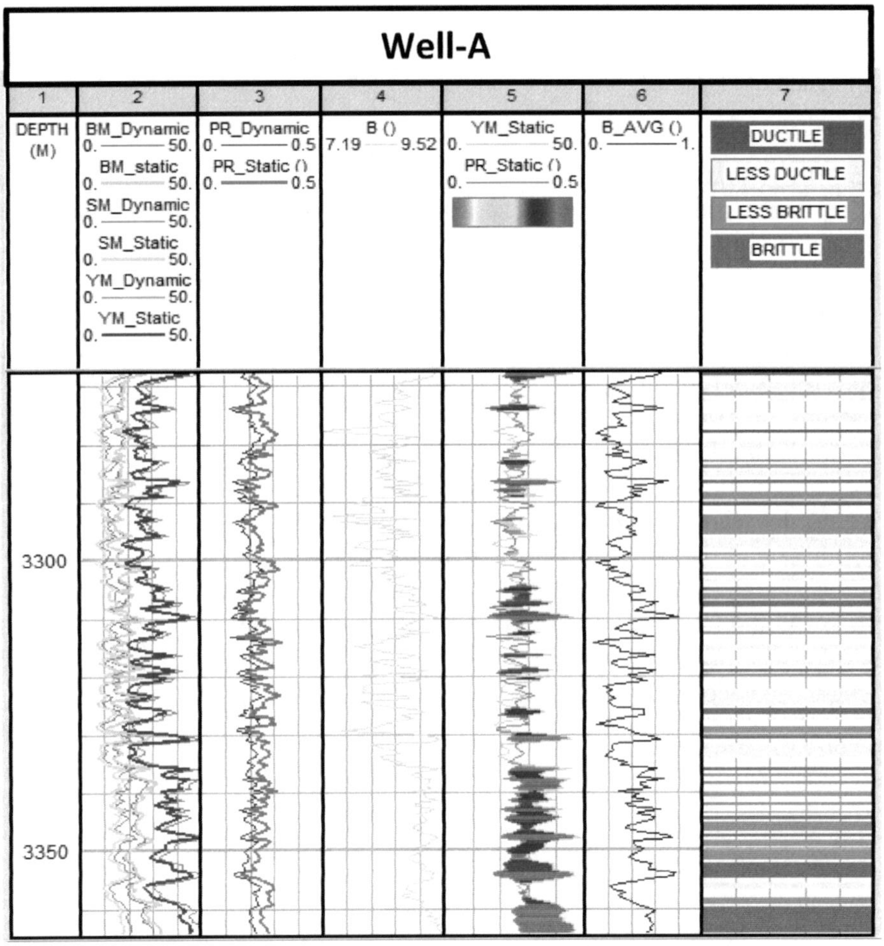

Figure 4. Discriminant analysis of Well-A; shows that studied formation has both brittle and ductile zones.

The erroneous values in Well-B were removed near 3300m, which caused a break in the trend of mechanical properties in all tracks of figure 6. High values of brittleness were seen in the lower part of the reservoir in Well-B, as was in Well-A. There was no indication of ductile or less ductile rock features in Well-B, which was clear from brittle-ductile color-coding and discriminant analysis. High brittleness layers were present in interbedded form from middle to lower part of the reservoir. The results of Young's modulus and Poisson's ratio color coding had a discrepancy with the discriminant analysis in the upper part of the reservoir as shown in figure 5. In figure 6, the cross plot of Young's modulus and Poisson's ratio shows that reservoir Zone has more of the data points falls in the brittle zone near to the brittle-ductile boundary. The results of this approach should also be checked against the mineralogical approaches for the studied Formation.

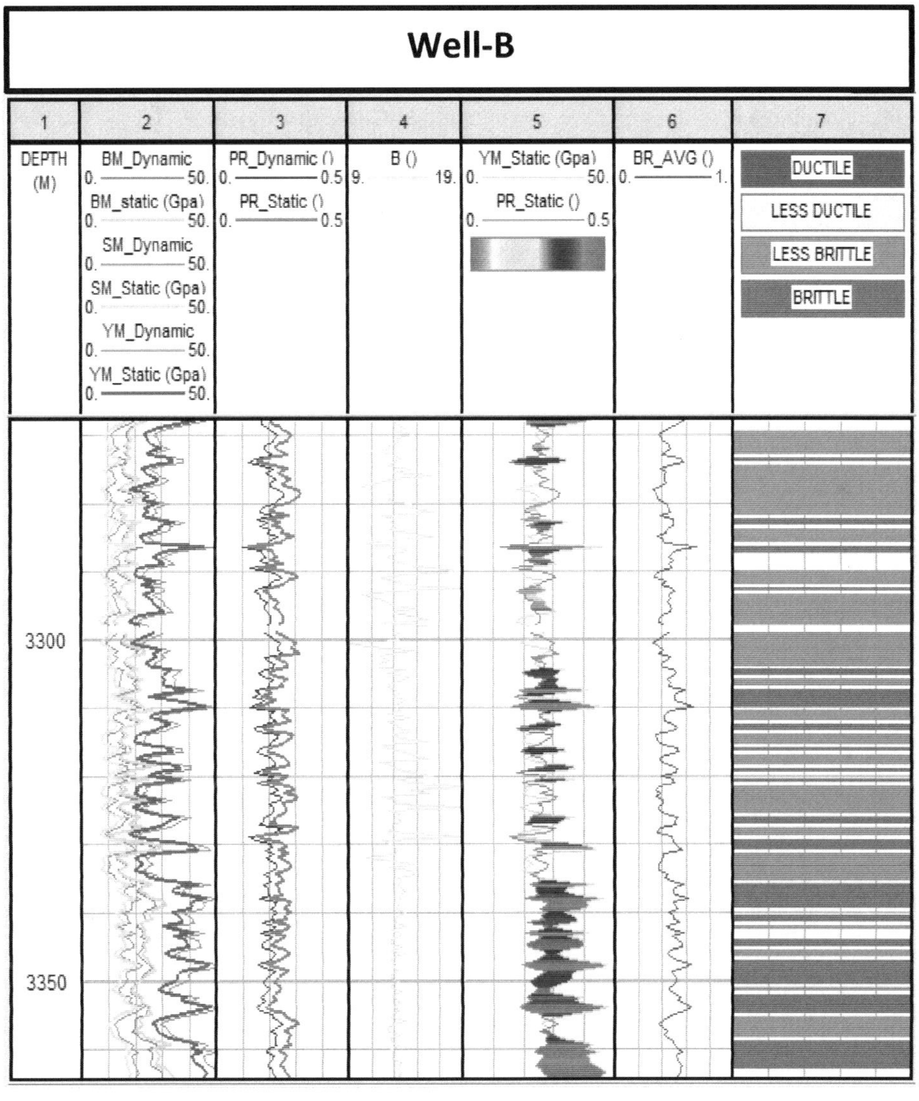

Figure 5. Discriminant analysis of Well-B; there is no indication of ductile or less ductile rock features in this well which is clear from brittle-ductile colour coding and discriminant analysis.

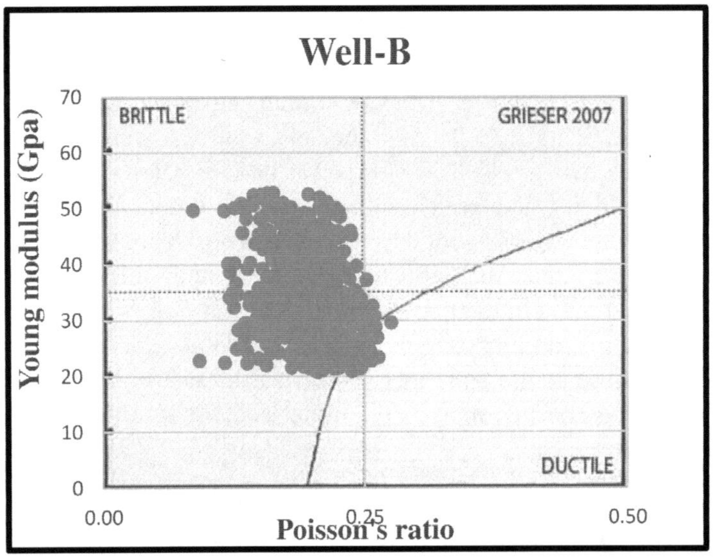

Figure 6. Cross plot of poison's ratio v.s Young's Modulus Well-B; shows that reservoir has more of the data points falls in brittle zone near to brittle-ductile boundary.

In Well-C, the brittleness estimated from the empirical equation of laboratory testing had very anomalous behavior along the reservoir depth as shown in figure 8. The same trend in the calculated mechanical properties were evident from other plots in the tracks as well. A ductile portion was identified in the Zone, which should be correlated to nearby offset wells to understand its lateral behavior. It was found that in all studied wells, there was high brittleness in the lower part of the reservoir, which is due to highly compressed shale formation. That part of the reservoir could be a good candidate for hydraulic fracturing if it satisfies all constraints of the prospective reservoir. Most of the data was falling in the low brittle zone to brittle zone ranging from 0.31 to 0.44 brittleness. Very small number of data points were present in the high brittleness zone. Ductile behavior was high in the studied formation as compared to brittle zone as shown in figure 7. Cross plot of Poisson's ratio and Young Modulus show that the higher the value of young Modulus and lower the value of Poisson's ratio it will be brittle and vice versa [10].

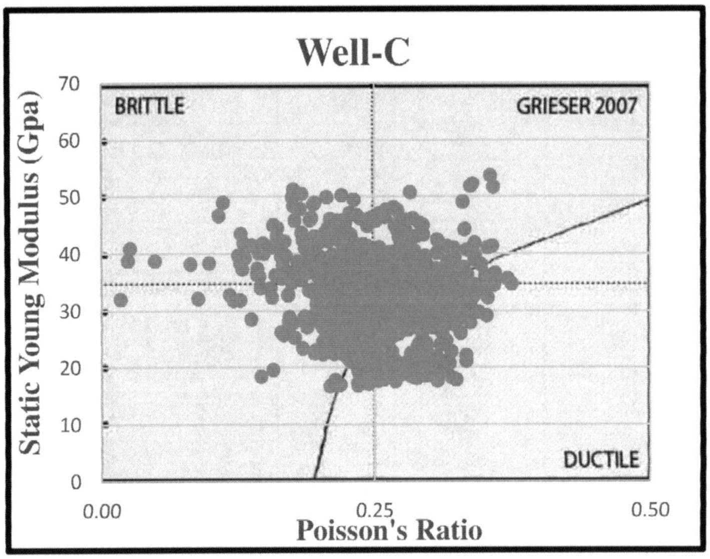

Figure 7. Cross plot of poison's ratio v.s Young's Modulus Well-C; Most of the data was falling in the low brittle zone to brittle zone ranging from 0.31 to 0.44 brittleness. Very small number of data points were present in the high brittleness zone.

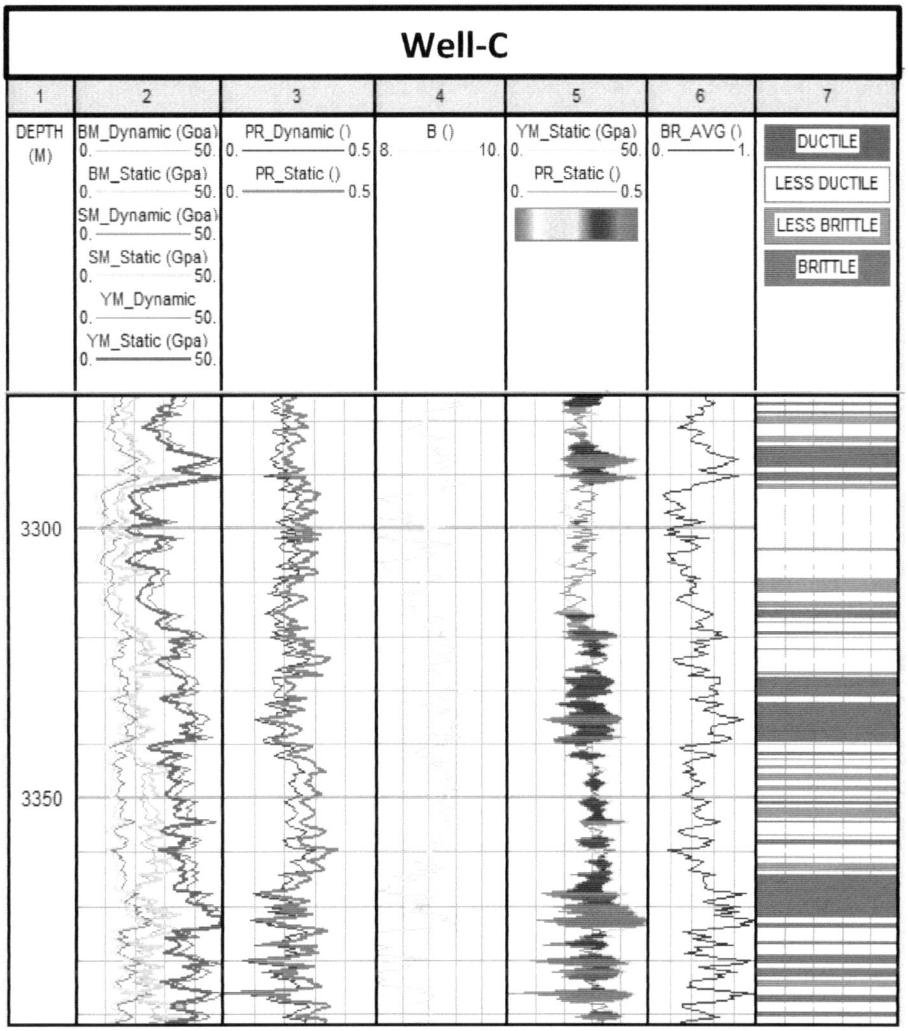

Figure 8. Discriminant analysis of Well-C;The brittleness estimated from empirical equation of laboratory testing has very anomalous behaviour along the reservoir depth. Ductile behaviour was high in the studied formation as compared to brittle zone.

5. Conclusions

- The understanding of brittle and ductile nature of the reservoir is the keystone for the successful execution of petroleum activities ranging from exploration to production.
- Brittleness and ductile behavior can be identified successfully using the empirical approaches and well log data.
- Brittle and ductile behavior depends upon the clay content as a part of the mineralogy of the rock.
- The upper part of the reservoir Zone-X has high sand content while the lower part is primarily based on shale interval in compressed morphology.
- High ranges of Young's modulus and corresponding low values of Poisson's ratio are an indicator of high brittleness in the formation.
- The behavior of mechanical properties is not consistent in Well-A, Well-B, and Well-C, due to heterogeneity in these wells.

IOP Conf. Series: Materials Science and Engineering **414** (2018) 012021 doi:10.1088/1757-899X/414/1/012021

References

[1] Charlez P, 1997. The impact of constitutive laws on wellbore stability: a general review. *SPE Drilling & Completion, 1997.* **12(02)**: p. 119-128.

[2] Rickman R. 2008. A practical use of shale petrophysics for stimulation design optimization: All shale plays are not clones of the Barnett Shale. in SPE Annual Technical Conference and Exhibition. *Society of Petroleum Engineers.*

[3] Song L, 2015. Prediction of continental shale brittleness based on brittle-ductile transition analysis. in 2015 SEG Annual Meeting. *Society of Exploration Geophysicists.*

[4] Jaeger, J.C., N.G. Cook, and R. Zimmerman, 2009. Fundamentals of rock mechanics. *John Wiley & Sons.*

[5] Valko, P. and M.J, 1995. Economides, *Hydraulic fracture mechanics.* Vol. **28.** Wiley Chichester.

[6] Zhang, D., P. Ranjith, and M. Perera, 2016. The brittleness indices used in rock mechanics and their application in shale hydraulic fracturing: A review. *Journal of Petroleum Science and Engineering.* **143**: p. 158-170.

[7] Hetenyi, M.I, 1950. *Handbook of Experimental Stress Analysis.*

[8] Morley A, 1944. *Strength of Materials*: with 260 Diagrams and Numerous Examples. New York: Longmans, Green and Company.

[9] Huang, X.-R, 2015. Brittleness index and seismic rock physics model for anisotropic tight-oil sandstone reservoirs. *Applied Geophysics*, **12(1):** p. 11-22.

[10] Grieser, W.V. and J.M. Bray, 2007. Identification of production potential in unconventional reservoirs. in Production and Operations Symposium. *Society of Petroleum Engineers.*

[11] Shi, X, 2016. Brittleness index prediction from conventional well logs in unconventional reservoirs using artificial intelligence. in International Petroleum Technology Conference. *International Petroleum Technology Conference.*

[12] Zaigham, N.A. and K.A. Mallick, 2000. Prospect of hydrocarbon associated with fossil-rift structures of the southern Indus basin, Pakistan. *AAPG bulletin*, **84(11)**: p. 1833-1848.

[13] Kadri, I. B,1995. Petroleum geology of Pakistan: Pakistan Petroleum Limited.

[14] Munir, K., Iqbal, M. A., Farid, A., & Shabih, S. M, 2011. Mapping the productive sands of Lower Goru Formation by using seismic stratigraphy and rock physical studies in Sawan area, southern Pakistan: a case study. *Journal of Petroleum Exploration and Production Technology*, **1(1)**, 33-42. doi: 10.1007/s13202-011-0003-9.

[15] Ismail, A, 2017. A comparative study of empirical, statistical and virtual analysis for the estimation of pore network permeability. *Journal of Natural Gas Science and Engineering.*

[16] Nygård, R, 2006. Brittle–ductile transition, shear failure and leakage in shales and mudrocks. *Marine and Petroleum Geology*, **23(2)**: p. 201-212.

[17] Jahandideh, A. and B. Jafarpour, 2016. Optimization of hydraulic fracturing design under spatially variable shale fracability. *Journal of Petroleum Science and Engineering.* **138**: p. 174-188.

[18] Altindag, R. and A. Guney, 2010. Predicting the relationships between brittleness and mechanical properties (UCS, TS and SH) of rocks. *Scientific research and Essays.* **5(16)**: p. 2107-2118.

[19] Bradford, I., 1998. Benefits of assessing the solids production risk in a North Sea reservoir using elastoplastic modelling. in SPE/ISRM. Rock Mechanics in Petroleum Engineering. *Society of Petroleum Engineers.*

[20] Hucka, V. and B. Das, 1974. Brittleness determination of rocks by different methods. in *International Journal of Rock Mechanics and Mining Sciences & Geomechanics Abstracts.* Elsevier.

[21] Perez Altamar, R. and K. Marfurt, 2014. Mineralogy-based brittleness prediction from surface seismic data: Application to the Barnett Shale. *Interpretation.* **2(4):** p. T255-T271.

[22] Mavko, G., T. Mukerji, and J. Dvorkin, 2009. *The rock physics handbook*: Tools for seismic analysis of porous media. Cambridge university press.

[23] McPhee, C. and C. Enzendorfer, 2004. Sand Management Solutions for High-Rate Gas Wells, Sawan Field, Pakistan in SPE International Symposium and Exhibition on Formation Damage Control. 2004. *Society of Petroleum Engineers*.

[24] Lal, M, 1999. Shale stability: drilling fluid interaction and shale strength. in SPE Asia Pacific Oil and Gas Conference and Exhibition. *Society of Petroleum Engineers*.

[25] Archer, S. and V. Rasouli, 2012. A log based analysis to estimate mechanical properties and in-situ stresses in a shale gas well in North Perth Basin. *WIT Transactions on Engineering Sciences*. **81**: p. 163-174.

Extraction of oil from algae for biodiesel production, from Quetta, Pakistan

Rizwan Ullah Baig[1], Abeera Malik[1], Khadim Ali[1], Sehar Arif[1], Sadam Hussain[1], Mazhar Mehmood[1], Kamran Sami[1],Ali Nawaz Mengal, Mohammad Najam Khan[1]

[1]Department of Chemical Engineering, Balochistan University of Information Technology, Engineering and Management Sciences Quetta, Pakistan

Email: abeera.malik@buitms.edu.pk

Abstract: The world is facing the energy crises and environmental issues in this century due to increased industrialization and overuse of natural resources for energy such as fossil fuels. The burning of fossil fuels generates greenhouse gases which aggravate the global warming. Researchers across the world are focusing on renewable, less CO_2 and NOx emissions fuels. These fuels are important for sustainability and green economy. Biodiesel production from Algae is emerged as the promising alternative fuel, technically and environmentally acceptable and easily available. In this study, different parameters of the oil extraction process from algae biomass were studied. Samples of algal species Spirogyra were collected from Chashma Achuzai, Quetta, Pakistan and employed as a feedstock for biodiesel production. Oil from macroalgae Spirogyra was extracted using n-Hexane as a solvent. Effects of n-Hexane to oil ratio, size of algal biomass and contact time on the percentage yield of extracted oil was studied and analyzed. It was concluded that maximum amount of oil was extracted from Spirogyra by using a greater ratio of solvent to algal biomass, maximum contact time, smaller algal biomass size.
Keywords. Spirogyra, solvent extraction, biodiesel, macroalgae, Quetta

1. Introduction

Presently, global warming effect, depletion in fossil fuel reserves, and higher petroleum prices are the main issues driving worldwide interest on the development of alternative renewable, biodegradable and sustainable biofuels [1]. Biofuels produced from algae considered to be a potential candidate to replace conventional fossil fuels [2]. Furthermore, it is the fastest developing alternative to petrodiesel fuel in many developed and developing countries worldwide [3]. Besides, its decreases particulate emissions, unburned hydrocarbons, and sulfur dioxide generated through its combustion process [4]. The use of Macroalgae for biodiesel production is the most promising alternative to fossil fuels depletion and pollution caused by these fuels. The certain species of algae contain more than 50% of oil content, which can be extracted and refined into transportation fuel easily using currently available technologies [5]. Microalgae have some benefits than other feedstocks for biodiesel as they have fast growth rate, permit the use of non-arable land, can be grown in wastewater, do not replace the edible traditional crops, they can be harvested daily, and it is not a seasonal production. Its waste can be used as a feed for animals or other purposes [6].

Algae can be used to generate energy in several ways. One of the most efficient ways is to produce biodiesel from algal oil by transesterification. One of the main obstacles to fully taking advantage of lipid-producing microalgae is the ability to successfully and efficiently extract oil from the biomass cells. A number of methods are used for extracting the algal oil i.e., mechanical extraction, chemical extraction [7]. Algal oil can be efficiently extracted using chemicals by solvent extraction method [8]. Solvent Extraction technique is favored than mechanical pressing method since it recovers all the oil leaving just 0.5–0.7% leftover oil in the algae biomass [9]. Benzene and di-ethyl ether can also use but the popular chemical for solvent extraction is hexane, which is less expensive and hazardous than another solvent [6]. Spirogyra is the green algae, available abundantly in springs, ponds and brackish water. It has a structure like a cellular hair and includes unbranched cylindrical fibers, which is about 1/10 mm in diameter and few centimeters long. Spirogyra produces lipids, carbohydrate, and proteins that can be utilized for the production of biodiesel, or bioethanol [10].

2. Materials and methods

The purpose of the study is to chemically extract oil from Macroalgae species collected from an open pond at Chashma Achozai site in Quetta, using n-Hexane as a solvent. Effects of n-Hexane to oil ratio, mesh size of algal biomass and contact time on the percentage yield of oil extracted were studied. The chemicals utilized, and the analysis procedure of study are discussed in this section.

2.1 Chemicals

The chemicals used n-Hexane, Methanol and NaOH were of analytical grade and purchased from Merck. The chemicals were used without any further purification. The complete experiment was carried out in the laboratory of Biotechnology, Department of Biotechnology, Balochistan University of Information Technology Engineering and Management Science Quetta, Pakistan.

2.2 Algae sample collection

The algae samples were collected from Chashma Achuzai Quetta, Pakistan. The samples collected from an open pond, illustrating in figure 1.

Figure 1. Collection of algae from the open pond

Figure 2. Algae identification and drying associated with the algal biomass.

2.3 Algae Identification and Preparation for Oil extraction

Collected Algae sample was examined under Eclipse E200 Compound Microscope. The photograph in figure 2 was taken by enlarging it 40 times and 100 times. The samples were spread under the sun in the roof of the hostel for 2 days (48 hours) to evaporate the amount of water. The dried samples were ground with the help of pestle and grinder and the fine powder was passed through different micron sieves, to get different mesh size algal biomass, figure 2. The ground algae were dried for 30 min at 80°C in an incubator for releasing leftover water. Then the algae powder was stored in different jars for extraction experiment in a sealed container.

2.4 Oil extraction from Algae

The algae samples collected was dried (100%) and powdered. Hexane was mixed with the dried ground algae to extract oil in separating funnel of 250 ml. Then the mixture was kept for 24 h for settling and for separation of the two layers in the funnel. The organic phase containing the algae oil was emptied in the pre-weighted 50 ml beaker. The Algal oil was separated from Algae biomass by filtration and weighted it by using electronic weight balance. The extracted oil was evaporated in a water bath to release hexane. All extraction was performed in triplicates for the different parameters solvent extraction process. The oil yield (wt. %) was then calculated by utilizing the equation-1 [4, 8].

$$\text{Extracted oil efficiency (wt. \%)} = \frac{\text{Mass of oil extracted (grams)}}{\text{The total mass of dried algae}} \times 100 \qquad (1)$$

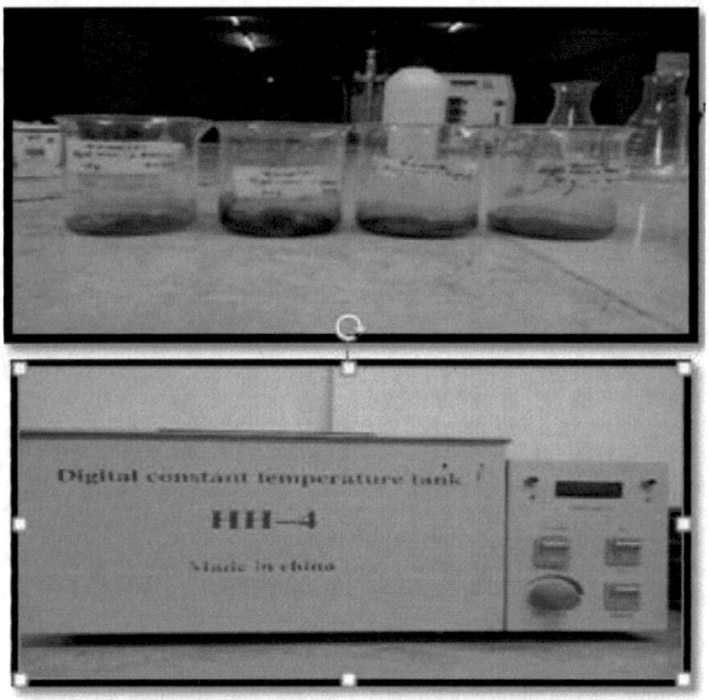

Figure 3. Solvent extraction in laboratory

3. Results and Discussion

Various Parameters of oil extraction from open pond algae by using solvent extraction method were studied. The study focused on to find out which parameter causes the increase in oil yield and improved the extracted efficiency.

3.1 Varying n-Hexane to Algae ratio

Experiments have been performed using the constant mass of dried Algae (Spirogyra) and by changing the volume of solvent. Derived results are then tableted in table 1. As it can be seen that increasing the solvent amount also improved the extracted oil efficiency.

Table 1. Amount extracted oil by varying volume of solvent used

Sample	Algae Biomass (g)	n-Hexane (ml)	Algae to n-Hexane ratio	Oil Extracted (g)	Extracted Efficiency %
01	30	30	1	0.79	2.63
02	30	40	1.33	0.92	3.07
03	30	50	1.66	1.57	5.23

Experiments have been performed using the constant volume of solvent and by changing the mass of dried Algae. Derived results are then tabulated in Table 2. It was observed that the percent yield of oil increased as the solvent to algae ratio increased. However, from the tabulated data, it is noted that extraction efficiency is good when the algae biomass to n-Hexane ratio is kept to 1:2. A very slight Increase is observed when the ratio is kept to 1:3 than the efficiency of 1:2 in extraction. It is concluded that the best optimum value for the biomass to n-Hexane ratio is 1:2.

Table 2. Amount of Oil extracted oil by varying mass of dried Algae

Sample	Algal Biomass (g)	n-Hexane (ml)	Algae to solvent ratio	Extracted Oil (g)	Extracted Efficiency %
01	20	30	1.5	0.7	4.1
02	15	30	2	0.8	5.33
03	10	30	3	0.82	7

The effect of solvent to algae ratio on percent yield of extracted oil is shown in figure-4. It was observed that the percent yield of oil increased as the n-Hexane to algae ratio increased. The higher yield at a solvent to algae ratio is attributed to the excess solvent available to extract oil from the algal biomass, so it is concluded greater will be the ratio between solvent to Algae biomass, greater will be the extraction efficiency [10, 11]. Increasing the algae to solvent ratio from 1:1 to 1:3 the extracted oil yield is 2.5 times more than the equal ratio.

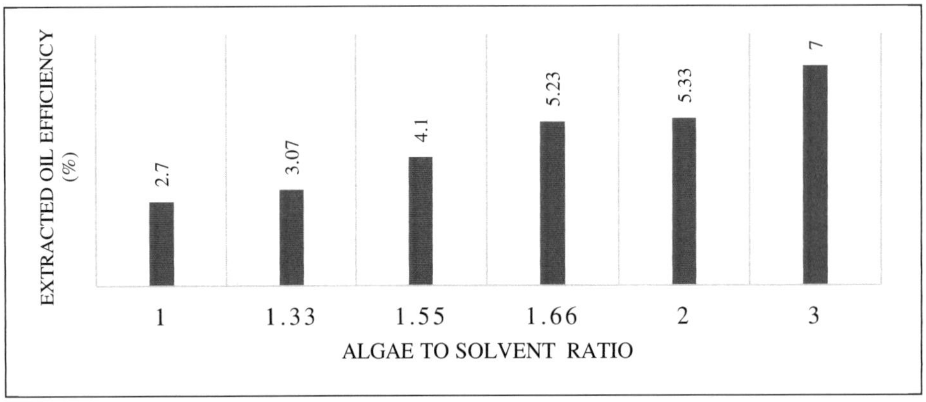

Figure 4. Effect of solvent to algae ratio on the extracted efficiency

3.2 Varying the algal biomass size

Experiments have been performed using different Algal biomass size by keeping the constant mass of Algal biomass and volume of solvent. The algae to solvent ratio were kept constant to 1: 1.66. Derived results are then tableted in table 3.

Table 3. Amount of Oil extracted using different size of Algal biomass.

Sample	Algal Biomass (g)	Mesh number	Algae biomass size (mm)	n-Hexane (ml)	Extracted Oil (g)	Extracted Oil Efficiency %
01	30	20	0.841	50	1.55	5.16
02	30	30	0.595	50	2.02	6.73
03	30	50	0.297	50	2.3	7.66

It was observed that (Figure-5) when the size of biomass decreased to 50 mesh from 20 mesh the oil extracted efficiency increased from **5.16% to 7.66%**. This can be defensible by the improved contact area between the algae biomass and solvent. The smaller sized particles have maximum interaction with solvent as compared to large particles of Algal biomass and thus increased the yield [10, 12].

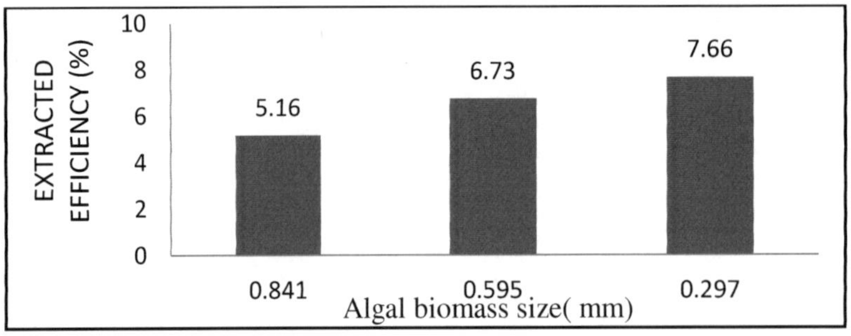

Figure 5: Effect of size of algal biomass on the extracted efficiency

3.3 *Varying contact time between n-Hexane and algae biomass*
Experiments have been performed at different contact time while the other parameters i.e. solvent to algae ratio and size of algae remains constant. Derived results are then tabulated in Table 4.

Table 4: Oil extracted at different contact time.

Sample	Algal Biomass (g)	n-Hexane (ml)	Times (Hours)	Extracted Oil (g)	Extracted Efficiency %
01	30	50	10	1.02	3.4
02	30	50	15	1.04	3.46
03	30	50	25	1.57	5.23

The contact time was varied from 10 to 25 hours. It was observed (Figure-6) that extracted oil efficiency increases as the contact time increases. Maximum yield at maximum contact time can due to enhanced interaction between the solvent and algal biomass, which lead to homogenous mixing and increased in solubility of oil by solvent [10, 13]. Hence oil is extracted from all portions of the algae species by increasing contact time.

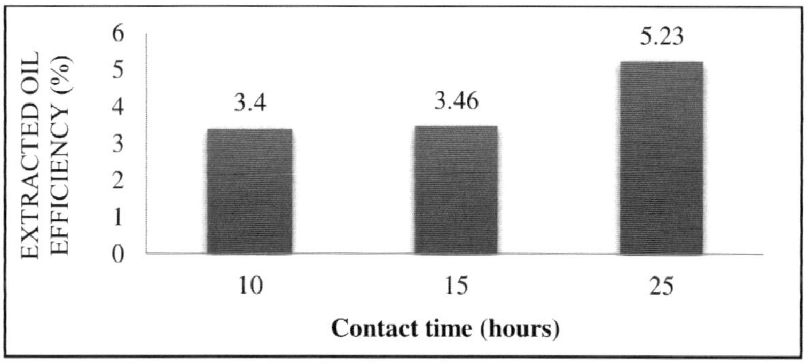

Figure 6. Effect of contact time on extracted efficiency

4. Conclusion

In this study, oil was extracted from Algae (Spirogyra) by using solvent extraction process. Also studied different parameters of the solvent extraction process. It was noted that the maximum amount of oil was extracted from algal biomass by using 3:1 ratio of n-Hexane to algal biomass, maximum contact time, smaller algal biomass size. Increasing the algae to solvent ratio from 1:1 to 1:3, the extracted oil yield is 2.5 times more than the equal ratio.

Biodiesel produced from Algae biomass can be considered an alternative choice as of easy availability and eco-friendly nature. A significant amount of biodiesel can be produced from macroalgae (Spirogyra). Spirogyra can be successfully used as feedstock for the producing biodiesel. The extracted oil can be converted to biodiesel by the trans-esterification reaction. Further study related to trans-esterification reaction and biofuel characterization will be needed.

5. References:

[1] Brennan, L. and P. Owende, 2010. Biofuels from microalgae—A review of technologies for production, processing, and extractions of biofuels and co-products. Renewable and *Sustainable Energy Reviews*,**14**(2): p. 557-577.

[2] Demirbas, A. and M. Fatih Demirbas, 2011. Importance of algae oil as a source of biodiesel. *Energy Conversion and Management*, **52(1)**: p. 163-170.

[3] Scott, S.A., M.P. Davey, J.S. Dennis, I. Horst, C.J. Howe, D.J. Lea-Smith and A.G. Smith, 2010. Biodiesel from algae: challenges and prospects. *Current Opinion in Biotechnology*, **21(3)**: p. 277-286.

[4] Arun, J., Shreekanth, S.J., Sahana, R., Raghavi, M.S., Gopinath, K.P. and Gnanaprakash, D., 2017. Studies on influence of process parameters on hydrothermal catalytic liquefaction of microalgae (Chlorella vulgaris) biomass grown in wastewater. *Bioresource technology*, **244**: p. 963-968.

[5] Varma, M.N., P.A. Deshpande, and G. Madras, 2010. Synthesis of biodiesel in supercritical alcohols and supercritical carbon dioxide. *Fuel*, **89(7)**: p. 1641-1646.

[6] Kumar, S.J., Kumar, G.V., Dash, A., Scholz, P. and Banerjee, R., 2017. Sustainable green solvents and techniques for lipid extraction from microalgae: A review. *Algal Research*, ***21***: p. 138-147.

[7] Barnwal, B.K. and M.P. Sharma, 2005. Prospects of biodiesel production from vegetable oils in India. *Renewable and Sustainable Energy Reviews*, **9(4)**: p. 363-378.

[8] Chen H, Zhou D, Luo G, Zhang S, Chen J, 2015. Macroalgae for biofuels production: progress and perspectives. *Renewable and Sustainable Energy Reviews*, **47(1)**: p. 427- 437.

[9] Santos, F.F.P., S. Rodrigues, and F.A.N Fernandes, 2009. Optimization of the production of biodiesel from sobean oil by ultrasound assisted methanolysis. *Fuel Processing Technology*, 90(2): p. 312- 316.

[10] Khan, S.A., Hussain, M.Z., Prasad, S. and Banerjee, U.C., 2009. Prospects of biodiesel production from microalgae in India. *Renewable and Sustainable Energy Reviews*, **13(9):** p. 2361-2372

[11] Topare, N.S., Raut, S.J., Renge, V.C., Khedkar, S.V., Chavanand, Y.P. and Bhagat, S.L., 2011.Extraction of oil from algae by solvent extraction and oil expeller method. *International Journal of Chemical Sciences*, **9(4)**: p. 1746-1750.

[12] Konga, A.K, A.S. Muchandi and G.P. Ponniah, 2017. Soxhlet extraction of Spirogyra sp. Algae: an alternative fuel. *Biofuels,* **8(1)**: p. 29 – 35.

[13] Fatima, N., Mahmood, M.S., Hussain, I., Siddique, F. and Hafeez, S., 2016. Transesterification of oil extracted from freshwater algae for biodiesel production. *Energy Sources, Part A: Recovery, Utilization, and Environmental Effects*, **38(15)**: p. 2306-2311.

Strategies to Manage Aquifer Recharge in Balochistan, Pakistan: An Overview.

Syed Mobasher Aftab[1], Rehanul Haq Siddiqui[1], and Muhammad A Farooqui[2]

[1]Balochistan University of Information Technology, Engineering and Management Sciences, Quetta, Pakistan
[2]COMSATS University Islamabad, Lahore Campus Defence Road Lahore, Pakistan

E-mail: syed.mobasher@buitms.edu.pk

Abstract. Balochistan plateau is dominated by the drainage system of eighteen river basins, where precipitation recharges both consolidated and unconsolidated aquifers. In eleven river basins, the groundwater levels are steadily depleting for the last three decades due to indefensible long-term groundwater extraction. The cumulative decline of water-table ranges from 2 to 3 meters/y. The most significant decrease of 60m in the last 12 years has been recorded in parts of Quetta Valley. The estimated total groundwater recharge in an average year of all river basins is 2,200 Mm^3, the withdrawal is 2,657 Mm^3. The groundwater overdraft in eleven river basins is 886 Mm^3, whereas 429 Mm^3 is available for sustainable development only in seven river basins. To enhance the groundwater recharge, many proposals have been advanced, and several studies have been carried out by different public and private sector organizations for the protection of natural resources. However, implementation of the recommendations of these studies remained limited with little progress causing severe shortages of groundwater resources. Assorted strategies have been developed by the concerned government departments for the protection of natural resources under their jurisdiction. A comprehensive provincial policy and subsequent strategy for the protection and sustainable development of natural resources in Balochistan is yet to be formed. In this paper the natural resource protection policies and strategies that are directly or indirectly associated with the surface and groundwater resources have been assessed. In addition, the federal and provincial policies, including water, biodiversity, climate change, environment, agriculture, forest, and rangeland management have been reviewed and summarized. In some of the previous strategies, certain measures have been proposed for the improvement of water resources. The measures include a ban on agriculture tube wells in urban areas, construction of storage, supply and delay action dams (DADs). As a consequence, 326 DADs with a storage capacity of 332 Mm^3 were constructed in different river basins. The studies represent that in current circumstances, the DADs are the appropriate means to recharge aquifers if supported by design modifications and catchment area specific watershed management plans. For the sustainable management of natural resources of the province a comprehensive, integrated watershed management strategy and a model are presented in this paper.
Keywords: Groundwater, recharge, strategies, Delay Action Dams, Balochistan

1. Introduction

Balochistan plateau is comprised of four distinct zones, the Upper-Highlands, the Lower-Highlands, the Kachi-Nasirabad-Makran Plain and the deserts. The Upper-Highlands where altitude ranges from 1,500 to 3,700 meters above sea level (masl), has warm summer and very cold winters. The precipitation ranges from 200-250 mm/y and characterized as a semi-arid zone. The Lower-Highlands, where altitude ranges from 600 to 1,200 masl, is the zone of hot and dry summers, but the winters are extremely cold to mild. This zone is characterized as arid where precipitation varies from 100-150 mm/y. The Kachi and small plains stretch from, Nasirabad to Dera Bugti and extends from Mekran coast to Iran border, where the summer temperature reaches up to 50 °C but winters are mild. The desert zone is located in western Balochistan mostly covered with sand-dunes where the summer is hot and arid, the windstorms make the area inhospitable. This desert zone is characterized as Hyper-arid and precipitation ranges from 0-100 mm/y. The distributions of 18 river basins of Balochistan and the location of Quetta Valley are shown in figure 1.

The geographical area of Balochistan is 44% with 5% of Pakistan's population. The provincial water assets can only be evaluated area-wise rather than population. The renewable water resources of Balochistan are about 30% per-annum of Pakistan as per geographical area. Pakistan's 87% available water is contributed by the Indus Basin Irrigation System. Balochistan receives yearly 3.75 billion m^3 of water from Indus Basin canal commands, which irrigates about 1% of the provincial lands that equal to 14% of the accessible waters. The remaining 99% of the provincial land area encountered exhausted environment and degraded natural resources, water scarcity and depleting groundwater (GW), resources. The water resources vary from basin to basin, some river basins have 1/2 - 1/3 of the average water potential. The floodwater is the largest resource of water in the province, about two-thirds of total available water, but only 40% is utilized. The GW constitutes around 4% of the total water available, but this is an over-utilized resource.

The main three water sources of the province; the largest is floodwater comprise of seasonal flood flows. The second is the IBIS consists of perennial and non-perennial water resources. The third is the GW abstracted by tubewells, dug-wells, *karizes* (underground water channels), and springs. The wasteful water usage, over-exploitation of GW and frequent droughts made the water management a complex and difficult task. Despite water scarcity, the provincial government does not have the capacity to manage the available water resources. To protect the natural resources, government agencies prepared several policies during the last two decades, but no policy has been implemented in its true spirit. The policies include protection and conservation of ecosystems, biodiversity, natural resources, environment effects of climate change, forests, rangelands, agriculture, watershed, surface water and GW resource management. The common aim of all policies is to conserve natural resources and to eliminate desertification and deforestation.

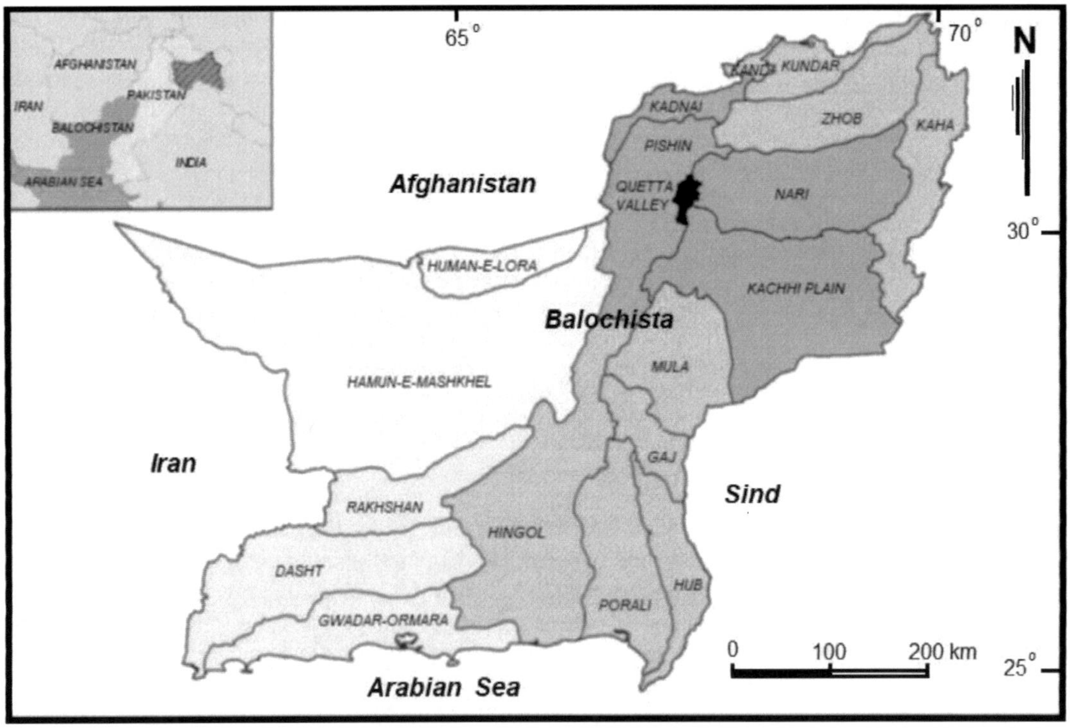

Figure 1. Balochistan plateau, 18 River Basins and Quetta Valley, [1].

2. Declining of Groundwater Resources

In Balochistan, GW is almost the only source of public supply, industrial and agricultural purposes. The river basins encompassed drainage areas from 1,115 Km^2 of Kand River to 84,916 Km^2 of Hamun-e-Lora. The cumulative average yearly precipitation ranges from 66-281mm. The GW occurs in consolidated and unconsolidated aquifers, which flows from the catchment boundaries to the axis of the valleys and follows the surface drainage pattern. In river basins, the depth of GW table ranges from 5-50 m below ground surface. In an average year, the estimated GW recharge of river basins varies between 10-300 Mm^3/y and the total GW recharge of all river basins is about 2,200 Mm^3/y. Its utilization varies from 19-566 Mm^3/y whereas its total utilization is 2,657 Mm^3/y, [2]. The amount of overdrawn GW in different basin ranges from 2-396 Mm^3/y and the total overdrawn from all river basins is 886 Mm^3/y. The safely exploitable in seven river basins ranges from 2-273 Mm^3 and the total available quantity for future exploration is 429 Mm^3/y, [2]. The estimated GW budget of all river basins is presented in table 1, [2].

The excessive unplanned pumping caused continuous depletion of GW resources at an alarming rate in eleven river basins out of eighteen. The cumulative decline of the GW table in the province is 2-3 m/y. The worst example of GW depletion is represented here from Quetta Valley. In Quetta Valley the GW table is declining at a distressing rate of 1-4 m/y in different areas, especially in the context of the continually increasing human population. The monitoring results of ten years from 2005-2014 in Quetta, represent that water table declining at a rate of 1.4-2.7 m/y and a total decline varies between 14-27 m. The eleven years monitoring results characterize that water table declining at the rate of 3.7 m/y, the total decline of the water table is 41 m. The monitoring results of thirteen years from 2003 to 2015 signify that water table decline at a rate of 1-4 m/y and the cumulative decline varies from 12-53 m. The declining trend of GW table at ten monitoring points in Quetta Valley is shown in table 2, [3].

Table 1. The GW Balance in an average year of River Bains, [2].

River Basins	Area Km^2	DAD's		Precipitation Ave. mm	GW Balance Mm^3			
		No	Storage Mm^3		Recharge	Utilization	Overdrawn	Surplus
Dasht	27,690	20	31	125.4	100	94	-	6
Gaj	6,025	4	8	203.1	70	72	2	-
Gawadar	17,065	5	95	65.7	40	25	-	15
H-e-Lora	84,916	1	0	129.2	40	141	101	-
H-e-Mashkel	8,260	14	33	120.9	300	27	-	273
Hingol	35,736	5	4	167.2	200	168	-	32
Hub	8,610	10	11	175.8	80	88	8	-
Kachhi	31,495	13	4	116.6	180	169	-	11
Kadanai	4,274	0	0	209.2	30	115	85	-
Kaha	11,995	9	5	281.3	190	319	129	-
Kand	1,115	0	0	245.6	10	19	9	-
Kunder	6,224	5	1	244.6	50	48	-	2
Mula	16,262	6	6	142.9	120	129	9	-
Nari	22,298	47	24	243.7	270	180	-	90
Pishin	18,133	132	45	199.9	170	566	396	-
Porali	18,540	17	21	171.5	140	146	6	-
Rakhshan	12,339	5	10	125.4	50	81	31	-
Zhob	16,425	33	34	235.2	160	270	110	-
Total	347,402	26	332	-	2,200	2,657	886	429

Table 2. The decline of groundwater in Quetta Valley, [3].

Monitoring			Decline (m)	
Points	Years		Per Year	Total
Labor Colony	2005-2014	10	2.7	27
Haji Saadullah CA	2005-2014	10	1.7	17
N-7 Old	2005-2014	10	1.4	14
N-3 Old	2005-2015	11	3.7	41
Al-Haded Quetta Cantt.	2003-2015	13	0.9	12
Agriculture College	2003-2015	13	1.9	25
Killi Nasaran Malezai	2003-2015	13	2.1	27
Chiltan Housing Scheme	2003-2015	13	1.9	25
GOR Colony	2003-2015	13	2.4	31
Killi Shabak K R	2003-2015	13	4.1	53

3. Strategies and Policies for Groundwater Management

To protect the ecosystem, biodiversity and natural resources different government agencies developed Strategies and Policies (SAP) for their respective domains and jurisdiction. Individual strategies encompass to protect and conserve the environment, effects of climate change, forests, rangelands, agriculture, poverty, river basins, watershed management and surface and GW resources.

The major aim of SAPs is to conserve the natural environment and to protect desertification and deforestation. Most of the SAP components and parameters are in replication, comparatively interconnected and overlapping. The SAPs devised in last two decades mostly from 1992 to 2015 at national and provincial levels. Here we present an overarching review and analyses of twenty national and provincial level SAPs. Most of the policies were prepared by respective federal and provincial ministries in collaboration with NGOs, national and international organizations, UN Agencies, and consultants. The conservation and protection of water resources has been the focal point of all individual plans, programs and SAPs. The SAPs are simplified and major components relevant to water resources of all individual policies are summarized in the following sections. The summary of the SAPs directly relevant to Balochistan is presented in table 3.

3.1 Federal Government's Strategies and Policies

In this section we will describe the different SAPs devised by the Federal Government. Some of the SAPs are devised simultaneously for all the provinces. While, some of the SAPs are specifically for individual provinces and separately for Balochistan. The Federal Government strategies are described in the following paragraphs.

3.1.1. National Biodiversity Strategy

International Union to Conserve Nature 2015 prepared the National Biodiversity Strategy, [4]. In Pakistan, the growth in human population puts extra pressure on natural resources. Likewise, excessive use of timber and non-timber forest products, use of wood, forage and expanding agriculture causing land degradation, desertification, loss of habitat and wild fauna, and the existence of wild species. The provincial rural lands are not managed by any government agency/department. It's a major reason for ecosystem degradation, overgrazing, and deforestation. The said areas are prone to air and water erosion, reduction in fertility of soil, waterlogging, and salinity. The degradation of the resource exceeds because of water scarcity, droughts, and mismanagement of available water and land resources. The loss of habitats is a grave hazard to biodiversity and a number of species confined to isolated pockets.

The National Biodiversity Strategy and Action Plan (NBSAP) provide an assessment of biodiversity and its trend in the country. The NBSAP presents specific objectives and goals and a plan of action along with coordinated arrangements and implementation measure. NBSAP has been prepared with the collaboration of national and provincial policy makers and planners, regional consultants and the World Conservation Union (IUCN). The NBSAP's major objective is to enhance biodiversity by the preservation of ecosystem, genetic diversity, and species. The major components of NBSAP covering different resources management and control regimes are presented in box 1.

Table 3. National and Provincial Strategies and Policies.

S#	Year	Strategies and Policies	Sponsored Agencies	
		National Strategies and Policies		
1	2015	National Biodiversity Strategy	GOP[a]	IUCN
2	2015	National Forest Policy	GOP	MO[c] Climate Change
3	2014	Millennium Development Goals	GOP	G-Water[d] Partnership
4	2012	National Sustainable Development Strategy	GOP	UNDP
5	2012	National Climate Change Policy	GOP	MO Climate Change
6	2010	National Rangeland Policy	GOP	MO Environment
7	2005	National Environmental Policy	GOP	MO Environment
8	2004	Agricultural Perspective and Policy	GOP	MO FA[e] & Livestock
9	2002	Pakistan Water Sector Strategy	GOP	MO Water & Power
10	1992	National Conservation Strategy	GOP	IUCN
		Provincial Strategies and Policies		
1	2014	Provincial Water Policy	GOB[b]	I&P[f] Department
2	2013	Poverty Alleviation Strategy	GOP	PPA[g] FUND
3	2011	Sustainable Development Policy	GOB	IUCN
4	2010	Participatory GW Management Strategy	GOB	Techno Consult Int.
5	2010	Ground Water Management Action Plan	GOB	Techno Consult Int.
6	2006	Integrated Water Resources Manag. Policy	GOB	I&P Department
7	2000	Balochistan Conservation Strategy	GOB	IUCN
8	2000	Strategy to Improve Governance	GOB	GKW Consult
9	2000	Water Sector Strategy	NGO	Taraqee Foundation
10	1992	Environmental Policy	GOP	Soil Survey Pakistan

[a]GOP = Government of Pakistan; [b]GOB Government of Balochistan; [c]MO = Ministry of; [d]G-Water = Global Water; [e]FA = Food & Agriculture; [f]I&P = Irrigation and Power; [g]PPA = Pakistan Poverty Alleviation

Box 1. The major components of NBSAP, [4].

1. Biodiversity awareness
2. Gender, poverty and biodiversity nexus
3. Biodiversity in national planning and policy
4. Terrestrial ecosystems, habitats, and species
5. Forest ecosystems
6. Inland wetland ecosystems
7. Coastal and marine ecosystems
8. Sustainable agriculture and agrobiodiversity
9. Sustainable production and consumption
10. Biosafety – Emerging issues and challenges

3.1.2. National Forest Policy

Ministry of Climate Change in 2015 prepared the National Forest Policy [5]. The country's land area under forests cover is only 5%. In most parts of the country the climate, water and soil conditions are not suitable for large-scale afforestation. Within their jurisdiction, the provincial governments developing and protecting forests. The country's vulnerability to climate change and its influences is quite evident. To protect the extreme environmental impacts, modification and adaptation procedures are the mainstay of the forest policy.

The forestry's scope should be extended immensely as it provides efficient measures to climate change as a global issue. It also tackles desertification together with air and water pollution and loss of biodiversity. Pakistan has various commitments and obligations as a signatory of international agreements and conventions. The commitments are related to forestry that requires national action plans, commitment and dedication of the provinces. An umbrella forest policy is prepared by the Federal Government for bolstering provincial forest policies in view of climate change. The objectives of the forest policy are highlighted in box 2.

Box 2. The major objectives of the forest policy; modified after MO Climate Change, [5].

1. Public awareness of socioeconomic, cultural and ecological standards of forests
2. Mass afforestation to increase forest coverage at global standards
3. Deforestation by regulating the trade and movement of timber
4. Establish protected areas and ecological corridors
5. Carbon footprint abating of energy in the economic sector
6. International convention's execution related to climate change, biodiversity & forestry
7. Consistent scientific planning of forests, education, and research

3.1.3. Millennium Development Goals

The Global Water Partnership has prepared Millennium Development Goals in 2014, [6]. The UN Conference on Sustainable Development organized in Rio de Janeiro in 2012, for a consultative exercise to consider post-2015 development framework. The outcome of the conference was to advance an agenda for member countries in the sustainable development through a consultative process for post-2015. It has been planned to create an intergovernmental procedure for the Sustainable Development Goals (SDGs) for General Assembly in 2015.

The SDG consultations covered eleven themes including water. The UN-Water, in collaboration with the Global Water Partnership (GWP), developed a document for the Open Working Group consultations held in 2014. A consultative procedure commenced in 30 countries, including Pakistan. The GOP and the Pakistan Water Partnership extended backing and possession to the UN Procedure and hold a consultation process. The Millennium Development Goals set for Pakistan, the progress achieved by 2013 and the processes to be achieved by 2030 was discussed. The outcomes of the workshop are outlined in box 3.

Box 3. Millennium Development Goals for drinking water, sanitation & hygiene, [6].

Access to safe drinking water, sanitation, and hygiene;

1. Investment with Public-Private Partnership to improve services for WASH
2. Enforce governance to control water thefts
3. Civil-Society contribution in delivering services to the poor communities

Water Resources;

1. Institutions to ensure sustainable withdrawal and uses of surface and GW
2. Develop pricing structure for domestic, agriculture, and industrial waters
3. Address water demand and equity in water accessibility and increase storages

Water Governance;

1. Acquire consensus for the formulation of National Water Policy
2. Strict water governance laws for the National Water Policy
3. Strengthen the water sector to accomplish targets

Managing Waste Water and Pollution;

1. Convert waste into wealth
2. Follow the principle of 3R: Reduce – Recycle – Reuse
3. Coordinate private sector for the management waste
4. In sub-sectors control and minimize pollution

IOP Conf. Series: Materials Science and Engineering **414** (2018) 012023 doi:10.1088/1757-899X/414/1/012023

3.1.4. National Sustainable Development Strategy

National Sustainable Development Strategy (NSDS), Pakistan's Pathway to a Sustainable and Resilient Future was developed by the GOP and United Nations Development Program (UNDP), [7]. This Strategy was established that climate change relates to sustainable development therefore careful and target-oriented planning is necessary.

Pakistan contributes 0.8% to global emissions and sufferer of worst casualties, a fact reflected in the various indexes which monitor climate change. The serious impacts are increasing in the frequency and severity of floods, droughts, cyclones and heat waves which cause severe damage to the lives, property, and agriculture. The NSDS implementation is a challenge to evolve, it may not impose. It can effectively reset the country's development and growth trajectory on to a sustainable pathway. A continued support and policy shift towards sustainable development concerns, through a network of institutions with the highlighted needs of the country and the global sustainable development program. The NSDS related to Environmentally Sustainable Development and the growing issue of Climate change and Sustainable Development are outlined in box 4.

Box 4. National Sustainable Development Strategy, [7].

Sustainable and inclusive economic growth
Social and human development
Environmentally sustainable development
1. Environmental sustainability
2. Air quality, water pollution, and water quality deterioration
3. Management of solid and hazardous waste
4. Forestry and biodiversity protection, and land degradation

The issue of sustainable development and climate change
1. Natural disasters and risk reduction
2. The economic costs of adaptation and sustainable development
3. Sustainable development and climate mitigation
4. Sponsoring sustainable development
5. Institutional response to climate change

3.1.5. National Climate Change Policy;

The National Climate Change Policy (NCCP) was developed by the Ministry of Climate Change, [8]. The Pakistanis are experiencing the impacts of climate change in all the provinces and across the country. The devastating floods, prolonged droughts, and extreme temperature waves are happening simultaneously in different parts. These climatic changes come along with extensive values and huge economic burden on effective communities.

The NCCP broadly covers the probable challenges and effects of climate change for future adaptation and mitigation. It offers an outline and an action plan for the climate change agenda and projects. The policy addresses the present and future expected climate change threats to various sectors in the country. The policy represents the adversative influences of climate change and extreme events and country's vulnerability. The implementation framework of NCCP document was prepared in 2013, [9]. The main objectives of Pakistan's Climate Change Policy are presented in box 5.

Box 5. Pakistan's Climate Change Policy, [8].

1. Sustainable economic development to tackle challenges of climate change
2. Gender-sensitive and pro-poor adaptation of cost-effective means
3. Ensure food, water, and energy security to eliminate challenges of climate change
4. Minimize extreme weather events effects like floods, tropical storms, and droughts
5. On climate change reinforce inter-ministerial coordination mechanisms
6. Increase the skills, awareness, and institutional capacity of communities
7. Sustainable conservation of natural resources

3.1.6. National Rangeland Policy;

The National Rangeland Policy prepared by the Ministry of Environment, 2010, [10]. The rangelands play an important role in the livelihoods of communities through the rearing of animals and gaining different products and facilities. The rangelands provide feed and forage to livestock. The ranges useful for water resources, recreation activities, wildlife, and fish habitat. The rangelands enhancing water infiltration, leading to GW recharge, sustained base-flow of streams and confine soil erosion. The social and economic parameters, ecological stability and shielding ecosystems are related to rangelands. In Pakistan, the rangelands span-over 60% of the total land area of the country, while in Balochistan about 79% of the total area, [11].

The ranges inhabitant a fragile ecosystem and due to overgrazing the productivity of rangelands and the biodiversity have declined considerably. Due to mismanagement of rangeland soil erosion augmented, land degradation accelerated, biodiversity and reduced productivity. The overall goal of the policy is to rehabilitate and restore the rangelands and pastures and increase productivity. Also, to enhance environmental and regulatory functions, conserve rangeland biodiversity and to mitigate negative impacts of climate change. The rangelands policy comprised eleven important thrusts described in box 6.

Box 6. The Rangelands Policy Thrusts; modified after MO Environment, [10].

1. Monitoring and assessment rangeland resource
2. Integrated rangeland resource planning
3. Rehabilitation and management of range resource
4. Increase forage production on other lands
5. Elevation of rangeland enterprises
6. Rangeland management for sustainable base-flow and curtail sedimentation
7. Rehabilitation ecosystem to conserve biodiversity
8. Global warming and climate change impact mitigation
9. Train farmers and capacity enhancement of relevant stakeholders
10. Introduce range research, technology development, and transfer
11. Generate financial resource

3.1.7. National Environmental Policy

The National Environment Policy (NEP) was prepared by the Ministry of Environment, [12]. The policy addresses the environmental issues facing the entire country. The environmental issues are related to pollution of fresh and coastal water resources. The other environmental issues are climate change, mismanagement of waste, loss of biodiversity, desertification, and deforestation, aridity and natural disasters. The policy addresses cross-sectoral issues and environmental degradation and meeting international obligations. The policy developed to conserve, protect, and restore the environment to improve the quality of people's lives through sustainable development. The major sectors of the NEP are summarized in the box 7.

Box 7. National Environment Policy, [12].

Sectoral guidelines

1. Water supply and management
2. Forestry, biodiversity and protected areas
3. Climate change and ozone depletion
4. Agriculture and livestock
5. Multilateral environmental agreements

Cross-sectoral guidelines

6. Population, poverty and environment
7. Health, gender and environment
8. Trade, local governance and environment
9. Natural disaster management

3.1.8. Agricultural Perspective and Policy

The Agricultural Perspective and Policy was prepared by the Ministry of Food, Agriculture and Livestock, [13]. The agriculture sector is the lifeline and the single largest sector of Pakistan's economy. It contributes 24% to the GDP and employs 48.4% of the labor force. It also contributes 70% of the foreign exchange earnings through the export of raw materials, semi-processed and processed agricultural products. About 68% of the population live in rural Pakistan and depends upon agriculture for sustenance. Pakistan is facing a comparatively low productivity level of crops due to gaps in yields of crops on a different type of farms and because of water availability in dry and wet years. To smoothen the supply of agriculture water it seems appropriate to tame the river flow system. A great irrigation system was set up to divert and distribute water supplies through canals from the Indus Basin Irrigation System established by link canals and water reservoirs. The recent droughts exposed the vulnerability of the IBIS and additional storage facilities felt to increase supply with operating flexibilities. Water supply systems are expensive, groundwater development is at its absolute limits, and water conservation may add 10-15% of total supply by the enhanced efficiency of an irrigation system which is 40%. To increase the agricultural productivity of high value crops, a number of policies and strategy measures were established. The Pakistan's Agricultural Policy concentrated on the sustainable domestic food security. Special emphasis was given to water management, conservation and increase in the productivity. The water management and conservation in agriculture is linked to many factors as listed in box 8.

Box 8. Water Management and Conservation, [13].

1. River water diversions system
2. Water supply situation and mobilization
3. Water requirements and conservation
4. Water vision and water issues
5. Water management strategy and program
6. Improvement of water productivity

3.1.9. Pakistan Water Sector Strategy

The strategy for the water sector was prepared by the Ministry of Water and Power, [14]. Pakistan did a lot of efforts for the development of surface and groundwater resources. In this regard, many projects accomplished to strengthen the infrastructure for water delivery system, constituting and world's biggest canal irrigation system developed. With the increase in population, the country's water demand is also increasing, ultimately the country becoming a resourceful country to a water scarce country. It's required to manage water resource to ensure sustainable supply to all sectors.

The government of Pakistan and the Asian Development Bank jointly introduced the Water Resources Strategy Study. The study was commenced by the Office of the Chief Engineering Advisor/Chairman, Federal Flood Commission, Ministry of Water and Power. The major objectives of the study are to formulate a roadmap for water sector development for a proficient delivery system and conservation of resources for future demands.

The government has addressed the water sector and took many incentives in this regard. The important incentives include Ten-Year Perspective Plan of Planning Commission, [15]; Vision 2025 of Water and Power Development Authority [16], and the National Water Policy of the Ministry of Water and Power [17]. The national water sector profile covers the sub-sectors including; development of water resources, rural and urban water supply and sanitation, pollution control, industrial water supply, irrigation and drainage, environmental and flood protection. The Pakistan Water Sector Strategy provides a roadmap for the sectors future needs based on planning approach. The strategy is applicable across the water sector and the main points are summarized in box 9.

Box 9. The Water Sector Strategy; modified after Ministry of Water and Power [14].

1. Strengthen agencies responsible for water resource delivery
2. Ensure private sector participation and community involvement in the water sector
3. Minimize spending in the water supply by obligatory duties on water services
4. Confirm financially self-sustaining services in all sub-sectors
5. Improve private sector investment in water supply, improve data in the water sector
6. Ensure public awareness and education on water issues
7. Adopt the principles of IWRM, develop water storage facilities
8. Promote managing watersheds for sustainable water resources
9. Ensure water utilization for socioeconomic growth
10. Progress conditions of equity in access to all sorts of water utilization
11. Improvement in water quality, coverage, and dependability of domestic supply
12. Increase water quality of water bodies, rivers, and GW by regulations
13. Ensure water access to fulfill the community needs, in all water sectors
14. Ensure sufficient water use for agriculture and food requirements
15. Improve productivity and expenditures in irrigation through Farmer Organizations
16. Develop hydropower, control waterlogging, salinity and disposal of saline drainage

3.1.10. National Conservation Strategy

The National Conservation Strategy (NCS) was developed by the International Union to Conserve Nature, during 1983-1992, [18]. The NCS is formulated for to environmental conservation of the country. In addition to protecting the natural resource and incitation of environmentally friendly projects. The guiding principles of the NCS ensure integrated exertion environment protection and to conserve the natural resources. The strategy provides wide-ranging references to relevant government departments, public and private sector corporations, donor agencies, financial institutions to work collectively for sustainable development in the respective field of studies. The NCS has eight core program areas for the priority implementation are given in box 10.

Box 10. Core programs of National Conservation Strategy, [18].

1. Improve soils in croplands, increase irrigation efficiency
2. Protecting watersheds, supportive forestry & plantations
3. Rangeland restores and livestock improvement
4. Shielding water bodies and biodiversity conservation
5. Developing and deploying renewable
6. Preventing pollution, manage urban wastes
7. Integrating population and environmental programs
8. Preserving cultural heritage

3.2 Provincial Government Policies

In the provinces, the provincial governments are responsible for the management of natural resources. The provincial government of Balochistan devised different strategies for the management of natural resources. The SAPs devised by the Balochistan Government are described in the following paragraphs.

3.2.1. Provincial Water Policy

The Provincial Water Policy was prepared by the Irrigation and Power Department (IPD), [19]. The provincial IPD, Government of Balochistan has a mandate and responsibility to ensure adequate and sustainable water resources management through the holistic development and efficient use of available and new water resources. The IPD is also engaged with rainwater harvesting through efficient watershed management, and to curtail uncontrolled withdrawal of GW by 2030. The improvised cropping patterns for suiting to the environment and water situation and the use of spate irrigation system is essential. Introduction of efficient irrigation technology is mandatory. The efficiently plan and develop water budgets to meet the requirement of different sectors on the actual requirement basis. In addition, institutional reforms, capacity building through advanced learning and research at all levels, effective response against the challenges are also the responsibilities of the IPD. To achieve the defined targets, the Irrigation and Power Department has formulated the Provincial Water Policy to effectively encounter the immense challenges. The basic document of water policy was approved by Provincial Cabinet, but the same has not been transformed into an Act for legal coverage. The provincial water policy is aimed to achieve certain objectives, presented in box 11, [19].

Box 11. Major objectives of the Provincial Water Policy, [19].

1. Efficient conservation and management of available water resources
2. Harnessing of flood and rainwater through efficient watershed and surface water management and development of surface water bodies
3. Efficient GW recharge measures to arrest sharp depleting GW levels
4. Increase of GW exploitation in areas where potential exists
5. Measures for water losses, equitable water distribution among beneficiaries
6. Efficient drainage system to mitigate waterlogging and salinity
7. Identification of technically feasible and economically viable and community-oriented schemes through professionals and timely completion of the projects
8. Institutional reforms and capacity building to make the organization and professionals equally dynamic and responsive future challenges ahead for assuring the sustainability of water resources

3.2.2. Balochistan Poverty Alleviation Strategy

The Balochistan Strategy organized by the Pakistan Poverty Alleviation Fund, [20]. Being least developed province, Balochistan needs integrated development approach through strengthening related institutions for the development and social progress. To address the current socioeconomic and political malaise in the province, Pakistan Poverty Alleviation Fund (PPAF) started consultative processes for community involvement, politicians, government departments and local NGOs. Ultimately, the PPAF developed a proficient organic development strategy for deprived communities of Balochistan. The main objective of the strategy is to improve the socioeconomic conditions of the community empowerment. It's imperative to bring them to the rural communities in the mainstream by transformation and involvement in decision making. The preservation and respect to the local culture, traditions, customs and identities are the focal point during the working norms with the communities. The PPAF had the experience to assist diverse communities to improve their skills, capabilities and loud their voices take part in the development process. The environmental supporting functions are outlined in box 12 along with strategy goals.

Box 12. Poverty Alleviation Strategy goals & environmental supporting functions, [20].

Gole-1	Hunger and poverty eradicate
Goal 2	Attain primary level education
Goal 3	Encourage gender equality and women's involvement
Goal 4	Diminish child mortality
Goal 5	Advance maternal health
Goal 6	Eliminate malaria, HIV/AIDS, and other diseases
Goal 7	Ensure sustainable environment
Environmental Supporting Functions	Water conservation, supply, and sanitation, and Kariz system Management of sustainable water resources & dispersing processes Disaster preparedness, flood water storage & drought mitigation Skill development and sustainable management practices

3.2.3. Sustainable Development Policy

The Sustainable Development Policy prepared by the International Union to Conserve Nature, [21]. The Integrated District Development Vision (IDDV) established by a consultative process. The existing situation, challenges, measures, issues, opportunities, and constraints are related to governance, natural resources, social sectors, disaster management and economic development. The prime objectives of the IDDV are to the sustainability of natural resources, environment protection, and sustainable development. The water supply, sanitation, healthcare, population, settlements, education, industry, commerce, energy, mining, wildlife, forests, poverty eradication, protected areas, and governance is the prime sectors.

The air, water, food and relevant products of daily consumption and trade are required for natural resources to sustain life. The socioeconomic and the survival of rural communities depend on natural resources. The mismanagement of natural resources and natural disasters such as droughts and floods may reduce the productivity which affects them badly. The climate, water, land-use, agriculture, livestock, rangelands, forests, wetlands, are the key elements of natural resources of rural communities. The development agencies are working to combat the depletion of natural resources and for the sustainable development. The optimistic characteristics of districts are people, forests, agriculture, rangelands, watersheds, biodiversity, mineral deposits and protected areas. The aridity, fast depleting GW, periodic and overgrazed and degraded rangelands, low productivity of livestock, rugged terrain and droughts are major problems. The population density is very high, it's required to stop and reverse the degradation and depletion of natural resources by promoting and utilizing alternate resources. The major issues projected to address on a priority basis for the sustainable development of the natural resources is presented in box 13.

Box 13. Key issues for protection and sustainable development of natural resources, [21].

1. Extreme aridity and periodical drought.
2. Fast depleting GW because of indiscriminate mining, subsidy on the power supply for agriculture, and changing the land-use in the recharge zones.
3. Depleted, overgrazed, low-productivity of livestock and rangeland mismanagement.
4. The degradation and decline in wildlife populations and habitats.
5. Degradation of the forest, deforestation due to overexploitation and mismanagement.
6. Communities' poverty related to degraded natural resources for their livelihood.
7. GIS Coordinators, GIS-based information databases, and lack of scientific research
8. Insignificant funds for natural resource management in livestock, rangelands, agriculture, wildlife, forests, and for protected areas.
9. Exploitation of water resources, gas, firewood, and electricity.

3.2.4. Groundwater Management Action Plan

The Groundwater Management Action Plan was prepared by the Techno-Consult International, [21]. The Balochistan Small Scale Irrigation Project (BSSIP) sponsored by the World Bank and GW Management Action Plan prepared for Pishin Lora Basin and implementation of a Pilot Program in Quetta Sub-basin. The study was conducted by the Techno-Consult International with the collaboration of Associated Consulting Engineers and CAMEOS Consulting Engineers. The consultants prepared an action plan that comprises several compounding factors which critically influence the management of provincial water resources and provide a firm basis for justifying the need for a GW management action plan. The prime goal of the management plan is to arrest the declining trend in water levels and deterioration of its quality and to stabilize the system at a certain level as determined with the consultation of stakeholders through structural and nonstructural measures. The management strategy includes exact assessment of the resource and prioritizes sectors and adoption of measures for management and utilization of the resource. The major GW management issues are given in box 14.

Box 14. The GW management action Plan, [21]

1. Water Resource Assessment and Monitoring
2. GW Level and Quality Monitoring
4. Institutional Limitations and Lack of Coordination
5. Poor Governance
6. Enforcement of Legislation
8. Stakeholder Consultation
9. Political Involvement and Malpractice
10. Advanced Technology
11. Poor Cost Recovery
12. Lack of Participatory GW Management

3.2.5. Participatory GW Management Strategy

The Participatory GW Management Strategy was drafted by the Techno Consult International, [22]. In Balochistan, no GW management system exists, at the watershed level, it is a local problem and needs to be analyzed and managed at local level. The communities should involve in managing GW with a holistic approach of a natural resource. The primary goal of the management is to arrest the declining trend in GW levels. The system must be stabilized at a certain level determined with the consultation of all the stakeholders through structural and nonstructural measures. A clear management strategy, including exact assessment of the resource, prioritization of use for specific sectors and adoption of measures for management/utilization of the resource specifically during drought should be adapted. The regulatory and enforcement mechanisms for management are deemed to fail in the developing economies. The last option is to involve local communities to educate and motivate the people to adopt sustainable means to use GW. The participatory GW management strategy must be adopted to tackle the problem at the grass root level. Farmer organizations formed for the small-scale irrigation schemes would also be motivated to start the GW management activities in the command and catchment areas with properly defined roles. Hydrologic principles, GW flow, monitoring of water table should be postulated with the assistance of farmers, and tentatively calculated monthly GW budgets. The major points of the participatory GW management strategy are given in box 15.

Box 15. Participatory GW Management Strategy; modified after Techno Consult Int., [22].

1. The Regulatory and Enforcement Mechanisms
2. User participation in GW Resource management
3. Social Awareness for GW Management
4. Knowledge of aquifer, GW situation and implications of GW depletion
5. Historical changes in land use, reliance on various water sources and pressure on GW
6. Participation rural communities for monitoring GW and preparing water budget
7. Encourage community to manage the GW resources as a common commodity
8. Communities in rural and urban areas are aware of the pressure on GW resources
9. Rural and urban tubewells operators and users understand the GW depletion rates
10. Communities develop and adopt mechanisms for community management of GW
11. Communities adopt measures to conserve water and economies in their use of water
12. Tubewells owners know the benefits of adopting high-efficiency irrigation systems
13. Urban communities understand the implications of GW exploitation for sustainability
14. Stakeholders understand why the power subsidy needs to be addressed
15. Participatory Hydrologic Monitoring and Micro Planning

3.2.6. Integrated Water Resources Management Policy Balochistan

Integrated Water Resources Management (IWRM) approach utilized to formulate the Water Policy by Irrigation and Power Department, [1]. The IWRM Policy comprising fifteen policy thrust areas, which is vital for improvement and sustainability of water resources. The policy thrust areas were recognized

IOP Conf. Series: Materials Science and Engineering **414** (2018) 012023 doi:10.1088/1757-899X/414/1/012023

by assessing sub-sectors, issues of water use. The assessment is a real-world approach as compared to the traditional one that focuses on water resources.

The effective implementation of the policy will help various stakeholders to implement the policy for their relevant sub-sectors. To evaluate issues and formulating policy and reforms, the water-poverty-environment framework utilized as evaluation criteria. The basinal approach is recommended to manage the water resources in a complete way. For basinal planning, the watershed-water-farming systems were followed to enhance water productivity and water use sustainability. The modified twelve-policy thrust area for the IWRM Policy are given in box 16.

Box 16. IWRM Policy Thrust Areas, [1].

1. Water availability and development potential
2. Water resource monitoring, assessment & demand management
3. Linking water development with IWRM approach for agriculture
4. Crops and cropping pattern adjustment with water availability
5. Environmental water management
6. Cost recovery of irrigation infrastructure
7. Cost effectivity of water conservation interventions
8. Endorsing inter-provincial cooperation
9. Community participation
10. Institutional strengthening and restructuring
11. High-efficiency irrigation systems
12. GW management and development

3.2.7. Balochistan Conservation Strategy

Balochistan Conservation Strategy (BCS), prepared by the Planning and Development Department of the Government of Balochistan in collaboration with IUCN, [23]. Balochistan faces challenges with major environmental issues and climate-change challenges. The shortage of water, degraded pastures and forests have been the subjects of concern for many decades. The settlements categorized by poor planning, contaminated water supplies, air pollution, inappropriate sanitation and waste management services. With the increase in population, the situation becomes worse as the urbanization accelerated.

The BCS devised to protect the natural environments of the province through community participation and involvement. The strategy designed to tackle the issues of socioeconomic development based on sustainability and conservation of natural resources. The overall goal of the strategy is the welfare of communities and protecting ecosystems, the entire strategy related to water resources of the province. The principles and objectives of the BCS are to provide a workable strategic plan for environment conservation and sustainable utilization of natural resources. The Strategy was also meant to promote social change for environmental protection, natural resources, and biodiversity. It's also facilitating access information for better decision-making and to improve the mechanisms for endorsing public awareness and support for the sustainability of natural resources. The modified action agenda of the BSC is comprised of seven core programs given in box 17.

Box 17. The core action agenda of the BCS, [23].

1. Improve governance, effective institutions, and authentic data management
2. Controlling depletion and pollution of GW increasing irrigation efficiency
3. Agricultural sustainability and poverty alleviation
4. Managing rangelands and enhancing the productivity of livestock
5. Sustainable planning and management of urban areas
6. Sustainable environment-friendly development
7. Conserving biodiversity, wetlands rehabilitating and forests sustainably

3.2.8. Strategy to Improve Governance
In Balochistan, all water supply schemes depend upon GW including Quetta Water Supply Scheme. Over-pumping from tubewells at a rate that exceeds the natural recharge, lead to a situation that the aquifers have exhausted in almost all parts of the province. If exploitation continues at current trends considerable water losses and shortcomings in resource management aggravate the situation. Responding to these conditions, the Government of Balochistan has decided to carry out the "Quetta Water Supply and Environmental Improvement Project", with Technical Assistance (TA), from the Asian Development Bank and GKW Consult, [24]. The TA is to prepare water resource management projects for conservation and development of Quetta. The TA assessed various policy options and institutional arrangements to encourage efficient water allocation by consumption-oriented pricing.

The actual institutional arrangements in Balochistan, characterized by a multitude of institutions interfering in the sectors concerned, and with overlapping responsibilities and insufficient coordination are the main reason. In the absence of coherent water resource and environmental management policies resulted in the inadequate performance of the operating agencies. Institutional strengthening, streamlining of responsibilities and capacity-building for political as well as an operational level are an important precondition to achieving rational water resource and environmental management. The TA proposed institutional setup at the upper level by the establishment of a single Water Regulatory Authority by merging BWASA & PHED. Like BIDA, an authority may be created as Public Health Engineering Authority to effectively manage the water resources in rural and urban levels. It is proposed that the Balochistan Water Resources Management Agency be established to coordinate the planning, evaluation, and monitoring of plans for the integrated development of the land water and the underground resources. The Balochistan Water Board should be established to prepare, implement, and monitor the plans. The proposed institutional setup to improve governance summarized in box 18.

Box 18. Proposed institutional setup to improve water governance, [24].

1. Balochistan Water Board, (BWB);
   Establish by reconstituting provincial water board to formulate policies and rules
2. Balochistan Water Resources Management Agency, (BWRMA);
   Establish by upgrading Bureau of Water Resources as an executing agency
3. Existing District Water Committee;
   Reconstituted for the implementation of policies at the district level
4. Water Resources Agency;
   Act as the board's executing agency, organized into three functional departments
   - *4.1 Water Resources;*
     For monitoring, research, planning, project and technical specifications
   - *4.2 Water Rights and Licenses;*
     Technical approvals, contracts, legal affairs and water conservation
   - *4.3 Administration and Finance;*
     Personnel, revenues, finance and general affairs

3.2.9. Water Sector Strategy
The water sector strategy formulated by Taraqee Foundation [25], Taraqee is a Quetta based local NGO, working in Balochistan since 1994. The Foundation has been actively engaged in grass root level activities in the community and physical infrastructure, microfinance, basic health services, gender and advocacy, education, child protection and the environment. The foundation recognized the fact that water due to its multi-faced roles in the economic and health sectors significantly affects the poverty levels of marginalized communities. Considering that water is a finite resource, the water strategy strives to create a more sustainable water resource base in the province. The strategy will encourage communities to take part to conserve and protect natural resources, diversity in shared habitats. The communities should learn from nature and natural processes to attain environmental health, social comfort, and economic benefits. The strategy helps to maintain the sustainability of

natural resources, maintaining ecosystem, ecological integrity and provide appropriate conditions to maintain self-sustaining communities in a better ecological environment. The water strategy of Taraqee Foundation is based on certain principles described in box 19.

Box 19. The guiding principles of the Water Sector Strategy, [25].

1. Promoting good governance in the water sector
2. Empowering communities to become self-sustaining
3. Ensuring equity in decision making
4. Enabling indigenous and traditional knowledge and water rights
5. Improving security and reducing conflict
6. Ensuring Women's participation

3.2.10. Provincial Environmental Policy

The provincial Environmental Policy was prepared by the Soil Survey of Pakistan, [26]. Environmental Profile of Balochistan presented a firsthand information about the gloomy situation of degraded environmental conditions. Desertification is well advanced and accelerated by human and small stock population explosion as well as by excessive pumpage of GW. The decline of biodiversity is also associated with desertification process. The policy stated that the GW decline and degradation of natural resources and pastures will demise the small stock industry and decrease the productivity of GW-based orchards. The predicted ecological disaster also implies the economic collapse of the natural resource-based economy. The existing forests and rangeland degradation, devegitation, GW depletion, salinization, erosion, and biodiversity decline are eminent in every part of the province. The effectiveness of legislation and Public Administration is achieving poor environmental sustainability. The first objective of the provincial environmental policy is a sustainable use of renewable natural resources and the safeguarding of the environmental quality. The second objective is the safeguard of environmental quality, the chemical and biological quality of water, air, and soil in relation to human health. The main points of the policy are summarized in box 20.

Box 20. Provincial Environmental Policy, [26].

1. Integrating population and environment programs
2. Deforestation of watersheds and depletion of aquifers
3. Desertification and de-vegetation of rangelands
4. Waterlogging and Salinization of irrigated lands
5. Exploitation of marine and coastal resources
6. Declining of wildlife habitats, wild flora and fauna
7. Damaging mining practices, water and air pollution
8. Unplanned urban settlement and industrial growth
9. Unmanaged urban and industrial solid wastes

4. Outcomes of previous strategies

The main objective of SAPs are the conservation, restoration and efficient management of environmental resources. The integration of natural resource considerations in policy making and planning processes. The capacity building of government agencies, NGOs and other stakeholders at all levels for better management of water resources. For sustainable natural resources management from water governess to community empowerment are integral part of SAPs. Specific measures have been proposed in previous strategies for the management of surface and GW resources. The measures included, ban on agriculture tube wells in urban areas, development of new water resources, construction of water storage and delay action dams (DADs) etc.

4.1 Groundwater Recharge

For groundwater recharge the construction of DADs has been proposed in different strategies, [27]. Subsequently, 326 DADs with a storage capacity of 336 Mm^3 were constructed as shown in table 1. The construction of 100 DADs is in progress and approximately 80% work has reportedly been completed. To assess recharge through DADs, studies were conducted on 14 and 25 dams during 1997 and 2008 respectively, [28]. The studies show that estimated seepage of GW through fourteen DADs was 5.46 Mm^3/y and 28.38 Mm^3/y from remaining twenty-five DADs. The estimated discharge through Karizes located at the downstream of 25 DADs increased from 6.86 Mm^3/y to 24.27 Mm^3/y, table 4. These studies represent that in current circumstances, the DADs are appropriate means to recharge aquifers if supported by integrated watershed management strategies. The effectiveness of DADs may further be enhanced by increasing the height of the dam and storage capacities to supply for downstream community utilization. The GW recharge may also be increased by introducing outlet systems in the dams by releasing sediment-free water for recharging through river beds.

Table 4. GW seepage from DADs and Kariz Flows, [28].

Delay Action Dams	Water Depth (m)	Seepage (000, m^3)		Kariz Flow (000, m^3)	
		Before	**After**	**Before**	**After**
1 Balozai	5.9	No seepage studies conducted before the construction of DADs	-	89	1,020
2 Gharagai	5.4		993	107	776
3 Ghazloona	0.8		927	Dry	Dry
4 Lalai	6.7		1,042	179	1,053
5 Nari Kach	11.0		983	Dry	1,539
6 Batogi	1.2		881	-	-
7 Chill Bagho	1.9		1,188	-	-
8 Sangar	0.5		-	268	1,062
9 Wulgi	2.0		1,088	-	-
10 Akhter Nika	10.9		2,047	2,233	5,771
11 Chinjann	1.7		1,284	268	853
12 Darloon	1.9		574	89	882
13 Tore Khezai	4.3		1,331	Dry	Dry
14 Amach	1.1		-	Dry	292
15 Isplangi	1.5		1,111	-	-
16 Piyyan	2.2		563	-	-
17 Brewary	2.3		359	-	-
18 Hangi & Takhtani	3.4		405	-	-
19 Tangi Dabari	10.2		1,103	Dry	980
20 Kapip	4.1		10,706	-	-
21 Babar Wulmai	8.7		748	500	953
22 Kawas	5.8		-	535	4,203
23 Zandra	2.1		-	Dry	1,024
24 Aghbargi	9.4		1,045	357	1,463
25 Kannar	0.5		-	2,233	2,401
Total	-		28,378	6,858	23,980

4.2 Mismanagement of Natural Resources

In Balochistan, overexploitation of natural resources and climate change affected the natural resources, communal life, and economy of the province. Low precipitation, fewer snowfall events and increased temperatures resulting in lengthier and severe droughts. Furthermore, depleting water resources, less productivity of agriculture and rangelands are the major issues. Agriculture is the main source of income in rural communities, it's not sustainable due to inefficient use of GW. The rangelands are overstocked, depleted and have low productivity. For the poverty eradication livestock and rangelands are important for rural communities. The GW is depleted because of over-exploitation, increase in population and by inefficient irrigation and agriculture. The poor planning, mismanagement, and overexploitation of natural resources substantially reduced the natural recharge of aquifers. Among other causes are, the lack of watershed management activities in the catchment areas and residential societies in potential water recharge areas. The IUCN reflected mismanagement of natural resources in the Integrated District Development Vision Balochistan, [21]. The major causes of mismanagement of natural resources are arranged in box 21.

Box 21. Causes of Mismanagement of Natural Resources; modified after IUCN, [21].

1. Inappropriate subsidy on the power supply and reduced recharge due to de-vegetation and land use in potential water recharge areas.
2. Depleted, overgrazed and low-productive rangelands and livestock due to negligence
3. The decline in wildlife population, ecosystems habitats, deforestation, degradation, and overexploitation of vegetation
4. The poverty of communities related to natural resources
5. Little funding for livestock, rangelands, agriculture, wildlife, forests & protected areas
6. Inefficient utilization of water, gas, electricity, and firewood

4.3 Improvement of GW Resources

The management of GW is a multifaceted association between communities, environment, integrated watershed management and the relevant SAP's. The overexploitation of environment and natural resources are among the major causes of GW depletion. The options for improvement include; watershed management, community participation, increasing rainwater harvesting, GW recharge, efficient irrigation systems, and water governance. In some of the previous studies, certain measures have been proposed for the improvement of water resources, which included an Integrated District Development Plan; IUCN, [22]. The outcome of the studies carried out by IUCN for the improvement of water resources is presented in Box 21, [22]. A firsthand water resources management strategy can be attained by the combination of the major causes of mismanagement of natural resources as outlined in Box 20, and the study's outcome for the improvement of water resources outlined in box 22.

Box 22. Studies Outcome for Improvement of Water Resources, [22].

1. Water loss reduction and leakage control of supply
2. A complete ban on agriculture tubewells in urban areas
3. Recycled wastewater in urban areas to farmers
4. Construction of DADs, storage and supply dams
5. Development of new water resources and recharge measures
6. Environmental Improvement Project for drinking water
7. Effective monitoring and metering of the system
8. Construct wastewater recycling plants to irrigate land
9. Drainage system's rehabilitation

5. Integrated Watershed Management Strategy

The Balochistan Plateau is comprised of 18 river basins that are divided into 73 sub-basins, numerous watersheds, and many-more subwatersheds. The river basins encompassed drainage areas from 1,115 Km^2 of Kand River to 84,916 Km^2 of Hamun-e-Lora. It's challenging and occasionally impossible to manage a river basin of 84,916 Km^2 in an arid climate with little resources. The subwatersheds require exceptional management scopes and needs, planning issues, priorities from monitoring to data collection and implementation. The integrated watershed management is a process to manage human activities and natural resources on a watershed basis. Further, to consider social, economic and environmental issues, and community interests to manage water resources sustainably, [30].

To protect and conserve natural resources the Integrated Watershed Management Strategy (IWMS) is developed. The IWMS is prepared with the critical assessment of the twenty SAPs related to federal and provincial governments. The twenty SAPs are revised, enhanced, reformed and presented in the boxes from 1-22. For the sustainable development of water resources appropriate planning, management, conservation and protection measures have been proposed in the subsectors. The IWMS is a flexible yet comprehensive strategy that may be updated anytime whenever additional data are available. The updates and modification can be made with the mutual discussion, consultation and participation of the concerned government agencies, water experts, NGOs, representatives of the civil society and end beneficiary communities. The strategy comprised of five leading sectors with 27 subsectors to conserve natural resource relevant to surface and GW management and is presented in box 23.

The water governance is a multi-level participation process beyond the government, where the decision is made by public institutions, civil society, the private sector, and by the communities. It encircles the relationships between governments and societies, including laws, regulations, institutions, and interactions that affect the ways in which governance systems function, [31]. The principles of water governance are imperative for the implementation of effective policies. The policies help for responsible groundwater use, equity, sustainability, and efficiency. In this background, the establishment of the Balochistan Water Board, Balochistan Water Resources Management Agency and upgrading the Bureau of Water Resources are recommended.

The environmental concerns are related to global warming and climate change. Appropriate policies and measures are required to minimize the effects of weather events of extreme nature such as storms, floods, and droughts. Environmental management is associated with water, air, solid and hazardous waste contamination and pollution. To mitigate the environmental impacts, the strategy ensures water and food security and challenges posed by the climate change. The sustainable rural environment for communities is an important issue to address with an increase in population.

The conservation and sustainable development initiated with institutional research, monitoring, and GIS-based information databases. It is also addressed to the management of urban areas, coastal, marine resources, improve resources security and reduce conflicts, sustainable utilization of water, improve water supply coverage and quality protection.

The natural resource management is related to water, soil, plants, animals and lands. For the successful management of natural resources, the cumulative work of land use planners, water management, biodiversity conservation, and the future sustainability of agriculture, mining, tourism, fisheries, and forestry is imperative. It's important to improve the water quality of rivers, water bodies and GW through legislation and laws. The management is imperative to control deforestation, halt desertification, management of protected areas and ecological corridors, and sustainable water flows. The social and economic awareness of GW management, conservation, and resource allocation is also an integral component of resource management.

The community empowerment is an effort to create a partnership between government and communities with an emphasis to improve the welfare of rural communities. It is to empower individual communities to achieve desired results to improve their quality of life and living standard. The communities identified their problems and probable solutions with the support of the government departments. The NGO's, individual experts, businesses and developmental agencies work collaboratively with communities to solve their problems. All rural development programs would be identified by individual communities and implement by the best means for attaining the set goals of development. In this regard, the empowering legislation will encourage communities for collaborative efforts for the betterment of the entire community in all aspects.

Box 23. Integrated Watershed Management Strategy; compiled by the authors.

Water Governance:
1. Balochistan Water Board; establish to formulate policies and rules
2. Balochistan Water Resources Management Agency; upgrading Bureau of Water Resources
3. District Water Committees; strengthen to implement policies at district level

Environmental Concerns:
4. Water, air, solid and hazardous waste pollution management
5. Integrating population, gender, poverty, health, trade and environmental programs
6. Mitigating impact of global warming and climate change, such as floods, droughts & storms
7. Sustaining an environment-friendly development and provide environmental education
8. Improve water quality of rivers, water bodies and GW through legislation and laws

Conservation and Sustainable Development:
9. Improve institutional research, monitoring, GIS-based information databases & coordinators
10. Management of urban areas, coastal, marine resources, improves security & reduce conflicts
11. Sustainable utilization of water; improve water supply coverage and quality protection
12. Better conditions of equity for irrigation, urban and rural domestic water
13. Sustainable agriculture, agrobiodiversity, production, consumption, efficiency and demand
14. Water vision, sustainable withdrawal, conservation, management and productivity
15. Improve data collection network, water management, conservation, construction of DADs

Natural Resource Management:
16. Ecosystem rehabilitation for biodiversity conservation, wildlife habitats, flora and fauna
17. Biodiversity protection and conservation; gender, poverty and biodiversity nexus
18. Rehabilitation terrestrial, inland wetland, coastal ecosystems
19. Controlling deforestation, establishing protected areas and ecological corridors
20. Conserving, rehabilitating, developing, forests sustainably, scientific planning and research
21. Re-vegetation rangelands, halt desertification for sustainable water flow & livestock
22. Afforestation, harnessing flood and rainwater through efficient conservation in watersheds
23. Management of watersheds for sustainability of surface and GW

Community Empowerment:
24. Empowering communities for self-sustaining, social awareness and women's participation
25. Improving physical infrastructure of communities in tune ecosystem of surrounding area
26. Community understanding of water issues, eradicate poverty dependent on natural resources
27. Promote IWMS principles, demand assessment, development, monitoring and cost recovery

To support the IWMS a conceptual strategic model is prepared. The model represents the flow of major activities and their linkages with each-other as well as subsectors. The model is the outcome of a systematic and methodological analysis of IWMS, and related SAPs. The conceptual model provides an association from water governesses to community empowerment.

The water governance is the prime body to oversee and regulate the water management activities in the province from the top level. It is encircled all agencies related to natural resources, all water-related issues, and activities from policy to implementation through communities. The water governance will also oversee the work of all government agencies related to water management and conservation.

The individual government departments will prepare SAPs with the consultation of all stakeholders for the sustainable management of natural resources. The concerned agencies also implement relevant SAPs and other activities with the coordination and assistance of rural communities. The sustainable socioeconomic development of the rural communities would be the focal point of the strategies. All field activities demonstrated would be carried out in consultation and ownership of the concerned communities. IWMS model compiled by authors is presented in figure 2.

Figure 2. Conceptual Model, Integrated Watershed Management Strategy.

6. Conclusions

In most parts of Balochistan plateau, the GW levels are declining due to increased abstraction through subsidized electrically powered and solar pumps. The GW is easily accessible in larger capacities and from greater depths by improved drilling techniques. The decline has been focused in irrigated areas of semiarid and arid regions. In arid areas, the present GW recharge is negligible in comparison with extraction. Consequently, the ongoing decline in GW storage is bound to cause deterioration of the GW dependent ecosystem and depletion of surface water resources. It also deteriorates GW quality, causes land subsidence and forces rural populations to migrate to urban centers.

To manage aquifer recharge, the conservation and utilization of available surface water resources wisely and efficiently along-with harness of flood and rainwater. The GW demand can be managed by modern irrigation techniques, low delta crops, water conservation strategies and policies, and higher pricing of electricity and water supply metering. The other factors are hydrogeological investigations, resource assessment, monitoring, resource allocations with community consultation and farmer-led GW management.

The IWMS is prepared with the assessment, analyses and summing up the twenty SAPs related to federal and provincial governments. The strategy comprised of five leading sectors and twenty-seven subsectors relevant to surface and GW resources. The leading sectors are; water governance, environmental concerns, conservation and sustainable development, natural resource management and the community empowerment.

For the sustainable development of water resources appropriate planning, management, conservation and protection measures have been proposed for the subsectors. The IWMS is a flexible yet comprehensive strategy that may be updated anytime with the consultation of all the stakeholders.

To support the watershed management policy a comprehensive strategic model is prepared. The model represents the flow of major activities and their linkages with each-other as well as subsectors. The model is the outcome of a systematic and methodological analysis of IWMS, and related SAPs and provides links between water governesses to community empowerment.

References

[1] Department of Irrigation and Power, Government of Balochistan, 2006. *Integrated Water Resources Management Policy*. Quetta, Balochistan, Pakistan

[2] Halcrow Pakistan (Pvt) Limited, Cameos, 2008. *Effectiveness of the Delay Action / Storage Dams in Balochistan*. Royal Netherlands Government, IPD, GOB, Pakistan.
[3] B-WASA, 2018. *GW monitoring reports of Quetta, unpublished*. GOB, Quetta, Pakistan.
[4] G. O. Pakistan, IUCN, 2015. National *Biodiversity Strategy and Action Plan*. Pakistan.
[5] Ministry of Climate Change, Government of Pakistan, 2015. *National Forest Policy*. Pakistan.
[6] Global Water Partnership, Pakistan Water Partnership, 2014. *UN process towards a Post-2015 Development Agenda, Report on Pakistan National Consultations*. Karachi, Pakistan
[7] Government of Pakistan, UNDP, 2012. *National Sustainable Development Strategy, Pakistan's pathway to a sustainable & resilient future*. Islamabad, Pakistan.
[8] Ministry of Climate Change, G. O. Pakistan, 2012. *National Climate Change Policy*. Pakistan.
[9] Climate Change Division, Government of Pakistan, 2013. *Framework for Implementation of Climate Change Policy*. Islamabad, Pakistan.
[10] Ministry of Environment, Government of Pakistan, 2010. *National Rangeland Policy (Draft)*. Government of Pakistan, Islamabad, Pakistan.
[11] Sarfraz Ahmad, Muhamad Islam and Sarwat N. Mirza, 2012. Rangeland degradation and management approaches in Balochistan, Pakistan. *Pak. J. Bot.*, **44**: 127-136.
[12] Ministry of Environment, GOP, 2005. *National Environmental Policy*. Islamabad, Pakistan.
[13] M. Hanif, Shakeel A. K, Fayyaz A. N, 2004. *Agricultural Perspective and Policy*. Ministry of Food, Agriculture and Livestock Islamabad, Pakistan
[14] Ministry of Water and Power Office of the Chief Engineering Advisor, Chairman Federal Flood Commission, 2002. *Pakistan Water Sector Strategy, National Water Sector Profile*. Government of Pakistan, Islamabad, Volume **1-5**.
[15] Planning Commission, Ministry of Planning, Development and Reform, 2001. *Ten-Year Perspective Plan of Planning Commission*. Islamabad, Pakistan.
[16] Ministry of Water and Power, WAPDA, 2001. *Vision 2025 of WAPDA*. Islamabad, Pakistan.
[17] Ministry of Water and Power, 2002. *National Water Policy, Pakistan*. Islamabad, Pakistan.
[18] Environment and Urban Affairs Division, IUCN, CIDA, UNDP, 1995. *The Pakistan National Conservation Strategy*. Islamabad, Pakistan
[19] Department of Irrigation and Power, Government of Balochistan, 2014. *Water Resources Development in Balochistan*. Quetta, Balochistan, Pakistan
[20] Pakistan Poverty Alleviation Fund, Government of Pakistan, 2013. *Balochistan Strategy, a road map to inclusive socio-economic development of the province*. Islamabad, Pakistan.
[21] Government of Balochistan, IUCN Pakistan, Embassy of the Kingdom of the Netherlands, 2011. *Integrated District Development Vision*. Quetta, Balochistan, Pakistan.
[22] Techno Consult International, Associate Consulting Engineers, and CAMEOS, 2010. *Ground Water Management Action Plan for Pishin Lora Basin & Implementation of Pilot Program in Quetta Sub-basin*. Balochistan, Pakistan.
[23] Planning and Development Department, Government of Balochistan, IUCN-The World Conservation Union, 2000. *The Balochistan Conservation Strategy*. Quetta, Pakistan.
[24] Government of Balochistan, Asian Development Bank, GKW CONSULT, 2000. *Quetta Water Supply and Environmental Improvement Project*. Quetta, Balochistan, Pakistan.
[25] Taraqee Foundation, 2000. *Water Sector Strategy*, Balochistan. Quetta, Pakistan
[26] Hein V G, Baig M S, 1992. *Environmental Profile Balochistan, Pakistan*. Land Resources and Urban Sciences Department, International Institute for Aerospace Survey and Earth Sciences Enschede, Netherlands. Ecology Unit, Soil Survey of Pakistan.
[27] Pakistan Engineering Services, Cameos Consultants, 2017. Detailed hydrological study report of surface/flood water resources of 18 major Basins of Balochistan. Quetta, Pakistan.
[28] Nippon Giken Inc., Sanyu Consultants Inc., 1997. Feasibility Study on the Irrigation Water Resources Dev with Delay Action Dams Project in Balochistan, Pakistan. IPD, GOB, JICA

Preparation of Phase Change Material on Polyester/Cotton Blend

Sidra Saleemi[1,2]*, Abdul Malik Rehan Abbasi[3], Zamir Ahmed Abro[1,3], Nadir Ali Rind[1,] Fu-Jun Xu[1]

[1]College of Textiles, Donghua University, Shanghai 201620, R. P. China
[2]Department of Textile Engineering and Technology, University of the Punjab, Quaid-e-Azam Campus, Lahore, Pakistan
[3]Department of Textile Engineering, BUITEMS, Quetta, Pakistan
[4]Department of Textile Engineering, Mehran University of Engineering and Technology, Jamshoro, Pakistan

Abstract. Fundamental principles of science are now increasingly employed for the manufacturing of innovative textile products which protect the consumer from dangers and risks; protection from extreme environment is one of the properties of textile products. This research is based on study of concentrations of phase change material (PCM) in textiles which can produce thermo-regulating characteristics to control body temperature useful for various daily wear and technical textiles. The latent heat which is absorbed during the process of phase change is studied and discussed in this research. Polyester cotton bland fabrics were treated with phase change material (microencapsulated paraffin wax) by a pad dry cure. These microcapsules are bounded onto the fabric surface by using a binder. The microcapsules and their effectiveness are tested by using DSC (Differential Scanning Calorie Meter). The testing is further extended to the physical property in term of air permeability, stiffness and tensile strength to ensure the performance properties of the textile fabric. All the properties of treated fabric were evaluated with respect to the add-on of binder

1. Introduction

Scientific development in various fields of quality and value of human life has certainly increased. It must be recognized, however, that technological advances have exposed us to greater risks and danger that involved extreme environments also. Fortunately, textile is the simple and effective way to protect against most of this danger. Fundamental principles of science are now gradually employed for the manufacturing of innovative textile products (1). The protection from extreme surroundings becomes most important in the areas where variation in temperature is more obvious such as, space wear textiles, medical textiles, sportswear textiles, automobiles and usual apparel. This increase the importance of thermal regulation (PCM) finishes providing the textiles with usual and functional properties. These finishes provide defences against extreme hot and cold environments (2).

Thermal regulation is obtained by the materials called "Phase Change," the process of going from one physical state to another, i.e. from a solid to a liquid and vice versa to regulate the temperature of the body in extreme situations. Fiber and textile which have automatic acclimatizing properties are recently attracting more and more attention (3).

Many PCM finishes used on textiles to protect wearer against space wearer where temperature may under -270 °C and thermal discomfort in usual and sports. Basically, active wear needs to provide the thermal balance between the heat released and to the environment, when the wearer engaging in sports (4). Polyester cotton blend used in sports and usual apparel textiles due to its comfort and strength but it is more subjected to thermal discomfort during doing activity (5).

The paraffin wax uses as thermal comfort element is not new. It has been used in building construction ceiling and wall for thermal comfort and thermal energy store. In textile for thermal comfort in sports and space wear. Micro-encapsulated paraffin wax is the strong element to give the comfort in ranges of extreme environment due to the better range melting and solidifying ranges.

Coating, lamination, finishing, melt spinning, bi-component synthetic fiber extrusion, injection molding, and foam manufacturing are some processes for PCMs' incorporation into the structure. Application of microencapsulated finish by pad dry cure is the simple and economic mean of achieving thermoregulation specifically in blended fabrics (6). Whereas other processes of finishing have some draw backs such as: coating is time consuming and cost-effective process, in lamination first preparing the sheet of polymer and then combine with fabric in other process so this process is also time consuming and cost effective, need of more experience and care, in melt spinning and bi-component synthetic fiber extrusion processes are only for the synthetic material,

Therefore, in this present study microencapsulated paraffin wax used as a phase change material for polyester cotton blend by simple pad dry cure process. Different concentration of chemicals was used to find the optimum recipe.

2. Materials

The material used in this work includes bleached P/C blend fabric having 155 GSM with plain weave kindly supplied by Popular Textile Mills, Karachi. ITO Finish PCM, ITO Binder and Silica base Softener were brought from Masood Textile Mill (MTM) Faisalabad. LJ-SPECIALITIES LTD is the manufacturer of these chemicals.

3. Methodology

Four recipes of ITO PCM Finish were planned for coating the samples with different blending concentration of Finish, Binder and Softener. These were prepared according to recipe given in Table 1, measured all the ingredients then put in a beaker and stirrer gently as all the chemicals are dissolve evenly.

Table 1. Recipes with variation in all the chemicals in (g/L)

Chemical Used	Recipe 1	Recipe 2	Recipe 3	Recipe 4	Recipe 5
ITO Finish	80	50	150	80	80
ITO Binder	40	30	50	30	50
Softener (silica base)	45	35	55	45	45
Acetic acid	2	2	2	2	2

The finish was applied with 70 % pickup and was controlled by pressure knobs in padding machine. After padding the samples were dried in Mini dryer. The rapid model type mini dryer machine was used to dry the samples. The samples were dried at 100 °C for 2 minutes. Mini dryer model was used to cure the fabric samples after drying. The samples were cured at 150 °C for 3 minutes.

3.1 Measurements

DSC graphs were used to study the thermal behaviour of the microcapsules. A Q200 instrument from TA Instruments was used to do the DSC analysis. Thermo gravimetric analyses were carried out in Punjab University TGA from 25°C to 250°C at a heating rate of 10°C/min flowing nitrogen (20 ml/min).

The maximum force required to extend a test piece to its breaking point is called as strength. Tensile strength of fabric was measured as per the standard ISO-13934-1 strip test method. Bending length and Flexural rigidity show the stiffness of fabric. Bending length was measured on Shirley stiffness tester before and after wash by ASTMD-96 R-02 standard method.

4. Results and discussion

4.1 Effect of different combinations of finish and binder on thermal stability

The DSC heating curve of microencapsulation paraffin wax shows exothermic in figure 1(a) and 1(b), through these graphs the amount of potential heat of fusion (melting) is find out, and the heat of latent heat of fusion of the fabric in phase change process are in the range of 60 – 128 J/g.

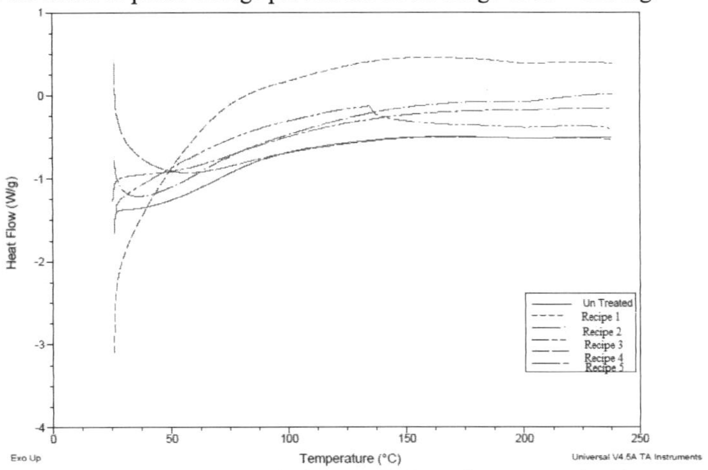

Figure 1(a) The microscopic image of Elastodien wrapped with PA6

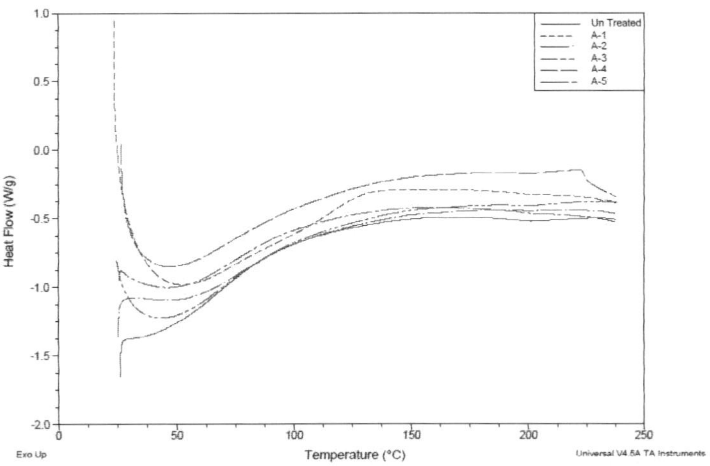

Figure 1(b). Effect of Different Concentration of Chemicals on TGA

It is observed from the results that the reduced amount of binder has slightly effect on the thermoregulation property and the increased amount of binder have the inverse effect on thermoregulation property. The thermal stability curves show the decreased amount of binder in standard recipe have no effect on the thermal stability of material, and the evaluation of increased amount of binder was inverse effect on the thermal stability of the material.

4.2 Effect of Different Concentrations of Finish and Binder on Tensile Strength

Figure 2 implies that tensile strength of the fabric gradually reduced as the concentration of the recipe decreased due to the coating of the finish. The tensile strength of the Recipe 1 was 576 N. Maximum tensile strength of fabric was obtained at Recipe 3 which is 621 N, and the minimum result was obtained at Recipe 4 which is 585.5 N, whereas the value of untreated fabric was 565.3 N. It is found from the above results that the variation in tensile strength of the fabric was due to the different concentrations of the chemicals make polymer film on the surface of the fabric and softener alter the tensile behaviour of fabric and increase the strength of fabric.

This figure also shows the durability of the finish after 15 industrial washes. The method used for washing the samples was AATCC-61. It is observed that the tensile strength of the fabric was slightly reduce by washing which indicates that our finish coating is durable.

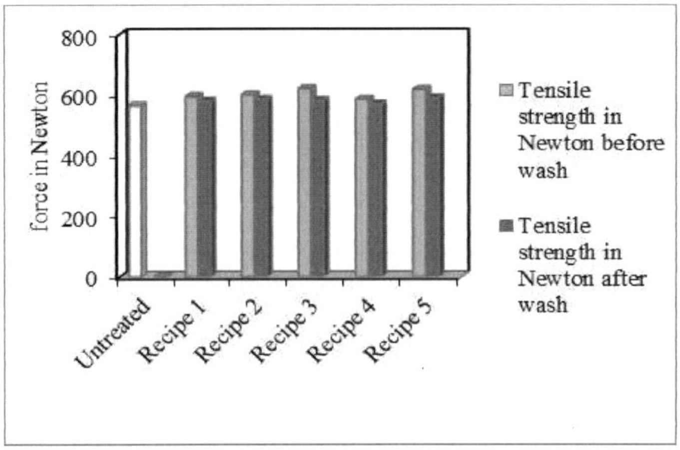

Figure 2 Effect of Different Concentration of Chemicals on tensile strength.

4.3 Effect of Different Concentrations of Finish and Binder on Air permeability

The air permeability of the fabric gradually reduced as the concentration of the finish and binder increased in recipe as shown in Figure 3. The air permeability of the Recipe 1 was 90 ml/sec. Maximum air permeability of fabric was obtained at recipe 2 which is 105 ml/sec and the minimum result was obtained at recipe 5 which is 85 ml/sec. It is noticed that it was due to the different concentrations of the chemicals which close the pours of the fabric due to which the air permeability of the fabric decreases. Air permeability of fabrics increased by washing shows that some of chemical rub-off during washing. However, this permeability still lesser than untreated.

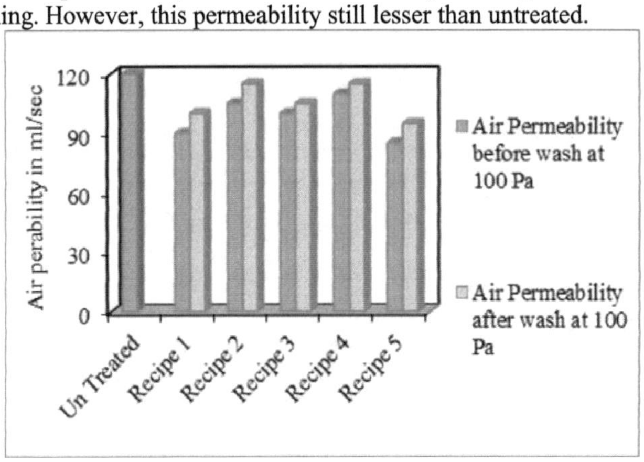

Figure 3. Effect of Different Concentration of Chemicals on Air Permeability.

4.4 Effect of Different Concentration of Finish and Binder on Stiffness

The bending length of the fabric gradually increased as the concentration of finish and binder increased in recipe as depicted in Figure 4.

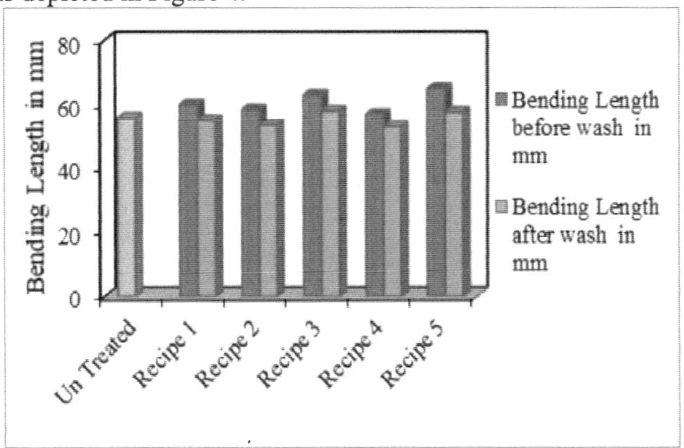

Figure 4. Effect of Different Concentration of Chemicals on Bending Length.

The stiffness in fabric is due the binder makes a polymer film. It is concluded that the variation in bending length of the fabric was due to the different concentration of the chemicals.

5. Conclusion

It is concluded that the low quantity of binder has slightly inverse effect on tensile strength and thermal decomposition of fabric as compare to standard recipe amount of binder. By decreasing the amount of binder, the value of tensile strength is 600.2 N the better results of tensile strength were taken from increased amount of binder. By comparing the results of air permeability and stiffness between higher and lower quantity of binder it was obviously observe that the lower quantity of binder has favourable than higher quantity of binder. By comparing the results before and after 15 industrial washes, the binder variation is not affected then lower amount of binder is preferable as good durability with lower cost.

6. References

1. Sharma A, Tyagi VV, Chen C, Buddhi D. Review on thermal energy storage with phase change materials and applications. Renewable and Sustainable energy reviews. 2009;13(2):318-45.
2. de Cortazar MG, Rodríguez R. Thermal storage nanocapsules by miniemulsion polymerization. Journal of Applied Polymer Science. 2013;127(6):5059-64.
3. Sarier N, Onder E. Organic phase change materials and their textile applications: an overview. Thermochimica Acta. 2012;540:7-60.
4. Shin Y, Yoo DI, Son K. Development of thermoregulating textile materials with microencapsulated phase change materials (PCM). II. Preparation and application of PCM microcapsules. Journal of Applied Polymer Science. 2005;96(6).
5. Ghali K, Ghaddar N, Harathani J, Jones B. Experimental and numerical investigation of the effect of phase change materials on clothing during periodic ventilation. Textile Research Journal. 2004;74(3):205-14.
6. Sánchez P, Sánchez-Fernandez MV, Romero A, Rodríguez JF, Sánchez-Silva L. Development of thermo-regulating textiles using paraffin wax microcapsules. Thermochimica Acta. 2010;498(1-2):16-21.

Extraction and Quantification of Chlorophyll from Microalgae *Chlorella* sp.

M Amin[1,2], P Chetpattananondh[1], M N Khan[2], F Mushtaq[2] and S K Sami[2]

[1]Department of Chemical Engineering, Faculty of Engineering, Prince of Songkla University,90112, Hat Yai, Thailand
[2]Department of Chemical Engineering, Faculty of Engineering and Architecture, BUITEMS, 87300, Quetta, Pakistan

Email: engr_amin63@yahoo.com

Abstract. Algal biomass emerged as a potential source of bioenergy and valuable derivatives in recent years. The major characteristics of microalgae such as high oil contents, carbon sequestration, high growth rate and valuable by-products are leading factors to compete with traditional resources. The aim of current study was (i) to extract and optimize chlorophylls (a and b) at temperature (30-40°C) and time (60-120 min) by ultrasonication assisted with methanol: hexane (2:1v/v) (ii) to find suitable dilution factor for chlorophyll analysis by UV spectroscopy considering 1:10, 1:15,1:20 ml/ml and (iii) to determine suitability of dissolving methanol/hexane extract in different solvents for analysis and quantification by simultaneous equations. The full factorial design RSM was employed and found that model is well fitted (0.99 R^2). Maximum recovery of total chlorophylls (a and b) was 17.15 µg/ml achieved at 30 °C and 120 min. The absorbance spectra peaks were found good with a dilution factor of 1:20 ml/ml. Dissolving the extract for analysis in same solvent is suitable choice even though acetone shows sharp peaks, but not in agreement with beer law. These pigments have a high market in pharmaceutical, dietary products, and food industry and recovery of these compounds can play an important role to make bioprocess industry more economical.
Keywords. Quantification, chlorophyll, Microalgae *Chlorella* sp.

1. Introduction

Microalgae are considered as 3rd generation resource for energy and received high focus in recent years. A microalga is unicellular organism, which possess the tendency to convert sun energy into chemical energy efficiently via photosynthetic process. A variety of strains available in algal class and *Chlorella* is one of them. Microalgae *Chlorella* sp. is single celled and green species with 2-10 µm [1]. It contains about 20, 45, 20 % fat, protein, carbohydrate [2,3], and 0.5-1 % pigments per weight of dry biomass [4]. The composition of biomass may vary due to different cultivation media and associated factors. *Chlorella* has oil contents of 28-32% [5] and abundantly available pigments due to photosynthetic action [6].

Chlorophyll is bioactive compound, which have wide application in pharmaceuticals, food and color industry [7]. Mainly there are two types of green pigments, namely chlorophyll-a and b, however excessive heat, light or air can destabilize the product [8].This destabilization can degrade the chlorophyll product. The structure of chlorophyll is porphyrin macrocycle with four pyrrole rings, while presence of single isocyclic with pyrrole ring built phorbin structure. There are four carbons and a nitrogen atom in pyrrole ring. The position of nitrogen atoms easily attracts Mg^{+2} ions for binding. In chlorophyll-b, formyl group take over methyl group in ring, which differentiate it from chlorophyll-a. This structure difference between two classes of chlorophyll makes identification process easier by their peaks at respective wavelength and region (665 for Ch.-a, and 652 for Ch.-b) [8].

The downstream process of cell disruption for microalgal cell to release associated products is necessary, which could be achieved via physical or chemical method [9]. Earlier, soxlet extraction procedure assisted with solvent was famous one, but it is time and energy intensive. The technological development in research and industrial sector introduced the new ways of disruption such as ultra-sonication and microwave. These methods do not only rupture the high dense cell very effectively but also conform short duration as well.

The use of ultrasonic as extractant method is of increased interest widely due to high efficiency and shorter time required [10]. The extraction process usually carried out with aid of chemicals such as chloroform, acetone, methanol and ethanol. A normal protocol for chlorophyll extraction use acetone as effective solvent, however this is one of toxic solvent. The solvent selected for extraction process must be less harmful. Normally cell disruption of algal biomass release intercellular products, which comprised of lipids and chlorophylls mixture known as crude extract. Lipids are precursor of biodiesel production, which is not an objective of current study. Various studies conducted on lipid extraction from microalgae but limited data is found on quantification of chlorophylls in crude extract using different solvent. Current study focused on optimization of process and quantification of chlorophylls in crude extract, extracted via solvent assisted ultrasonic technique.

2. Material and Methodology

2.1. Material and Chemicals

A marine microalgae *Chlorella* sp. was cultivated in 20 m^3 open aerated pond at National Institute of Coastal Aquaculture (NICA), Songkhla, Thailand as shown in Figure 1. The culture growth was attained using CO $(NH_2)_2$ and 16-16-16 fertilizer (16% N_2, 16% P and 16% K) as growth medium. Due to open cultivation scheme the other sources such as Light and CO_2 were provided naturally for photosynthetic action. Commercial grade n-hexane, methanol, ethanol and acetone were purchased from ACI Labscan Ltd., while alumunium sulphate was obtained from Saim chemical company Ltd. Thailand.

Figure 1. Marine *Chlorella* sp. cultivated in 20 m^3 open pond system

2.2. Biomass harvesting and post processing

The cultured biomass attained peak growth phase in 7 days with cell density of 0.8 g/l was harvested using aluminium sulphate as flocculant agent. After harvesting biomass slurry was filtered using cheese cloth to remove any present contamination, filled in 20 L gallons and stored at 4 °C. The stored algal biomass slurry was washed thrice with DI water to remove salinity and pH was recorded as 7.5.

Cleaned biomass slurry was subjected to dewatering using vacuum filtration with GENVAC vacuum pump (PVL 3, 0.5 mbar, Italy). The slurry was stirred gently to enhance filtration rate. The wet paste as shown in Figure 2 was collected in zip locked air tight bags and stored at 4 °C for short period of time before starting extraction. 1gallon (20 L) slurry yields about 1 kg wet paste.

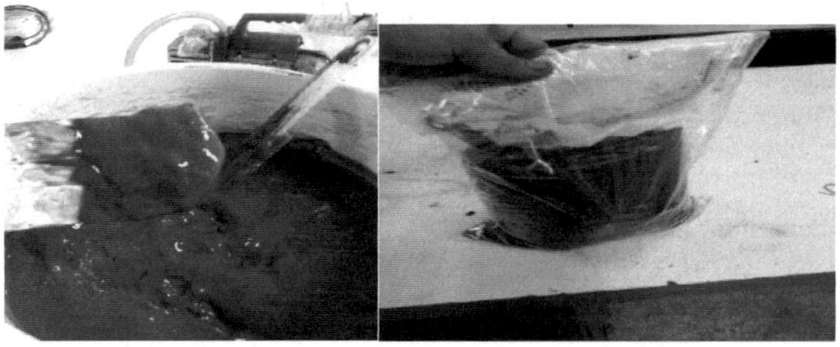

Figure 2. Fresh wet algae paste (10-12% dw solid biomass) after vacuum filtration

Extraction process was performed using ultrasonic bath (Crest power sonic, 45 kHz, CP 2600, USA). Experiment was conducted at different temperature (30-40 °C) and time (60-120 min) values. The Stored wet paste biomass was brought to room temperature firstly, then 10 gm (dw% basis) was placed in 250 ml Duran bottle. Organic solvent methanol: hexane (2:1 v/v) was mixed with sample and stirred for 1 minute. To prevent chlorophyll degradation due to light, sample bottles were covered with aluminium and placed in ultrasonic extraction bath at predefined condition. The extract after different time intervals were taken out and filtered using whatman 4 by vacuum application.

Residual biomass was stored for further processing, while filtrate was evaporated using Heidolph Laborta 4000 rotary evaporator for solvent recovery. The extract obtained after solvent recovery was solid at ambient temperature. This solid was resuspended in methanol initially, instead of solvent mixture used for extraction, because developed simultaneous equation only deals with single solvent. Methanol was selected because of its higher proportion as extractant and have polar nature as well, which can easily dissolve green pigments. Optimum condition for chlorophyll quantity (previously using methanol) was determined, after that extract was dissolved in different solvents such as ethanol and acetone to perform analysis.

2.3. Analysis

The compositional analysis of wet paste biomass was performed according to standard procedures of Association of Official Analytical Chemist (AOAC). The AOAC protocols 991.20, 920.39,942.05 were adopted to find out protein, crude fat and ash. Moisture contents were determined on the weight loss basis at 90-95 °C, while ANKOM ^{200} analyzer was used to measure crude fiber contents at ADCET, PSU, Thailand. The carbohydrate and energy contents were determined by calculation method.

The UV-spectrophotometer (Agilent 8453, USA) was used for chlorophylls analysis. The sampling quartz of 1 cm was used in spectrophotometry. UV system was calibrated prior to analysis and blank reading was taken. Initially different dilutions 1:10, 1:15, 1:20 and 1:25 ml/ml (sample: solvent) were performed to get clear peaks at full spectrum range (100-800 nm). Further measurements were taken at best found dilution factor. Green pigments (chlorophyll a and b) absorb light in red and blue region. The developed simultaneous equations [Equation (1-3), for methanol, ethanol and acetone respectively] were used to quantify chlorophylls from plants [11].

$$\begin{aligned} Ch-a &= 16.29A_{665.2} - 8.54_{652.0} \\ Ch-b &= 30.68A_{652.0} - 13.58A_{665.2} \end{aligned} \quad (1)$$

$$\begin{aligned} Ch-a &= 13.36A_{664} - 5.19_{649} \\ Ch-b &= 27.43_{649} - 8.12A_{664} \end{aligned} \quad (2)$$

$$\begin{aligned} Ch-a &= 12.25A_{663.6} - 2.55_{646.6} \\ Ch-b &= 20.31_{646.6} - 4.91A_{663.6} \end{aligned} \quad (3)$$

The samples absorbance was recorded at desired wavelength as per above equation and quantity (μg/ml) of chlorophylls were calculated. The total quantity of chlorophyll a and b was determined based on summing up the individual chlorophyll values and modeled as yield.

The 3^2 (2 factors 3 levels) full factorial response surface experimental design with 2 additional runs at centre point was selected to observe the effects on response (total quantity of chlorophylls). The design of the experiment, parametric analysis *(*ANOVA*)* by multiple regression and optimization was performed using Statistica version 10.0. Two factors temperature as X_1 (30, 35 and 40 °C) and time as X_2 (60, 90 and 120 min) were coded into three level as -1 (low), 0 (center) and +1 (high). The relationship between coded and actual value is presented in Equation (4).

$$X_i = \frac{x_i - x_o}{\Delta x} \quad (4)$$

Where X_i is coded value of independent variable; x_i is original factor; x_o is base value at center point and Δx is step change between low and high level. The experimental data were fitted according to Equation (5), which is the general form of the proposed model.

$$y = \beta_o + \sum_{i=1}^{k} \beta_i X_i + \sum_{i=1}^{k} \beta_{ii} X_i^2 + \sum_{i>j}^{k} \beta_{ij} X_i X_j + \varepsilon \quad (5)$$

Where y is response and β_o, β_i, β_{ii} β_{ij} are linear, quadratic and interaction terms of model. The predictive response was studied for dependent variables and correlation equations were developed.

3. Results and discussion

3.1. Wet paste biomass compositional analysis

The compositional analysis of wet paste was determined by standard AOAC procedures. The composition of different microalgae strains is different, even same strain could have different composition due to diversity in culture. Moisture contents were found 90%, which means only 10% (dw) solid contents of algal biomass available. Fang et al., (2016) reported the findings for *Chlorella* sp. Wet biomass [12]. It is in normal range with respect to vacuum filtration application to concentrate biomass from 5-15%. Protein, crude fat, ash, carbohydrates and energy (kcal) contents were found 0.76, 0.8, 1.36, 1.59 and 9.43 respectively.

3.2. Ultrasonic extraction and optimization

The full factorial response surface methodology was adopted to design and optimize the ultrasonic extraction experiment. Response surface methodology is one of the most attractive tool for optimization of process [13]. The details of experiment such as coding and actual factors scheme, experimental and predicted response are presented in Table 1. The response fitted model regression analysis (ANOVA) is shown in Table 2. The data shows that experimental and predicted response is in good agreement and residual is minimal. Regression analysis was performed at 95% confidence level.

The model F value is good and making it significant, only 0.01% chance is there that error could be due to some unavoidable factor. Both linear, interaction and squared interaction terms are significant having p-value less than 0.05. Lack of test is non-significant with p-value of 0.91, which confirms the best fit of model. The R-squared of 0.99 very close to 1 is presenting good fit of model as well at 95% confidence.

On the other hand, adjusted and predicted R-squared value of 0.98 and 0.97, which are very close to R-squared value of 0.99 are well justifying the model. The standard deviation of 0.18, and 1.26% C.V leading to reliability of model. By backward elimination as setting confidence level at 0.95 a refined equation (6) in terms of coded factors was developed for response variable.

$$Y = 13.80 - 1.57X_1 - 0.95X_2 + 0.43X_1^2 - 1.17X_1X_2 + 1.12X_1^2X_2 \quad (6)$$

The selected solvent methanol shows high response in extracting these value-added products, as it was confirmed in our preliminary analysis as well. Some studies also reported methanol as consistent solvent for excellent capability of chlorophyll extraction compared to other solvents [14,15].

Response surface and contour plot of factors with respect to yield are given in Figure 3. Both factors showing significant effect on response, at 30 °C yield increased linearly with time increment but at 35 °C and 40 °C response yield trend was found in decreasing order as time increased. Kong et al.[16] study states that chlorophyll extraction via ultrasound increased linear with time at specific temperature, after that it started to decrease. This could be due to fact that chlorophylls start to degrade, if excessive heating applied [17]. The optimum condition was observed 30 °C and 120 min at which recovery was 17.19 (μg/ml). Two trials were performed at optimized condition to check accuracy and found ± 0.02.

Table 1. Full factorial design runs with actual and model predicted response

Run	Actual Factors		A (665.2)	A (652.0)	Coded Factors		Total chorophylls (μg/ml)	
	Temperature (°C)	Time (min)			X_1	X_2	Experimental	Predicted
1	30	120	0.85	0.67	-1	1	17.20	17.15
2	35	120	0.62	0.51	0	1	12.90	12.8
3	35	90	0.70	0.54	0	0	13.80	13.85
4	35	90	0.75	0.54	0	0	14.0	13.80
5	30	60	0.80	0.56	-1	-1	14.50	14.45
6	40	120	0.59	0.46	1	1	11.70	11.67
7	35	60	0.91	0.56	0	-1	14.80	14.75
8	40	60	0.76	0.53	1	-1	13.70	13.67
9	40	90	0.64	0.49	1	0	12.60	12.67
10	35	90	0.82	0.51	0	0	13.50	13.8
11	30	90	0.98	0.59	-1	0	15.70	15.8

3.3 Dilution factor and dissolving of methanol/hexane extract in different solvents.

The dilution factor for spectroscopy analysis was investigated and 1:20 ml/ml factor is reasonable to produce clear peaks. Dissolving the methanol/hexane solvent extract in different single solvents (as per simultaneous equations) in methanol, ethanol, acetone was solely based on fact to plug data in available equations, and in case of suitability choice of ethanol (non-toxic) as friendly solvent.

It has been found that dissolving extract in different solvent (instead of extractant) is not good choice, however peaks were good but absorbance was more than 2, which sounds not good, because spectrophotometer linear absorbance range is between 0.1-1. Due to this fact chlorophyll calculated amount and absorbance values of sample diluted in ethanol and acetone are not presented here.

Table 2. Regression analysis (ANOVA) of fitted model.

Source	**SS**	**MS**	**F**	**df**	**p**	**significance**
Model	22.69	4.54	144.82	5	<0.0001	significant
X_1	14.73	14.73	470	1	<0.0001	significant
X_2	1.80	1.80	57.61	1	0.0006	significant
X_1X_2	5.52	5.52	176.25	1	<0.0001	significant
X_1^2	0.51	0.51	16.34	1	0.0099	significant
$X_1^2 X_2$	1.69	1.69	53.86	1	0.0007	significant
$X_1 X_2^2$, X_2^2						Not significant
Residual	0.16	0.031		5		
LOF	0.30	$1e^{-2}$	0.16	3	0.91	Not significant
Pure error	0.13	0.063		2		
Total	22.85			10		

Precession	**R^2**	**R^2 Adj.**	**Std. Dev**	**Pred. R^2**	**C.V %**
41.9	0.99	0.98	0.18	0.97	1.26

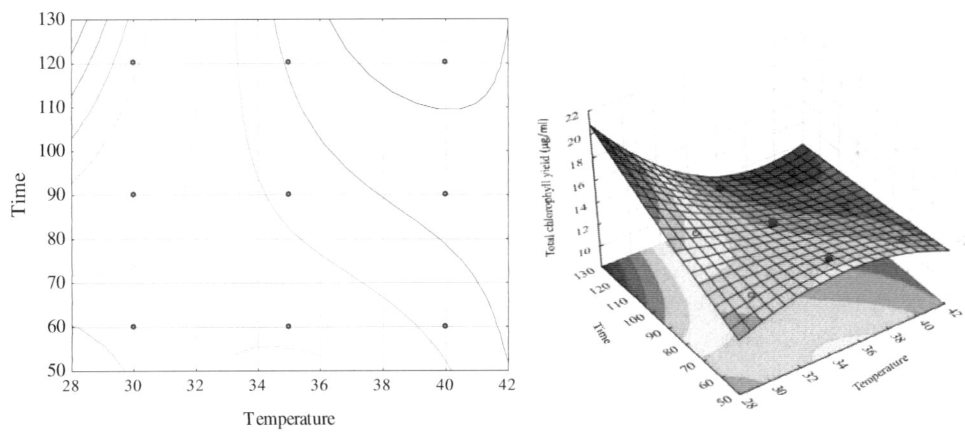

Figure 3. Surface and contour plots of fitted response for total chlorophyll yield ((µg/ml).

4. Conclusion

Current study presents that marine *Chlorella* sp. possess considerable amount of chlorophylls, which could be extracted and purified for wide range applications. High temperature was found not suitable for extraction as it might degrade pigments. Dissolving the extract (methanol/hexane) in methanol was found good, but it's dissolving in ethanol and acetone was not accurate despite of producing clear peaks. It is concluded that same solvent (methanol in current study because of higher proportion with respect to hexane) utilized for extraction is best for dissolving again for analysis. More work is required in purification and developing simultaneous equation for mixed solvent.

Acknowledgment
The corresponding author is grateful to Graduate School Prince of Songkla University, Hat Yai, Thailand for the provision of Thailand's Education Hub for Southern Region of ASEAN Countries (THE-AC) scholarship, 2016.

References

[1] Kelaiya S V., Chauhan P M and Akbari S H 2015 Fuel Property of Biodiesel Made From Microalgae (Chlorella sp.) *Curr. World Environ.* **10** 912–9
[2] Phukan M M, Chutia R S, Konwar B K and Kataki R 2011 Microalgae Chlorella as a Potential bio-energy feedstock *Appl. Energy* **88** 3307–12
[3] Kay R A and Barton L L 1991 Microalgae as Food and Supplement *Crit. Rev. Food Sci. Nutr.* **30** 555–73
[4] D'Alessandro E B and Antoniosi Filho N R 2016 Concepts and studies on lipid and pigments of microalgae: A review *Renew. Sustain. Energy Rev.* **58** 832–41
[5] Chisti Y 2008 Biodiesel from microalgae beats bioethanol *Trends Biotechnol.* **26** 126–31
[6] Nurachman Z, H H, Rahmaniyah W R, Kurnia D, Hidayat R, Prijamboedi B, Suendo V, Ratnaningsih E, Panggabean L M G and Nurbaiti S 2015 Tropical marine Chlorella sp. PP1 as a source of photosynthetic pigments for dye-sensitized solar cells *Algal Res.* **10** 25–32
[7] Borowitzka M A 2013 High-value products from microalgae-their development and Commercialisation *J. Appl. Phycol.* **25** 743–56
[8] Halim R, Hosikian A, Lim S and Danquah M K 2010 Chlorophyll extraction from microalgae: A review on the process engineering aspects *Int. J. Chem. Eng.* **2010**
[9] Lee J Y, Yoo C, Jun S Y, Ahn C Y and Oh H M 2010 Comparison of several methods for Effective lipid extraction from microalgae *Bioresour. Technol.* **101** S75--S77
[10] Parniakov O, Apicella E, Koubaa M, Barba F J, Grimi N, Lebovka N, Pataro G, Ferrari G and Vorobiev E 2015 Ultrasound-assisted green solvent extraction of high-added value Compounds from microalgae Nannochloropsis spp. *Bioresour. Technol.* **198** 262–7
[11] Sumanta N, Haque C I, Nishika J and Suprakash R 2014 Spectrophotometric Analysis of Chlorophylls and Carotenoids from Commonly Grown Fern Species by Using Various Extracting Solvents *Res. J. Chem. Sci. Res. J. Chem. Sci* **4** 2231–606
[12] Yang F, Cheng C, Long L, Hu Q, Jia Q, Wu H and Xiang W 2015 Extracting lipids from Several species of wet microalgae using ethanol at room temperature *Energy and Fuels* **29** 2380–6
[13] Li Y, Cui F, Liu Z, Xu Y and Zhao H 2007 Improvement of xylanase production by Penicillium oxalicum ZH-30 using response surface methodology *Enzyme Microb. Technol.*
[14] Riemann B and Ernst D 1982 Extraction of chlorophyll a and chlorophyll b from Phytoplankton using standard extraction techniques *Freshw. Biol.*
[15] Molnár É, Dóra Rippel-Pethő ! and Bocsi R 2013 Solid-Liquid Extraction of Chlorophyll From Microalgae From Photoautotroph Open-Air Cultivation *Hungarian J. Ind. Chem. Veszprém* **41** 119–22
[16] Kong W, Liu N, Zhang J, Yang Q, Hua S, Song H and Xia C 2012 Optimization of ultrasound-Assisted extraction parameters of chlorophyll from Chlorella vulgaris residue after lipid Separation using response surface methodology *J. Food Sci. Technol.* **51** 2006–13
[17] Schumann R, Häubner N, Klausch S and Karsten U 2005 Chlorophyll extraction methods for the quantification of green microalgae colonizing building facades *Int. Biodeterior. Biodegrad.* **55** 213–22

Effect of Steel Confinement on Axially Loaded Short Concrete Columns

H Farooq[1], M Usman[1], K Mehmood[2], M S Malik[1], A Hanif[2]

[1]School of Civil and Environmental Engineering, National University of Science and Technology, Sector H-12, Islamabad, Pakistan
[2]Department of Civil Engineering, Mirpur University of Science and Technology, Allama Iqbal Road, Mirpur AJ&K, Pakistan

E-mail: asadhanif@must.edu.pk

Abstract. This paper presents experimental investigations of Reinforced Concrete columns strengthened with steel strips and steel jacket under axial compression loading. Total of six specimens were prepared in the laboratory, out of which; 1) two columns were unconfined columns, 2) two columns were strengthened with steel strips and 3) remaining two columns were strengthened with a steel jacket. All specimens were subjected to axial compression loading in increment until the failure of specimens using UTM at NUST. It was concluded after the experimental work that the steel strips enhanced axial load carrying capacity by a factor up to 1.54 times and steel jacketing enhanced the capacity by a factor of 2.38. It has also been observed that the steel confinement increased the cracking load by a factor up to 1.5 to 2.66 for strips and jacketing respectively. The failure occurred due to the crushing of concrete and buckling of steel strips and jacket between the fasteners. The steel strips and jackets are economical and feasible options for confining the concrete to increase its axial load carrying capacity.
Keywords: Aging, steel, concrete, reinforced columns

1. Introduction

Aging and adverse conditions are increasing the vulnerability of many old deteriorated concrete buildings [1]. The need to improve the behavior and safety of un-confined concrete buildings is becoming increasingly important [2]. Some of the methods to strengthen load bearing structures are to apply fiber-reinforced polymers (FRP), Steel Jacketing [3]. Taghdi [4] strengthened the masonry walls for improving their shear behavior by diagonally applying the steel strips and stated them to be quite effective. Despite the effectiveness of diagonal steel strips, the problems associated with the modification in aesthetics and augmented mass were not addressed, K C Voon [5] employed vertical and horizontal arrangement of steel rebars to enhance the shear strength of walls, and eventually stated them to be effective in their results. The same research also concluded that the employed arrangement of rebars also improved the post-tensioned behavior of walls. The research conducted by Farooq et. al. [6, 7] also stated improvements in the compressive and shear behavior of masonry by means of lightweight steel strips, mounted on their surface. The research also concluded that no significant improvement in compressive strength was observed by reducing the spacing of horizontal strips. Strengthening is essentially a requirement for reinforced concrete (RC) columns, manufactured with low quality of materials, with inadequate reinforcements, and insufficient splice lengths. In addition to improvement in the load-bearing capacity, strengthening is often executed in a manner that it results in altering the failure pattern from brittle to ductile, and for strengthening, numerous techniques are available and utilized, as indicated by the literature [8].

Each strengthening methodology has its own suitability along with particular limitations. One such prominent technique is the use of steel jackets in which versatile configurations i.e. steel caging; wrapping, etc. are adapted to confine the columns. The most common form among these is the steel caging in which four steel angles, followed by the welding of horizontal steel straps at particular intervals along a column's height, are positioned at the corners of an RC column. Such a technique is usually employed during the strengthening process of square and rectangular X-Section columns. Tarabi and Albakry [9] has stated that such a method is typically considered swift and economically feasible. Moreover, the worthy contribution towards the enhancement of dynamic performance, improvement in the axial load bearing behavior, and increased ductility are other advantages attributed to this strengthening technique [9-13]. The technique is broadly employed and practiced in the United States, Japan, and Taiwan [11, 14], and has been regarded as highly effective for post-earthquake retrofitting of axial structural elements [15]. From the literature, it can be stated that columns, after being strengthened by means of steel jackets, may exhibit two different failure configurations. The first pattern of failure indicates the yielding of angles [9, 16, 17], while the second configuration of failure may be observed when yielding occurs in horizontal strips. This particular failure pattern is witnessed when the length of a cross-sectional dimension of a column is bigger than the length of angles. Thus, after the application of an axial load, the lateral dilation of concrete exposes the horizontal strips to additional pressures [9, 10]. The structural capacity of a column is greatly influenced by the spacing of strips. In this regard, [16] made an extensive study to assess the variation in the load carrying capacity of strengthened columns by varying the spacing of the horizontal steel strips throughout the columns' height. The research concluded that increased spacing eventually resulted in decreasing the structural capacity of columns. Similarly, the behavior of steel-jacketed RC columns was also studied by Belal [18]. In his research, C-sections, steel angles, batten plates, and steel plates with specific intervals were used for strengthening purposes. The experimental test revealed that strengthening increased the failure load and load carrying capacity by 18% - 45%.

Considering the cost and difficult implementation of available concrete strengthening procedures, the need for some cost-effective technique was highly desirable which should involve local materials and nominal skill [19-21]. To cater to this issue, the technique used by Farooq et. al. [7] for strengthening masonry was used for the confinement of the un-strengthened concrete columns. The concrete columns were strengthened using lightweight Steel Strips and compared with the already available Steel Jacketing technique to study the efficacy of lightweight steel strips strengthening on existing concrete short columns. The application of steel strips has effectively addressed the issues of alterations in aesthetics and the augmented mass and is found to be quite cost-effective. This research is, however, limited to investigate the axial strengthening of columns only. The behavior of the strengthened columns will be carefully studied with due importance to the stress-strain behavior, composite action and the confinement effect. Experimental results revealed that steel strengthening resulted in significant increase in axial capacity, modulus of elasticity and improved compressive behavior.

2. Specimen preparation and material properties

This study is focused on the experimental testing of strengthened short columns with the geometry of 152.4 x 152.4 x 355.6 mm with 4#3 longitudinal bars and 3#2 ties @ 100 mm c/c with 25 mm cover in the laboratory. The Specimen columns were prepared in the laboratory for carrying out compression test using the universal testing machine. Total six short columns were prepared. Out of six specimens; two specimens were reference specimen, two were strengthened with steel strips and two were strengthened with steel jacket as shown in table 1.

The specimens were prepared using 1:2:4 cement-sand-aggregates with w/c ratio of 0.55 and the mechanical properties of the material is shown in table 2. To achieve uniform compaction up to the best possible level vibratory compactor was used. After a moist curing period of 28 days, samples were taken out of the curing tank and placed in the laboratory at normal conditions of temperature and moisture until their surface became dry. The materials were tested in the concrete laboratory of Military College of Engineering Risalpur, Pakistan. Galvanized steel sheet was used for jacketing of columns. For both categories, steel strips as well as steel jackets the same 17-gauge steel with a thickness of 1.41 mm was used. The cross-sectional dimension of steel strips was 50.8 x 1.41 mm. The steel strips were anchored all around the columns using Hilti-Impact Anchors. Each anchor had a

IOP Conf. Series: Materials Science and Engineering **414** (2018) 012026 doi:10.1088/1757-899X/414/1/012026

capacity to resist a force of 0.25 KN. These anchors were 45mm long with 6mm diameter. Six columns were prepared all having a height of 355.6 mm with same aspect ratio as shown in figure 1.

Table 1. Designation of concrete specimens.

Specimen Type	ID	No of Specimen	Cross section (mm)	Height (mm)	Steel confinement
Reference specimen	RCC	2	152.4x152.4	355.6	nil
Strip strengthened column	SSC	2	152.4x152.4	355.6	Horizontal strips; spacing 101.6 mm, thickness1.41 mm, cross section 50x1.41 mm Vertical strips; spacing 101.6 mm, thickness 1.41 mm, cross section 50x1.41 mm
Steel jacket strengthened column	JSC	2	152.4x152.4	355.6	Steel jacket Thickness 1.41 mm, Cross section 152.4x1.41 mm

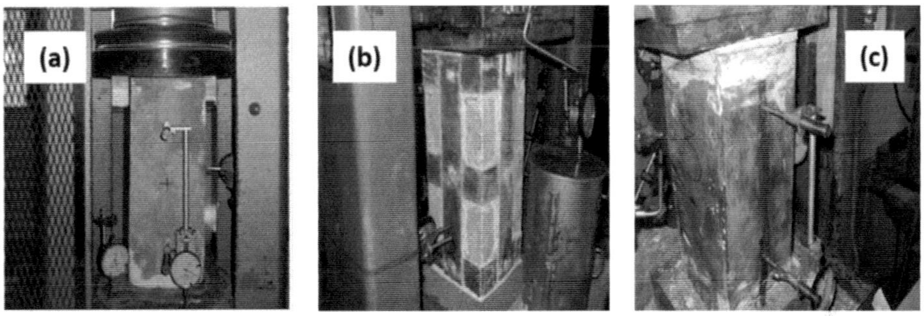

Figure 1. Test specimens; (a) Reference Concrete Columns- RCC, (b) Stripped strengthened Columns-SSC, and (c) Jacket strengthened columns–JSC.

Table 2. Results of the material characterization test.

Item	Properties	Values	Standard
Fine Aggregate	Fineness Modulus	2.42	ASTM-127/128
	Bulk Specific Gravity	2.328	
	Avg Absorption	1.56%	
	Unit weight	1489.53 Kg/m^3	
Coarse Aggregate	Specific Gravity	2.665	ASTM-127/128
	Avg Absorption	1.066%	
	Unit weight	1225.63 Kg/m^3	
	Crushing value	24.46%	BS882:1965
	Impact value	15.42	BS812:1967
Confining Steel	Yield strength	235 MPa	-
	Ultimate strength	303 MPa	-

2.1. *Test specimens*

Six columns were prepared all having a height of 355.6 mm with same aspect ratio. These specimens were divided into two groups. Each column accompanied 3 concrete cylinders for its 28 days compressive strength (f'c).

2.2. *Instrumentation of specimens*

Universal Testing Machine was used to subject the columns to tests. Strain and deformations were measured by employing strain gauges, mechanical gauges and LVDTS. The two faces of column contained strips in which the horizontal and vertical strains were recorded by using four strain gauges. Similarly, the horizontal and vertical strains in concrete were recorded by using four mechanical gauges, while for measuring the out of plane deflections in the columns, three LVDT were eventually employed, as depicted by the figure 2.

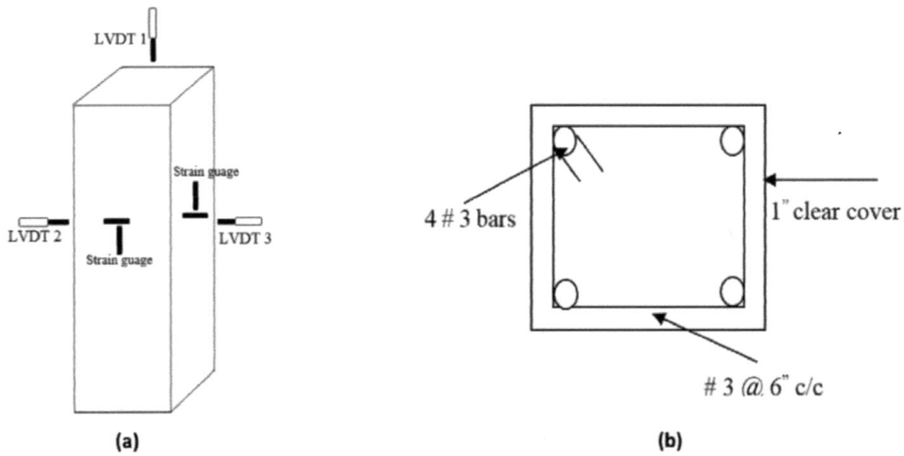

Figure 2. (a) Instrumentation of test specimens, **(b)** Cross section with reinforcement details

3. Results and discussion

3.1. *Stress-strain relationship*

Stress-axial strain relationship is shown in Figure 3a. The un-strengthened column failed at a stress value of 16.63 MPa and strain recorded was 0.0087 whereas, steel strip confined column failed at a value 54 % greater and jacket strengthened columns at a value 138 % greater than the un-strengthened column. This shows that the increase in the amount of confinement affects significantly the load carrying capacity that is evident from the proportion of the increase in strength with the amount of confinement. The cracking stress (stress at which the first crack appeared in the masonry during loading) and strain for strengthened columns have increased considerably. The increase in cracking stress for strengthened columns as compared to specimen RCC is by a factor of 1.5 for SSC, 2.66 for JSC column specimens as summarized in table 3.

Table 3. Summary of Test results.

Specimens	28days f'c (MPa)	Compressive Stress (MPa)	Normalized Stress	Ultimate Strain	Cracking Stress (MPa)	Normalized Stress
RCC	19.72	16.63	-	0.0087	11.52	-
SSC	18.7	25.58	1.54	0.00943	17.91	1.50
JSC	21.45	39.65	2.38	0.0089	30.69	2.66

The obtained stress versus strain relationships in axial and lateral directions of the specimens are presented in figure 3b. In almost all the cases, the curves come out to be bilinear, possessing no descending branch. The ultimate stress has been increased by the virtue of confinement provided by the steel strips; nevertheless, the unconfined columns exhibited bigger ultimate lateral strains in comparison with those of confined columns. For confined columns, with a substantial increase in stress, the escalation of lateral strain is less. Thus, it is observed that the ultimate stress of SSC and JSC specimens are 1.54 and 2.38 times higher than those of RCC specimen respectively. Similarly, the cracking stress and strain have also increased considerably. The increase in the lateral strain value in comparison to the axial strain value in a column for confined ones was very less than the unconfined specimens causing the increase in strength. These less lateral strains were definitely arrested by the steel strips. The unconfined columns showed contraction initially and then dilated because of the crushing of unconfined concrete. The very fewer strain values will also be there in steel strips which give the direct contribution of steel in strengthened is less and the main enhancement comes from the confinement effect and the composite behavior of the columns. Stress-axial Strain behavior of steel strips indicated that steel strips maintained their elastic limit and did not step into plasticity. All confined columns manifested extremely small axial strains. The smaller values of strain essentially indicate that the strength gain is attributed to the confinement provided by the strips and steel jacket. It is evident from the results that the axial strength of columns confined with steel strips has increased considerably due to confinement and the composite behavior but with variations.

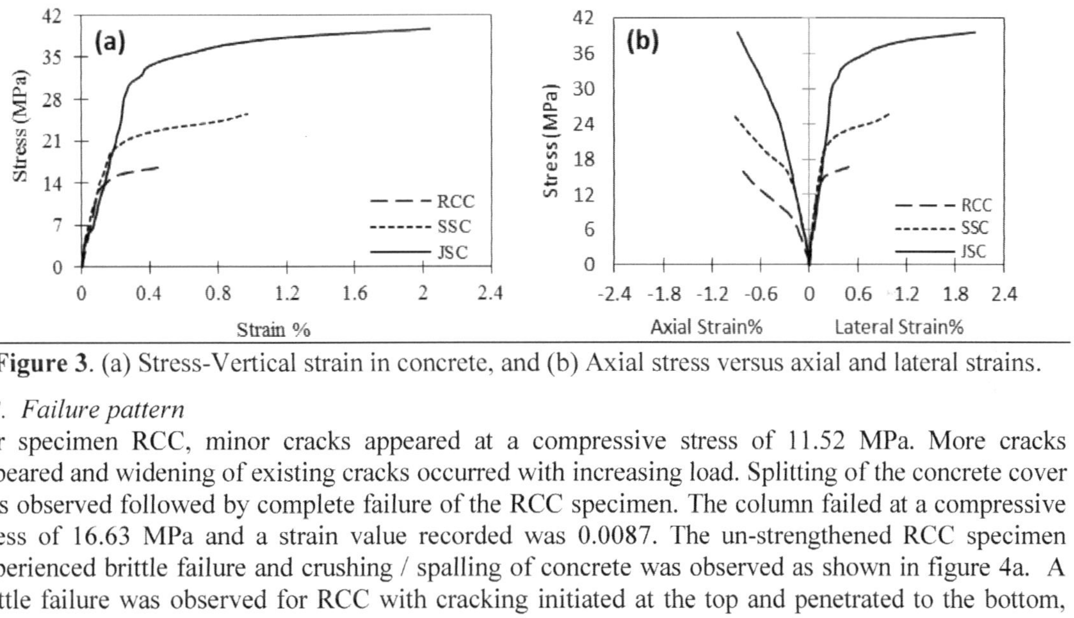

Figure 3. (a) Stress-Vertical strain in concrete, and (b) Axial stress versus axial and lateral strains.

3.2. Failure pattern

For specimen RCC, minor cracks appeared at a compressive stress of 11.52 MPa. More cracks appeared and widening of existing cracks occurred with increasing load. Splitting of the concrete cover was observed followed by complete failure of the RCC specimen. The column failed at a compressive stress of 16.63 MPa and a strain value recorded was 0.0087. The un-strengthened RCC specimen experienced brittle failure and crushing / spalling of concrete was observed as shown in figure 4a. A brittle failure was observed for RCC with cracking initiated at the top and penetrated to the bottom, eventually resulting in the crushing of concrete.

Figure 4. Pictorial view of test specimens after failure; ((a) Reference Concrete Columns RCC, (b) Stripped strengthened Columns SSC, and (c) Jacket strengthened columns JSC.

Minor cracks appeared at a compressive stress of 17.91 MPa was observed in case of strip-strengthened concrete columns specimen SSC and failure of the column occurred at a compressive stress of 25.58 MPa with strain value of 0.00943 at the failure. Both confined columns exhibited an analogous pattern of failure. After the development of cracks in concrete, they became wider, eventually causing the crushing of the material inside the columns. The confinement of concrete columns with steel resulted in a considerable increase in cracking stress. The cracks were well distributed and both the confined specimens experienced ductile failure. The bending and pulling out steel strips due to crushing of concrete was observed at failure in case of specimen SSC (figure 4b). In case of specimen JSC, at failure bulging and crushing of concrete was observed causing opening of steel jacket at corner (figure 4c). At a stress value of 30.69 MPa minor cracks appeared in case of specimen JSC. The column failed at a compressive stress of 39.65 MPa with a strain value of 0.0089 was recorded at the failure as summarized in table 3.

3.3. Poisson's ratio

Poisson's Ratio plays an important role in describing the behavior of unconfined and confined reinforced concrete columns. Figure 5 and Figure 6 show the relation between Poisson's ratios versus the axial strain for the unconfined and confined reinforced concrete columns. The failure of the concrete tied column is characterized by the development of hairline cracks which make the compression member, incapable of resisting tensile stresses. This abrupt failure of columns with outward buckling of longitudinal bars between ties is changed to gradual and ductile failure by confining patterns of steel strips and jacketing. As shown in Figures, Poisson's Ratio of confined columns is higher in all of the two groups as compared to unconfined columns. In case of group 1, the values of confined columns are greater than unconfined columns by a factor of 2 to 3 in case of SSC and JSC which indicates the effectiveness of steel strips and steel jacket in strength enhancement and improving the behavior of specimens. In addition, for the strengthened specimens the value of lateral/axial strain stabilized at approximately the same value regardless whether the strengthening was by steel strips of steel jacket.

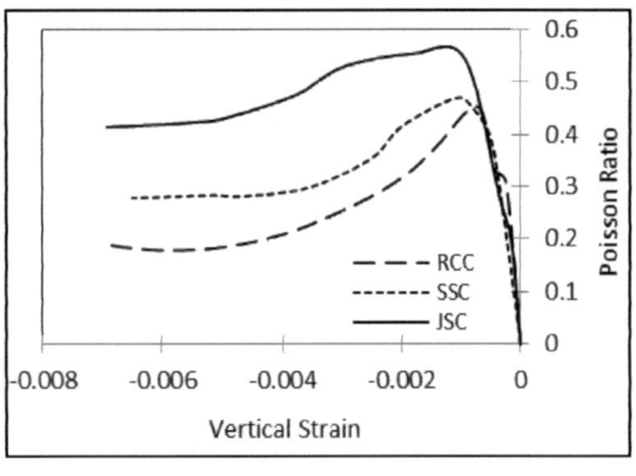

Figure 5. Poisson's Ratio vs vertical strain plot.

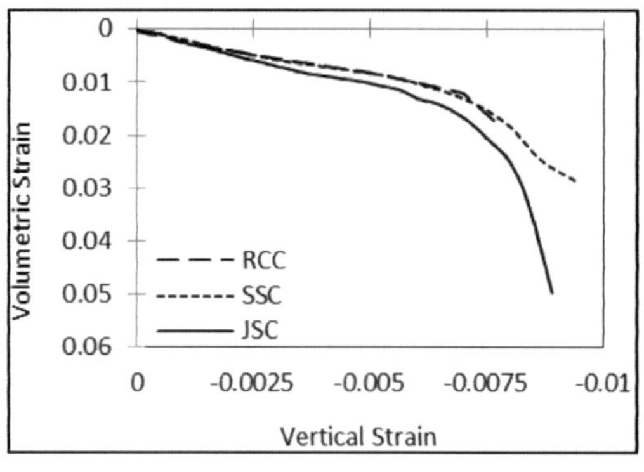

Figure 6. Volumetric Strain vs Vertical Strain.

3.4. Volumetric strains

Volumetric strain for all the specimens was calculated using the equation 1 and figure 6 shows the axial strain / volumetric strain ε_v relationship. The negative value designates contraction, whereas a positive value specifies expansion or the dilation. Figure 6 shows that the phenomenon of contraction has been observed by all the columns with no dilation during the test and volumetric strain value is increasing with increase in axial strain. This particular behavior is mainly attributed to the elastic response rendered by the steel strips and steel jacketing systems respectively which caused the rupture of the specimen because of the consistent increase in confining pressure. For all confined columns, the contraction has been observed which is followed by no recovery, while for unconfined specimen RCC, initially the contraction was observed which was followed by the recovery branch. Due to the compressive load, the contraction was observed in specimen RCC. Eventually, the lateral strain escalated after the initiation of cracking in comparison with the axial strain. This increased in lateral strain ultimately caused the expansion and final failure of columns. While in confined columns, steel jacketing and steel strips effectively countered the proliferation of lateral strain which ultimately resulted in increasing the strength and eventually postponed the cracking itself. Equation 1 was used to calculate volumetric strains and maximum values recorded were 0.018, 0.028 and 0.049 respectively for RCC, SSC, and JSC.

$$\varepsilon_A + 2\varepsilon_l \quad (1)$$

where ε_A = axial strain; ε_l = lateral strain.

3.5. *Axial – lateral strain relationship*
Axial verses lateral strain nexus for concrete is shown in figure 7, which is based on the absolute values of compressive axial strains. For the case of unconfined columns, the almost linear relationship is observed between axial and lateral strains from 0.1% to 0.2% of axial strain values. Afterward, the concrete begins to laterally dilate at a higher rate because of the accretion of damage, which ultimately renders the softening in the relationship of axial/ lateral strain. For the case of confined columns, the axial/lateral strain relationship remained immune to the effects of steel jacket and the steel strips. This essentially implies that response of unconfined columns and confined columns were almost similar initially because of the insignificant concrete lateral dilation. After that, the imposed axial strain initiates to increase at a higher rate than that of the lateral strain during a transition regime where the concrete dilation is counteracted because of the presence of steel jacketing and steel strips. Finally, the strengthened specimens dilated laterally because of the accumulated damage due to an increase in axial strain.

The Modulus of Elasticity values of confined specimen columns were 42% and 133% higher than the specimen RCC for specimens SSC and JSC respectively as shown in figure 8. Higher values of modulus of elasticity is due to additional strengthening being provided by steel confinement. The steel jacket was the most effective in increasing the strength.

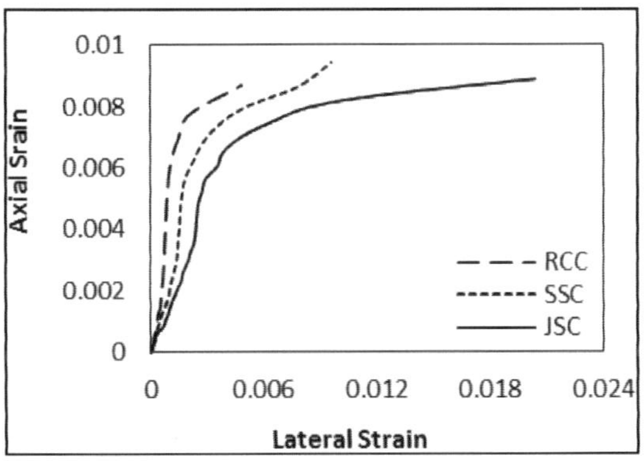

Figure 7. Axial strain vs lateral Strain.

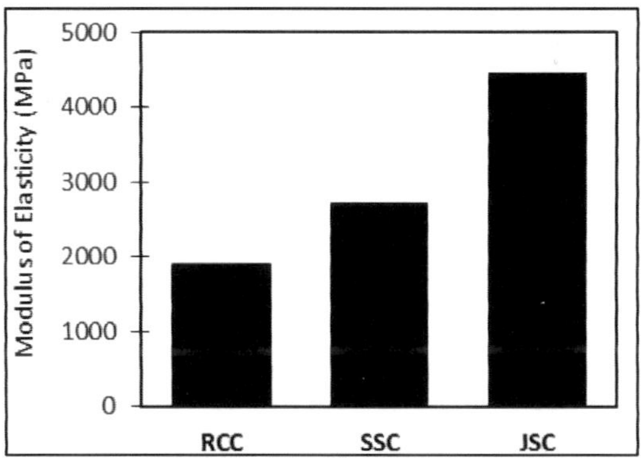

Figure 8. Comparison of E values for columns.

3.6. *Confinement effects*
Steel strips were responsible to provide passive confinement to concrete columns; however, the horizontal strips remained essentially unstressed for the small axial load. The lateral strain in concrete got elevated and the additional strain was taken by the steel strips, thus forming a confinement pressure at axial loads ranging from 11.52 to 14.06 MPa. Vertical stripes, due to their confinement effects, significantly improved the axial strength of the specimens, as projected by the figures above. Equation 2 is used to calculate the axial strength of the specimens.

$$P = P_m + P_c + P_{st} \tag{2}$$
P = Axial strength of the specimen
Pm = Strength due to concrete
Pc = Strength due to confinement
Pst = Steel strips contribution to the axial strength

The following equation number 3 was used to calculate the rendered impact of the steel strips to the axial load:

$$Pst = m\varepsilon_{st}EA \tag{3}$$
n = total number of steel strips
ε_{st} = strain in steel strips
E = Young's Modulus of the steel strips = 16875 MPa
A = Cross-sectional area of a steel strip = 76.2 mm2

Substituting these values in Eq. 3, the contribution of the steel strips to the axial load can be calculated. According to one of the module by Triantafillou is derived on the basis of Mander's Model. The model was originally proposed to evaluate the effective confinement pressure for tangular concrete cross sections, confined with steel. In Mander's model, the effect of confinement can be calculated and the same model was used by Triantafillon for calculating the effect of steel confinement on compressive strength of unconfined concrete. The relation between the compressive strength of unconfined concrete f'co and confined concrete f'cc is evaluated using the following equation.

$$\frac{f'_{cc}}{f'_{co}} = 2.254\sqrt{1 + 7.94\,f'_1/f'_{co}} - 2\,f'_l/f'_{co} - 1.254 \tag{4}$$

Where 〖f'〗_1 is the effective confining lateral pressure. The percentages of confining pressure, i.e., 〖f'〗_l/〖f'〗_co for SSC and JSC are 0.094, 0.33 respectively.

4. Conclusions

The conclusions are drawn and observations made from the study of confinement of short concrete columns are as follows:

- Axial strength of concrete columns is greatly enhanced by using steel strips. The steel strips also proved to be effective in increasing the axial strength of confined columns.
- The significant increase in axial strain with respect to lateral strain shows increased ductility and when coupled with high compressive strength, it justifies the used of additional confinement in retrofitting and new construction.
- The strength enhancement was mainly attributed to the confining effect provided by the steel for a composite behavior, while the direct contribution made by the steel for increasing the strength was less.
- Considering the easy availability of raw materials for the proposed strengthening technique, it offers a great advantage over more sophisticated strengthening techniques.
- Additionally, with the minimum additional cost, this study will help the strengthening and retrofit of rural structures.
- The use of steel strips can further study for other types of loadings also in order to develop it into a more comprehensive retrofitting technique

References

[1] Bhasker P T 2015 Aging Effects on Structural Concrete and Long-term Storage of Spent Nuclear Fuel in DCSS at *ISFSIs Transactions, SMiRT-23* Manchester, United Kingdom - August 10-14, 2015

IOP Conf. Series: Materials Science and Engineering **414** (2018) 012026 doi:10.1088/1757-899X/414/1/012026

[2] Wewin I 2018 Strengthening method of concrete structure IOP Conf. Series: Earth and Environmental Science **126** 012051

[3] Kim S W 2017 Strengthening methods for reinforced concrete infrastructure using FRP composites *Proc. of the Inst. of Civil Engrs. – Str.and Bldg.* 1-10

[4] Mustafa T et al 2000 Seismic retrofitting of low rise masonry and concrete walls using steel strips *J. of Str. Engg.* **126** 1017-1025

[5] Voon K C and Ingham J. M. 2007 Design Expression for the In-Plane Shear Strength of Reinforced Concrete Masonry *J. of Str. Engg.* **133** 706-713.

[6] Farooq S H. et al 2006 Technique of Strengthening Masonry Wall Panels using Steel Strips *Asian J. of Civil Engg.* **7** 625-642.

[7] Farooq S H. et al 2008 Effect of Horizontal Reinforcement in Strengthening of Masonry Members *Mehran Uni. Res. J.of Engg. & Tech.* **27** 49-62.

[8] Shri B. and Pravin W 2011 Materials and Jacketing Technique for Retrofitting of *Structures Int. J.of Adv.Engg Res. and Stud.* I 5-19.

[9] Tarabi A M and Albakry H F 2014 Strengthening of RC columns by steel angles and strips *Alexandria Engg J* **53** 615–626.

[10] Badalamenti V et al 2010 Simplified model for compressive behavior of concrete columns strengthened by steel angles and strips *ASCE J. Eng. Mech.* **136** (2) 230–238.

[11] Choi E et al 2013 Effect of steel wrapping jackets on the bond strength of concrete and the lateral performance of circular RC columns *Engg. Str.* **48** 43–54.

[12] Areemit N et al 2015 Strengthening of Deficient RC Columns by Steel Angles and Battens under Axial Load Department of Civil Engineering, *Khon Kaen University, Thailand.*

[13] Adam J M et al 2009 Axially loaded RC columns strengthened by steel caging. Finite element modeling *Const. and Bldg Mat.* **23** 2265–2276.

[14] Lin M L et al 2010 Seismic steel jacketing of rectangular RC bridge columns for the mitigation of lap-splice failures Earthquake *Eng.Struct. Dyn.* **39** 1687–1710.

[15] Wasti S T and Ozcebe G 2006 Advances in earthquake engineering for urban risk reduction *Nato Science Series: IV: Earth and Environmental Sciences* **66** 1-552.

[16] Cirtek L. 2001 RC columns strengthened with bandage experimental programme and design recommendations *Const. and Bldg Mat.* **15** 341-349.

[17] Calderón P A et al 2009 Design strength of axially loaded RC columns strengthened by steel caging *Mat.and Des.***30** 4069–4080.

[18] Belal M F et al 2015 Behavior of Reinforced Concrete Columns Strengthened by Steel Jacket *HBRC J.* **11** 201-212.

[19] Hsu C T et al 2003 Flexural strengthening of reinforced concrete beams using carbon fibre reinforced polymer strips *Mag. Conc.Res.* **55** (3) 279-288.

[20] Das A K et al 2018 Design of propping schemes of 60-storey reinforced concrete building: a case study *Proc. of the Inst. of Civil Engrs. – Str.and Bldg.* **171** (5) 409-424.

[21] Lu C et al 2018 An investigation on the behavior of a new connection for precast structures under reverse cyclic loading *Engg. Str.* **169** 131–140.

Comparative analysis of heat transfer coefficient for wood and non-wood pulp fiber

G Khan[1], S N Kazi[2], M A Syed[2], S Nasar[1] and A Kakar[3]

[1] Department of Chemical Engineering, Faculty of Engineering & Architecture, Balochistan University of Information Technology, Engineering & Management Sciences, Quetta, Pakistan
[2] Department of Mechanical Engineering, Faculty of Engineering, University of Malaya, Kuala Lumpur, Malaysia
[3] Department of Economics, Faculty of Management Sciences, Balochistan University of Information Technology, Engineering & Management Sciences, Quetta, Pakistan

Email: ghulamullahkhan73@gmail.com

Abstract. The comparative study for obtained heat transfer coefficient for Acacia Mangium (wood) and Kenaf bast (non-wood) pulp fiber suspension were performed in a close loop circular pipe heat exchanger. During the test study it was observed that the magnitude of heat transfer coefficient strong influenced by the parameters including: fiber type, pulp concentration, characteristics as well as fiber pulping method. At a low fiber suspension concentration and higher flow rates, it was observed that higher values of heat transfer coefficient were achieved than that of water, in contrast, a decrease in heat transfer values were found at increasing concentration and lower pulp suspension. It was found that values of heat transfer coefficient decreases with increasing the fiber flexibility for both Acacia Mangium and Kenaf bast respectively. In the present research work, the fiber dimensions and fiber flexibility are inter-related along with values of heat transfer coefficient of both wood and non-wood fiber suspensions.
Keywords. Comparative analysis, heat transfer coefficient, pulp fiber

1. Introduction

There are diverse application of fiber suspension in various process industries like textile industries, food processing, packaging and most significantly the major and most applicable use of the huge volume of fiber suspension in pulp and paper industry. The raw materials for paper manufacturing are pulp comprises of natural cellulose fiber mostly obtained from wood and non-wood plants. Trees are the core source of wood fiber which constitute both softwood (Pine, Fir, spruce etc.) and hardwood (eucalyptus, ash, Aspen, etc.). Due to the limitations, demand as well as environmental concern of wood source, nonwood plant fibers become a substitute source of natural occurring wood fibers [1-2]. Among them Kenaf is one of the most economical plant-base fibers commonly used for paper and pulp production [3]. Moreover, Kenaf Bast (outer portion) fiber is characterized as long fiber like softwood fibers and commonly used for the production of high grade pulping not only for paper and pulp industry but also for as for textile, filter and composite materials as well [2,4-6].

The studies of heat transfer in the suspension forms have two fold benefits: for the Advancement about substantial models that might control the rejected paper processing and secondly for Different range about industries and can be utilized for planning about the equipment designing for pulp handling and the pipelines for supplying pulp in the paper and pulp industries. In 1994, the heat transfer effects of natural wood and synthetic fibers suspension in water were investigated for the first time. Later on these investigations were further improved for a variety of concentration and at a different flow rates in a pipe-line test setup to achieve more useful results for correlating heat transfer coefficient (h_C) to different fiber properties [7-8]. Furthermore, these finding in the form of h_C correlation to specific fiber properties could be significant in order to retard low quality paper production. Kazi and Duffy [9] further investigated the correlations and observed that h_C were increases by decreasing fiber concentration and increasing flow velocity using annular pipe heat exchanger. Additionally, heat transfer coefficient (h_c) was affected by various pulping method including: mechanical and chemical fiber pulping as well as degree of fiber beating [10-11].

Present research work focus on the comparative analysis of heat transfer coefficient (h_c) for Acacia mangium (wood fiber) and Kenaf Bast (non-wood) pulp fiber at various pulp flow and different fiber concentration parameters. Moreover, the effect of different fiber processing on heat transfer coefficient (h_c) may also be considered for deep insight monitoring of h_c for hindrance of paper production loss.

2. Methodology

2.1 Materials

Both Kenaf Bast and Acacia mangium fiber samples were prepared by mechanical pulping and chemical pulping process by soda pulping by bleaching acacia mangium and Kenaf bast. All the investigated pulp samples were provided by Forest Research Institute of Malaysia (FRIM).

2.2 Experimental set-up

The experimental test setup consists of storage tank of 100 Liters, a close loop pipe line with an aluminum heat transfer test section having dimensions of (900 mm length and 41.4 mm inner diameter) along a centrifugal pump for pulping the suspension. A magnetic flow meter at the inlet section and pressure drop transducer was installed at the inlet and outlet ends of the test section for measuring the pressure drop. A chiller was connected in parallel to stock tank for maintaining the desired temperature inside the tank. A schematic diagram of the whole set-up has been shown in Figure 1.

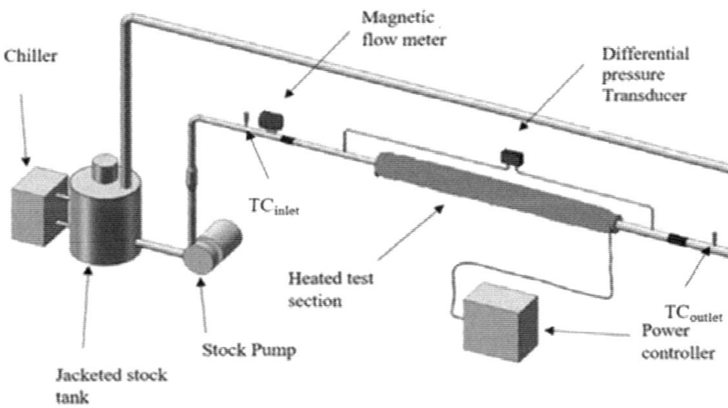

Figure 1. Experimental test section diagram

2.3 Data acquisition

A data logger, Magnetic flow meter and Differential Pressure cell were connected for measuring the inlet, outlet and test section wall temperature, flow rates and pressure drop across the test section. Power of the installed heater at the test section was calculated from the supplied current and voltage to the heater. The obtained data were taken the conditions of steady state, bulk temperature, heating flux and instantaneous conditions.

2.4 Experimental procedures

All the investigated wood and nonwood fiber samples were supplied by FRIM. A known quantity of the fiber sample were disintegrated in the disintegrator equipment up to 3500 rpm until the whole fiber bunches were uniformly dispersed, before sending to the experiment test rig [12]. All the experimental settings are tabulated in Table 1.

Table 1. Experimental conditions

Test variable parameters	Desire range
Velocity	0.4 - 2.8 m/s
Pulp bulk temperature	30 °C
Fiber concentration	0.2 - 0.6 wt.%
Heat flux	34 k Watt/cm^2

The surface temperatures (Ts), wall resistances (λ/δ), temperature of thermocouples (Ttc), and heat flux ($q^{\cdot}$), relation was calculated using equation (1).

$$T_s = T_{tc} - q^{\cdot}/(\lambda/x) \quad (1)$$

Where λ is thermal conductivity of test section and δ represents thermocouple distances of thermocouple from the inner surface of the pipe.

The value of heat transfer coefficient h_c is calculated from the test section wall temperature Ts, the pulp bulk temperature T_b and heat flux q, using the following equation (2).

$$h_c = q^{\cdot}/(T_s - T_b) \quad (2)$$

Equation (3) and equation (4) are used for the evaluation of bulk temperature (T_b) and Nusselt number (Nu), and obtained by the following equation (3) and equation (4).

$$T_b = (X/L)\,[T_{out} - T_{in}] + T_{in} \quad (3)$$

$$N_u = (h_{avg}\,D)/K \quad (4)$$

Where, X and L represent the distances of thermocouples tips in the flow direction and the heated length-wise respectively.

3. Results and discussions

3.1 Effect of fiber concentration on heat transfer coefficient (h_C)

Heat transfer coefficient (h_c) data at three different weight concentrations (0.2 wt.%, 0.4wt.%, 0.6 wt.%) of acacia mangium and kenaf bast fiber suspensions together with water are shown in Figure 2. The obtained results in a comparison with water run and the three concentrations of both samples suggested that flow of fiber suspension at velocities lower than 1 m/s results in a lower heat transfer coefficient than water. Moreover, concentration effects on heat transfer coefficient are more dominated when the velocity increases beyond 2 m/s. However, at the middle and high velocities range the suspensions with the lowest concentration (0.2 wt.%) tends to show the heat transfer coefficient higher than water in the case of acacia mangium only.

The present research work reveals that the mechanically pulped kenaf have h_c values lower than the corresponding values of water during the course of velocity as well as concentration excluding for a lower fiber concentration of 0.2 wt.% which exhibit a higher hc values than that of water data at the velocity higher than 0.8m/ s. It was investigated that heat transfer coefficient significantly decreases with increasing the fiber suspension concentration and increases with the decrease in the concentration of kenaf bast fiber pulp.

In fact, the thermal conductivity of fiber suspension have a thermal conductivity lower than water, due to this particular fiber behavior, the enhancement of h_C contributed by fibers by the virtue of modification of turbulent eddies.

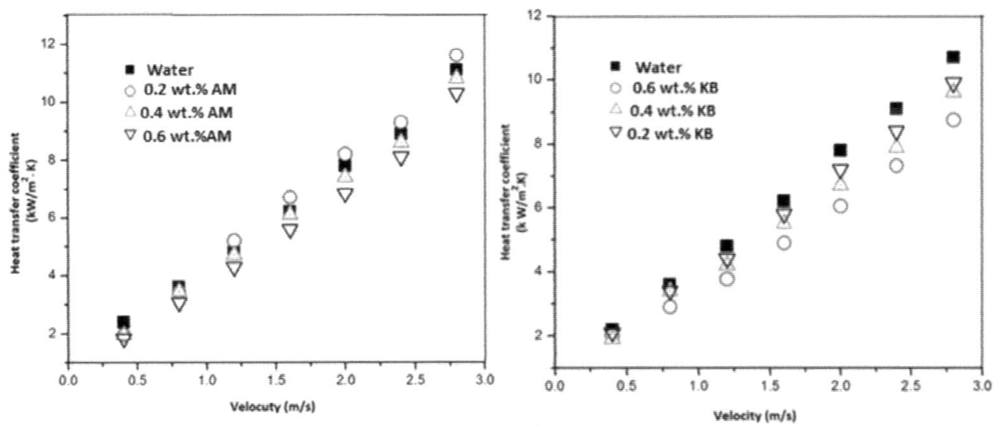

Figure 2. Heat transfer coefficient as a function of velocity for different concentrations

3.2 Effect of fiber bleaching on heat transfer coefficient (h_C)

The effect of fiber beaching on h_C were examined by AMB and KBB pulps, and prepared by AM and KB pulp fiber. Figure 3 represents the heat transfer coefficient (h_C) over a variety of velocity for water, bleached AM and bleached KB at fiber concentration 0.6 wt% and at bulk temperature 30 °C.

The experimental data indicate that the values of h_C for both bleached fiber pulp are lower than the corresponding values for water throughout the range of velocities. The result of bleaching lead a net increase in fiber flexibility, which further lead a lower h_C value as compared to unbleached pulp fiber. Moreover the bleached KB are more flexible as compared to bleached AM, therefore h_C values for bleached KB are lower to the corresponding bleached AM pulp fiber as obvious in Figure 3. Same results were achieved by the other researcher as well [13]. During chemical pulping, fiber swelling and fiber detachment within fiber and from fiber wall occur due to lignin removal (after bleaching) from pulp fiber producing a more flexible fiber and lower h_C values.

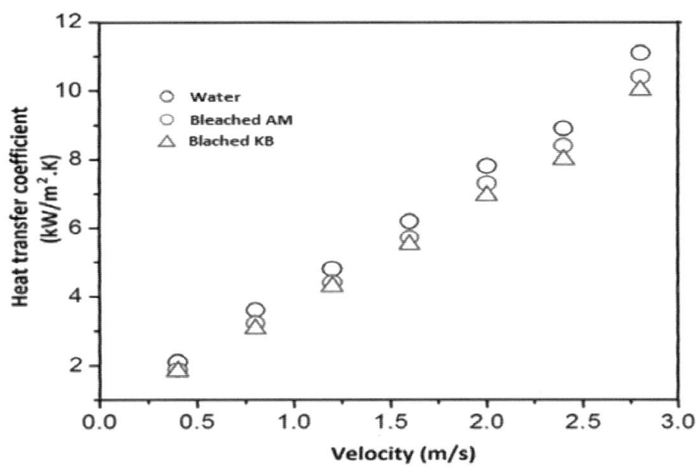

Figure 3. Heat transfer coefficient as a function of velocity for water, bleached AM and bleached KB fiber pulp suspensions

4. Conclusion

Significant differences in the values of h_C were observed after fiber pulping process including: fiber concentration and fiber pulp bleaching of both wood (acacia mangium) and nonwood (kenaf bast) pulp fiber suspensions. Up to certain limits of fiber pulp suspension concentration, a considerable amount of h_C values have been observed. Moreover, bleaching of pulp suspension provides a more flexible and more uniform fiber surface with a lower h_C values. By adopting proper fiber pulping processes, fiber quality can be improved by changing fiber pulp characteristics for the production of especial products.

References

[1] Kaldor A F, Karlgren C and Verwest H 1990 Kenaf-a fast growing fiber source for papermaking *Tappi Journal* **73** 205-8

[2] Mossello A A, Harun J, Shamsi S R F, Resalati H, Tahir P M, Rushdan I and Mohmamed A Z 2010 A review of literatures related to kenaf as a alternative for pulpwoods *Agricultural Journal* **5** 131-8

[3] Han J S and Rowell J S 1997 Chemical composition of fibers *Paper and composites from agro-based resources* 83-134

[4] Villar J, Revilla E, Gómez N, Carbajo J and Simón J 2009 Improving the use of kenaf for kraft pulping by using mixtures of bast and core fibers *Industrial crops and products* **29** 301-7

[5] Dutt D, Upadhyay J, Singh B and Tyagi C 2009 Studies on Hibiscus cannabinus and Hibiscus sabdariffa as an alternative pulp blend for softwood: An optimization of kraft delignification process *Industrial crops and products* **29** 16-26

[6] Ahmed S, Kazi S, Khan G, Dahari M, Zubir M, Ahmad P and Montazer E 2017 Experimental investigation on momentum and drag reduction of Malaysian crop suspensions in closed conduit flow. In: *IOP Conference Series: Materials Science and Engineering*: IOP Publishing p 012065

[7] Kazi M S N, Duffy G G and Chen X D 1999 Heat transfer in the drag reducing regime of wood pulp fibre suspensions *Chemical Engineering Journal* **73** 247-53

[8] Duffy G, Kazi S and Chen X 2000 New Way to Monitor Wood Pulp Fibre Quality Using Heat Transfer Coefficient. In: *Proceedings of the 54th Annual Appita Conference, Melbourne, Australia, Apr,* pp 3-6

[9] Kazi S, Duffy G and Chen X 2012 Validation of heat transfer data for fibre suspensions in coaxial pipe heat exchangers *Experimental Thermal and Fluid Science* **38** 210-22

[10] Kazi S, Duffy G and Chen X 2014 Validation of heat transfer and friction loss data for fibre suspensions in a circular and a coaxial pipe heat exchanger *International Journal of Thermal Sciences* **79** 146-60

[11] Kazi S, Duffy G and Chen X 2015 Heat transfer coefficient of flowing wood pulp fibre suspensions to monitor fibre and paper quality *Applied Thermal Engineering* **78** 172-84

[12] Tappi T 2002 222 om-02: Acid-insoluble lignin in wood and pulp *2002–2003 TAPPI Test Methods*

[13] Kazi S, Duffy G and Chen X 2015 Heat transfer coefficient of flowing wood pulp fibre suspensions to monitor fibre and paper quality *Applied Thermal Engineering* **78** 172-84

Numerical Evaluation of Slope Stability measures: A Case Study of Birham Landslide, Murree, Pakistan.

Shamsher Sadiq[1], Urwah Khan[2], Mehtab Alam[3], Saeedullah Jan Mandokhail[4], Maqsood-ur-Rehman[5], Nizamani Zubair Ahmed[1], Ayaz Ikram[6]

[1] Graduate Student, Department of Civil and Environment Engineering, Hanyang University, Seoul, South Korea
[2] Graduate Student, Nagoya University, Japan
[3] Graduate Student, Institute of Mountain Hazards and Environment, Chinese Academy of Sciences, Chengdu, P.R China
[4] Assistant Professor, Department of Civil Engineering, BUITEMS, Quetta, Pakistan
[5] Senior Engineer, National Engineering Services of Pakistan (NESPAK)
[6] Junior Engineer, Mott Macdonald Pakistan (pvt) Ltd. (MPP)

Email: ershrajput@hotmail.com

Abstract. Slope failures endanger the public safety and one of major hazard considered in mountainous terrain. In this paper, slope stability measures have been evaluated using the limit equilibrium method (LEM). A case study of slope failure from Birham land slide, Murree, Pakistan has been modelled using SLOPE/W program. In-situ boring tests are performed to collect laboratory test specimens. Geotechnical properties (i.e. shear strength and stiffness parameters) for the idealised slope sections are based onlaboratory tests performed on the collected in-situ specimens. Slope stability measures are evaluated in terms of factor of safety (FoS) for static and dynamic loading. Pseudo-static method is used for dynamic analysis, considering the Peak ground acceleration (PGA) from seismic hazard map of Pakistan. Results shows that static and dynamic FoS of original slope are inadequate therefore resistance to slip surface of original unstable slope is provided by slope geometry modification with pile reinforcement. Based on these remedial measures, computed FoS of modified slope are found to be sufficient for static and dynamic slope stability.
Keywords. Stability, birham landslide, numerical evaluation, Murree, Pakistan.

1. Introduction

Murree is a hilly resort town located in northern area of Pakistan. The area of presented case study is located in Murree formation consisting of sandstone, shale and weakly bounded slope forming material. Moreover, it lies in seismically active region due to active Murree Boundary Fault (MBT) Figure 1. Due to highly fragile hilly terrain and seismically active region, this provides a favorable failure zones for faults during seismic event [1, 2].

There are several number of land sliding events along Islamabad Murree Expressway because of rapid urbanization and widening of roads and so on. Menal Z. [1] studied the landslide susceptibility for Murree using Geographic Information System (GIS) and remote sensing techniques. The proposed susceptibility maps depicted in figure 2 shows very low, low, moderate, high and very high landslides with percentage 15%, 27%, 29%, 21% and 8% respectively.

In this paper, a case study of slope failure from Birham land slide that was triggered 10th August, 2012, in Murree, Pakistan has been modelled using Limit Equilibrium (LE). [3] Slope/W program is used to perform LE analysis to calculate static and dynamic Factor of Safety (FoS) for the original and modified improved slopes.

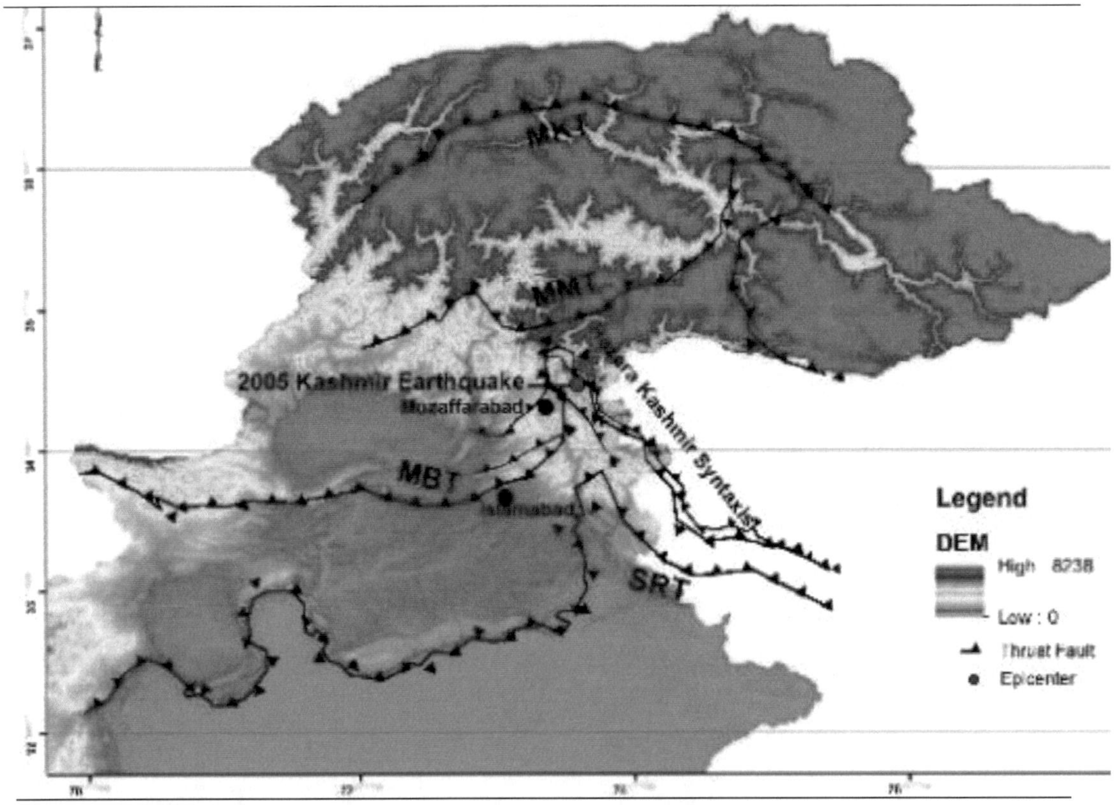

Figure 1. Map showing major fault systems in Northern Pakistan and Kashmir Earthquake, 2005 epicentre location, [2].

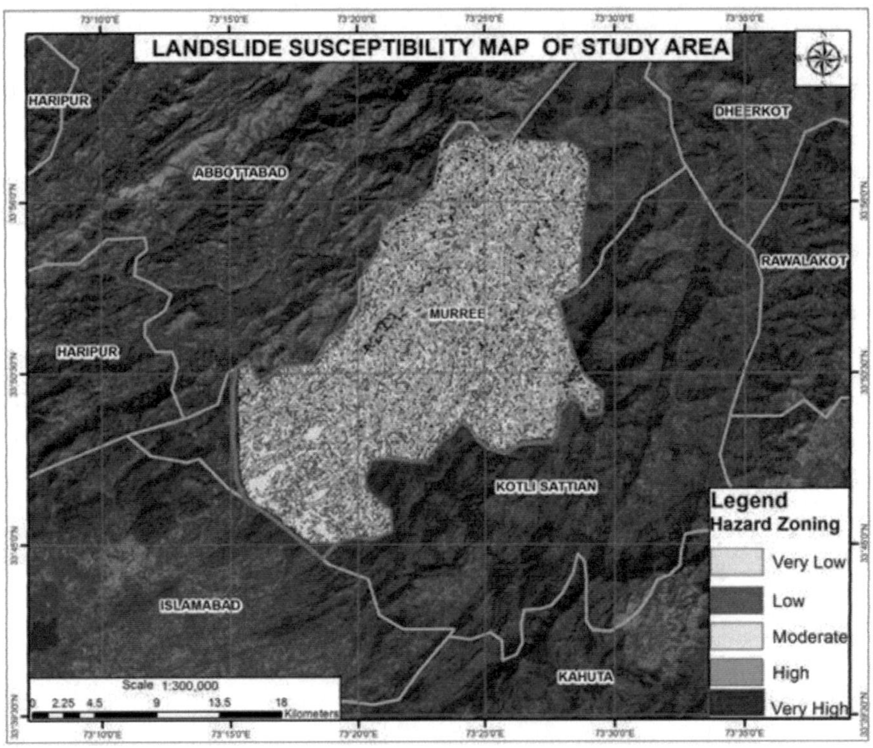

Figure 2. Landslide Susceptibility map, Murree [1].

2. Case Study

The town of Murree in Himalayan foothills lies about 50km to Northeast of Islamabad at an elevation of approximately 2200m. The lithology comprised of shale, siltstone and sandstone belonging to the Oligocene-Miocene Murree and Kuldana formations.

Slope stability analysis requires the stiffness and strength properties of soil/intact rock to be known. Geotechnical investigations were performed at the three locations as shown in layout plan in Figure 3, to collect the samples from top, middle and bottom of slope, details of sample and laboratory test results are presented in Table 1. The slope strata mainly comprise of shale, siltstone and sandstone. Based on careful extrapolation of boreholes, slope stratigraphy is idealized as shown in Figure 4.

Table 1 Summary of Geotechnical properties

Borehole No.	Run No.	Depth (m)	Rock type	Point load strength (MPa)	Specific gravity	Bulk Density (kPa)	Moisture content (%)
BH-1	R-1	10.7-11.1	Claystone	-	2.63	22.9	2.02
	R-2	13.5-13.7	Siltstone	4.4	2.72	23.9	4.54
BH-2	R-3	5.4-5.6	Sandstone	8.1	2.75	25.1	1.37
	R-4	6.0-6.4	Shale	-	2.67	25.8	1.19
	R-5	14.1-14.2	Claystone	-	2.65	24.6	3.45
	R-6	15.9-16.1	Sandstone	16.1	2.74	24.3	0.30
BH-3	R-7	17.3-17.4	Shale	-	2.62	24.2	5.98
	R-8	21.5-21.6	Claystone	2.8	2.69	22.1	6.22
	R-9	26.9-27.2	Shale	-	2.63	23.1	4.80

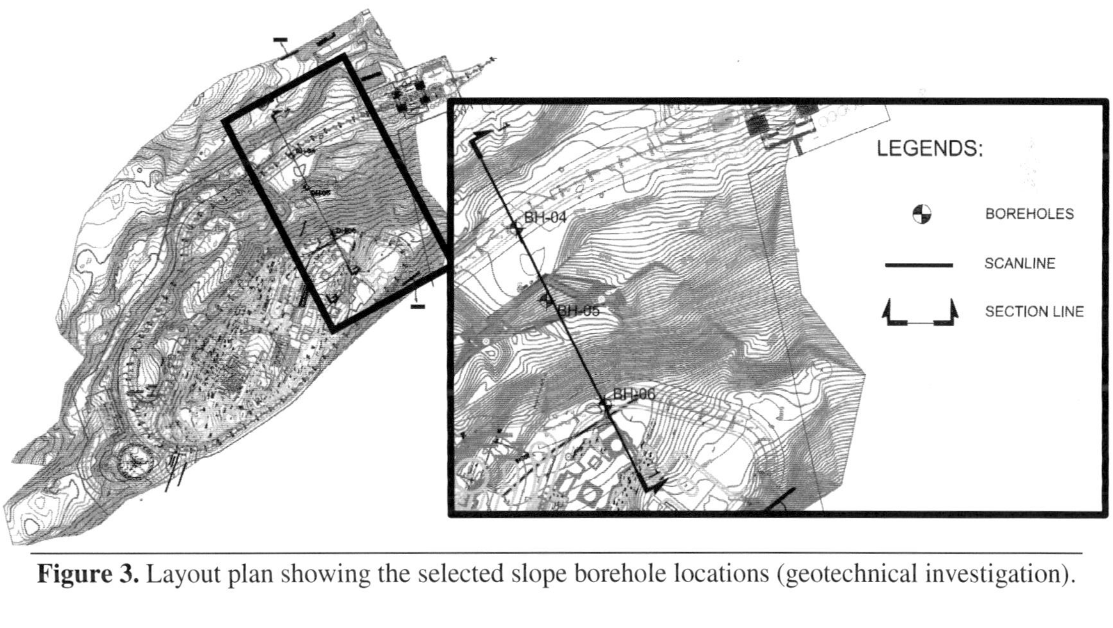

Figure 3. Layout plan showing the selected slope borehole locations (geotechnical investigation).

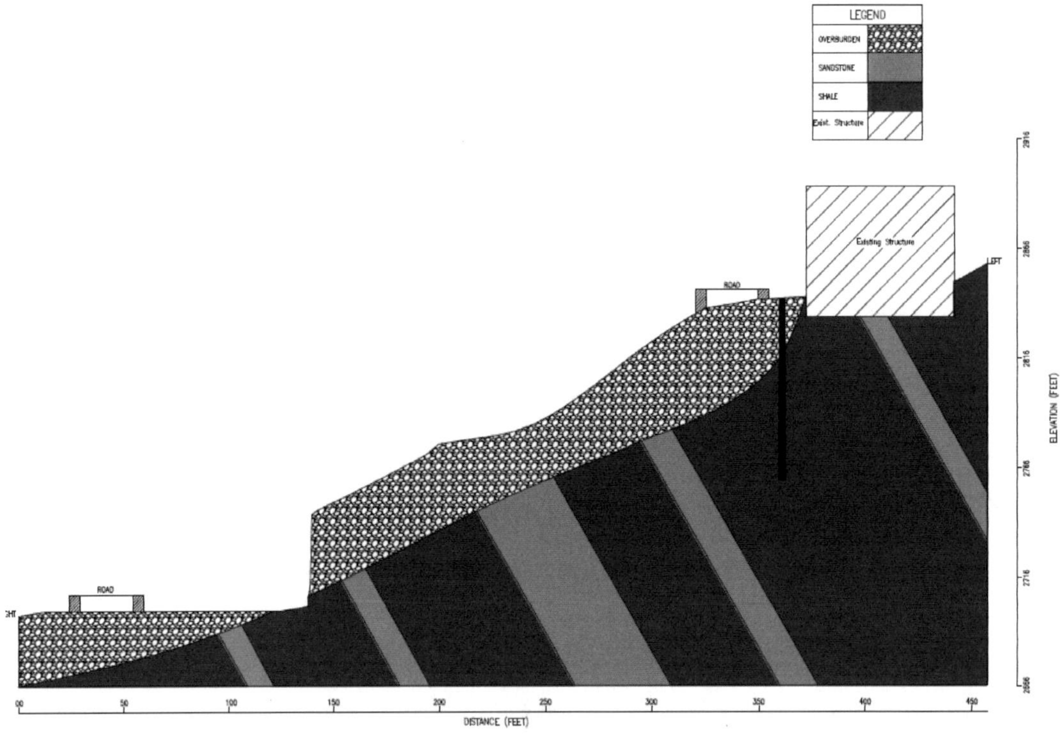

Figure 4. Idealized slope geometry and stratigraphy of selected slope section 1-1'.

3. Numerical modelling

Based on layout plan and selected section 1-1' shown in Figure 3 and idealized slope geometry and stratigraphy shown in Figure 4, two dimensional plane strain (2D) static and pseudo static analysis to calculate factor of safety (FoS) are performed using SLOPE/W program. The soil and rock properties used for simulations are summarized in Table 2. The constitutive behaviour of soil and rock both are simulated using the elastic-perfectly plastic assumption of Mohr-Coulmn model. The boundary conditions in numerical model were applied such that the vertical face was fixed in horizontal direction and base was fixed in both horizontal and vertical directions.

Table 2. Summary of soil and rock properties used in numerical modelling

Layer	Soil/rock type	Unit weight (kN/m^3)	Cohesion (kPa)	Friction angle (Deg.)
1	Overburden	19.66	32.60	20.0
2	Shale	23.96	369.2	23.5

Pseudo static analysis is one of simplest methods of idealizing the slope stability problem subjected to dynamic loading. Although earthquake force is dynamic, however if the strength is reduced less than 15 percent by cyclic loading, pseudo static analysis of the seismic loading can be used [4]. In pseudo static analysis, the Peak Ground Acceleration (PGA) generated by earthquake shaking is considered to create inertial forces. These inertial forces act in horizontal and vertical directions [5].

Building code of Pakistan (BCP) [6], provides the seismic hazard level in terms of horizontal Peak Ground Acceleration (PGA) for the four seismic zones of Pakistan Figure 5, the level of PGA corresponding to each zone is presented in Table *3*. The selected site shown in Figure 5 is located in Zone 2B, level of hazard in terms of horizontal PGA is 0.2g. Dynamic FoS is calcuted by performing pseudo static analysis, in which 0.2g PGA value is used to introduce the seismic loading.

Table 3. Peak horizontal acceleration for different seismic zones

Zone	**1**	**2A**	**2B**	**3**	**4**
PGA (g)	0.075	0.15	0.20	0.30	0.40

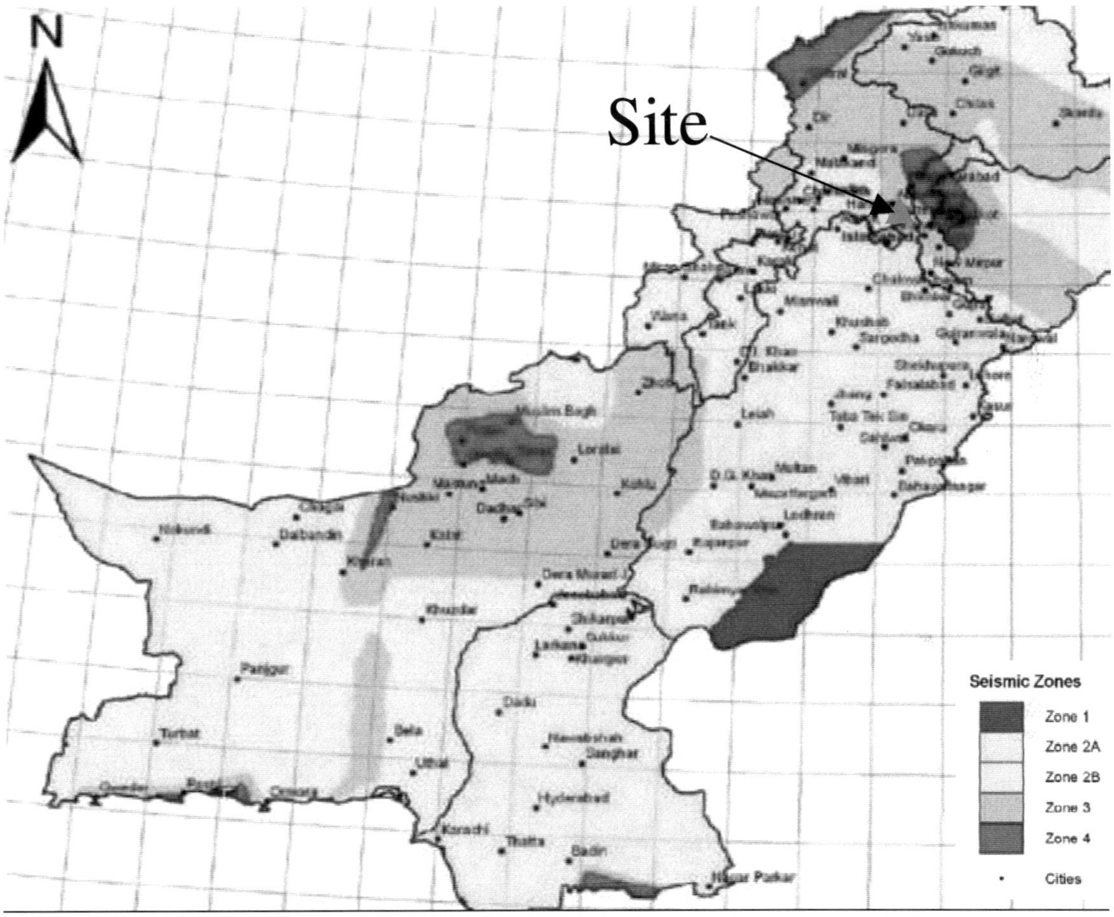

Figure 5. Seismic zonation of Pakistan, Building code of Pakistan (BCP) [6]

4. Results and Discussion

The stability of slope is evaluated in terms of FoS. The computed static and pseudo static FoS for both the original Figure 6 and improved slopes Figure 7 are presented in Table 4. The calculated minimum FoS of original slope for both static and pseudo static analysis is less than the required [6] code guidelines. Slope geometry modification and pile reinforcements are the remedial methods usually used for improvement of unstable slopes in Pakistan. Pile shear strength and arching between the rows of these piles are two type of resistances provided to critical slip surface of unstable slope. Pile location, length, diameter and centre to centre spacing are selected based on parametric study (not presented here for brevity). Calculated FoS of modified slope Figure 7 are adequate and slope is stable under static and dynamic loading.

Table 4 Calculated Factor of Safety

Slope	Analysis	Factor of Safety (FoS)	Comments
Original	Static	1.000	<1.5 (Unstable)
	Dynamic	0.718	<1.1 (Unstable)
Improved	Static	1.807	>1.5 (Stable)
	Dynamic	1.184	>1.1 (Stable)

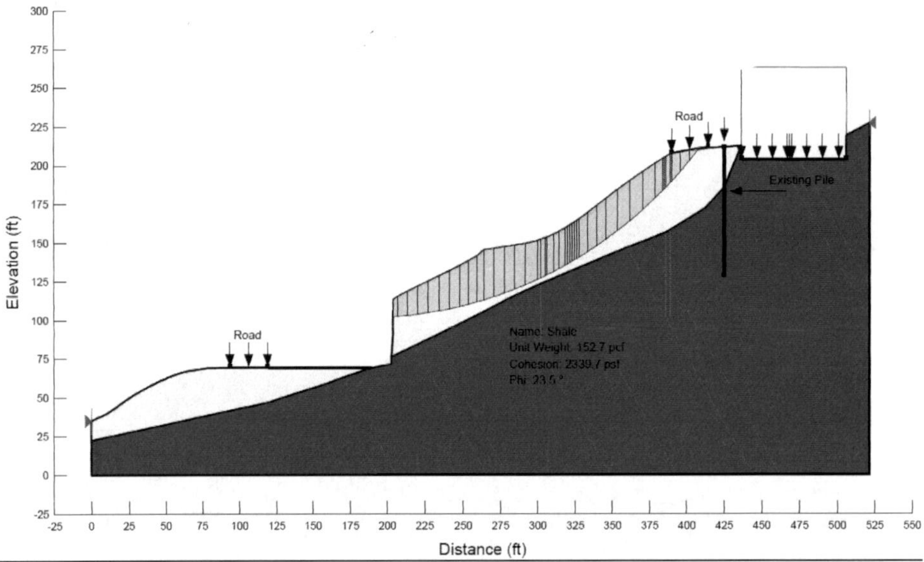

Figure 6. Slope stability analysis for original section under static and dynamic loading.

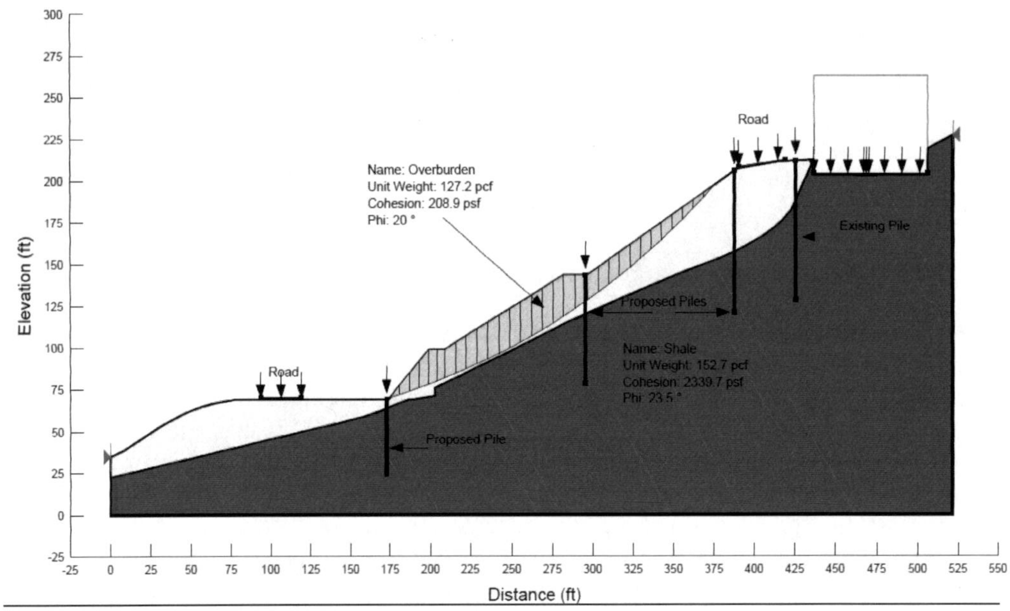

Figure 7. Slope stability analysis for the modified geometry and proposed shoring (RCC pile walls) under static and dynamic loading

5. Conclusions

Murree, Pakistan is susceptible to major landslide events due to its location in seismically active region and weak loosely bound slope forming material. A case study of slope failure from Birham land slide has been numerically modelled using SLOPE/W program. Slope stability in terms of FoS is evaluated for original and modified slope under static dynamic loading. Dynamic loading is considered as a pseudo-static by defining horizontal PGA of 0.2g from seismic hazard map of Pakistan. Results shows that static and dynamic FoS of original slope are inadequate therefore resistance to slip surface of original unstable slope is provided by slope geometry modification with pile reinforcement. Based on these remedial measures, computed FoS of modified slope are found to be sufficient for static and dynamic slope stability.

References

[1] Menal Zaheer, S.R.A., Zohaib Fayyaz, Landslide Susceptibility Mapping using Geomatics; A Case Study of Murree Area, Pakistan. Proceedings of the 19th International Conference on Soil Mechanics and Geotechnical Engineering, Seoul 2017

[2] Mahmood, I., et al., Analysis of landslides triggered by October 2005, Kashmir earthquake. PLoS currents, 2015. 7.

[3] Krahn, J., Stability modeling with SLOPE/W: An engineering methodology. GEOSLOPE/W International Ltd. Calgary, Alberta, Canada, 2004.

[4] Duncan, J.M., S.G. Wright, and T.L. Brandon, Soil strength and slope stability. 2014: John Wiley & Sons.

[5] Kramer, S.L., Geotechnical earthquake engineering prentice hall. Upper Saddle River, NJ, 1996.

[6] Building Code of Pakistan, Seismic Provision, SP-2007.

Engineered Silica Nanoparticles and silica nanoparticles containing Controlled Release Fertilizer for drought and saline areas.

Ayesha Mushtaq[1], Nelofer Jamil[1], Sabeena Rizwan[1], Fazeela Mandokhel[1], Musarrat Riaz[1], G L Hornyak[2], M Najam Malghani[3] and M Naeem Shahwani[4]

[1]Department of Chemistry, SBK Women's University, Quetta, Pakistan.
[2]Center of Excellence in Nanotechnology, AIT, Thailand.
[3]Facultyof Biotechnology & Informatics, BUITEMS, Quetta. Pakistan.
[4]Faculty of Chemical Engineering, BUITEMS, Quetta. Pakistan.

Corresponding author email: muhammad.naeem@buitms.edu.pk

Abstract. Silicon supplementation was explored to increase salinity tolerance in plants, in the beginning of this cenutury, therefore silica nanoparticles (SiO_2 Nps) can also be used as a trial to improve salinity tolerance in plants. In this study SiO_2 Nps have been synthesyzied by some modification in the conventional method of Stober by thermal hydrolysis of Tetraethylorthosilicate. A compound Controlled Release Fertilizer (CRF) was synthesized that carried NPK and SiO_2 Nps inside the core and Chitosan as the first semi-permeable coating and Sodium Alginate and Kaolin as an outer most superabsorbent coating. The synthesized SiO_2 Nps were characterized by TEM and XRD. The water absorbency of CRF beads showed that they can absorb large amounts of water and double their weight. The Nutrient released rate from CRF beads was very slow and sustained for six months at room temperature. The SiO_2 Nps containing superabsorbent CRF was capable of releasing the nutrients slowly, withhold large amounts of water therefore can help plants control salinity and survive better in drought and saline conditions without harming the environment. The synthesized compound fertilizer is biocompatible, biodegradable and nontoxic so ideal for growing plants in drought and salt effected areas.
Keywords: Silica nanoparticles, saline, drought, Sol-Gel, controlled release fertilizer

1. Introduction

Silica nanporous materials occupy analytical interest due to their extreme use in agriculture, drug delivery, enzyme encapsulation, cell markers, chromatography, chemical sensors and biological images [1]. Due to broad applications of silica nanoparticles in multidiscipline their synthesis by various methods is getting researcher's utmost attention. Silica nanoparticles can be used in agriculture, as a source of silicon to alleviate salinity stress in plants growing in drought and saline environments. Silica nanoparticles (SiO_2 Nps) have been effectively prepared by both inorganic and organic colloids [2, 3]. SiO_2 Nps can be prepared up to 10 nm by the sol–gel method using TEOS by controlling the conditions in the Stober reaction [4]. The latest ongoing research activiteies have established and emphasized the capibility of ultrasonication techniques for the synthesis of many nanoparticles [5]. Uultrasonic-based synthesis of SiO_2 Nps resulted in tiny patrticles of about 13 nm [6]. Due to advantages of ultrasonication, this study has been carried out by ultrasonication based synthesis of SiO_2 Np by modified Stober process (Sol-Gel). These SiO_2 Nps were further encapsulated in a compound-controlled release fertilizer to be used for improving agriculture in drought and saline areas.

Excessive use of pesticides and chemical fertilizers for increasing agricultural yield has polluted the environment and water resources [7]. Nanoparticles like Zinc Oxide & Zinc stannite has been used for photocatalytic degradation for purification of polluted water [8-10]. The application of CRF is getting popularity over the conventional fertilizer, as the conventional fertilizer releases excessive residues randomly that persist in the soil and can leach down to the water reservoirs. CRF can release the nutrients slowly, at the required amount and time needed by the plant, and reduces the toxic wastes and remains.

Application of Controlled Release Fertilizers is a highly beneficial technology that encapsulates or coat the conventional fertilizers in the form of granules, of incorporated polymer matrix, which have slow-release properties, allowing controlled flow of nutrients at the time and amounts needed for plant [11]. Chitosan is one of few natural polymer that is the biocompatible, biodegradable, non-toxic towards plant tissue, which offers excellent opportunity to improve controlled release fertilizer behavior and with the combination of kaolin superabsorbent polymer, it can broaden its application especially in drought and marginal areas [12]. Therefore, a compound Controlled Release Fertilizer can be synthesized that can carry NPK fertilizer and SiO_2 Nps inside the core and it can be coated with chitosan as the first semi-permeable coating and Sodium Alginate and Kaolin as an outer most superabsorbent coating. These beads will be capable of releasing the nutrients slowly that can withhold large amounts of water, help plants control salinity and survive better in drought conditions.

2. Materials and Methods

2.1 Synthesis of Silica Nanoparticles

TEOS (1.8 ml) was added slowly to the co-solvent (water & ethanol) and mixed by ultrasonication for 1 hour. Equal amount (1.4 ml) of NH_4OH and DI water were mixed, separately on stirrer for 30 minutes and added dropwise and carefully to the TEOS solution during sonication by a micro-pipette. The mixture was sonicated for two hours until it turned milky and gel like. It was centrifuged at 4000 rpm for 20 minutes and SiO_2 Nps were collected and dried in the oven.

2.2 Synthesis of Controlled Release Fertilizer

2.2.1. Preparation of CS-NPK-SiO_2 Nps beads for controlled release fertilizer

Chitosan (3 %) solution was used as dope while TPP (5 %) solution as a cross-linker. Dissolved granules of NPK (20 %) and SiO_2 Nps (0.1 g) were added to chitosan solution as dope. Under continuous stirring, slowly the dope was dropwise added with the help of a syringe, into the TPP solution. After cross-linking, the formed CS-SiO_2 Nps -NPK beads were collected & washed with DI water to remove the remaining TPP on the beads surface.

2.2.2. Double coating of CS -NPK-SiO_2-NPs beads with SA-Kaolin superabsorbent

Superabsorbent polymer coating of Sodium Alginate (SA) was made by its continuous stirring at 65 °C, for two hours. This solution was divided into two parts, one part remained with SA to provide a second absorbent coating on CS-SiO_2 Nps -NPK beads to make SA-CS-SiO_2 Nps -NPK beads that can withhold more water.

Another part contained SA and 5 % kaolin (w/v) to coat beads with Kaolin (K) superabsorbent as the outermost coat (K-SA-CS- SiO_2 Nps -NPK). CS-SiO_2 Nps -NPK beads (a) were then placed inside each treatment for 20 minutes, under continuous stirring. Coated beads were then dried in the oven at 70 °C for 24 hours for further use.

2.3 Methods for characterization of SiO_2 Nps & CRF beads

The particle size of SiO_2 Nps were determined by Transmission Electron Microscope (JEM-2100 plus) with operating voltage of 200 KV to determine the images. X-Ray diffraction spectrum was taken by Bruker, 2nd Gen, D2, Phaser. The X-Ray Diffaraction spectrum of precipitated SiO_2 Nps was recorded with an intensity of 30 KV.

2.4 Water Absorbency of CRF beads

Absorbency of synthesized beads (CRF) was measured by putting small number of dried beads (0.1 g) in 100 ml water for 90 minutes. Beads were filtered and dried for calculating absorbency. Absorbency was calculated by the formula;

$W = \frac{M-M*}{M*}$ (where M=weight of wet beads,M*= weight of dried beads)

2.5 Nutrient Release Rate of CRF beads

Release of nutrients from different types of coatings (Chitosan, Sodium alginate, Kaolin) was checked by measuring the TDS daily, of different types of beads by placing them (0.1 g) in 100 ml water for six months at room temperature.

3. Results and Discussion

3.1 Synthesis and characterization of silica nanoparticles

Although various methods have been reported for the synthesis of silica nanoparticles including microemulsions, stober and sonochemical [13-15], but the Stober method is the pioneer and the most promising simplest method that has been extensively used and also modified by many researchers. Ultrasound assisted synthesis of silica nanoparticles by modified Sol-gel (Stober method) has been carried out in some recent research [6, 16]. Ultrasonication is recommended to an agglomeration phenomenon that effects the shape of suspended nanoparticles therefore it can be used to contol the shape and the size of the nanaoparticles. The ultrasonication process is proved to eradicate the primary structure of the silica gel and modifies the size and the shape of the formed clustures with particle re-agglomeration [5]. The particle size of silica nanoparticles synthesized by the conventional method was 200 nm, while the particle size reduced from 100 nm to 50 nm by ultrasonication method. Therefore it can be concluded that increasing the duration or time of ultrasonication has a significant decrease in the particle size of the nanoparticles. There is no particular technique which can characterize the physiochemical properties (size and composition) of nanoparticles [17]. Therefore in this study various techniques like TEM and XRD, have been used to characterize the physiochemical properties like size, morphology and composition of Silica nanoparticles.

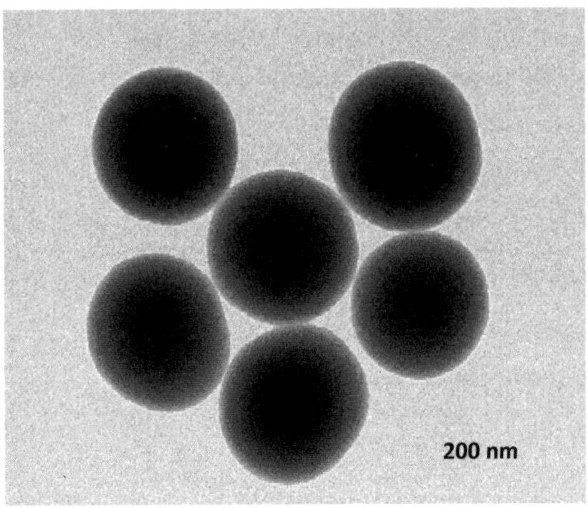

Figure 1. TEM image of SiNps of 200nm

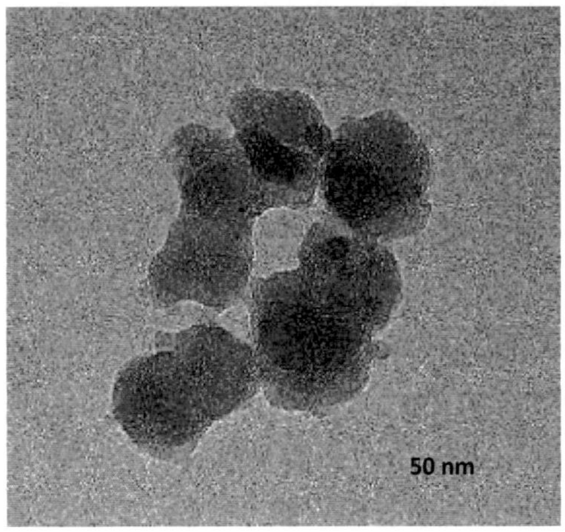

Figure 2. TEM images of SiNps of 50nm

TEM images (Figure 1 & 2) illustrated that silica nanoparticles prepared by conventional Stober method were spherical and larger in size about 200 nm (Figure 1), while the silica nanoparticles by ultrasonication method reduced to size of 50 nm (Figure 2) . Our synthesized SiO_2 Nps looked same like colloidal SiO_2 Nps arranged into clustures that could also be a result of drying mechanism in the sample preparation stage [5]. The dffraction patterns of precipitated SiO_2 Nps were obtained in 2θ range from 10-70°. The characteristic peak of SiO_2 Nps prepared is presented as Figure 3. The spectrum appeared as a broad halo peak centered at 2θ angle of 22°, that confirm amorphous nature of silica nanoparticles. These results are similar to the spectrum obtained as a broad band with the equivalent Bragg angle at 2θ = 22 ° [18-20]. The broad diffraction band centered at 2θ angle of about 22.5 ° is a known typical characteristic of silica [5, 21].

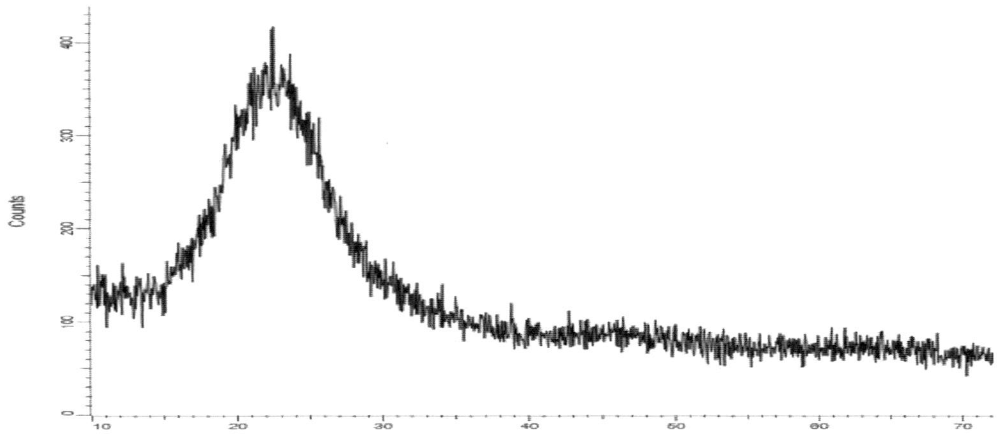

Figure 3. XRD pattern of SiNps

3.2 Water Absorbency & Nutrient Release Rate of CRF beads

Water absorbency is the capibility of dried beads to absorb water at room temperature without any cycling and incubation in water bath [22]. It is useful for the determination of water holding capacity of the prepared beads. The absorbency of synthesized CRF beads with different coatings and incorporation of SiO_2 Nps in water was recorded. Water absorbency of chitosan coated beads < sodium alginate < Kaolin-SA which prooved that Kaolin-SA polymer has unique superabsorbent properties and its application can be useful in drought, prune and marginal areas with less rainfall or water stress conditions. Sodium alginate is naturally able to absorb 200-300 times its own weight, once it was attached on the beads surface, it improved the swelling and absorbance of beads. Kaolin effected the surface of dried beads by lowering the content of hydrophilic group and decreased the osmotic pressure differences [22]. The synthesized beads (K-SA-CS-SiO_2 Nps -NPK) in the present study are capable of withholding large amounts of water and can double their original weight after absorption of water.

The nutrient release rate (Figure 4 & 5) of dried beads (K- SA-CS-NPK, SA-CS-NPK, CS-NPK), with SiO_2 Nps (red lines) and without SiO_2 Nps (black lines) has been recorded for six months in summer (Average temperature ranged 18°C - 24°C) and six months in winter (Average temperature ranged 4°C - 15°C). The rate of nutrient released was measured as TDS (Total Dissolved Solids) which was higher in summer, as compared to the winter, certainly due to high temperature. It took about three months (90 days) in summer, to reach peak nutrient released which gradually slowed down and completed in six months. Whereas nutrient released rate reached to maximum after six months (about 180 days) as it was very slow due to low temperature in winter and it decreased gradually when the nutrients released completed. This experiment has also shown release rate of nutrients in different types of coatings. The first coating of Chitosan coated beads (CS-SiO_2 Nps -NPK) released higher amounts of nutrients which decreased by second coating of sodium alginate (SA-CS-SiO_2 Nps -NPK) and become lowest with the application of kaolin coating (K-SA-CS-SiO_2 Nps -NPK). This result has proved the effect of superabsorbent (sodium Alginate + Kaolin) that can slow down the release of nutrient.

The synthesized SiO_2 Nps containing controlled released fertilizer (K-SA-CS-SiO_2 Nps -NPK) is capable of releasing the nutrients slowly and has higher water absorbing capacity. Therefore, the synthesized CRF can be highly useful for drought and marginal areas due to its absorbency and slow release properties. Moreover SiO_2 Nps , can help plants to survive better in the saline environment. SiO_2 Nps priming have been found effective in improving growth of wheat under salt stressed [23].

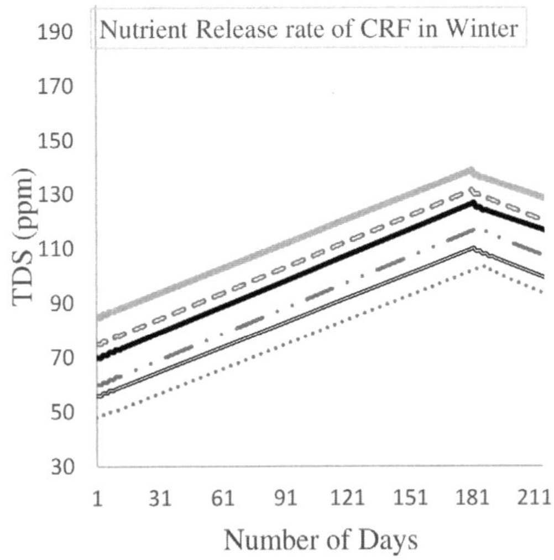

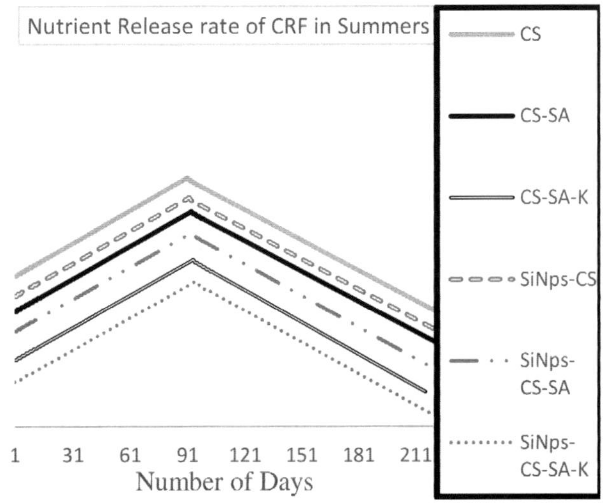

Figure 4. TDS record for 6 months in winter **Figure 5**. TDS record for 6 months in summer

In conventional fertilizers most of the nutrients are lost in the environment. About 40% to 70% Nitrogen, 50 to 70% Potassium and 80 to 90% Phosphorous get leached and hence lost and becomes unavailable to plants within few days. Whereas this synthesized CRF is capable of releasing N, P, and K for plants from 3 to 6 months that can be very helpful for healthy growth of plants. Secondly in drought and rain-fed areas less water is available for plants. This novel CRF can withhold large amounts of water, prevent leaching of excess water and keep the roots moist and water available for plants survival.

The novelty of this synthesized controlled release fertilizer is its coating by natural, non-toxic materials like Chitosan, Sodium alginate and Kaolin, that are biodegradable and non-toxic therefore the problem of toxic wastes in the soil has been reduced. Moreover, in this study, the combination of sodium alginate and kaolin, the superabsorbent polymer has improved the absorbency & controlled release fertilizers behavior and broaden its application especially for drought and marginal areas. Although conventional polymers like, polyethylene, polyvinyl and polyacrylic acid has been used for CRF synthesis but they remained in the environment for many years and produced toxicological effects on the environment.

The introduction of clays in CRF has improved the polymer performance, due to the ability of clay particles to absorb water or lose water in response to simple-humidity-content changes in the surrounding environment. Most of the clay mineral is hydrophilic, that swells with the absorption of water and greatly increase its volume. Kaolin as a mixture with polystyrene-starch has been used for coating the fertilizer [24]. Our findings are in agreement for using kaolin clay, as the polymer (sodium alginate and kaolin) increased the water withholding capacity by many folds. Together with sodium alginate, kaolin served to inhibit too much water diffuse inside the chitosan matrix, thus maintaining low release rate of nutrients. Besides improving the release rate, the outer layer also functions to protect the CRF from physical damage. Our synthesized CRF used chitosan, sodium alginate and kaolin all of which are completely degradable which makes this CRF novel, biocompatible and environment friendly along with its excellent controlled release properties.

4. Conclusion

The results of this study indicated that ultrasonication is very helpful tool in reducing particle size in the synthesis of SiO_2 Nps. SiO_2 Nps can be incorporated in a controlled release fertilizer for growing plants in drought and saline areas. SiO_2 Nps and NPK were coated by Chitosan as inner layer and superabsorbent Sodium alginate & Kaolin polymer as outer coating. This CRF was capable of withholding large amounts of water and releasing the nutrients very slowly that is very useful for plants grown in drought and saline areas. The novel CRF is biocompatible, biodegradable and non-toxic especially aimed to prevent negative effect on the environment as compared to the conventional fertilizer that releases nutrients abruptly, that remain in the soil and probably leach down to the water reservoirs.

5. Acknowledgement

The authors acknowledge the financial support of SBKWU and HEC, Pakistan. Special thanks to Dr. G. L. Hornyak for facilitating the synthesis and TEM of SiO_2 Nps in, AIT Thailand.

References

[1] Gorji B, Ghasri MA, Fazaeli R, and Niksirat N 2012 Synthesis and characterizations of silica nanoparticles by a new sol-gel method. *Journal of Applied Chemical Research* **6** 22-26

[2] Matijevic E 1993 Preparation and properties of uniform size colloids. *Chemistry of materials* **5** 412-426

[3] Hammouda A, Gulik T, and Pileni M 1995 Synthesis of nanosize latexes by reverse micelle polymerization. *Langmuir* **11** 3656-3659

[4] Tadanaga K, Morita K, Mori K, and Tatsumisago M 2013 Synthesis of monodispersed silica nanoparticles with high concentration by the Stöber process. *Journal of sol-gel science and technology* **68** 341-345

[5] Jafari V, Allahverdi A, and Vafaei M 2014 Ultrasound-assisted synthesis of colloidal nanosilica from silica fume: Effect of sonication time on the properties of product. *Adv Powder Technol* **25** 1571-1577

[6] Edrissi M, Soleymani M, and Adinehnia M 2011 Synthesis of Silica Nanoparticles by Ultrasound-Assisted Sol-Gel Method: Optimized by Taguchi Robust Design. *Chemical Engineering & Technology* **34** 1813-1819

[7] Baruah S, Khan MN, and Dutta J 2016 Perspectives and applications of nanotechnology in water treatment. *Environmental chemistry letters* **14** 1-14

[8] Khan MN, Al-Hinai M, Al-Hinai A, and Dutta J 2014 Visible light photocatalysis of mixed phase zinc stannate/zinc oxide nanostructures precipitated at room temperature in aqueous media. *Ceramics International* **40** 8743-8752

[9] Khan MN and Dutta J 2015 Comparison of photocatalytic activity of zinc stannate particles and zinc stannate/zinc oxide composites for the removal of phenol from water, and a study on the effect of pH on photocatalytic efficiency. *Materials Science in Semiconductor Processing* **36** 124-133

[10] Khan MN, Jaisai M, and Dutta J 2015 Photocatalytic Inactivation of Escherichia Coli Using Zinc Stannate Nanostructures under Visible Light. *Advanced Materials Research* **1** 131 203

[11] Wu L and Liu M 2008 Preparation and properties of chitosan-coated NPK compound fertilizer with controlled-release and water-retention. *Carbohydrate Polymers* **72** 240-247

[12] Wu L, Liu M, and Rui L 2008 Preparation and properties of a double-coated slow-release NPK compound fertilizer with superabsorbent and water-retention. *Bioresour Technol* **99** 547-54

[13] Pileni M-P 2003 The role of soft colloidal templates in controlling the size and shape of inorganic nanocrystals. *Nature materials* **2** 145-150

[14] Stöber W, Fink A, and Bohn E 1968 Controlled growth of monodisperse silica spheres in the micron size range. *Journal of Colloid and Interface Science* **26** 62-69

[15] Van Helden A, Jansen JW, and Vrij A 1981 Preparation and characterization of spherical monodisperse silica dispersions in nonaqueous solvents. *Journal of Colloid and Interface Science* **81** 354-368

[16] Gholami T, Salavati-Niasari M, Bazarganipour M, and Noori E 2013 Synthesis and characterization of spherical silica nanoparticles by modified Stöber process assisted by organic ligand. *Superlattices and Microstructures* **61** 33-41

[17] Behzadi S, Ghasemi F, Ghalkhani M, Ashkarran AA, Akbari SM, Pakpour S, Hormozi-Nezhad MR, Jamshidi Z, Mirsadeghi S, and Dinarvand R 2015 Determination of nanoparticles using UV-Vis spectra. *Nanoscale* **75** 134-5139

[18] Venkateswaran S, Yuvakkumar R, and Rajendran V 2013 Nano silicon from nano silica using natural resource (RHA) for solar cell fabrication. *Phosphorus, Sulfur, and Silicon and the Related Elements* **188** 1178-1193

[19] Rida MA and Harb F 2014 Synthesis and characterization of amorphous silica nanoparitcles from aqueous silicates uisng cationic surfactants. *Journal of Metals, Materials and Minerals* **24**

[20] Stanley R and Nesaraj AS 2014 Effect of surfactants on the wet chemical synthesis of silica nanoparticles. *International Journal of Applied Science and Engineering* **12** (1) 9-21

[21] Gao G-M, Zou H-F, Gan S-C, Liu Z-J, An B-C, Xu J-J, and Li G-H 2009 Preparation and properties of silica nanoparticles from oil shale ash. *Powder technology* **191** 47-51

[22] Liang R, Liu M, and Wu L 2007 Controlled release NPK compound fertilizer with the function of water retention. *Reactive and Functional Polymers* **67** 769-779

[23] Mushtaq A, Jamil N, Riaz M, Hornyak G, Ahmed N, Ahmed SS, Shahwani MN, and Malghani MNK 2017 Synthesis of Silica Nanoparticles and their effect on priming of wheat (Triticum aestivum L.) under salinity stress Biological Forum **9** 150-157

[24] Wang Q, Zhang J, and Wang A 2009 Preparation and characterization of a novel pH-sensitive chitosan-g-poly (acrylic acid)/attapulgite/sodium alginate composite hydrogel bead for controlled release of diclofenac sodium. *Carbohydrate Polymers* **78** 731-737

Domestic three stage water-treatment option for harvested rainwater in water-stressed communities

M H Nawaz[1] and M A Baig[2]

[1]Institute of Environmental Sciences and Engineering, National University of Sciences and Engineering, Islamabad, Pakistan
[2]Institute of Environmental Sciences and Engineering, National University of Sciences and Engineering, Islamabad, Pakistan

Email. mehwishhaqnawaz@gmail.com

Abstract. Rainwater harvesting (RWH) practice can be traced back millennia, the degree of its modern implementation varies greatly across the world. And is the best option for water-stressed communities. A harvested rainwater quality monitoring study was undertaken at 3 lakes constructed in NUST Islamabad, Pakistan for a period of 10 months over two seasons i.e. wet and dry periods. Overall, harvested rainwater was of good quality, falling within the recreational water quality as per WHO standards with the exceptions for pH, color, turbidity, total coliforms and total bacterial count. A large number of samples tested positive for the total bacterial count and total coliforms, showing that disinfection of harvested rainwater is mandatory prior to use and its direct consumption without treatment may pose a health risk. For its treatment, an indigenously designed (SwissPak) water filter employing physicochemical methods were tested for making harvested rainwater fit for potable use. This filter contains silica sand for pre-filtration, granulated chlorine for disinfection and charcoal for removal of taste, odor and dissolved organic while alum was used as coagulant initially. This filter successfully improved the harvested rainwater quality and proved itself a suitable option for water treatment.
Keywords: Rainwater harvesting, water quality, physicochemical water treatment

1. Introduction

Human health, livelihood, and continuous development is directly and indirectly linked to the provision of safe drinking water [1]. The ratio between total water withdrawals and available renewable surface water at a sub-catchment level is termed as water stressed defined by World Water Resources Institute, it also highlights the alarming situation that Pakistan will have to face being a most water-stressed country in the region by the year 2040.Recent conditions threatening Pakistan and making it 23rd most water-stressed country in the world, if not tried to implement water conservation strategies [2]. Pakistan being a victim of water stress, facing poor water availability, in villages ponds are a source of water for animal and human beings as well. As considerable predictions open the gate for alternatives especially in developing countries where water withdrawal will increase by 50 % as compared to 18 % in developed countries. This will directly push almost 800 million people into absolute water scarcity [3]. About 44 % of Pakistan's population is lacking the safe water resources reported by Pakistan Council of Research in Water Resources (PCRWR). The surface and groundwater use has reached the upper limits in most parts of the country. Rainwater harvesting is the answer to mitigate the upcoming challenges linked with rapid economic growth and population burden. Mostly rainwater is pure from extensive contamination, easily treatable which makes it eye-catching option in the water-stressed condition [4, 5]. But health risks are also associated with the consumption of untreated rainwater [6, 7]. Recent study was designed to examine low-cost, indigenously designed water filter integrated coagulation, flocculation, filtration and disinfection water treatment system for harvested rainwater.

2. Methods and material

Lakes of New campus of National University of Sciences and Technology Islamabad, Pakistan were taken as the study site where rainwater harvesting is in practice (figure 1). These lakes 1, 2 and 3 have area 1.5, 2 and 2.25 acre with height 25, 16 – 20 and 25 feet respectively. And they have a storage volume of 0.17, 0.16 and 0.74 GL respectively. Composite water samples were collected from all three lakes of NUST. It was done with seasonal variation before and after the rain (figure 2). All the sampling and preservation methods carried out for the quality analysis in water samples were in the light of Standard Methods for the Examination of Water and Wastewater [8].

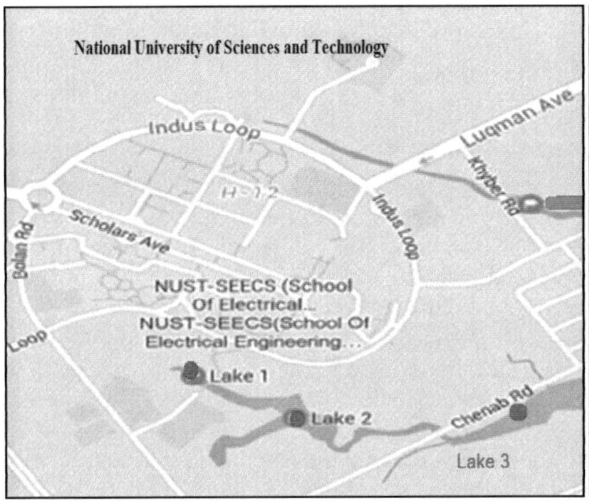

Figure 1. Red circles showing location of NUST Lakes. **Figure 2.** (a) Lakes in wet, (b) Lakes in dry season.

Table.1 shows water quality parameters associated with units and methods of analysis. The temperature of water samples was measured at the sampling points by a mercury thermometer. In the laboratory, all the samples were analyzed for different physicochemical and biological parameters. pH, electrical conductivity (EC), dissolved oxygen (DO) and turbidity was measured by pH, EC, DO and turbidimeters respectively. TDS were determined gravimetrically at 180 °C. EDTA complexometry titration, with indicator Eriochrome Black T for total hardness. Chlorides were determined by the argentometric method. The total alkalinity was measured by titration, with indicators phenolphthalein and methyl orange. Colour, phosphorus and nitrates were measured by using a spectrophotometer. Total bacterial count was done by spread plate count and T. Coliform by membrane filter technique.

2.1. Treatment methodology

Water treatment methodology is shown in (figure.3). Composite water samples were collected from all the three lakes, and coagulation was done by alum stick (1 turn of alum stick/L of water). Coagulation involves adding a chemical, which has a positive charge. This chemical binds (flocculation) with the negative charged dissolved and suspended particle in raw water and neutralize these. Two hours were given for flocculation. Then this water was improved by pouring it into the three-stage SwissPak portable water filter. It is simple, portable, reliable, lightweight and low-cost household water treatment option. This point of use water filter is specifically designed for rural households for treating turbid and biological contaminated water. It is based on modern filtration technologies of large filter plants that are simplified for use by rural household level. It is robust and has a high output flow. It keeps water cool and fresh without using energy. It improves the filtration process and significantly prolongs the life of replaceable filter cartridges. It has provision for coagulating very turbid waters.

Table 1. Water quality parameters associated with abbreviations, units and analytical methods used.

Variables	Abbreviations	Units	Methods
Temperature	Temp	°C	Thermometer
PH	pH	pH unit	pH meter
Turbidity	-	NTU	Nephelometric
Dissolved Oxygen	DO	mg L^{-1}	DO meter
Total dissolved solid	TDS	mg L^{-1}	Gravimetric
Electrical conductivity	EC	µS cm^{-1}	Electrometric
Colour	-	TPC	Spectrometer
Chloride	Cl	mg L^{-1}	Titrimetric
Total Hardness	T.Hard	mg L^{-1}	Titrimetric
Alkalinity	ALKY	mg L^{-1}	Titrimetric
Phosphorus	P	mg L^{-1}	Spectrometer
Nitrates	NO_3	mg L^{-1}	Spectrometer
T. Bacterial Count	TBC	CFU/ml	SPC
T. Coliform	TC	CFU/100ml	MF Technique

Table 2.: Daily range and total monthly rainfall values for Islamabad- Pakistan during the study period (October 2015 to July 2016)

Month/Year	Daily Range (mm)	Total monthly rainfall (mm)
Oct. 2015	0-136	190.2
Nov. 2015	0-10	18.7
Dec. 2015	0-25.8	25.9
Jan. 2016	0-36	54.8
Feb. 2016	0-36	86.9
Mar. 2016	0-72	202.1
Apr. 2016	0-4	12.1
May. 2016	0-17	27.4
Jun. 2016	0-19	47.5
Jul. 2016	0-75.4	368.6

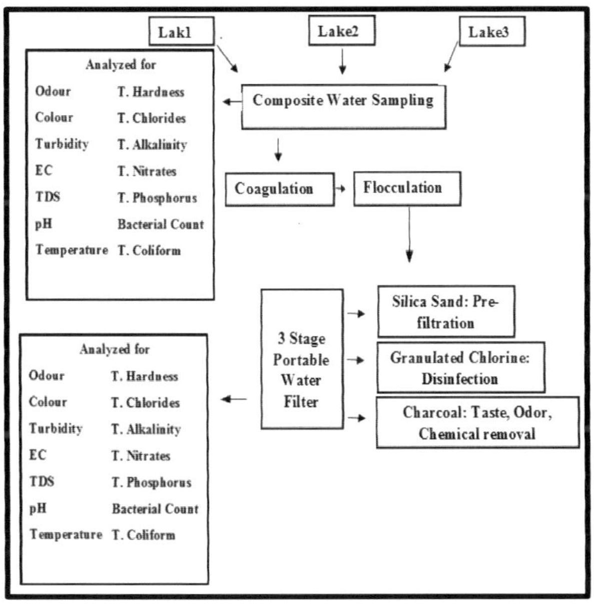

Figure 3. Flow diagram of study

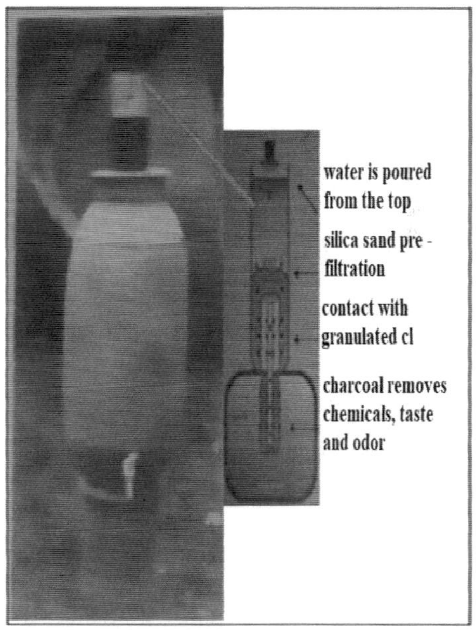

Figure 4. Schematic diagram of three stage portable water filter.

The filter is powered by gravity flow. Its upper part is made of plastic pipe through which water is poured. Its outflow is 3-4 liters per minute. When water is poured from the top, silica sand acts as pre-filtration, and then have contact with the granulated chlorine for disinfection (simple chlorine dosing and dispelling) and carbon block removes the chemical, taste, and odor (figure 4). Its main body is made of mud which can store 30 liters of water. Physicochemical and biological analysis (Total Bacterial Count, Total Coliform, Colour, Odour, Turbidity, DO, TDS, pH, EC, Total Hardness, Chloride, Alkalinity, Phosphorous and Nitrates) of water was done before and after treatment. The daily rainfall range and the total monthly rainfall values for each month of the study are presented in table 2. The total monthly rainfall varied from month to month and ranged from 12.1 mm to 368.6 mm. The values reflected the rainfall pattern of rainy and dry seasons in the year. July 2016 had the highest monthly total rainfall of 368.6 mm followed by March 2016. November 15, April, May and June 16 were typically dry months with very low amounts of rainfall. Water sampling was done Oct 15 to July 16, all the data regarding rainfall, the pattern of rainy and dry seasons in the year is shown in table 2. Monthly rainfall ranged from 12.1 mm to 368.6 mm. Highest monthly rainfall was recorded 368.6 mm in July 2016 followed by March 2016. Low amount of rainfall was recorded in dry months of November 15, April, May and June 16.

3. Results and discussion

Harvested rainwater quality found reliable when compared with drinking water guidelines except for color, turbidity, pH, Bacterial count and T Coliform in dry and wet season table 3 and 4. The analytical result of physicochemical and biological water quality parameters before and after treatment was significantly different for all the three lakes. And the efficiency of the three-stage portable water filter was evaluated based on the 14 water parameters and will be represented in the form of graphs for each parameter. Quality of rainwater samples was already under the guidelines for most of the parameters, so water filter was tested for academic purposes while considering the removal of turbidity, color, bacterial count and T. coliform.

Presence of microorganisms in water is directly linked with its temperature [9]. Bacterial growth rate, the decay of the disinfection residual is affected by water temperature [10]. More than 25 °C temperature can be a problem and change the taste, odor, and color. The temperature of raw water ranged before treatment 21.3 – 21.7°C and 27- 28 °C in wet and dry seasons respectively. After treatment, a decrease in temperature was found as its body is designed with mud texture which keeps water cool and fresh without using energy. It is said that pH should range from 6.5 to 8.5. As it does not cause any harm when intake. pH of lake water samples was from 7.3 – 7.6 and 8.1 – 8.2 in wet and dry seasons respectively. These recorded values indicating that rain from studies area is not acidic in nature [11]. Another study reported similar pH values 7.63 – 8.80 from harvested rainwater [12]. After treatment, a slight decrease in pH was found. It means chemical did play a minor role in changing pH. It was recorded that lake water had EC 227 – 254 and 571 – 1100 µs/cm in a wet and dry season respectively (figure 5a). Similarly, the low range of EC is reported from harvested rain water [12]. And 20 – 30 percent removal was achieved by three-stage portable water filter treatment. The physicochemical quality of harvested rainwater in Spain appeared to be better than the average quality found in the literature review [10].

These lakes were originated from outside and inside of NUST and bring both point and non-point source pollution from surrounding sectors. These sources add turbidity as well. It was also found that the value of turbidity increases every time after precipitation due to the water turbulence and soil erosion. The turbidity of water samples ranged 14-30 and 8.9-25 NTU in a wet and dry season respectively (figure 5b). The used water filter is specifically designed for rural household. The SwissPak treatment proved very efficient for the removal of turbidity 92-94 percent. So, it has provision for coagulating very turbid waters. DO is affected by temperature level in a water body and found to be critical for the survival of aquatic organisms for aerobic respiration. A relation between DO and temperature was found, when the low temperature in lakes DO was more and less DO was recorded in higher temperature. It was found that DO remained the same, as before the treatment 9.2-9.3 and 8.3-8.5 mg/L respectively in wet and dry weather (figure 5c). Electrical conductivity increases an amount of dissolved salts, showing a direct relationship

[13]. Total dissolved solids were 116-161 and 296-563 mg /L in a wet and dry season respectively (figure 5d).

Table 3. Descriptive statistics of rainwater quality of lakes in dry season along with national standards of drinking water quality*

Parameters	Site No	Observations	Minimum	Maximum	Mean	SD	*NSDWQ
	1	7	75	258	163	105	15
Color(TCU)	2	7	64	295	123	90	15
	3	7	43	240	99	74	15
	1	7	10.2	33.5	24	8	5
Turbidity(NTU)	2	7	9.1	35.7	10.4	10	5
	3	7	7.5	33.7	15.8	8.9	5
	1	7	7.5	10	8.87	1.02	-
DO (mg/L)	2	7	7.6	10	8.9	1.02	-
	3	7	7.5	10	8.87	1.04	-
	1	7	392	1233	15.8	8.9	-
EC (µs/cm)	2	7	380	1121	719	313	-
	3	7	306	684	536	133	-
	1	7	170	655	419	198	1000
TDS (mg/L)	2	7	185	570	371	158	1000
	3	7	142	382	274	82	1000
	1	7	7.5	8.4	7.89	0.41	6.5-8.5
pH	2	7	7.3	8.3	7.84	0.42	6.5-8.5
	3	7	7.2	8.7	7.84	0.59	6.5-8.5
	1	7	16.17	30.3	23	5.84	25
Temperature (◦C)	2	7	16.5	30.4	22.99	5.85	25
	3	7	15.8	30.2	22.99	6.1	25
	1	7	151	337	252	70.2	500
Hardness (mg/L)	2	7	145	299	228	59	500
	3	7	133	271	206	56	500
	1	7	39	129	90.14	34.26	250
Chlorides (mg/L)	2	7	33	125	79.43	32.91	250
	3	7	29	115	68	29.31	250
	1	7	109	345	208.71	101.97	1000
Alkalinity (mg/L)	2	7	102	281	189.71	82.45	1000
	3	7	102	192	144.57	42.17	1000
	1	7	4.22	5.9	4.88	0.6	-
Phosphorus(mg/L)	2	7	2.17	5.7	3.92	1.37	-
	3	7	1.32	3.51	2.09	0.7	-
	1	7	6.8	28.97	16.22	9.09	50
Nitrates (mg/L)	2	7	10.2	28.02	21.07	8.3	50
	3	7	3.75	31.5	13.74	9.58	50
	1	7	90000	227000	15551	51074	0
Bacterial count/ml	2	7	89000	286000	154000	70114	0
	3	7	71000	150000	102428	24905	0
	1	7	20200	29800	25871	3926	0
T Coliform100ml	2	7	20000	28100	23557	3463.07	0
	3	7	16900	22600	19328	1949.97	0

The SwissPak treatment removed 19-37 percent TDS. Color of raw water ranged from 34-55 and 43-173 TCU in wet and dry seasons respectively (figure 5e). Mostly this color was due to the algae present in it. It was found that the color of the raw water was low every time after precipitation. And found higher in all the study months, exceeding the permissible limits. Efficient results were found by SwissPak treatment. Color removal was 94 – 97 percent. Total alkalinity of raw water ranged from 89 – 142 and 192 – 290 mg/L in a wet and dry season respectively (figure 5f). By SwissPak treatment, alkalinity removal was 22 – 32 percent in both types of water samples. Total hardness of raw water ranged from 109 – 131 and 269 – 301 mg/L in a wet and dry season respectively (figure. 5g). And another study reported hardness in harvested rainwater, where the highest value was found 74mg/L [12]. By SwissPak treatment 29 – 43 percent hardness removal was achieved. Total chlorides of water ranged from 21 – 27 and 69 – 122 mg/L in a wet and dry season respectively (figure 5h). In a SwissPak treatment 22 – 34 percent removal was achieved. Total nitrates of water ranged from 3.72 – 11.43 and 3.16 – 29.08 mg/L in a wet and dry season respectively (figure 5i). In SwissPak treatments nitrates removal was 67 – 97 percent from both types of samples. Total phosphorus of water ranged from 2.01 – 2.62 and 1.32 – 5.23 mg/L in a wet and dry season respectively (figure 5j). It was found that SwissPak treatment improved the water quality by removing 42 – 64 percent phosphorus. Water used for drinking purpose must follow the standards which state bacteria should be absent from 1mL of water [14]. However, all the tested water samples confirmed the presence of bacteria. Harvested rainwater quality is dependent on the location and the distance from pollution source [15]. If harvested, rainwater is surrounded by vegetation, plantation, trees and other types of vegetation it provides a suitable environment for a variety of avian, mammal and reptiles fauna [16]. As bacterial count and coliform were high is lake water clearly suggesting the source of pollution was of animal residing in or near lakes and human feces. This analysis clearly suggested the need for water disinfection for making it suitable for use.

There are a number of treatment methods according to the type of raw water [17]. The microbiological quality of harvested rainwater was assessed in terms of total bacterial count and T coliform and tested positive. The high bacterial count and total coliforms was attributed to wastewater coming from outside, groundwater seeping into the porous, as soil act as the medium for bacterial contamination [18]. Similar results were found from harvested rainwater in Jordan, one of the water pollution indicator fecal coliform tested positive [19]. One of the studies regarding domestic rainwater harvesting quality revealed that water born disease risk was associated with the use of water [20].

In Greece, rainwater samples were tested in terms of chemical and biological parameters. Water sound chemically safe from any impurity but total coliforms, *Escherichia coli* and enterococci were detected in 80.3%, 40.9% and 28.8% of the rainwater samples showing pollution [12]. Bacterial count ranged 50,000 – 108,000 and 90,000 – 127,000/ml in wet and dry season respectively (figure 5k). The improving efficiency of lakes water by three-stage portable water filter was successfully determined by before and after treatment analysis of water. In dry season water pollution was higher as compared to wet after precipitation. Similar results are recorded from harvested rainwater with seasonal variation [12].

The results of study matched with previous studies [22, 23] in which microbial contamination was linked with season and weather condition. Number of similar studies have been conducted in different countries [11, 22, 24-27]. But filter was found efficient to improve the quality of lakes by 100 percent removal. Similar results were found for T Coliform, it ranged from 13900 – 21000 and 20100 – 24800/ 100ml in wet and dry conditions respectively (figure 5l). Used treatment proved efficient by removing 100 percent coliform and making water clean and suitable for drinking. Similarly, efficient results were found, while using mineral pot filter in removing bacteria and water turbidity [21].

IOP Conf. Series: Materials Science and Engineering **414** (2018) 012030 doi:10.1088/1757-899X/414/1/012030

Table 4. Descriptive statistics of rainwater quality of lakes in rainy season along with national standards of drinking water quality (NSDWQ)*

Parameters	Site No	Observations	Minimum	Maximum	Mean	SD	*NSDWQ
	1	4	35	88	58	36.55	15
Color(TCU)	2	4	32	88	52	37.87	15
	3	4	21	35	41	32.58	15
	1	4	15.7	37.5	24.8	15.48	5
Turbidity(NTU)	2	4	19	210.6	82.68	109.72	5
	3	4	14.2	64.8	43.05	35.81	5
	1	7	7.5	10	8.87	1.02	-
DO (mg/L)	2	7	7.6	10	8.9	1.02	-
	3	7	7.5	10	8.87	1.04	-
	1	4	289	582	365	236	-
EC (μs/cm)	2	4	232	524	317	217	-
	3	4	220	372	277	140	-
	1	4	129	288	186.5	116.39	1000
TDS (mg/L)	2	4	128	261	186.25	105.16	1000
	3	4	116	165	134.5	56.49	1000
	1	4	7.2	8.7	7.8	2.67	6.5-8.5
pH	2	4	7.2	8.7	7.9	265	6.5-8.5
	3	4	7.2	8.1	7.7	2.35	6.5-8.5
	1	4	15	33.6	25.4	15.02	25
Temperature(◦C)	2	4	15.02	33.2	25.26	14.89	25
	3	4	15.05	33.2	25.23	14.89	25
	1	4	115	178	137.5	62.19	500
Hardness(mg/L)	2	4	120	138	125.75	39.7	500
	3	4	98	128	112.25	40.51	500
	1	4	6.6	105	42.7	53.66	250
Chlorides(mg/L)	2	4	8.5	95	37.88	48.26	250
	3	4	16.8	42	24.95	17.74	250
	1	4	92	180	120.8	70.3	1000
Alkalinity (mg/L)	2	4	82	159	108.75	62.8	1000
	3	4	48	108	78.75	52.93	1000
	1	4	2.12	11.2	6.5	6.72	-
Phosphorus(mg/L)	2	4	1.7	14.36	7.85	8.58	-
	3	4	2.01	16.84	8.25	9.51	-
	1	4	3.16	10.5	6.7	4.91	50
Nitrates (mg/L)	2	4	5.79	25.87	13.92	12.32	50
	3	4	3.29	29.02	17.07	15.59	50
	1	4	108000	300000	183500	138871	0
Bacterial count/ml	2	4	64000	236000	96375	119957	0
	3	4	50000	155000	93500	73631	0
	1	4	11800	28800	21150	12344	0
T Coliform100/ml	2	4	10200	26700	17200	11203	0
	3	4	5900	21500	13875	1949	0

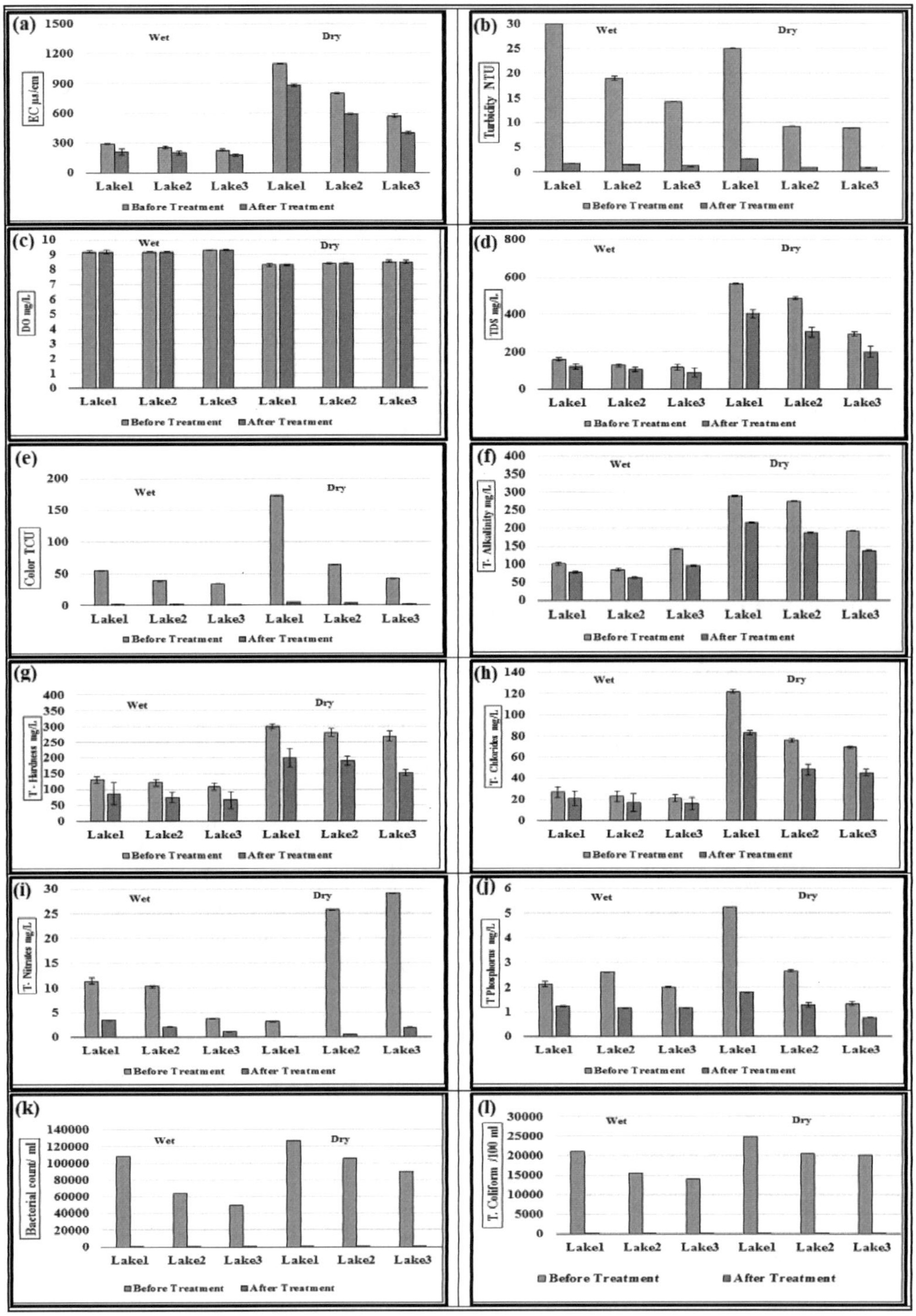

Figure 5. (a-l) SwissPak treatment of harvested rainwater in dry and wet season.

4. Conclusions

Rainwater harvesting becomes the suitable solution in the absence of fresh water for the livings. Quality of harvested rainwater in Islamabad is satisfied in terms of physicochemical accept microbiological parameters. So this harvested water need some treatment before use. Three stage portable water filter is an efficient option. This water filter is specifically designed for rural households for treating turbid and biological contaminated water. Therefore, rainwater harvesting should be in practice to overcome the water shortage in areas like Islamabad, where water shortage becomes the prominent problem in summers. And three stage portable water filter should be used to overcome the shortage of safe water. In dry areas, rainwater harvesting is one of the options for overcoming the need of water for survival, so three-stage water treatment filter can be successfully applied in any remote, dry lands like Thar, Thal, and Balochistan, and in other arid sites in rainy days. Because it is simple, reliable, lightweight and low-cost household water treatment option. This can also be the best option in flooded condition to provide drinking water.

References

[1] Grey, D and Sadoff, C W 2007 Sink or swim? Water security for growth and development. *Water Policy,* **9(6)**, 545-571.
[2] Maddocks A, Young R and Reig P 2015 Ranking the world's most stressed countries in 2040. As found online in http://www.wri.org
[3] Global Water Partnership (GWP) 2010 As found online in http://www.gwp.org/en/The-Challenge/The-Urgency-of-Water-Security/
[4] Meera V and Ahammed M M 2006 Water quality of rooftop rainwater harvesting systems: A review. *Journal of water supply: Research & Technology*. **55(4)**, 257-268.
[5] Che-Ani A I, Shaari N, Sairi A, Zain M F M and Tahir M M 2009 Rainwater harvesting as an alternative water supply in the future. *European Journal of Scientific Research*. **34(1)**, 132-140.
[6] Ahmed W, Huygens F, Goonetilleke A and Gardner T 2008 Real-time PCR detection of pathogenic microorganisms in roof-harvested rainwater in Southeast Queensland, Australia. *Applied and environmental microbiology,* **74(17),** 5490-5496.
[7] Ahmed W, Brandes H, Gyawali P, Sidhu J P and Toze S 2014 Opportunistic pathogens in roof-captured rainwater samples, determined using quantitative PCR. *Water Research*. **15 (53)**, 361–369.
[8] APHA 2005 Standard Methods for the Examination of Water and Wastewater. 21st ed., American public Health Association, Washington D. C.
[9] Ramteke P W, Bhattacharjee J W, Pathak S P and Kalra N 1992 Evaluation of coliforms as indicators of water quality in India. *Journal of applied microbiology*, ***72*(4)**, 352-356.
[10] Farreny R, Morales-Pinzon T, Guisasola A, Taya C, Rieradevall J and Gabarrell X 2011 Roof selection for rainwater harvesting: quantity and quality assessments in Spain. *Water research*, ***45*(10),** 3245-3254.
[11] Zhu K, Zhang L, Hart W, Liu M and Chen H 2004 Quality issues in harvested rainwater in arid and semi-arid Loess Plateau of northern China. *Journal of arid environments*, ***57*(4)**, 487-505.
[12] Kelin H, Yuang H F, Hong L and Robert W E 2005 Spatial variability of shallow ground water level, electrical conductivity and nitrate concentration and risk assessment of nitrate contamination in North China plain. *Environment International,* **31**, 896–903.
[13] Sazakli E, Alexopoulos A and Leotsinidis M 2007 Rainwater harvesting, quality assessment and utilization in Kefalonia Island, Greece. *Water research*, ***41*(9)**, 2039-2047.
[14] WHO, UNICEF, Guidelines for Drinking Water Quality, 4th ed., World Health Organization, Geneva, Switzerland, 2011 http://whqlibdoc.who.int/
[15] Forster J 1996 Patterns of roof runoff contamination and their potential implications on practice and regulations of treatment and local infiltration, Water Sci. Technol. **33**, 39–48.
[16] Appan A 1997. Roof water collection systems in some Southeast Asian countries: status and water quality levels, *J. R. Soc. Health* **117**, 319–323.

[17] Ahmed W, Richardson K, Sidhu J P S, Jagals P and Toze S 2013. Inactivation of faecal indicator bacteria in a roof-captured rainwater system under ambient meteorological conditions,*J. Appl. Microbiol.* **116**,199–207.

[18] Coombes P J, Argue J R. and Kuczera G 2000 Figtree Place: a case study in water sensitive urban development (WSUD). *Urban Water* **1**, 335-343.

[19] Abdulla F A and Al-Shareef A W 2009 Roof rainwater harvesting systems for household water supply in Jordan. *Desalination,* ***243*(1-3),** 195-207.

[20] Kahinda J M M, Taigbenu A E, and Boroto J R 2007. Domestic rainwater harvesting to improve water supply in rural South Africa. *Physics and Chemistry of the Earth, Parts A/B/C*, ***32*(15-18)**, 1050-1057.

[21] Karim M R, Rahman S, Hossain M A, Islam M A, Mahmud S G and Mahmud Z H 2016. Microbiological effectiveness of mineral pot filters as household water treatment in the coastal areas of Bangladesh. *Microbial Risk Analysis*, **4,** 7-15.

[22] Simmons G, Hope V, Lewis G, Whitmore J and Gao W 2001 Contamination of potable roof-collected rainwater in Auckland, New Zealand. *Water Research*, ***35*(6)**, 1518-1524.

[23] Evans C A, Coombes P J and Dunstan R H 2006 Wind, rain and bacteria: The effect of weather on the microbial composition of roof-harvested rainwater. *Water research*, ***40*(1)**, 37-44.

[24] Dillaha III T A and Zolan W J 1985 Rainwater catchment water quality in Micronesia. *Water Research*, ***19*(6)**, 741-746.

[25] Yaziz M I, Gunting H, Sapari N and Ghazali A W 1989 Variations in rainwater quality from roof catchments. *Water research*, ***23*(6)**, 761-765.

[26] Mantovan P, Pastore A, Szpyrkowicz L and Zilio-Grandi F 1995 Characterization of rainwater quality from the Venice region network using multiway data analysis. *Science of the total environment*, ***164*(1)**, 27-43.

[27] Chang M, McBroom M W and Beasley R S 2004 Roofing as a source of nonpoint water pollution. *Journal of environmental management*, ***73*(4)**, 307-315.

In-plane shear strength analysis of Basalt Fiber-Reinforced Epoxy Laminates with Biowaste Catalyst Free Carbon

A N Mengal[1], S Karuppanan[2], Z N Qaisarani[3], M Siddique[3], M Kashif [4], M Najam[3]

[1]Department of Mechanical Engineering, Balochistan University of Information Technology Engineering and Management Sciences, Quetta, Pakistan.
[2]Department of Mechanical Engineering, Universiti Teknologi PETRONAS, Malaysia
[3]Department of Chemical Engineering, Balochistan University of Information Technology Engineering and Management Sciences, Quetta, Pakistan.
[4]Coventry University, 16, Burnsall Road, Coventry, CV5 6BP, UK.

Email: aliract@gmail.com

Abstract. This study aims to analyse the effects of carbon nanospheres (CNSs) on the in-plane shear properties of basalt fiber-reinforced epoxy composite laminate (BFR). The CNSs were obtained from an economical fibrous residue attained from the sago palm tree, which is known as biowaste sago bark. Hand lay-up method was used to fabricate the unidirectional basalt fiber-reinforced epoxy composite laminates. The epoxy resin was mixed with carbon nanosphere particles (i.e., 0.6 wt% - 1 wt%). In-plane shear tests have been conducted as per ASTM standards. In addition, emphasis on the microstructural investigation using Scanning Electron Microscopy (SEM) is given, in order to study the fracture surfaces of the composite laminates. The results demonstrated significant improvement in in-plane shear strength when carbon nanosphere particles were included in the basalt fiber-reinforced epoxy composite laminate. The best result was obtained at 1.0 wt% CNSs. It displayed an increment of 37.1% in in-plane shear strength, and 36.4% increment in modulus of rigidity, respectively, in comparison to neat basalt fiber-reinforced epoxy composite laminate. The improved accomplishment of CNSs/ basalt fiber-reinforced epoxy composite laminate is due to good distribution of CNSs particles in the epoxy matrix.
Keywords. Shear strength, fiber reinforced, biowaste

1. Introduction

Fiber-reinforced polymer composites (FRCs) have been used extensively in many applications, especially in areas such as construction, aerospace, recreational equipment, automotive, marine and other high performance applications. This is due to the advantages of having high stiffness to weight ratio and their ease of processing [1, 2]. To enhance the overall properties of the composite laminate, fibers, for instance carbon fiber, glass fiber, and basalt fiber are used as filler for epoxy-based laminates. Recent research demonstrated the advantages of utilizing natural fibers in the reinforcement of polymers in order to improve the environmental safety [3]. As a waste material, the use of natural fibers as the reinforcement for the composite of the polyester is a good way move towards eco-reusing [4]. Furthermore, the benefits of natural fibers over glass fibers are in terms of the cost, density, strength-to-weight ratio, resistance to breakage during processing, energy content and recyclability [5].

Among the natural fibers, basalt fiber (BF) has been proven as a potential reinforcement of composite materials because of its good mechanical properties. Basalt fibers originated from basalt rocks through melting process. The basalt rocks can be broken down into smaller particles to produce fibers characterized by high strength, excellent fiber/resin adhesion, ability to be easily processed using conventional processes and equipment [6], sound insulation properties, low water absorption [7], good resistance to chemical attack, ecologically friendly, free from carcinogens and health hazards, high operating temperature range, and can be produced at considerably lower cost than carbon fiber [8-

11]. In addition, BFs do not contain any other additives in the production process, which makes them cost effective. Furthermore, BFs are known to have higher tensile strength than E-glass fiber and they have larger strain to failure than carbon fiber [9]. The excellent heat resistance and low water absorption capability make them suitable as thermal insulation materials and hot fluid conduits. BFs also have advantages over the glass and asbestos fibers in terms of environmental cleanness [12]. Moreover, the basalt fiber has high chemical stability and sound mechanical properties as well as being non-toxic and non-combustible [13].

Basalt fibers which are extracted from the volcanic rock are preferably used for reinforcing polymer matrices [14]. The chemical composition of BF is approximately identical to glass; the basic components of both elements are SiO2, Al2O3, CaO, MgO, K2O, Na2O, Fe2O3 and FeO [15]. It is also important to highlight the melting temperature of the basalt fibers which ranges between 1350 and 1700°C. Basalt will solidify as a partial crystalline structure when it undergoes slow cooling process. In fact, basalt fibers can be used from 200 to 600°C without any significant losses of the mechanical properties. Basalt fibers can be applied for structural strengthening of material based on the advantages that they possess. Besides that, they also have great potential as transportation and construction materials. Before using fiber-reinforced polymer (FRP) materials as a strengthening material, preferably the mechanical and durability issues of the material are tested first.

Similarly, carbon-based nanomaterials have potential function in ultracapacitors [16-18], nanoelectronics, catalysis, microelectrical devices, sensors, and electrochemistry. Among the various forms of carbon nanomaterials, carbon nanospheres are getting more attention. This is because, in its spherical form, they are usually unclosed shells with little waving flakes that follow the curvature of the sphere. Many studies found that the mechanical, electrical and thermal properties of polymer composites are enhanced when CNSs are used as fillers in the resulting composites.

In the present study, carbon nanosphere particles will be used to enhance the properties of basalt fiber/epoxy composite laminates. The carbon nanospheres (CNSs) are obtained by a simple environmentally benign catalyst-free pyrolysis procedure from biowaste sago bark, which is a low-priced fibrous residue acquired cheaply from the local sago palm tree. The sago bark is rich in cellulose and hemicellulose, contain a small percentage of lignin and have a porous structure. The basalt fiber/epoxy laminates dispersed with 0.6–1.0 wt% carbon nanosphere particles will be fabricated by hand lay-up method. To the best knowledge of the authors, no one has yet reported on the use of carbon nanospheres particles for the reinforcement in basalt composite laminates. The present work aims to study the new approach to investigate the effect of the incorporation of inexpensive carbon nanosphere particles in a basalt fiber-reinforced-epoxy matrix on the mechanical properties (in-plane shear properties) of the composite laminates. The fractured surfaces will then be characterized by scanning electron microscopy (SEM).

2. Experimental Investigation

2.1 Materials

Basalt fiber (300 g/m2 unidirectional 0°), was supplied by Suretex Composite International China. The epoxy used was epocast (Bisphenol) and the curing agent was amine based epoharden (cycloaliphatic amine). Both epocast and epoharden were supplied by Portal Trading Malaysia. The resin hardener mix ratio was 2:1 by weight. The prepared carbon nanospheres were in powder form. Scanning electron microscopy (SEM) was used to present the structure of the carbon nanospheres particles. Figure 1 shows the formation of carbon nanoparticles. The average particle size was 45-60 nm.

Figure 1. The particle size distribution of CNSs using SEM

2.2 Fabrication of composite laminates

First, the epoxy resin was dispersed in the acetone (10 wt% to resin). The CNSs powder was then mixed with the epoxy acetone solution (without hardener). The suspension was mixed for 24 hours using automatic mixer. After that, the hardener was added to the solution. The solution was stirred manually for 15 minutes to homogenies it. The mixture was then degassed for 1 hour in a dessicator to remove the entrapped air bubbles. CNSs powder concentrations added varied from 0.6 wt% to 1 wt% with respect to the mixture of the epoxy/hardener. The basalt fiber reinforced composite laminates were fabricated by the hand lay-up method. A total of four plies of basalt fibers fabric with a size of 300 mm x 300 mm were stacked together. Three laminates were fabricated for this work. First laminate was of neat basalt, second laminate was of basalt with 0.6 wt% of CNSs particles, and third laminate was of basalt with 1.0 wt% of CNSs particles. The laminates were cured at room temperature for 24 hours. Then the specimens for in-plane shear test were cut according to the ASTM standards from the prepared laminates. The schematic layout of fabrication of BFRP laminates is shown in Figure 2.

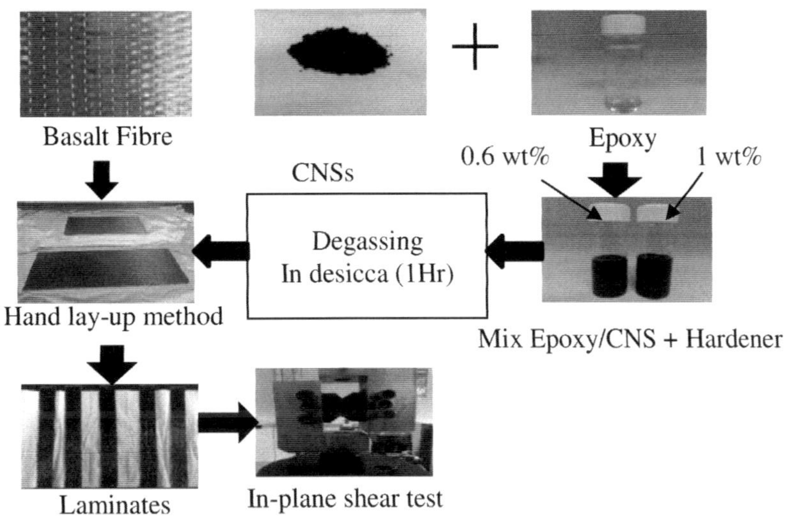

Figure 2. Schematic layout of the preparation and fabrication process of the BFRP and composite laminates.

2.3 Material testing

In-plane shear test was carried out to determine the shear strength of the laminate composites. The in-plane shear tests were performed at room temperature under standard laboratory conditions (23±5°C and 50±10% relative humidity) according to ASTM D7078, respectively. The average in-plane shear stress was defined by the ratio of maximum applied load P to the cross section area A (w x t) between the two notches, where w is the distance between upper and lower notch tip and t is the thickness of the specimen.

$$\tau xy = Pmax/A \quad (1)$$

Scanning electron microscopy (SEM, JEOL JSM-5900) was utilized to identify the morphological structure of the carbon nanospheres particles and the cross-section of the fracture surface of the composite laminates.

3. Results and Discussion

In-plane shear tests were performed to observe the effect of CNS particles on the stress concentration. It is due to the existence of the geometry and material discontinuities at the free edge notches of the composite laminates. The obtained stress-strain curves are presented in Figure 3. The obtained results for in-plane shear tests are reported in Table 1. The in-plane shear strengths of the composite laminates increased by 24.3% and 37.1% for basalt with 0.6 wt% CNS and basalt with 1.0 wt% CNS, respectively. The results also show significant improvement in the shear modulus. The modulus of neat basalt was 2.8 GPa, but the modulus of composite laminates increased to 3.14 GPa and 3.82 GPa for basalt with 0.6 wt% CNS and basalt with 1.0 wt% CNS, respectively. Similar to the results in tensile test, basalt with 1.0 wt% CNS showed the highest improvement in in-plane shear properties among all the conditions. This indicates that the increase of the mechanical properties of the composite laminates is caused by the homogenous dispersion and the strong interactions between epoxy matrix and CNS particles, and basalt fiber-epoxy interfacial interaction. This brings to the efficient stress-transfer from epoxy to CNS particles and basalt fibers. A good dispersion of filler particles is necessary for an effective load transfer of polymer matrix to filler particles. Studies showed that the load-transfer mechanism happens due to the mobility of filler particles. The particles orient themselves under tensile stress, forming temporary cross-links with polymer matrix, and creating a local region of enhanced strength. The improvement of in-plane shear modulus of the basalt fiber-epoxy composite laminates is because of the increase in shear modulus of the nanospheres matrix.

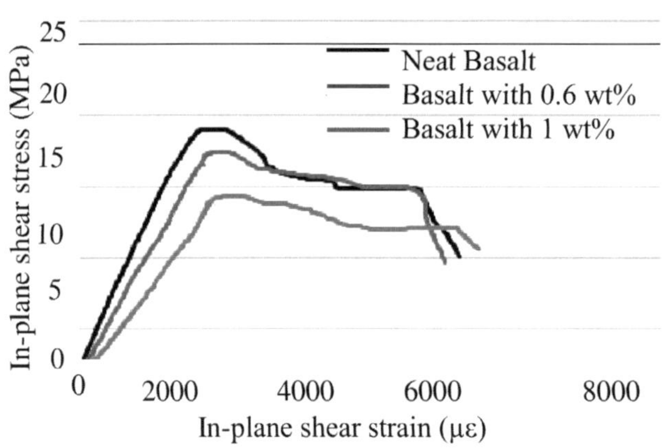

Figure 3. Typical in-plane stress-strain curves of the laminate

Table 1. Summary of the in-plane shear properties of BFRP and CNSs

Sample	In-Plane Shear Properties			
	In-plane shear strength (MPa)	% Gain in strength	Modulus (GPa)	% Gain in modulus
Neat Basalt	14	0	2.8	0
Basalt with 0.6wt% CNS	17.4	24.3	3.14	12.1
Basalt with 1.0wt% CNS	19.2	37.1	3.82	36.4

3.1. Fractography

Figure 4 shows the SEM images of the prepared samples after in-plane shear test. The detailed fracture process such as matrix debonding and fiber pull outs were observed in each of the samples. In the case of neat basalt fiber laminates, fiber pull out was observed as shown in Fig 4 (a). The fracture surfaces showed improvement with the inclusion of CNS particles (see Fig 4 (b) and 4 (c)). Some single fibers were seen on the surface rather than completely pulled-out fiber bundles, as shown in Fig. 4 (b) and 4 (c). Basalt with 0.6 wt% CNS (Fig 4 (b)) displays matrix debonding when the in-plane shear test was carried out

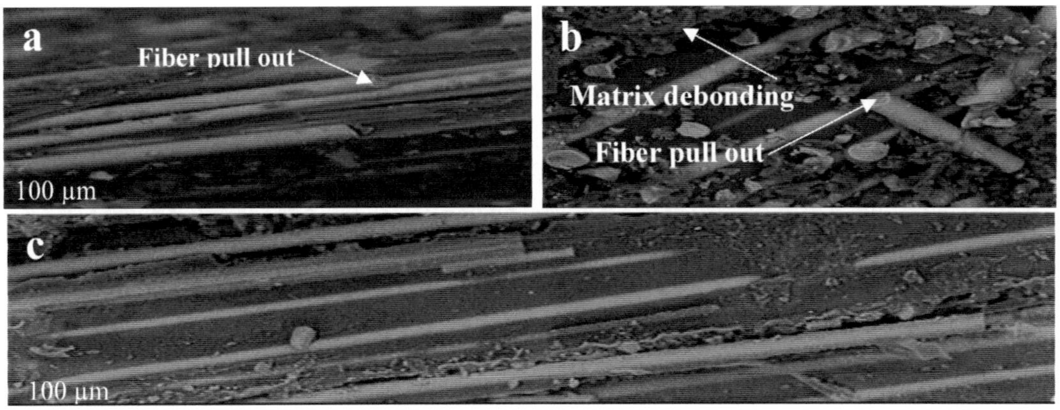

Figure 4. Representative SEM images of the fracture surfaces after in-plane shear test of: (a) Neat basalt, (b) Basalt with 0.6 wt% CNS, and (c) Basalt with 1.0 wt% CNS.

4. Conclusion

In this study, we have demonstrated that in-plane shear strength and modulus of rigidity are enhanced by reinforcing the basalt fiber with carbon nanospheres particles. To summarise, this paper examined the result of carbon nanospheres particle-loading of 0.6 wt% and 1.0 wt%, on the mechanical properties of basalt fiber-reinforced epoxy composite laminate. The composite laminates with different CNS loading showed better in-plane shear properties compared to the neat basalt fiber-epoxy composite (BFRP). The best result was found for 1.0 wt% addition of CNS particle, where the in-plane shear strength increased by 24.3% and 37.1% with the addition of 0.6wt% and 1wt% of CNS particles, respectively. Furthermore, the modulus of rigidity increased by as high as 36.4% when added with 1.0 wt% of CNS particles.

SEM images displayed good epoxy-fiber bonding with carbon nanospheres particles. The improvement in the mechanical properties is associated to the good dispersion of CNS particles in the epoxy matrix. This paper indicates that the In-plane shear strength of neat basalt fiber can be improved by the homogeneous dispersion of inexpensive CNS particles obtained from the biowaste sago bark in the epoxy matrix. With proper dispersion of micro/nano fillers, it will enhance the mechanical properties of the basalt composite laminate. The applications of basalt fiber in the industries would benefit from this study and could contribute to more knowledge on the exploitation of micro/nano particles to further improve the properties of epoxy-based composite materials.

References

[1] Dean D, Obore A M, Richmond S and Nyairo E 2006 *Compos. Sci. Technol* **66(13)** 2135-42
[2] Liu L, Huang Z, He C L and Han X J, 2006 *Ma.t Sci. Eng A -Struct* **435** 309-17
[3] Manikandan V, WinowlinJappes J T, Suresh S M and Amuthakkannan P 2012 *Compos. Part B. Eng.* **43(2)** 812-18
[4] Bodros E, Pillin I, Montrelay N and Baley C 2007 *Compos. Sci. Technol* **67** 462-70
[5] Wambua P, Ivens J and Verpoest I 2003 *Compos. Sci. Technol* **63** 1259–64
[6] Wang B P and Zhang W *Build 2002 Mater. Technol. Appl* **4** 17-18
[7] Sim J, Park C and Moon D Y 2005 *Compos. Part B. Eng* **36(6-7)** 504-12
[8] Lopresto V, Leone C and De Iorio I 2011 *Compos. Part B. Eng* **42(4)** 717-23
[9] Wei B, Cao H L and Song S.H 2011 *Corros. Sci* **53(1)** 426-31
[10] Colombo C, Vergani L and Burman M 2012 *Compos. Struct* **94(3)** 1165-74
[11] Liu Q, Shaw M T, Parnas R S and McDonell A M 2006 *Polym. Compos* **27(5)** 475-83
[12] Olexandr M and Yuriy T 2005 *Filtr Separat* **42**, 33-3
[13] Berozashvili M 2001 *Adv. Mater. Com. New.s Compos. Worldwide* **6**, 5-69
[14] Goldsworthy W B 2000 *Compos. Technol* **8** 15
[15] Militky J, Kovacic V and Rubnerova J 2002 *Eng. Fract. Mech* **69**, 1025–33
[16] Lee J, Yoon S, Hyeon T, Oh S and Kim K *Chem. Commun* **21**, 2177-78
[17] Yang H, Shi Q, Liu X, Xie S, Jiang D, Zhang F, Yu C, Tu B and Zhao D 2002 *Chem. Commun* **23**, 2842-43
[18] Stroller M D, Park S, Zhu Y, An J and Ruoff R S 2008 *Nano Lett.* 2008, **8 (10)** 3498-3502

Modeling and Simulation of a 100 MW Concentrated Solar Thermal Power Plant Using Parabolic Trough Collectors in Pakistan

K Liaqat[1], M Anss[2], A Ali[3] and A Nawaz Mengal[4]

[1,4]Department of Mechanical Engineering, Balochistan University of Information Technology, Engineering and Management Sciences (BUITEMS), Quetta, Pakistan.
[2]UET Lahore, Pakistan.
[3]Chairperson, Department of Mechanical Engineering, BUITEMS, Quetta. Pakistan.

Email: kashif29725@gmail.com

Abstract. The increasing energy demands of the modern era has created a critical situation. The world is now moving towards renewable and sustainable methods of producing energy, among them the most abundant renewable energy resource is solar energy. As the lifetime and efficiency of Concentrated Solar Power (CSP) is more as compared to photovoltaic (PV) and considering the solar potential of Pakistan, the design of a 100 MW Concentrated Solar thermal power plant using Parabolic Trough Collectors and a 6-hour thermal energy storage is proposed. The CSP plant is modeled and simulated using System Advisor Model (SAM). Based on certain parameters, a location receiving an annual Direct normal irradiance (DNI) of 1955 KWh/m^2/year near Nawabshah is selected for the hypothetical solar thermal power plant. The plant consists of 189 solar collector loops with 8 parabolic trough collectors in each loop and HITEC solar salt as HTF. The simulation results show that the plant can generate 245,688,560 kWh (245.68 GWh) of electricity annually with a capacity factor of 28.1% and 93.8% gross to net conversion. The results of PTC Power plant encourage further investigation and development of CSP technologies for electricity generation in Pakistan.
Keywords: Simulation, solar, power plant, collectors

1. Introduction

The modern developing world's energy requirements are growing rapidly. The use of fossil fuels to overcome these requirements has created a critical situation because the fossil fuels are causing serious environmental problems and in addition, the resources are in limited supply. To overcome this problem the focus is shifted towards the use of new sustainable electricity generation options, which take advantage of renewable energies and are economical. Solar Energy is one of the most promising renewable energy technology that can reduce the fossil fuel consumption and CO_2 Emissions.

Concentrated solar power (CSP) is the most promising way to convert solar energy into electricity. In contrast, Photovoltaic (PV) technology is less efficient and has a shorter life. In CSP technology, the sun rays coming from the sun are reflected a receiver by using collectors (mirror), the receiver carries fluid that absorbs the heat. There are different methods to concentrate the solar radiation depending upon the applications and desired temperatures. Parabolic trough collectors (PTC) as compared to flat plate solar collector requires less surface area decreasing concerns for high thermal losses. PTC shows least development risks and is the best one to achieve high temperature with commercially suitability, as compared to other concentrating technologies like Fresnel collector, solar tower, and dish sterling.

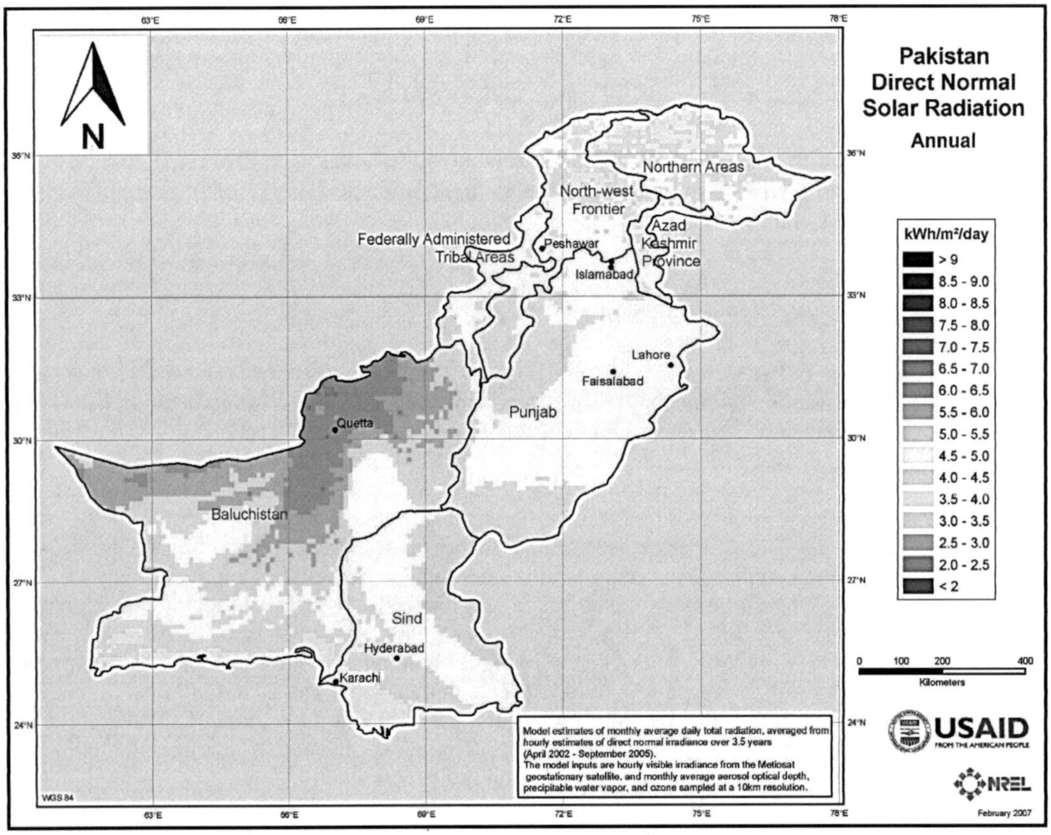

Figure 1. Annual DNI map of Pakistan

Pakistan is undergoing an unprecedented energy crisis since long time back. The demand and supply shortfall is around 5000MW in the country and an electrification rate of 73% (90% in urban areas and 68% in rural areas) [1]. Electricity generation in Pakistan is dominated by fossil fuels. Pakistan has limited fossil fuel resources and needs to import to fill this gap. Due to this energy shortfall, urban areas are facing 10–12-hour load shedding while in rural areas electricity remains unavailable for 16-18 hours[2]. There is a 2,900,000 MW solar energy potential due to Pakistan's geographical location with more than 300 sunshine days, 26-28°C average annual temperature and 1900-2200 kWh/m^2 annual global irradiance [3]. The country receives on average global insolation of 19 megajoules per square meter of solar energy daily. These figures are ideal for a Concentrated Solar Thermal Power Plant. As stated in the Green Peace International report, the desert areas of the country are among the most suitable locations for CSP plants[4]. Hence solar thermal power generation is one of the ways to solve the energy crises of the country without harming the environment as well.

2. Site Selection

The primary phase of CSP power plant is site selection. Many parameters are considered while fulfilling the requirements of a CSP site. These factors are essential for the plant to generate electricity economically and to make it work best.

2.1. Site Requirements

The essential requirements for a CSP site are availability of Direct Normal Irradiation (DNI) > 1800 kWh/m^2/year (5.2 kWh/m^2/day) [5]. One axis sun tracking mechanism to maximize the incident solar radiation. Flatland area with an overall slope of less than 1 - 3% [6]. Lands with higher slope reduce the efficiency of a CSP plant, due to the reason that slop tends to give rise to self-shadowing and shading cast by nearby collectors. This can reduce the significant amount of radiation intercepted by the collector [7]. The wind speed of the location should be less than 15.64 m/s [8], higher wind speed can cause structural damage to the collector assembly. Other than these requirements some other required factors are convenient grid connection, water resources, Good transportation facilities and Back up fuel.

2.2. Site assessment for Power plant in Pakistan

Most of the area of Pakistan receives an average DNI of 4.5 to 6 kWh/m^2/day. Values of just over 6.4 kWh/m^2/day are reached in the southwestern region of Balochistan decreasing gradually towards the northeast of the country to 4.0 kWh/m^2/day [1]. Figure 1 shows the annual DNI map of Pakistan.

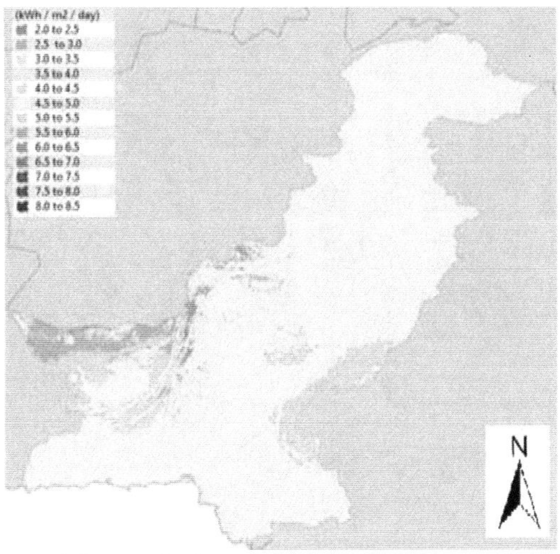

Figure 2. Geospatial Toolkit Results

The site selection is carried out by using Solar radiation maps of Pakistan, Geospatial Toolkit by NREL (National Renewable Energy Laboratory) and Google earth. Initially, the primary criteria are considered. Figure 2 shows the results of Geospatial toolkit with parameters adjusted as, minimum DNI of 5.2 kWh/m^2/day, the maximum land slope of 3% and the land is selected as barren or sparsely vegetated. The shortlisted sites from these results are given in table 1.

Table 1. Shortlisted Sites [9]

Balochistan	Punjab	Sindh
Quetta	D.G.Khan	Nawabshah (Nara)
Qila Abdullah	Raheem Yar Khan	Sanghar
Panjgur	Bahawalpur	Khairpur

Considering the remaining requirements Nara desert, a location near district Nawabshah located at latitude 26.55 °N and longitude 69.25 °E with annual DNI of 1955 kWh/m^2/year, is the most optimum site for the plant [9]. The details specifications of the site are given in table 2.

Table 2. Selected site specifications [9]

Requirement	Availability
Water Resources	Nara Canal (8 km) Chotiari Dam (37 km)
Grid Connection	132 kV Sanghar (~50km)
Wind Speed	0 - 5.4 m/s
Transportation	Nawabshah Railway station (62 km), Sanghar road (12 km)
Backup Fuel (Natural Gas)	Sawan gas field (60.82 km), Kadanwari gas field (77.18 km)

3. Modeling and Simulation

The parabolic trough power plant works on the Rankine cycle. The heat is absorbed by the heat transfer fluid and is then transferred to water to generate steam. The integration of thermal storage helps to provide electricity during the period when solar radiation is not available. Figure 3 shows the schematics of a Parabolic Trough Power plant.

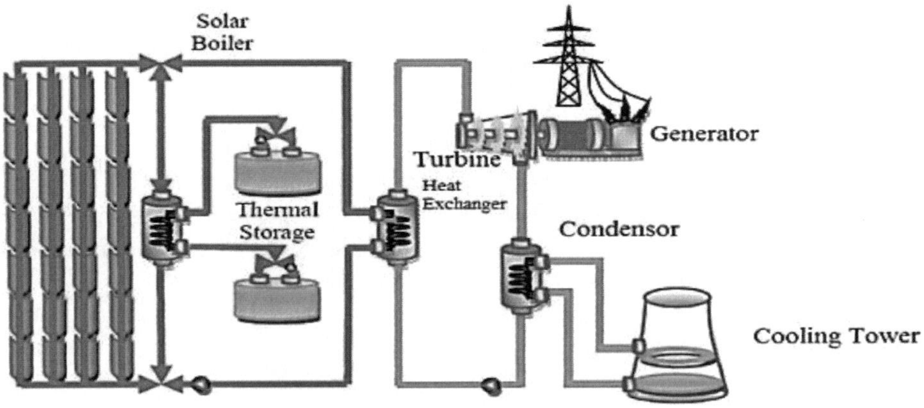

Figure 3. Schematics of the Plant

3.1. Heat Transfer Fluid (HTF)

Synthetic oils and molten salts are the most commonly used HTF for CSP power plants. Among them, Therminol VP-1 and Hitec solar salt are suitable options due to their higher specific heats, good heat carrying capacity and high density. Hitec Solar salt can achieve the highest temperature as compared to others and has a higher specific energy over Therminol VP-1. A detailed comparison of HTF properties is given in table 3.

Table 3. HTF properties [10]

Name	Type	Min Operating Temp °C	Max Operating Temp °C	Freeze Point °C
Hitec Solar Salt	Nitrate Salt	238	593	238
Caloria HT 43	Mineral Hydrocarbon	-12	315	-12 (pour point)
Therminol VP-1	Mixture of Biphenyl and Diphenyl Oxide	12	400	12 (crystallization point)

3.2. Mathematical Modeling

Duffie and Beckman[11] presented a model to express useful heat gain Q_u for a single concentrated collector given as:

$$Q_u = F_R A_a \left[\frac{S - A_r}{A_a U_L (T_i - T_a)} \right] \qquad (1)$$

Where F_R is collector heat removal factor, A_a is aperture area of the concentrator, A_r is receiver area, S is absorbed solar radiation, U_L is heat loss coefficient, T_i is fluid inlet temperature, T_a is ambient temperature and output temperature of a collector is given by.

$$T_o = T_i + \frac{Q_u}{mC_p} \quad (2)$$

T_o is outlet temperature, m is the mass flow rate of HTF and C_p is the specific heat capacity of HTF.

$$No\ of\ loops\ required = \frac{\left(\frac{P_{th}}{Output\ Power\ collector}\right)}{No.of\ collectors\ per\ loop} \quad (3)$$

3.2.1. Solar Field and Solar Multiple. The first factor for CSP plant is a solar field, an area where solar collectors are to be installed. The field aperture area expressed as a multiple of the aperture area required to operate the power cycle at its design capacity and is known as Solar Multiple (SM)*[10]*.

$$SM = \frac{Power\ Cycle\ Capacity}{Solar\ solar\ field\ Capacity} \quad (4)$$

3.2.2. Design thermal output of Solar Field. Thermal energy obtained from the solar field under design condition at a given solar multiple, and is expressed as:

$$Q_{sf,des} = SM\left(\frac{W_{pb,des}}{\eta_{\ des}}\right) \quad (5)$$

Where $Q_{sf,des}$ is solar field design heat output, $W_{pb,des}$ design work out from power block and $\eta_{\ des}$ is design efficiency.

3.3. Solar Radiation Data and System Advisor Model (SAM)

SAM is a financial and performance model developed by the National Renewable Energy Laboratory (NREL) with the grants funded by U.S. Department of Energy, for the field of renewable energy. For annual performance prediction and cost estimation, it uses hourly solar radiation data of a typical year obtained from the concerned location. SAM uses the SAM CSV format for solar resource data, and can also read data from files in the TMY3, TMY2, and EPW formats. SAM comes with built-in radiation data for many locations including India and Bangladesh[10]. Unfortunately, it doesn't include data for locations in Pakistan, thus for this work, the hourly radiation data of a year of the hypothetical plant location has been obtained from NREL National Solar Radiation Database (NSRDB) and is then integrated to SAM.

3.4. Plant Specifications and Configurations

The input parameters used in SAM for the simulation of the hypothetical plant are given in table 4.

Table 4: Plant Specifications and Configurations

Parameter	Value	Parameter	Value
Solar Field Parameters		**Design Point**	
Solar Multiple	2	Actual No. of loops	189
Design Irradiation	887 W/m^2	Single loop Aperture	5248 m^2
Design ambient Temp	150 ^{o}C	Total Aperture Reflective Area	991872 m^2
Design wind Velocity	5 m/s	Field Thermal output	623.596 MWt
No. of field subsections	2	Water usage Per wash	0.7
Row Spacing	15 m	Washes per year	63 L/m^2, Aper
Stow Angle	170°	Number of Assemblies per loop	8
Deploy Angle	10°	**Power Cycle**	
HTF Pump Efficiency	0.85	Design Gross Output	111
Heat Transfer Fluid	Hitec Solar Salt	Estimated Gross to Net Conversion	0.9
Design Loop Inlet Temp	293	Estimated net output at Design	100
Design loop outlet Temp	525	Rated Cycle Conversion Efficiency	0.356
Min Single loop flow rate	1 kg/sec	Design Inlet Temperature	525 ^{o}C
Max Single Loop Flow rate	12 kg/sec	Design Outlet Temperature	293 ^{o}C
Min field velocity	0.153546 m/s	Boiler Operating Pressure	100 Bar
Max Field Velocity	1.99737 m/s	Cooling Condenser Type	Evaporative
Header design Min field velocity	2 m/s	**Thermal Storage**	
Header design Max field velocity	3 m/s	Full load Hour of Thermal Storage	6 hr
Receiver	Solel UVAC 3	Storage Volume	10482.8 m^3
Absorber Tube Inner Diameter	0.066	**Parasitic**	
Absorber Tube Outer Diameter	0.07	Balance of Plant Parasitic [5]	0.02467 MWe/Mwcap
Total Receiver Losses	215.5 W/m	Auxiliary heater, Boiler Parasitic [5]	0.02273 MWe/Mwcap
Total Land Area	858 Acres	Piping Thermal loss coefficient [12]	0.45 W/m^2-k

4. Results and Performance Assessment

The simulation of the plant is carried out for one year i.e. 0 to 8760 Hours. The plant starts to generate electricity around 8 A.M on average. The highest amount of energy is generated during the month of May (26129.1 MWh) and the minimum electricity is generated during the month of December (12112.8 MWh). Figure 4 shows the monthly variation of the generated electricity. Maximum cycle electrical power output and maximum field thermal power incident were recorded as 124.586 MWe and 749.855 MWt respectively. The total thermal output of the field is recorded as 745534 MWht. The hypothetical plant is estimated to generate (245.68 GWh) of electricity annually. The gross to net conversion is estimated to be 93.8%. Furthermore, the plant operates with a capacity factor of 28.1% which is a promising value compared to the previously available value of 20.5% using Therminol VP-1[9].

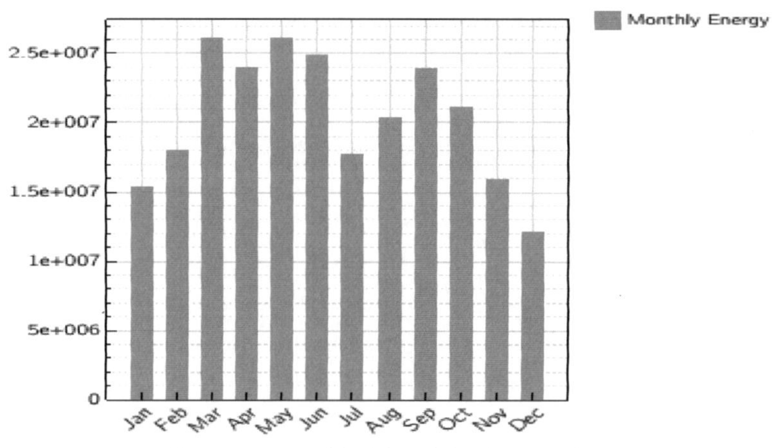

Figure 4. Monthly Electricity Generation

4.1. Performance Comparison

The results obtained for the CSP plant are compared to similar work available in literature and the comparison is given in table 5. For this comparison, it shows that the CSP power plant using Parabolic Trough collectors in Nawabshah is technically viable and has the feasibility to work in this environment.

Table 5. Performance Comparison with similar work

Author	Plant Capacity (MW)	Annual Energy Generation (GWh)	Capacity Factor
This Work	100	245	28.1 %
Bishoyi[13]	100	285	32.6 %
Mohamed et al.[14]	100	237	21.1 %

5. Conclusion

Based on the criteria of CSP power plant a location within Pakistan is selected and simulation of a 100MW Power plant using Parabolic Trough Collectors is performed. The results obtained provide the necessary data for designing a CSP power plant in Pakistan. The thermal performance of the plant showed good results and this work can be utilized for further investigation of CSP technologies in Pakistani Climate. The research also showed that Hitec Solar salt can increase the performance of the CSP Plant. However, further research and investigations are required to analyze and improve parameters such as Loop Configuration and Thermal Capacity. This work only investigates the thermal performance of the plant, in addition to this, financial models are also needed to be developed to check the economic viability of the technology. This technology has more life and efficiency than PV plants and can be utilized as a solution to energy concerns of Pakistan.

References

[1] "Sustainable Development Policy Institute ("SDPI") (Pakistan Energy Vision 2035)."

[2] "Asian Development Bank (Pakistan Solar Rapid Assessment June 5, 2017)."

[3] "Farooq M, Shakoor A. Severe energy crises and solar thermal energy as a viable option for Pakistan. J Renew Sustain Energy 2013;5(1):013104."

[4] "Solar Thermal Power 2020 | Greenpeace International.".

[5] "K. Kaygusuz, Prospect of concentrating solar power in Turkey: the sustainable future, Renew. Sustain. Energy Rev. 15 (2011) 808–814."

[6] "Bin Yang; Jun Zhao; Wenbin Yao; Qiang Zhu; Hang Qu, Feasibility and Potential of Parabolic Trough Solar Thermal Power Plants in Tibet of China, APPEEC 2010, Asia-Pacific Power and Energy Engineering Conference 2010."

[7] "Richard G. Allen, Ricardo Trezza, Masahiro Tasumi, Analytical integrated functions for daily solar radiation on slopes, Agricultural and Forest Meteorology 139 (2006) 55–73."

[8] "Nusrat Kamal Raja, M. Shahid Khalil, Design and Manufacturing of Parabolic Trough Solar Collector System for a Developing Country Pakistan, Journal of American Science, 2011."

[9] "S. Saleem, A. ul Asar, Analysis & Design of Parabolic Trough Solar Thermal Power Plant for Typical Sites of Pakistan, IOSR J. Electr. Electron. Eng., e-ISSN: 2278-1676,"

[10] "System Advisor Model Version 2017.9.5 (SAM 2017.9.5). National Renewable Energy Laboratory. Golden, CO." .

[11] Duffie, J.A. and Beckman, W.A, Solar Engineering of Thermal Processes, 3rd Edition New York, USA 2005.

[12] "Jay Prakash Bijarniya, K. Sudhakar, Prashant Baredar, Concentrated solar power technology in India: a review, Renew. Sustain. Energy Rev. 63 (2016) 593–603."

[13] D. Bishoyi, "Modeling and performance simulation of 100 MW PTC based solar thermal power plant in Udaipur India - ScienceDirect." .

[14] "Mohamed Abbas, et al., Assessment of a solar parabolic trough power plant for electricity generation under Mediterranean and arid climate conditions in Algeria, Energy Procedia 42 (2013) 93–102."

Hydrothermal Alteration in Porphyry Cu-Mo-Au Mineralizations of the Chagai Arc, Balochistan, Pakistan

Rehanul Haq Siddiqui[1], Syed Mobasher Aftab[1] and Asif Hanif Chaudhry[2]

[1]Balochistan University of Information Technology, Engineering and Management Sciences, Quetta, Pakistan,
[2]Geoscience Advance Research Laboratories, Geological Survey of Pakistan, Islamabad, Pakistan

E-mail: rehanul.haq@buitms.edu.pk

Abstract. Subduction-related east-west trending Chagai Arc located in the western part of Pakistan. This arc is the western extension of Tethyan Magmatic Belt which is about 5000 km long and extends from central Europe, Turkey, Iran, and Pakistan. During the last four decades, several porphyry copper settings were reported from the Chagai Arc. Seven more famous occurrences are selected for the present study owing to more detailed work carried out on them. From west to east these deposits include Saindak, Reko Diq (formerly known as Koh-e-Dalil), Humai, Missi, Durbanchah, Dasht-e-Kain and Ziarat Pir Sultan. Hydrothermal alterations in the porphyry copper setting of Chagai Arc are mostly associated with tonalite porphyry stocks, except Durbanchah and Humai prospects which are hosted in dacite porphyry stocks, whereas Missi Prospect occurs in a granodiorite batholith. The alteration is normally developed in a concentric zonal pattern as observed in the majority of the world deposits, except that absence of a regular argillic and peripheral zone. In most of the occurrences the potassium silicate alteration zone (K-alteration zone) occurs usually within the intrusive porphyry stock, but in Saindak and Reko Diq deposit, some of the adjacent wall-rock sediments, and in Durbanchah setting the microdioritic country rock, has also undergone the K-alteration. Quartz sericitic or phyllic alterations zones are developed in all the occurrences as continuous or discontinuous haloes around the K-alteration zone except in Durbanchah prospect. In Humai prospect an advanced argillic alteration zone is developed around the K-alteration zone. Propylitic alteration has also developed in almost all settings and generally encircles the quartz sericitic alteration. The oxide mineralization is generally represented by goethitic jarositic, hematitic limonites and malachite, with minor pitch limonite, chrysochola, neotocite, brochanthite and molybdite. Hypogene mineralization is represented by pyrite, chalcopyrite and minor bornite. A regular supergene sulfide enrichment is only reported from the Reko Dik deposit and represented by chalcosite and covellite in other deposits restricted incipient type supergene sulfide enrichment is encountered. In terms of ore reserves the Riko Dik deposit is on the top with 2,200 million tons of copper followed by Saindak with 412 million tons and Dasht-e-Kain 350 million tons. Grade wise Reko Diq Deposit is again on the top with 0.53% Cu, 0.01% Mo and 0.3g/t Au, followed by Saindak with 0.4% Cu, 0.015% Mo and 0.35 g/t Au. Dasht-e-Kain deposit contains 0.35% Cu and 0.015% Mo. Gold has not been reported from the Dasht-e-Kain deposit. From other deposits of the Chagai Arc ore reserves and grade are not reported so far.
Keywords: Hydrothermal Alteration, Porphyry Copper, Tethyan Magmatic Belt, Chagai Arc, Balochistan

1. Introduction

The Chagai Arc located in the western part of Pakistan. This arc is the western extension of the Tethyan Magmatic Belt which is about 5000 km long that extends from central Europe, Turkey, Iran, and Pakistan. During the last four decades, several porphyry copper settings were reported from the Chagai Arc. The main part of the Arc present in the western part of Pakistan. A small component of the arc terrain extends northwards in Afghanistan and towards the west in Iran. The Chagai Arc trends in EW direction, it is around 500 km long and 140 km wide. It is terminated by the Chaman and

Harirud Fault Zones in the east and west respectively and convex towards the south, figure 1 and 2. Several porphyry copper occurrences have been found in the arc, out of which, seven are selected for the present study because relatively more geologic work has been done on them. These include well known Saindak porphyry copper deposit and lesser-known porphyry copper prospects of Reko Diq (formerly known As Koh-e-Dalil), Durbanchah, Humai, Ziarat Pir Sultan, Missi and Dasht-e-Kain. The eastern part of the Chagai belt between the Chagai Hills and Siah Koh is covered by Recent and Sub-recent alluvial deposits and may also be hosting a few hitherto unknown porphyry prospects [1].

The hydrothermal alteration is generally developed in all the porphyry copper systems and serves as an important tool for the recognition of porphyry copper mineralization in the area [2]. Three hydrothermally altered zones, namely potassium silicate, quartz sericitic (phyllic) and propylitic, have been identified in almost all the porphyry copper occurrences in the region. However, argillic, advanced argillic and a regular peripheral zone as described by Lowell and Giulbert [2], are not developed anywhere in the Chagai Arc. In this paper, an attempt has been made to describe the alteration zones in various deposits/prospects of the Chagai Arc.

2. Regional Geology

The current geotectonic setting of the Chagai Arc is represented by a Quaternary subduction zone dipping towards the north, an active Makran accretionary zone that is followed by the Chagai-Raskoh magmatic arc and Afghan Block, shown in figure 1. Apparently, this subduction zone setting reflects a continental margin typesetting. The majority of the preceding investigations have preferred this tectonic setting about the origin not only for the Quaternary volcanics but also for the older volcanic rocks up to Late Cretaceous [3, 4, 5, 6, 7, 8, 9, 10]. Siddiqui [11, 12], and Siddiqui et al. [13, 14, 15, 16, 17, 18, 19, 20], have opposed this perspective and documented an oceanic island arc setting for the Late Cretaceous to Paleocene volcanics of the Chagai Arc, transitional character for Eocene and continental margin or Andean type settings for Oligocene to Pleistocene volcanics.

The oldest rock suite developed in the Chagai Arc is a submarine stratified volcanic and volcanoclastic, suite known as Sinjrani Volcanic [21], which is Late Cretaceous in age and is composed mainly of andesitic flows, agglomerates, volcanic conglomerate, tuffs and subordinate amounts of limestone, shale, and sandstone. The total thickness of the Group is about 10,000 m [7], and it has been intruded by the Chagai intrusions [21], which range in size from cupolas and stocks to batholiths and include granite, quartz-monzonite, granodiorite, monzonite, diorite, quartz diorite, and gabbro. These intrusions are as old as Pre-Late Cretaceous [21], and as young as Lower Miocene [22, 23].

Multiple episodes continued in pulsatory fashion were responsible for the emplacement of the volcanic and plutonic rocks of the region [7]. Siddiqui et al. [15, 19], has proposed a link between calc-alkaline volcanism during the Oligocene to Miocene with the formation of porphyry copper deposits in the Chagai Arc.

The Chagai Arc has got tremendous importance for its economic mineral potential. The magmatic rocks arc intimately associated with deposits of copper, and iron including the porphyry copper, skarn copper and/or iron, manto-type copper, vein-type copper and volcanogenic sulfur and volcanogenic-stratiform iron. In addition, zinc, copper and silver-rich Kuroko type sulphide deposits have also been reported.

IOP Conf. Series: Materials Science and Engineering **414** (2018) 012033 doi:10.1088/1757-899X/414/1/012033

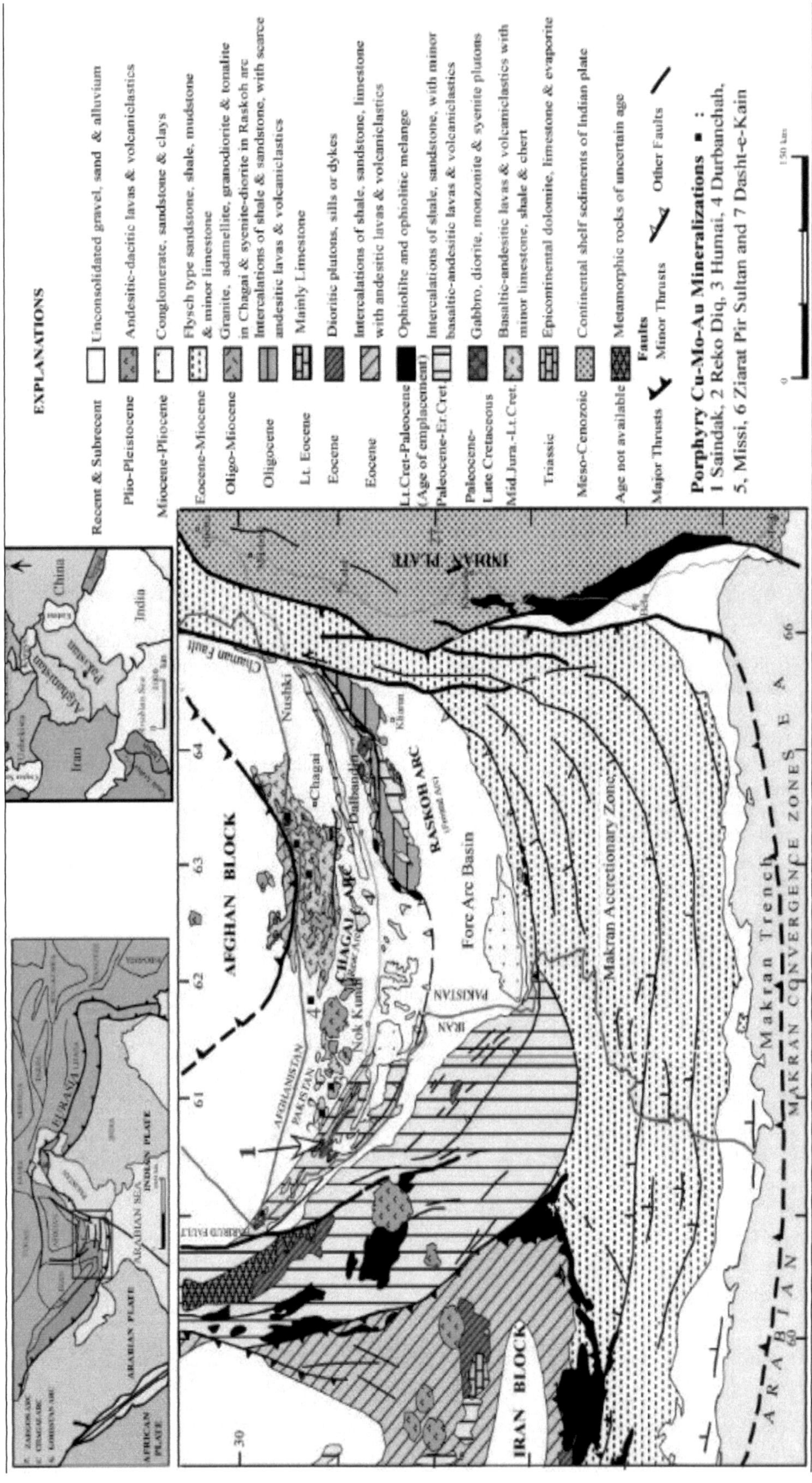

Figure 1. Geological map of the Chagai-Raskoh Arc terrain and adjoining area showing porphyry copper deposits, after Siddiqui et al., [19].

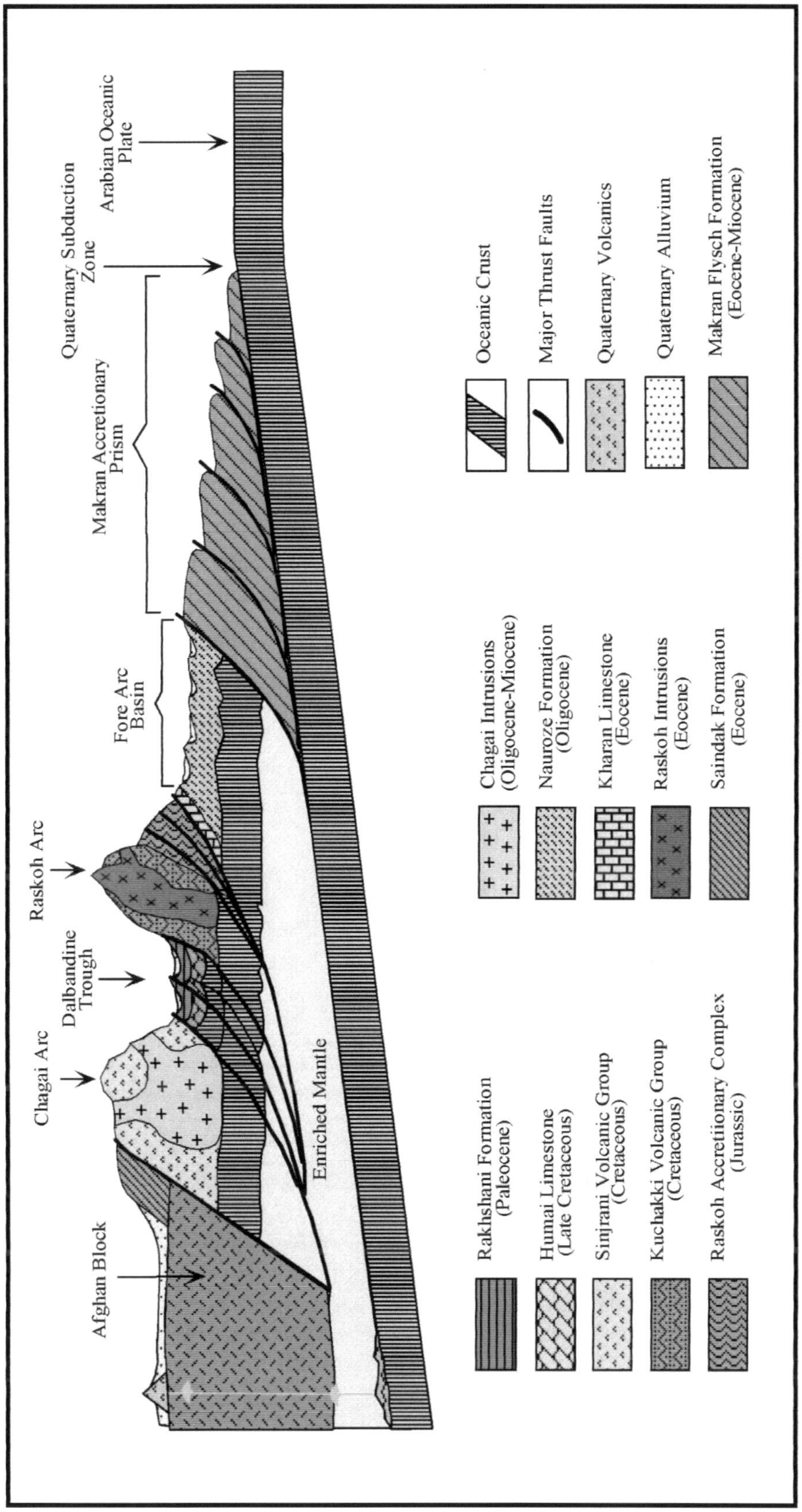

Figure 2. Schematic cross-section (not to scale) across the Chagai-Raskoh Arc terrain and adjoining area, after Siddiqui, [12].

3. Hydrothermal Alteration

Hydrothermal alteration in each of the porphyry copper occurrences is developed in a concentric manner. In each occurrence, the potassium silicate alteration zone (K-zone) occurs in the center, followed outwards by the quartz sericitic and the propylitic alteration zones, but none of the occurrences have regular argillic and peripheral zones proposed by Lowell and Guilbert [2], figure 3. The zones developed are discussed in the following paragraphs.

3.1. Potassium Silicate Alteration Zone (K-zone)

This type of alteration zones occurs centrally within each intrusive porphyry ore prospect; but at Saindak and Reko Diq, some of the adjacent wall rock sediments and, at Durbanchah, microdioritic wall rocks have also undergone such alteration [23, 24], table 1.

At Saindak, K-alteration is mainly associated with three comagmatic tonalite porphyry stocks, named respectively, north, south and east stocks according to their positions. Potassium silicate alteration at Saindak is characterized by the biotite, K-feldspar, quartz anhydrite, epidote, chlorite, tourmaline, pyrite, chalcopyrite, magnetite and minor molybdenite, Biotite occurs disseminated or as replacement of primary hornblende, in veinlets with other minerals and alone on partings. The southern body has a quartz-magnetite type of K-alteration in the center whereas, in the eastern ore body, K-alteration is superimposed by retrograded quartz sericitic alteration [22].

At Reko Diq, potassium silicate alteration is generally associated with five tonalite porphyry stocks, some neighboring host-rock sediments have also undergone the same alteration, K-alteration. In this prospect, is characterized by the biotite, K-feldspar, quartz, tourmaline, magnetite, pyrite, chalcopyrite and minor molybdenite. The main eastern body is centered by quartz-magnetite type of K-alteration, and retrograded quartz sericitic alteration has developed towards its eastern boundary. At Reko Diq deposit, the K-zone is partially destroyed by the emplacement of post-mineralization dacite porphyry intrusions [25, 26]. At Humai prospect, potassium silicate alteration is restricted to a dacite porphyry stock and is represented by chlorite after biotite, quartz, magnetite and minor pyrite and chalcopyrite [5].

The largely concealed Durban Chah porphyry copper prospect has a small outcrop of K-alteration, restricted in a marginal part of alluvial depression bordered by andesites of Sinjrani Volcanic Group. Potassium Silicate alteration at this prospect is associated with a dacite porphyry stock and is represented by biotite, K-feldspar, quartz, chlorite, and minor pyrite, chalcopyrite and molybdenite [27, 28].

At Ziarat Pir Sultan, the alteration is mainly associated with two tonalite porphyry stocks and K-alteration is exhibited by biotite and quartz with minor pyrite, chalcopyrite and molybdenite [28]. At Missi prospect, potassium silicate alteration zone has not been observed on the surface, but it is expected to occur below the quartz sericitic zone [5].

At Dashte Kain, the alteration is mainly associated with two tonalite porphyry stocks, named eastern and western stocks by position. The later stock is invaded by an intrusive breccia pipe. K-alteration is centered in each stock and is characterized by biotite, K-feldspar, anhydrite, chlorite, carbonate, sericite, and kaolinite with minor magnetite, pyrite, chalcopyrite, pyrrhotite, molybdenite, bomite, and enargite. A considerable part of the K-zone of the western stock has been superimposed by retrograded quartz sericitic alteration [29].

3.2. Quartz Sericitic or Phyllic Alteration Zone

The quartz sericitic alteration zone has developed as continuous and/or discontinuous halos around potassium silicate alteration zone in all the porphyry Cu settings of the Chagai belt, except at Durbanchah where it has not been encountered, whereas at Humai an advanced argillic zone has been found around K-zone, table 1.

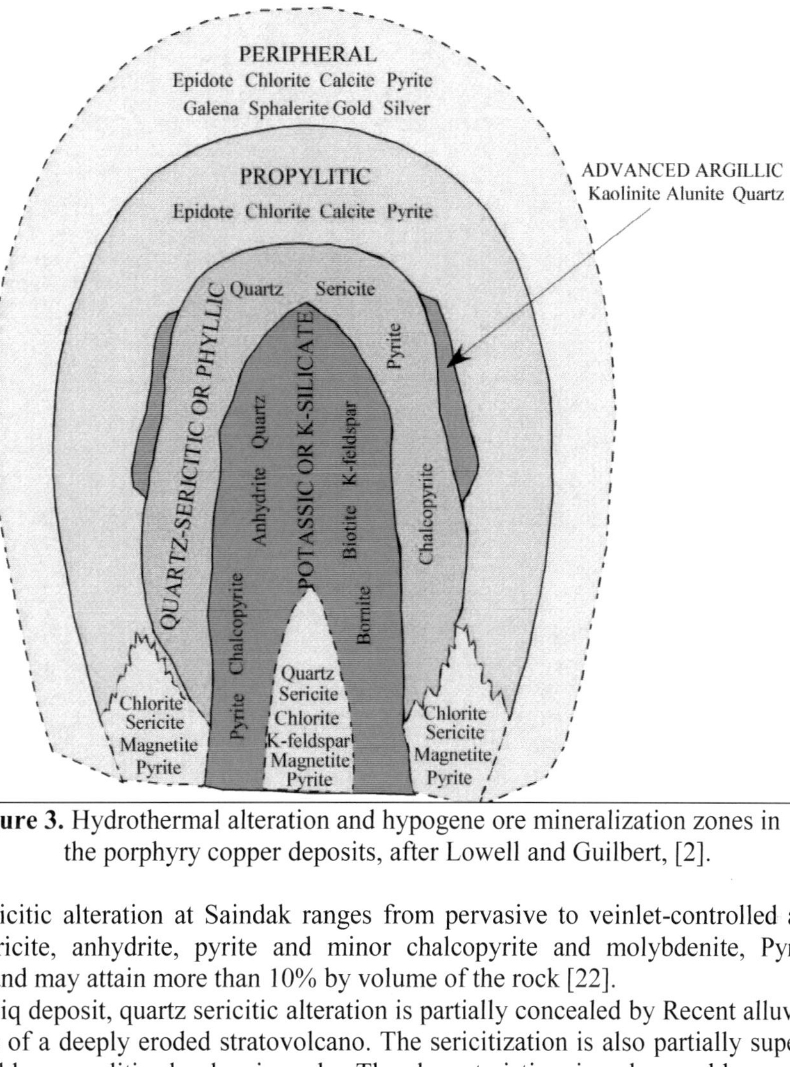

Figure 3. Hydrothermal alteration and hypogene ore mineralization zones in the porphyry copper deposits, after Lowell and Guilbert, [2].

Quartz sericitic alteration at Saindak ranges from pervasive to veinlet-controlled and comprises quartz, sericite, anhydrite, pyrite and minor chalcopyrite and molybdenite, Pyrite is locally abundant and may attain more than 10% by volume of the rock [22].

At Reko Diq deposit, quartz sericitic alteration is partially concealed by Recent alluvial deposits in the middle of a deeply eroded stratovolcano. The sericitization is also partially superimposed and surrounded by propylitized volcanic rocks. The characteristic mineral assemblages of the prospect are quartz, sericite, pyrite, and chlorite [25, 26, 30].

At Humai prospect, an advanced argillic alteration zone is developed around K-zone instead of quartz sericitic alteration, which is represented by alunite, quartz, sulphur, and pyrite [5]. There is no quartz sericitic alteration zone at Duroanchah, at least on the surface and K-alteration is directly surrounded by propylitic zone. At Ziarat Pir Sultan, quartz sericitic alteration occurs as a discontinuous halo around K-zone and is exhibited mainly by sericite, quartz, and pyrite [28]. Quartz sericitic alteration at Missi prospect is widely developed and characterized by sericite, quartz, and pyrite with minor chalcopyrite and molybdenite.

At Dashte Kain, quartz sericitic alteration zone partly encircles the K-alteration and is represented by quartz, sericite, and pyrite. At places, this alteration is also superimposed over a hydrothermal intrusion breccia. The northern portion of quartz sericitic alteration at Dasht-e-Kain is alluvium covered.

Table 1. Different porphyry Cu-Mo-Au settings of the Chagai Arc, Balochistan, Pakistan; [5, 15, 22, 23, 25, 27, 30, 31, 49].

	GEOLOGY AND AGE				HYDROTHERMAL ALTERATION ZONES				ORE GRADE	ORE RESERVE
NAME OF PORPHYRY COPPER SETTING	PRE-ORE HOST ROCKS	ARTICLE AGE	HOST ROCK	AGE OF MINERALIZATION[a]	POTASSIUM SILICATE	PHYLLIC (QUARTZ SERICITIC)	PROPYLITIC	Area of Alteration	(a) Copper (b) Molyb. (c) Gold	(Million tons)
SAINDAK	Siltstone, shale, sandstone andesitic lava flows and tuffs.	Oligocene	Tonolite porphyry (3-bodies)	20.3 ± 0.8 Ma***	Biotite, K-feldspar, quartz, anhydrite, pyrite, chalcopyrite, molybdenite	Quartz, sericite, clay minerals, pyrite, chalcopyrite, molybdenite	Chlorite, epidote, albite, calcite, pyrite, anhydrite	3 Km^2	(a) 0.4 % (b) 0.015% (c) 0.23 Os/T	412
REKO DIQ (Former Koh-e-Dalil)	Shale sandstone limestone, andesitic, tuff and lava flows	Palaeocene To Miocene	Tonolite porphyry. (12 bodies)	10-13 Ma	Biotite, magnetite, pyrite anhydrite, K-feldspar, Chalcopyrite, molybdenite	Quartz, sericite, pyrite and chalcopyrite	Chlorites, epidote, calcite	18 Km^2	(a) 0.6 % (b) 0.015% (c) 0.23 Os/T	6800
HUMAI (Chagai Arc)	Andesites, andesitic agglomerates and tuffs	Late Cretaceous To Miocene	Dacite to andesite porphyry (One body)	Not determined	Quartz, Pyrite magnetite, chlorite, chalcopyrite	Aphanitic silica, pyrite, alunite, sericite	Chlorite, Epidote, pyrite	1.1 Km^2	Not determined	150
DURBANCHAH	Granodiorite & diorite (Lt.Cre.-Miocene) Andesites (Lt. Cre.)	Late Cretaceous To Miocene	Dacite porphyry. (One body)	10.9 x 0.7 Ma**	Biotite, quartz h-feldspar, minor pyrite, chalcopyrite, molybdenite	Absent	Chlorite, epidote, calcite	9.75 $Km.^2$	Not determined	300
MISSI	Granodiorite and diorite	Late Cretaceous To Miocene	Granodiorite (One body)	Not determined	Absent	Sericite, quartz, pyrite minor, chalcopyrite, molybdenite	Chlorite, epidote, calcite, tourmaline, quartz, pyrite	1 Km^2	Not determined	100
ZIARAT PIR-SULTAN	Granodiorite & diorite, (Lt. Cre. To Miocene) Andesites (Lt. Cre.)	Late Cretaceous To Miocene	Adamillite & tonolite porphyry. (2-bodies)	21.0 x 0.7 Ma**	Biotite, quartz, pyrite and chalcopyrite	Quartz, Sericite, pyrite	Chlorite, Epidote, pyrite	2.5 Km^2	Not determined	200
DASHT-E-KAIN	Hornblende diorite with minor gabbro (Paleocene) Andesite, andesitic tuff, agglomerates	Late Craterous to Paleocene	Tonolite porphyry. (2-bodies)	31.6 x 1.2 Ma*	Biotite, chlorite k-feldspar, quartz pyrite, magnetite chalcopyrite, bornite, energite	Quartz, sericite, kaolinite, pyrite	Chlorite, epidote, carbonates, pyrite	3 Km^2	(a) 0.35 % (b) 0.015 %	350

[a]Age of Mineralization, [22, 23, 31, 49].

3.3. Propylitic Alteration Zone

The propylitic alteration zone is widely developed in all the porphyry copper settings so far found in the Chugai bell and generally encircles the quartz sericitic alteration zone, Table 1. At Saindak, the area beyond the quartz sericitic zone represents the propylitic alteration and its outer limits were defined as the outer limits of pervasive pyritization. This alteration at Saindak is typified by chlorite, epidote, albite, calcite, anhydrite, and pyrite [22].

The Reko Diq deposit shows this zone around the quartz sericitic alteration zone, mainly associated with andesitic flows and characterized by the presence of chlorite, epidote, and calcite [25, 26]. At Humai propylitic alteration zone is marked by chlorite, epidote, and calcite [5]. The propylitic alteration at Durbanchah is represented by the extensive development of chlorite, epidote, and calcite. At the contact zone, the andesite is completely replaced by epidote and thus is a good indicator of mineralized zone in the vicinity [27]. At Ziarat Pir Sultan, this alteration is characterized by chlorite, epidote and pyrite. At Missi prospect, the propylitic alteration was identified by the presence of chlorite, epidote and calcite.

Propylitic alteration at Dashte Kain is restricted in the pre-ore dioritic and andesitic rocks and is represented by epidote, chlorite, calcite, and pyrite. Epidote generally occurs in the ground mass replacing the plagioclase and ferromagnesian minerals and chlorite have developed after hornblende and biotite. Calcite generally occurs as open space fillings. At places, intense epidotization is observed replacing the whole rock [23]. The propylitic alteration zones are the most widespread alteration footprints that commonly surround phyllic alteration zones in the porphyry deposits of western Chagai Arc [24, 31].

4. Discussion

All the porphyry copper prospects of Chagai Arc are associated with tonalitic to dacitic porphyritic rocks except Missi prospect, which is associated with granodioritic rocks. Hydrothermal alteration has been developed in concentric zonal patterns. The intensity of alteration generally decreased from the center in each case as proposed by Lowell and Guilbert [2].

The presence of porphyritic texture in the host rocks of all the porphyry Cu settings suggests that they have gone through two different physico-chemical environments during their course of emplacement. In deeper environments and at high P, T conditions, larger crystals formed and when the partially crystallized magma further rose up and reached relatively in the shallower environments having low T, P conditions, smaller crystals of the groundmass were formed and consequently a porphyritic texture was developed contemporaneously due to decrease in confining pressure over the magma, a vapor phase was released as hydrothermal solution, which is responsible for hydrothermal alteration in the porphyry systems.

The Rose [32] proposed that hydrothermal alteration is generally controlled by the temperature and HCl/KCl ratio in the brines. The alteration in the porphyry copper system is mainly metasomatic/hydrothermal in nature and is in accordance with the mesothermal zone of Lindgren [33] but more recent studies report a wide range of fluid-inclusion homogenization temperatures (200°–760°C) from porphyry copper deposits [34, 35]. Bodnar [36] has reported magmatic fluid inclusions of high-salinity and halite bearing, which homogenized at a relatively high temperature (> 500°–600°C). Therefore, it is suggested that the alteration at the central zones might have been caused by primary magmatic, highly saline, Na, K, Ca, Cl, brines at about 500-750°C, while the alteration in the outer zone may be produced by the circulation of less saline meteoric water at about 250°C [37, 38]. Cooke et al. [39, 40], Richards [41], Sinclair [42], Sillitoe [43]; Sillitoe and Perelló [44] have suggested magmatic-hydrothermal fluids generated from subduction-related calc-alkaline igneous rocks for the formation of porphyry copper deposits. The source of metals in porphyry deposits is magmatic as evidenced by magmatic fluid inclusions found in the early quartz in veins [36].

The source of metal of porphyry copper setting of the Chagai belt may be the upper basaltic part of the subducted oceanic lithosphere in which massive copper sulphide mineralization, was emplaced during oceanic spreading along Mid-Tethyan ridge and subsequently partial melting of the above basalt along with massive copper sulphides, later on during the subduction under the southern margin of Afghan microcontinent produced calc-alkaline magmatism and associated porphyry copper mineralization in the Chagai belt [45]. Small quantities of both copper and molybdenum might have been derived from the wedge of mantle overlying the Benioff zone [46, 47]. Richards et al. [48], and Richards [41] revealed a compositional distinction between productive intrusions of porphyry deposits with fractionated REE patterns and depletions of manganese, thorium, yttrium, and the heavy-REE compared to barren intrusions in the Andes arc. Similarly, Siddiqui et al. [15, 19], have noted enrichment of large ion lithophile elements (LILE) and light REE in the pre-ore host volcanic rocks associated with porphyry copper bearing plutons of the Chagai Arc. This reflects that larger amount of subduction-related fluids (enriched in LILE and LREE) had contributed to the sub-arc mantle source of porphyry copper bearing magmatic rocks.

5. Conclusion

Description of hydrothermal alteration in the foregoing pages reveals that almost all the prospects exhibit a similar pattern of hydrothermal alteration which is quite comparable with the model proposed by Lowell and Guilbert [2]. Out of the above porphyry copper settings, only Saindak, Reko Diq and to some extent Dasht-e- Kain have been explored and evaluated so far.

At Saindak, the proved copper ore reserves are about 412 million tons, at Reko Diq 2,200 million tons and at Dashte Kain estimated reserves are about 350 million tons. Other porphyry copper prospects of the Chagai Belt have smaller alteration areas than Saindak.

References

[1] Ahmad, M. U., Siddiqui, R. H., and Chaudhary, M. A. (1986). Geological and geochemical exploration and preliminary evaluation of Dasht-e-Kain porphyry copper molybdenum prospect Chagai District, Balochistan Pakistan, *Geological Survey. Pak., Rec.* V. **72**

[2] Lowell, J.D. and Guilbert, J.M., 1970, Lateral and vertical alteration mineralization zoning in porphyry ore deposits. *Econ. Geol.* **65**, pp. 373-408.

[3] Stoneley, R., 1974. Evolution of a continental margin bounding a farmer southern Tethys, In Burk, C. I. and Drake, C. L., (eds.), *The Geology of Continental Margins*. Springer, New York, 873-887.

[4] Stocklin, J., 1974. Possible ancient continental margin in Iran. In Burk, C. I. and Drake, C. L., (eds.). *The Geology of Continental Margins*. Springer, New York, 889-903.

[5] Sillitoe, R. H., 1978, Evolution of mineral prospect and suggestion for exploration in Balochistan, Pakistan. *U.N. Report,* unpublished.

[6] Dykstra J D, (1978). A geological study of Chagai Hills Balochistan, Pakistan using LANDSAT digital data. *Ph.D. Thesis, Dartmouth College*, Hanover, USA

[7] Arthurton, R. W., Ahmed, S.A. & Iqbal, S., 1979, Geological history of the Alem Reg-Mashki Chah area, Chagai District, Baluchistan. In: Farah, A. & De Jong, K.A. (eds.) "*GEODYNAMICS OF PAKISTAN*" Geol. Surv. Pakistan, p. 361.

[8] Farah, A., Lawrence, R. D. and DeJong. K. A., 1984. An overview of the tectonics of Pakistan, In Haq, B. U. and Milliman, J. D., (eds.), Marine geology and oceanography of Arabian Sea and coastal Pakistan. *Van Nostrand Reinhold Company*, New York, 161-176.

[9] Arthurton R S, Farah A, Ahmed W (1982) The Late Cretaceous-Cenozoic history of western Balochistan, Pakistan - the northern margin of the Makran subduction complex. In: Leggett JK (ed) Trench Fore-Arc Geology. *Geol. Soc. London Spec Publ,* **10**: 343-385

[10] Kazmi, A. H. and Jan, M. Q., 1997. Geology and Tectonics of Pakistan. *Graphics Publishers, Karachi*, Pakistan, 554.

[11] Siddiqui R H, 1996. Magmatic evolution of Chagai-Raskoh arc terrane and its implication for porphyry copper mineralization. *Geologica* **2**: 87-119

[12] Siddiqui R H, 2004. Crustal evolution of the Chagai-Raskoh arc terrane, Balochistan, Pakistan. *PhD Thesis, Centre of Excellence in Geology*, University of Peshawar, Pakistan

[13] Siddiqui R H, Khan W, and Haque M, 1986. Petrological & petrochemical studies of northcentral Chagai Belt and its tectonic implications. *Acta Mineralogica Pakistanica,* **2**:12-23

[14] Siddiqui R H, Khan M A, and Jan M Q, 2005. Petrogenesis of Eocene Lava flows from the Chagai Arc, Balochistan, Pakistan and its tectonic implications. *Journal of Himalayan Earth Sciences,* **38**:163-187

[15] Siddiqui R H, Khan M A, and Jan M Q, 2007. Geochemistry and Petrogenesis of Miocene Alkaline and Sub-alkaline volcanic rocks from the Chagai Arc, Balochistan, Pakistan and their Implications for the Emplacement of Porphyry Copper Deposits. *Journal of Himalayan Earth Sciences.* **40**: 1-23

[16] Siddiqui R H, Khan M A and Jan M Q, 2009. Petrogenesis of Plio-Pleistocene volcanic rocks from the Chagai Arc Balochistan. *Journal of Himalayan Earth Sciences*, **42**: 1-24

[17] Siddiqui R H, Khan M A, Jan M Q, and Brohi I A, 2010. Paleocene Tholeeitic volcanism and Oceanic Island arc affinities of the Chagai Arc, Balochistan, Pakistan. *SURJ,* **42** (1): 83-98

[18] Siddiqui R H, Jan M Q and Khan M A, 2012. Petrogenesis of Late Cretaceous Lava Flows from a Ceno-Tethyan Island Arc: The Raskoh Arc, Balochistan, Pakistan. *J Asian Earth Sci.* **59**: 24-28

[19] Siddiqui R H, Khan M A, Jan M Q, Kakar M I, Kerr A C, 2015. Geochemistry and petrogenesis of Oligocene volcaniclastic rocks from the Chagai Arc: implications for the emplacement of porphyry copper deposits. *Arab J Geosci.* **8 (10)**: 8655-8667

[20] Siddiqui, R. H. Jan, M. Q., Khan, M. A., Kakar M. I., Foden, J. D. 2017, Petrogenesis of the Late Cretaceous Tholeiitic Volcanism and Oceanic Island Arc Affinity of the Chagai Arc, Western Pakistan. *Acta Geologica Sinica* (English Edition). **91** No. 4, 1248–1263)

[21] Jones A G, ed., 1961. Reconnaissance Geology of Part of West Pakistan. *A Colombo Plan cooperative project,* Government of Canada, Toronto: 550p

[22] Sillitoe, R.H. and Khan, S.N., 1977, Geology of Saindak porphyry copper deposit, Pakistan. *Trans. Inst. Min Metall.* Sec. B. **86**, pp. 27-42

[23] Siddiqui, R.H., 1984. Petrographic and ore microscopic study of Dashte Kain porphyry copper/molybdenum prospect Chagai District, Baluchistan, Pakistan. *Inf. Rel. Geological Survey of Pakistan* **213**, 26p.

[24] Razique A, Tosdal R M, and Creaser R A (2014) Temporal Evolution of the Western Porphyry Cu-Au Systems at Reko Diq, Balochistan, Western Pakistan. *Eco. Geol.* **109 (7)**: 2003-202

[25] Khan, F. and Ahmed, W., 1982, Reconnaissance geology of Reko Diq porphyry copper prospect, Chagai District, Baluchistan. *Geol, Surv. Pakistan, Unpublished report*, 9p.

[26] Bhutta, A. M., 2004, Porphyry type systems-configuration and distribution in the Chagai metallogenic belt with a reference to the Reko Diq Copper deposit, Balochistan, Pakistan. *Geol. Bull. Univ. Peshawar*, **37**, 179-189.

[27] Khan, A.L., 1975. A preliminary report on the discovery of sulphide mineralization in Durbanchah and Mashki Chah areas Chagai District, Baluchistan. *Inf. Rel. Geol. Surv.* Pakistan pp. 90, 6p.

[28] Sillitoe, R.H., 1975, Metallogenic evolution of a collisional mountain belt of Pakistan. *Rec. Geol. Surv. Pakistan*. **34**. 16p.

[29] Siddiqui R H, Hussain S A, and Haque M (1987) Geology and petrography of Eocene mafic lavas of Chagai island arc, Balochistan, Pakistan. *Acta Mineralolgica Pakistanica* **3**: 123-128

[30] Khan, F. and Ahmed, W., 1981. Geological map of Reko Diq Quadrangle, Chagai District, Balochistan, Pakistan. Geol. Surv. Pakistan, Quetta.

[31] Perello J, Razique A, Schloderer J, Rehman A U (2008) The Chagai Porphyry Copper Belt, Baluchistan Province, Pakistan. *Eco. Geol.* **103**:1583-1612

[32] Rose, S.W., 1970, Zonal relation of wall rock alteration and sulphide distribution at prophyry copper deposits. *Econ. Geol.* **65**, pp. 920-36.

[33] Lindgren, W., 1933, *Mineral Deposits*, McGraw Hill, New York.

[34] Klemm, L.M., Pettke, T., and Heinrich, C.A., 2007. Hydrothermal evolution of the El Teniente deposit, Chile—Porphyry Cu-Mo ore deposition from low-salinity magmatic fluids: *Economic Geology*, **102**, 1021–1045.

[35] Ulrich, T., Gunther, D., and Heinrich, C.A., 2001. The evolution of a porphyry Cu-Au deposit, based on LA-ICPMS analysis of fluid inclusions: Bajo de la Alumbrera, Argentina: *Economic Geology*, **96**, p. 1743–1774.

[36] Bodnar, R.J., 1995, Fluid inclusion evidence for a magmatic source of metals in porphyry copper deposits, in Thompson, J.F.H., ed., Magmas, fluids, and ore deposits: *Mineralogical Association of Canada,* Short Course. **23**, p.139–152.

[37] Sheppard, S.M.F., 1981, Identification of the origin of ore-forming solutions by the use of stable isotopes, In: Mitchel, A.H.G. & Garson, M.S. (eds.) *MINERAL DEPOSITS AND GLOBAL TECTONIC SETTINGS*. Academic Press, London. 405p.

[38] Henley, R.W. and McNabbe, A., 1978, Magmatic Vapour Plumes and groundwater interaction in porphyry copper emplacement. *Econ. Geol*. **73**, pp. 1-20.

[39] Cooke, D.R., Heithersay, P.S., Wolfe, R., and Calderon, A.L., 1998, Australian and western Pacific Porphyry Cu-Au deposits: *Journal of Australian Geology and Geophysics*. **17**, no. 4, p. 97–104. Accessed at: http://www.ga.gov.au/about/corporate/ga_authors/ajagg_17_4 /DR_Cooke_etal_ajagg174.jsp

[40] Cooke, D.R., Hollings, P., and Walshe, J.L., 2005, Giant porphyry deposits—Characteristics, distribution, and tectonic controls*: Economic Geology*. **100**, p. 801–818.

[41] Richards, J.P., 2003, Tectono-magmatic precursors for geophysical data over a copper gold porphyry Cu-(Mo-Au) deposit formation: *Economic Geology*. **96** p. 1419–1431.

[42] Sinclair, W.D., 2007. Porphyry deposits, in Goodfellow, W.D., ed., Mineral deposits of Canada: *Geological Association of Canada*, Special Publication. **5**, p. 223–243.

[43] Sillitoe, R.H., 1998, Epochs of intrusion-related copper mineralization in the Andes: *Journal of South American Earth Sciences*. **1**, p. 89–108.

[44] Sillitoe, R.H., and Perelló, J., 2005, Andean copper province—Tectonomagmatic settings, deposit types, metallogeny, exploration, and discovery: *Society of Economic Geologists, Economic Geology 100th Anniversary Volume*, p. 845–890.

[45] Sillitoe, R.H., 1972. A plate tectonic model for the origin of porphyry copper deposits, *Econ. Geol.* **67**, pp. 189-97.

[46] Oxburg, E.R. and Turcotte, D.L., 1970, Thermal structure of island areas, *Bull. Geol. Soc. Amer.* **81**, pp. 1665-88.

[47] Jenson, M.L., 1970, Province of cordillerian intrusives and associated metals. *Econ. Geol*. **66,** pp. 34.42.

[48] Richards, J.P., Boyce, A. J., and Pringle, M.S., 2001, Geological evolution of the Escondida area, northern Chile. A model for spatial and temporal localization of porphyry Cu mineralization: *Economic Geology*. **96**, p. 271–305.

[49] Britzman, L., 1979, Fission track ages of intrusives of Chagai District Baluchistan Pakistan. *MA. Dissertation Dartmouth College*, Hanover New Hampshire USA, unpublished.

Efficiency and Cost Analysis of Power Sources in Impressed Current Cathodic Protection System for Corrosion Prevention in Buried Pipelines of Balochistan, Pakistan

Muhammad Saddam Khan[1], Faizullah Khan Kakar[2], Surat Khan[2], Syed Owais Athar[2]

[1]Department of Electrical Engineering, BUITEMS, Quetta.
[2]Faculty of Information and Communication Technology, BUITEMS, Quetta.

E-mail: engr.saddam@outlook.com

Abstract. Corrosion is the physical deterioration of metals, including buried metal structures and gas pipelines, resulting from chemical reactions between metals and the surrounding environment. Cathodic protection of these underground metal structures & pipelines can be achieved either by the Impressed Current method or by the sacrificial anodes method. This research paper focuses on the corrosion prevention of buried metal structures in general and gas pipelines in particular, by the novel method of Impressed Current Cathodic Protection (ICCP) which uses electrical current from an external DC power source for its operation. Three major DC power sources i.e. Transformer Rectifier, Thermoelectric Generator and the Solar System are selected for this purpose and are projected over a period of time in a pre-designed ICCP for high transmission underground gas pipeline in Balochistan province of Pakistan. The efficiencies of these three power sources are then analysed and compared according to the climate effects, the versatility effects, the intensity of power output, their operational limitations and their initial, running and maintenance costs. The measured results will aid the selection of efficient and robust DC power source for ICCP design and will contribute to mitigating and controlling the corrosion rate in underground pipelines. Based on analysis of the measured results for the three available DC sources, the Solar System was found as the most efficient DC power source for Impressed Current Cathodic Protection of buried gas pipelines. The selection of efficient corrosion protection system will result in a smooth flow of oil and gas products through these buried pipelines which may otherwise lead to huge monetary and accidental losses when get corroded.
Keywords. Corrosion, solar system, impressed current cathodic protection

1. Introduction

The metals are extracted from their respective ores and are then structured in a reformed state after adding an enormous amount of heat into it. When these structured metals are exposed to the environment, the environmental contaminants such as oxygen, H_2O moisture, etc. react with these structured metals. Due to this reaction, the structured metals lose their energy and returns to their original state [1]. Uniform corrosion assaults uniformly at all areas of metals at the same rate, whereas localized corrosion assault some areas of metals at a different rate and in pitting corrosion pits are formed in the material [2].

1.1 Corrosion in Underground Pipelines

In case of underground iron pipelines, corrosion is a spontaneous or electrochemical process in which current leaves a structure at anode site and penetrates the structure at cathode site after passing through an electrolyte in which it is placed.

To analyze the mechanism of corrosion in iron it is observed that oxidation reaction take place at anode causing Fe^{2+} ions to release from the anode and OH^- ions are released from the cathode by a reduction reaction. The Fe^{2+} and OH^- ions combine to form a white precipitate of solid iron + hydroxide [1].

IOP Conf. Series: Materials Science and Engineering **414** (2018) 012034 doi:10.1088/1757-899X/414/1/012034

Equation (1), (2) and (3) shows oxidation reaction, reduction reaction and their net equation respectively.

$$Fe(s) \rightarrow Fe^{2+}(aq) + 2e^- \quad (1)$$

$$O_2(g) + 2H_2O(l) + 4e^- \rightarrow 4OH^- \quad (2)$$

$$Fe^{2+} + 2OH^- \rightarrow Fe(OH)_2(s) \quad (3)$$

This $Fe(OH)_2$ further undergoes oxidation in the presence of moisture and oxygen leading to the formation of brown solid called rust (see Equation (4)).

$$4Fe(OH)_2(s) + O_2(g)2H_2O(l) \rightarrow Fe_2O_3.nH_2O(s)(rust) \quad (4)$$

1.2 Effects of Corrosion on Underground Pipelines

Transportation of petroleum products and natural gas over long distances through underground and submerged pipelines is considered as most efficient, reliable and cost-efficient way of transporting these products. But these underground pipelines are vulnerable to mechanical damage, third party factors and corrosion.

The occurrence of corrosion in these pipelines can be understood by its adverse effects on industries and consumers alike.

- Corrosion may lead to the plant shutdowns and shutdown of process plants.
- Corrosion may cause loss of products and results in leakage of liquid from transportation pipelines, with a capacity to generate severe accidents and hazards. Replacement of corroded pipes causes huge monetary losses to companies and national economies [1].

1.3 Corrosion Prevention Techniques in Underground Pipelines

Methods widely used for the prevention of corrosion are:

- Use of material that is corrosion resistive such as stainless or plastic
- By changing the environment such as electrolyte soil
- By using a coating that isolates the structure electrically from the electrolyte e.g. epoxy paints, coal tar enamel and polyethene coating.
- Use of cathodic protection techniques [2]

The first line of defence against corrosion protection in underground and submerged structures (pipelines) is insulation coating which can either be of coal tar enamel, polyethene, tape or asphalt coating. Up to 50% – 99% of corrosion protection is achieved if the proper coating is ensured. But due to third party and mechanical damages during commissioning of these pipelines, this percentage reduces and achieving 100% corrosion protection through the primary line of defense is difficult. Therefore, the remaining corrosion protection is accomplished by a well-designed cathodic protection system also called secondary corrosion control techniques [3, 4]. In cathodic protection techniques, a direct current by means of an external source is utilized to make the structure (to be protected) at one potential and eliminating the anode and cathode site. The principal utilization of cathodic protection was traced in 1824 by Sir Humphrey Davy [5].

1.4 Cathodic Protection Techniques

Cathodic protection techniques are mainly classified into two types:

a. Sacrificial anode cathodic protection (SACP) method (Galvanic Anode Method).
b. Impressed Current Cathodic Protection (ICCP) method.

The choice of selecting these two methods for cathodic protection design depends upon the area of the structure to be protected (current requirement) and electrolyte (soil) resistivity. In both methods, the fundamental components are the anode, the cathode, DC power source (in ICCP an external power source), and electrolyte. The electrolyte is an electrically conductive solution from which electrons flow

IOP Conf. Series: Materials Science and Engineering **414** (2018) 012034 doi:10.1088/1757-899X/414/1/012034

from anode to cathode. The type of electrolyte differs in context of moisture content, oxygen level, and soil resistivity level.

1.4.1. Sacrificial anode cathodic protection system

The

Figure 1 illustrates the utilization of a metal *(anode),* which is normally more electro-negative *(magnesium, zinc and aluminium etc.)* than the metal *(cathode)* being protected *(mild steel or carbon steel etc.).* Current streams from the metal that is more electrically negative to the metal that is more electrically positive. This current stream brings about a fast utilization of the anode. In SACP the voltage difference is limited to anode and cathode, usually 1 volt or less, contingent upon the anode material and the environment. Oxidation and reduction reaction takes place at anode and cathode respectively [6].

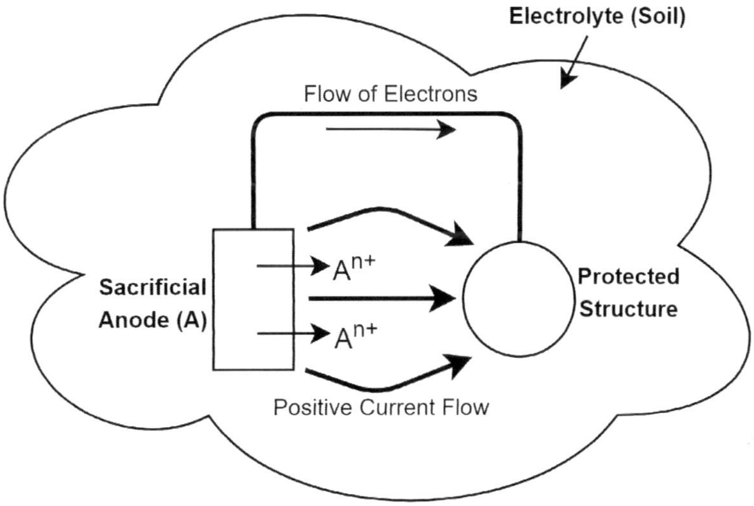

Figure 1. Schematic of Sacrificial Anode Cathodic Protection Method

1.4.2. The impressed Current Cathodic Protection system

This system is comprised of an external DC power source, an auxiliary or impressed current electrode, the corrosive solution and the structure to be protected. The DC power source delivers positive current from the impressed current electrode on to the structure (being protected) through the corrosive solution. The structure is cathodically polarized (its potential is lowered) and the positive current returns through the circuit to the DC power supply, as shown in Figure 2 [6].

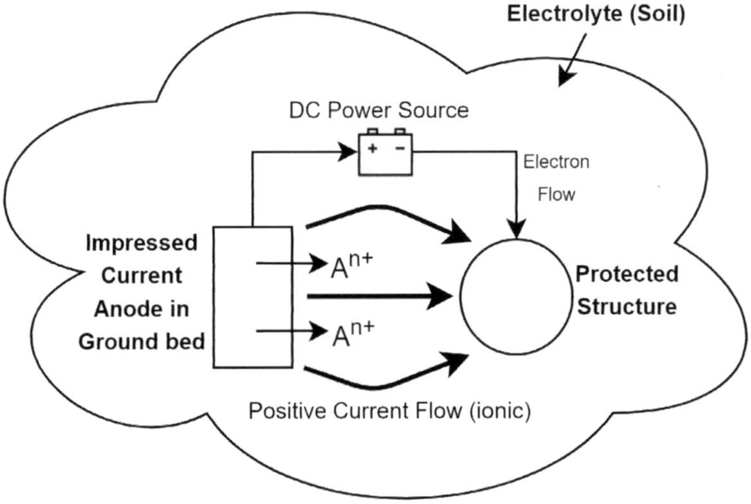

Figure 2. Schematic of Impressed Current Cathodic Protection Method

The utilization of the external power source allows an impressed current cathodic protection system to generate comparatively higher current output and has a tendency to use larger voltage differences than any sacrificial anode system. The larger voltage available with impressed current permits remote anode location, delivering more efficient current distribution pattern along the protected structure. Specially in low conductive environments, these larger voltages are found useful [1]. The selection of right type of external DC power source is crucial in ICCP which has a tendency to be versatile, robust, and effective and cost-efficient [6].

2. State of the Art

The problem of corrosion takes more serious attention when underground metallic pipelines are considered. The problem of corrosion in buried gas pipelines and their remedial action through impressed current cathodic protection has been studied by various researchers. The research studied in [7] concluded that the Energy crisis and corrosion are two major problems for the 21st century and proposes Thermoelectric Device as a promising solution for addressing those two problems. The application of Rectifier DC power source is presented in a patent paper [8]. Another research study [1] is based on prevention of corrosion in iron pipelines by utilizing the novel method of Impressed Current Cathodic Protection. Instead of deploying a conventional AC power source, solar energy is utilized to drive the system. Adly G S [2] concluded that to minimize the risk of corrosion and overprotection is caused by an increase in DC current in underground structures. An automatic regulating cathodic protection system is hence proposed. This is to overcome problems associated with conventional cathodic protection system. The research [3] analyzes and calculated the parameters of PV solar system for ICCP system to protect tanks from corrosion.

None of the researchers have discussed, analyzed and compared more than one DC sources after implementing it over a time period by means of real-time in a pre-designed ICCP system for high transmission underground gas pipelines. The objective of this research work is to analyze and compare climate effects, versatility effects, intensity of power output, operation limitation, as well as initial running and maintenance cost of three different DC sources by implementing it in a pre-designed ICCP system over a period of time at high transmission gas pipelines.

3. Methodology

To achieve the research objective, a 12-inch diameter underground high transmission natural gas pipeline made up of mild steel material, at two different locations (Sibi and Bolan) differing in resistivity of soil and climate in the Balochistan province of Pakistan, was chosen. The pipeline has coal tar enamel coating having coating efficiency of 75% analyzed after closed interval potential survey at both locations. Three number of trials have been performed at each location in three different months of winter and summer seasons respectively. The ground bed efficiency at both locations was observed as 50% to 60% as per ground bed resistance and commissioning year.

Three DC sources i.e. thermoelectric generator (TEG), transformer rectifier (TR) and a solar PV system are implemented and analyzed in a pre-determined impressed current cathodic protection system at both locations, one at a time respectively. The results for Bolan and Sibi sites are given in table 1 and table 2 respectively.

4. Analysis and Results

4.1 Technical Analysis

4.1.1. Analysis of the climate effects

The climate effects comprise of two parameters, i.e. specific resistivity of electrolyte and the temperature.

While analysing the climate effect of all three DC sources selected and their performance in Quetta Bolan region, the constant variable is coating efficiency, ground bed efficiency, soil resistivity and capacity of the DC power source (220 Watts), whereas the dynamic variable is temperature.

During winter: it has been observed that during the winter season while considering the constant and dynamic variable conditions at a similar location, the three DC sources energizes the ground bed at different levels. It is noticeable that the power drawn from solar system energizes the anode bed greater than that of TEG and TR.

During Summer season: it has been observed that during summer season where the weather conditions become dry and the moisture content in soil reduces, the output current and voltage ratio drew from DC source also changes. More voltage is required to carry out current and to maintain pipe to soil potential under specified protection level defined by National Association for Corrosion Engineers, NACE standard RP-01 i.e. at least 0.85 Volts, measured from CU CUSO4 reference electrode. It is noticeable that more power is delivered by solar system during the summer season. The result of a pipe to soil potential is also depicted in table 1.

While analysing the climate effect of all three DC sources selected and their performance in SIBI region, the constant variables are coating efficiency, ground bed efficiency, soil resistivity and capacity of the DC power source (220 Watts), whereas the dynamic variable is Temperature. Values are mentioned in table 2. In high resistive soil more, voltage is needed whereas in low resistive soil more current is needed to maintain the pipe to soil potential level.

During winter: The power delivered by individual DC source as well as the potential level of pipe to soil can be clearly analyzed from the values of the obtained results, mentioned in table 2. Solar PV systems can be seen as delivering more power subsequently energizing the available anode bed to ensure the Pipe to Soil Potential level greater than that of TEG and TR.

During summer: During the summer when the temperature rises the TEG loses its power by 0.8W / C° above 25 C°. But due to increase in solar irradiations during summer solar PV panels will deliver more power. The output power, voltage and current values to utilize the ground bed anodes and PSP readings are shown in table 2. One can easily compare the output power of all three DC sources and the PSP values of the pipeline.

Table 1: Test Results of Location 1 (Bolan)

QUETTA BOLAN SECTION (WINTER SEASON)																		
Specified DC Power Source	**Soil Type**	**Location**	**Soil Resistivity (Ωcm)**	**Month**	**Ambient Temperature (Min - Max) (°C)**	**Output Voltage**	**Output Current**	**Power Output (Watt)**	**Grounbed Efficiency (%)**	**Coating Efficiency (%)**	**PSP using CU CUSO4 Reference Electrode (Volt)**			**Intensity of the Power output in terms of effectiveness over distance (kilo-meter)**			**Versatility Effect After connecting to a 300ft Microwave System (kilo-meter)**	
											Down-Stream End	**Feeding Point**	**Up-stream End**	**Down-Stream End**	**Up-Stream End**		**Down-Stream End**	**Up-Stream End**
TEG	Rocky Gravel	Spezand, Balochistan	10,000 Ωcm	Dec	0 to 9	13 V	5A	65 W	50 to 60 %	75%	1.00 V	1.52 V	1.00 V	6 km	6 km	**Moist Weather Conditions**	4.5 km	4.5 km
TR				Jan	1 to 8	18 V	8A	144 W	50 to 60 %	75%	1.04 V	1.55 V	1.04 V	7.5 km	7.5 km		6.5 km	6.5 km
PV				Feb	1 to 11	20 V	10.5A	210 W	50 to 60 %	75%	1.10 V	1.57 V	1.10 V	11 km	11 km		10.5 km	10.5 km
QUETTA BOLAN SECTION (SUMMER SEASON)																		
Specified DC Power Source	**Soil Type**	**Location**	**Soil Resistivity (Ωcm)**	**Month**	**Ambient Temperature (Min - Max) (°C)**	**Output Voltage**	**Output Current**	**Power Output (Watt)**	**Groundbed Efficiency (%)**	**Coating Efficiency (%)**	**PSP using CU CUSO4 Reference Electrode (Volt)**			**Intensity of the Power output in terms of effectiveness over distance (kilo-meter)**			**Versatility Effect After connecting to a 300ft Microwave System (kilo-meter)**	
											Down-Stream End	**Feeding Point**	**Up-stream End**	**Down-Stream End**	**Up-Stream End**		**Down-Stream End**	**Up-Stream End**
TEG	Rocky Gravel	Spezand, Balochistan	10,000 Ωcm	June	13 to 35	16 V	3.8A	60.8 W	50 to 60 %	75%	0.98 V	1.49 V	0.98 V	5 km	5 km	**Dry Weather Conditions**	3 km	3 km
TR				July	18 to 39	21 V	6.3A	132.3 W	50 to 60 %	75%	1.00 V	1.52 V	1.00 V	7 km	7 km		5.5 km	5.5 km
PV				Aug	21 to 40	23 V	8.9A	204.7 W	50 to 60 %	75%	1.05 V	1.55 V	1.05 V	10 km	10 km		9 km	9 km

Note: TEG Stands for 'Thermoelectric Generator', TR stands for 'Transformer Rectifier', PV stands for 'Photovoltaic System', PSP stands for 'Pipe to Soil Potential

IOP Conf. Series: Materials Science and Engineering **414** (2018) 012034 doi:10.1088/1757-899X/414/1/012034

Table 2: Test Results of Location 2 (Sibi)

SIBI SECTION (WINTER SEASON)																		
Specified DC Power Source	**Soil Type**	**Location**	**Soil Resistivity (Ωcm)**	**Month**	**Ambient Temperature (Min - Max) (°C)**	**Output Voltage**	**Output Current**	**Power Output (Watt)**	**Groundbed Efficiency (%)**	**Coating Efficiency (%)**	**PSP using CU CUSO4 Reference Electrode (Volt)**			**intensity of the Power output in terms of effectiveness over distance (kilo-meter)**			**Versatility Effect After connecting to a 300ft Microwave System (kilo-meter)**	
											Down-Stream End	**Feeding Point**	**Up-stream End**	**Down-Stream End**	**Up-Stream End**		**Down-Stream End**	**Up-Stream End**
TEG	Hard Clay	Shohri Balochistan	2500 Ωcm	Dec	8 to 25	5.5 V	14A	77 W	50 to 60 %	75%	0.89V	1.38V	0.89V	7 km	7 km	**Moist Weather Conditions**	6 km	6 km
TR				Jan	4 to 17	7.5 V	19.9A	149.25W	50 to 60 %	75%	0.94V	1.45V	0.94V	8.5 km	8.5 km		7.5 km	7.5 km
PV				Feb	9 to 23	8.5 V	25A	212.5W	50 to 60 %	75%	1.03V	1.53V	1.03V	12 km	12 km		11.5 km	11.5 km
SIBI SECTION (SUMMER SEASON)																		
Specified DC Power Source	**Soil Type**	**Location**	**Soil Resistivity (Ωcm)**	**Month**	**Ambient Temperature (Min - Max) (°C)**	**Output Voltage**	**Output Current**	**Power Output (Watt)**	**Groundbed Efficiency (%)**	**Coating Efficiency (%)**	**PSP using CU CUSO4 Reference Electrode (Volt)**			**Intensity of the Power output in terms of effectiveness over distance (kilo-meter)**			**Versatility Effect After connecting to a 300ft Microwave System (kilo-meter)**	
											Down-Stream End	**Feeding Point**	**Up-Stream End**	**Down-Stream End**	**Up-Stream End**		**Down-Stream End**	**Up-Stream End**
TEG	Hard Clay	Shohri Balochistan	2500 Ωcm	June	28 to 41	4V	17A	68W	50 to 60 %	75%	0.88V	1.31 V	0.88 V	5.5 km	5.5 km	**Dry Weather Conditions**	4 km	4 km
TR				July	30 to 47	5.9V	23A	135.7W	50 to 60 %	75%	0.90 V	1.42 V	0.90V	8 km	8 km		7 km	7 km
PV				Aug	31 to 48	7.4V	27A	199.8W	50 to 60 %	75%	1.00 V	1.56V	1.00V	11.5 km	11.5 km		11 km	11 km

Note: TEG Stands for 'Thermoelectric Generator', TR stands for 'Transformer Rectifier', PV stands for 'Photovoltaic System', PSP stands for 'Pipe to Soil Potential

4.1.2. Analysis of Versatility effects

The Versatility effects mean that how much larger structure of the ICCP system can handle when relating to three different DC sources having similar power capacity i.e. 220 W. The 12-inch diameter underground pipeline is coupled with 300 feet high iron made a tower for microwave system at both selected locations. The PSP of the pipeline is recorded at feeding point, upstream end and at downstream ends of the pipeline. In the light of results recorded the PV Solar system takes clear-cut advantage over TEG and TR in context of versatility effects at both locations.

4.1.3. Analysis of intensity of the power output

This effect means that up to how long the magnitude of the output reaches, that is delivered by the DC source in ICCP to energize the pipeline potential. The analyses were done by integrating three DC sources respectively, one at a time in ICCP system at both said locations and then recorded the PSP of pipeline up to several kilometres in the up and downstream ends. The solar system has proved its efficient magnitude in both soil types by maintaining the PSP up to several kilometres greater than that of TEG and TR. Analyzed results are depicted in table 1 and table 2.

4.2. Cost Analysis

The cost comparison of three DC power sources for ICCP was carried out using present worth or net present value (NPV) analysis. For economic analysis, constant discount rate, exchange rate, annual maintenance and operation costs throughout the useful life of the external DC power sources were assumed. The following parameters were considered as common for all the three DC power sources: Useful life =20 Years, Discount rate = 6%, USD to PKR exchange rate as on 21/2/2018 = Rs110.76

4.2.1. Thermoelectric generator

Initial cost = $3160
Maintenance cost = $412 every five years
Operational cost = $17526 per year
Salvage value = $472 at the end of 20 years.

The economic formula for calculating present worth or present value (P) of a single payment and equal payment series are given in Equations (5) and (6) respectively [9].

$$\mathrm{P} = \mathrm{F}\,(1+i)^{-N} \tag{5}$$

$$\mathrm{P} = \mathrm{A}\left[\frac{(1+i)^{N}-1}{i\,(1+i)^{N}}\right] \tag{6}$$

Where 'P' is the present worth (PW) or present value, 'i' is the discount rate or interest rate, 'N' is the number of discounting periods and 'A' is the annual payment. Figure 3 shows the cash flow for the thermoelectric generator as per calculation.

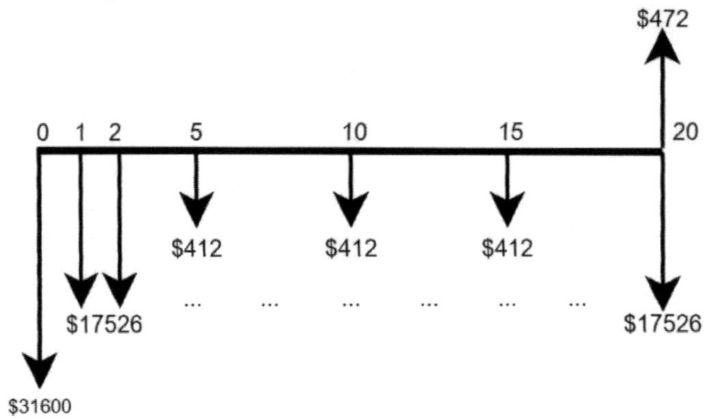

Figure 3. Cash Flow Diagram of Thermoelectric Generator

$$PW = 31600 + \frac{412}{(1+0.06)^{5}} + \frac{412}{(1+0.06)^{10}} + \frac{412}{(1+0.06)^{15}} + 17526\left[\frac{(1+0.06)^{20}-1}{(0.06)(1+0.06)^{20}}\right] - \frac{472}{(1+0.06)^{20}} \tag{7}$$

$$PW = \$\,233184.5 \tag{8}$$

4.2.2. Transformer Rectifier

Initial cost = $1120
Maintenance Cost = $73/year
Operational cost = $ 242/year
Annual O&M Cost = $242 + $73 = $315
Salvage value = $187 at the end of 10 Years

The cash flow diagram for Transformer rectifier as per computation is shown in Figure 4.

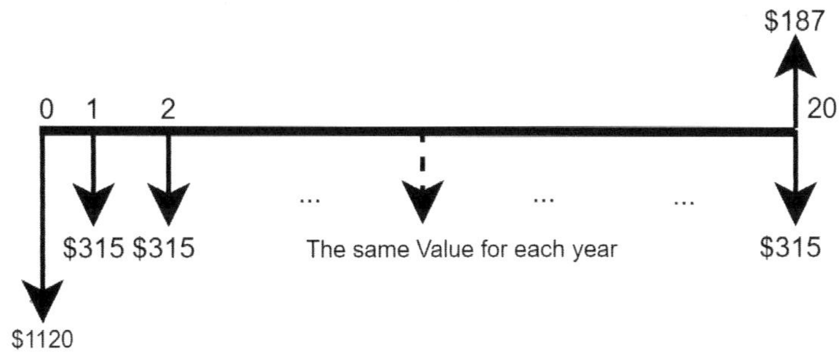

Figure 4. Cash Flow Diagram for Transformer Rectifier.

$$PW = 1120 + 315\left[\frac{(1+0.06)^{20}-1}{(0.06)(1+0.06)^{20}}\right] - \frac{187}{(1+0.06)^{20}} \quad (9)$$

$$PW = \$\,4674.74 \quad (10)$$

4.2.3. Solar PV System

Initial cost = $8902
Maintenance cost = $7327 after 10 years
Operational cost = $0
Salvage value = $283 every 10 years.

The cash flow diagram for solar PV system after the calculation is shown in Figure 5.

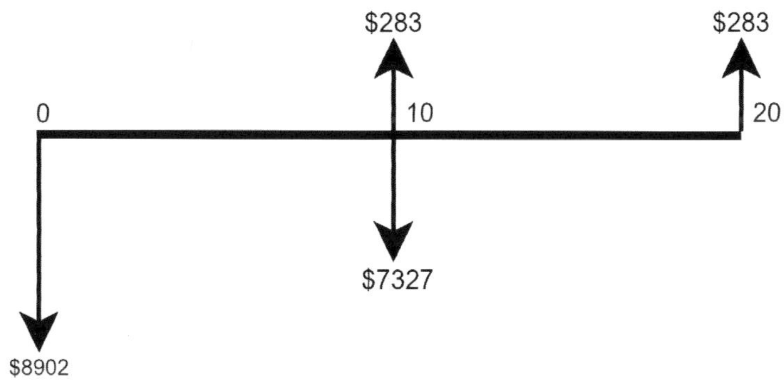

Figure 5. Cash Flow Diagram for Solar PV System.

$$PW = 8902 + \frac{7327}{(1+0.06)^{10}} - \frac{283}{(1+0.06)^{10}} - \frac{283}{(1+0.06)^{20}} \quad (11)$$

$$PW = \$\,12747.09 \quad (12)$$

5. Conclusion

Cathodic protection systems are used to protect buried and submerged metallic structures from corrosion. Impressed Current Cathodic Protection (ICCP) system is used for protecting larger metallic structures where the current requirement is high and external DC source is required. The focus of this research work is to mitigate the rate of corrosion in underground metallic pipelines in general and natural gas pipelines, by integrating three different DC sources for a specified period of time at two different locations in the province of Balochistan with an ICCP system. The selected locations are distinct from each other according to climate & soil resistivity. The efficiency of each DC source has been thoroughly analyzed after comparing their technical and cost efficiencies. Based on measured results the technical efficiencies of the Solar PV system was found a promising solution in mitigating the rate of corrosion, since it has tendency to deliver more constant DC power to an ICCP system both in hot and cold temperatures as well as in high and low resistivity environments, as compared to other two DC Sources used i.e. Thermoelectric generator and transformer rectifier. Further, unlike TEG and TR the solar PV system is not dependent on commercial utility supply, for its operation, which make it more preferable, especially in areas where there is no availability of commercial utility supply but enriched solar irradiations. However, the computed results found that Transformer rectifier is an economical DC source for an ICCP system compared with that of the Thermoelectric generator and Solar PV systems. The calculation and measured results are quite straightforward based on real-time data. The proposed research will aid companies and industries in selecting the right type of external DC source while designing the ICCP system according to climate, versatility, output power intensity and cost efficiency for corrosion protection of underground pipelines.

References

[1] B S S, Pranav K P and Raj R. K 2014 Solar powered corrosion prevention in iron pipelines using Impressed Current Cathodic Protection *2014 14th Int. Conf. Environ. Electr. Eng. EEEIC 2014 - Conf. Proc.* 372–7
[2] Adly G S, Anis W R and Hafez I M 2017 Design Of Photovoltaic Powered Cathodic Protection Systems *2017* International Journal fo Scientific & Technology Research **6** 246-253
[3] Janowski M and Wantuch A 2016 ICCP cathodic protection of tanks with photovoltaic power supply *2016 E3S Web of Conferences 10* **29** 1–4
[4] Al-Faiz M Z 2012 Cathodic Protection Remote Monitoring Based on Wireless Sensor Network *Wirel. Sens. Netw.* **4** 226–33
[5] Ezekiel Enterprises, LLC Ee C 2015 Introduction to Electrical Design for Cathodic Protection Systems
[6] Bashi S M, Mailah N F and Radzi M A M 2003 Cathodic protection system *Power Engineering Conference, 2003. PECon 2003. Proceedings. National* pp 366–70
[7] He W, Zhang G, Zhang X, Ji J, Li G and Zhao X 2015 Recent development and application of thermoelectric generator and cooler *Appl. Energy* **143** 1–25
[8] Benham R A 2007 Cathodic protection system for metallic structures
[9] Park C S 2001 *Contemporary engineering economics: A Canadian perspective* (Addison Wesley Longman)

IOP Conf. Series: Materials Science and Engineering **414** (2018) 012035 doi:10.1088/1757-899X/414/1/012035

Stability Analysis of Flood Bunds: A Study on Geotechnical Health Evaluation of Embankments

Sumayya Noshin[1], Mehtab Alam[2], Shamsher Sadiq[3], Saeedullah Jan Mandokhail[4]

[1]Graduate Student, Department of Civil Engineering, UET, Lahore, Pakistan
[2]Graduate Student, Institute of Mountain Hazards and Environment, Chinese Academy of Sciences, Chengdu, P.R China
[3]Graduate Student, Department of Civil and Environment Engineering, Hanyang University, Seoul, South Korea
[4]Assistant Professor, Department of Civil Engineering, BUITEMS, Quetta, Pakistan

Email: engr_mehtab@yahoo.com

Abstract. Flood constitutes one of the world's most serious environmental hazards. Flood bunds are the earthen hydraulic structures which are constructed along the River to control the Flood water to avoid damages to the infrastructures, crops, livestock and loss of human lives. Pakistan which lies in the Indus Basin has been facing severe threats and losses from the floods since histories. About 6807km length of flood embankment has been constructed to safeguard against the floods in the country. Punjab has been worst hit province by heavy floods and rains causing heavy loss. Geotechnical Evaluation is vital for proper functioning of such structures. In this study four flood bunds susceptible to potential embankment breaching during flood have been selected along river Chenab in district Muzaffar Garh Irrigation zone. Suite of analysis using GeoStudio software (SLOPE/W and SEEP/W for stability and seepage analysis respectively) has been performed by considering four different critical scenarios, (1) steady state at highest flood level (2) rapid drawdown from highest flood level (3) steady state at extreme condition with 3 feet free board (4) rapid drawdown from extreme condition with 3 feet free board. The safety of the flood bunds is evaluated in terms of River Embankment Breaching Vulnerability Index (REBVI), safe exit gradient and factor of safety against slope failure. Based on these calculated factors from the numerical analysis it is found that embankment at RD:25,29 are marginally safe, whereas at RD: 147, 157 are susceptible to piping failure. Therefore, it is recommended to make embankment impervious using cohesive material and to install cut-off walls or berms to lengthen the seepage path.
Keywords. Stability, Stability analysis, flood bunds, embankments

1. Introduction

EMBANKMENT / LEVEE / DYKE is an earthen bund extending generally in more or less parallel direction of river course and is designed to protect the area behind it on the country side from the over run of the flood water. Nature has blessed Pakistan with plenty of surface water which originates from Himalayans in the north and flows toward Arabian Sea through river Chenab and its tributaries. A number of barrages and head works have been constructed on different rivers, mostly during British rule in the Subcontinent. Major purpose of these structures was to divert the river water to the canals for irrigating the fertile lands of Chenab plane. These structures on the rivers cause the hindrance to the natural regime of the rivers which lead to the heavy floods, especially during monsoon season owing to heavy rainfall particularly in the upper Chenab basin. To control the floods and keep the river waters within limits earthen dykes have been constructed from time to time and again along the rivers in Pakistan. At present about 6803 km flood bunds are existing on different rivers in Pakistan out of which, 3332 km fall in Punjab [1]. Unfortunately, less attention has been paid to proper stability and seepage analysis based on extensive geotechnical investigations during planning and construction of these structures due to which the failure of flood bund occurs causing heavy loss to infrastructures, crops, livestock and human lives.

Punjab has been the worst-hit province due to rains and floods, where more than 1.7 million of the population is affected [1], [2].According to the NDMA, nearly 17,000 houses have been fully destroyed. Over 1,900 schools across the province were also damaged and remained non-operational during flood 2014 [1].

The geotechnical inspection and assessment of flood control embankments were studied by Dyer et al. [3] wherethey discussed that the failure was due to seasonal wetting and drying with potential cracking on the crest and side slopes. Morris et al. [4] suggested that reason of embankment failure can be broadly divided into two modes depending on where the failure generates. First mode is founding strata (settlement, sliding, seepage and piping and uplift pressure) and second mode is embankment structure (slope instability, internal seepage, erosion of outward face and toe). Mondal et al. [5] Investigated the function of geotechnical, geometrical and hydraulic properties on embankment breaching of Moyna drainage basin area using a group analysis based upon vulnerability index for the indicators. Failures in embankments of Irrigation canals in Sindh were overtopping, internal erosion, structural defects and piping and those were found out during study conducted by Bhanbhro et al. [6]. The recurrent history of overtopping and stability issues in the embankments provided the motivation for developing a better understanding of the behavior of this compacted fill under different loading and wetting conditions by Mountassir et al. [7]. Flood bunds can be protected using electrical resistance tomography (ERT) methods. Protecting these vulnerable slopes requires monitoring systems capable of identifying and alerting to asset operators changes in the internal conditions that precede failure [8]. Further, extensive literature regarding the flood bunds have been presented [9], [10].

In this paper, embankment stability is evaluated as a case study for the four embankment selected along Chenab river in district Muzzafar Garh, Punjab, Pakistan. Potential to embankment breaching is calculated in term of River Embankment Breaching Vulnerability Index (*REBVI*) suggested by Mondal et al. [5] and safe exit gradient using limit equilibrium program SEEP/W while factor against failure is evaluated using SLOPE/W (Geo Studio 2012) [11]. Remedial measures are recommended to avoid the embankment failure due to seepage.

2. SELECTION OF FLOOD BUNDS AND DESK STUDY FOR THE SELECTED SITES

2.1 Study Area

A study on four flood bunds of Chenab river was carried out to evaluate the health of flood bunds by investigating their geotechnical, geometrical and hydraulic properties to ascertain their strength against breaching during floods. The location map for selected flood bunds is given in Figure 1.

2.2 Field Survey and Subsurface investigations

For study purpose, four sites were selected on Chenab river which have potential for breaching, asshown in Figure 1. On two sites (RD 147+000 and RD 157 + 000) natural breaching was occurred during Flood 2014, while the other two sites (RD25+ 000 and RD29 +000) were remain safe in the vicinity of that breaching Flood Bunds.

IOP Conf. Series: Materials Science and Engineering **414** (2018) 012035 doi:10.1088/1757-899X/414/1/012035

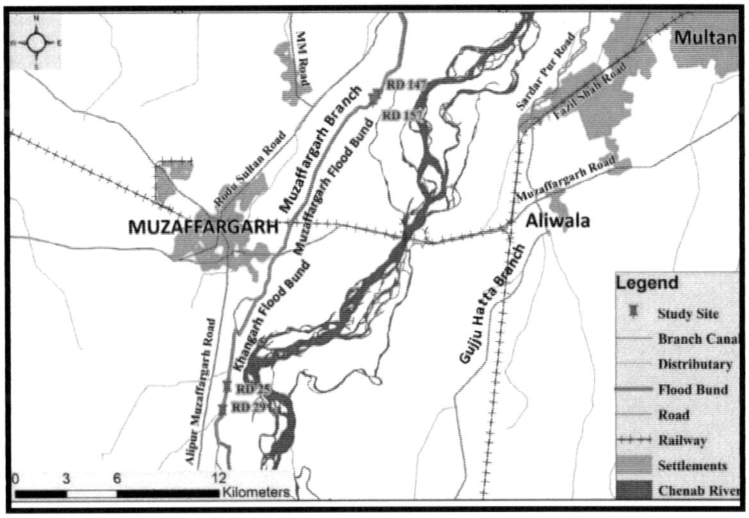

Figure 1. Location Plan of selected bund sites along Chenab river

Geotechnical investigations (SPTs and field permeability tests) were carried out. Disturbed and undisturbed soil samples were collected for laboratory testing. Based on direct shear, unconfined compression, atterberg limit, modified proctor and laboratory permeability (constant and falling head) tests the soil properties selected for embankment as well as foundation are shown in Figure 2.

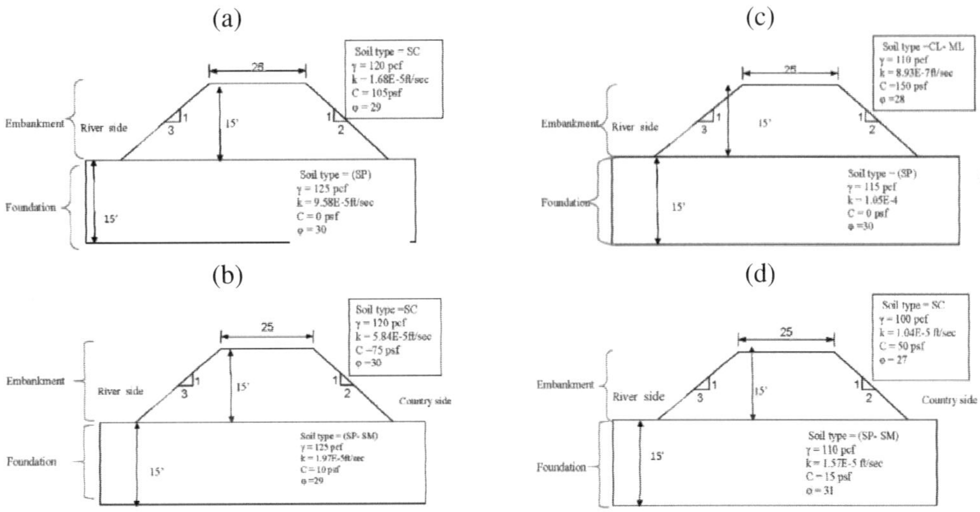

Figure 2. Selected embankment dimension and soil properties at R.D a) 25 b)29 c)147 d)157

3. Methodology Adopted for Health Evaluation of Selected Flood Bunds

3.1. Evaluation of River Embankment Breaching Vulnerability Index (REBVI)

A River Embankment Breaching Vulnerability Index (*REBVI*) was derived based on weightings of bank material to delineate the risk of vulnerability of the embankment. Transformation of absolute values into weighted values (W) for sites of flood bund is shown in Table 1.

Table 1: Methodological approach of River embankment breaching vulnerability index [5]

Transformation of Absolute values into Weighted value (W)									
Bank Materials and Geotechnical Analysis						Embankments Structure			Hydraulic pressure
Weight value	**Soil texture**	**Bulk density (g/cm^3)**	**Plasticity Index PI PI=LL-PL**	**Compressive strength (kg/cm^2)**	**Safety Factor**	**Top height (Bank Height**	**Base Width (m)**	**Bank Slope**	**Water height (m)**
0	loam	<1.3	>40(very high	>4 (hard)	0.2	>9	>20	<30	<4
1	Sandy clay	1.3-1.4	20-40(high)	2-4 (very stiff)	0.2-0.4	8-9	15-20	30-35	4-5
2	Clay loam	1.4-1.5	10-20 medium	1-2 (stiff)	0.4-0.6	7-8	10-15	35-40	5-6
3	Silt loam	1.5-1.6	5-10 (low)	0.5-1 (medium)	0.6-0.8	6-7	5-10	40-45	6-7
4	Sandy loam	>1.6	<5 very low	<0.5 soft	>0.8	<6	<5	>45	>7
Rank value	1	2	3	4	5	1	2	3	1

i. Ranking of different parameters. (1-7):1 means highly ranked and 7 means lowest ranked
ii. Weighting of values within parameter. (0-6):0 means very less vulnerable and 6 means very highly vulnerable.

After deriving the normal weights and ranks of all individual parameters, they were integrated with one another in a linear model to demarcate *REBVI* in the study area. The equation of *REBVI* is as follows:

$$REBVI = (R_{ST} \times W_{ST}) + (R_C \times W_C) + (R_X \times W_X) + (R_{PI} \times W_{PI}) + (R_P \times W_P) + (R_{UCS} \times W_{UCS}) + (R_\Phi \times W_\Phi) \quad (1)$$

where R= Rank Value, W= Weight Value, ST = Soil Texture, C= Compaction, X = Cross Section, PI = Plasticity Index, p= Permeability, UCS= Unconfined Compressive Strength, Φ = Angle of internal Friction

3.2. Numerical Analyses

The selected flood bunds are modelled under different critical condition in SEEP/W and SLOPE/W to study the failure of flood bunds under seepage pressure and safety against slope failure. Following four (04) critical Scenarios were selected, (i) Steady state at highest Flood Level (ii) Rapid drawdown from highest Flood Level (iii) Steady state at extreme condition with 3ft free board and (iv) Rapid drawdown from extreme condition with 3ft free board. The constitutive behaviour of both embankment and foundation were simulated using the elastic-perfectly plastic assumption of Mohr-Coulmn model. The boundary conditions in numerical model were applied such that vertical face is fixed in horizontal direction and base is fixed in both horizontal and vertical directions.

4. Results and Discussions

The result of the *REBVI* showed that the highest score is less potential to prevent from embankment breaching or bank erosion. Lowest score is more potential to prevent from embankment breaching or bank erosion. Calculated *REBVI* values at RD 25, 29, 147, 157 are 45, 47, 55 and 66 respectively. Based on the results of SEEP/W, seepage analysis the values of exit gradient for four critical conditions is more than the safe exit gradient (0.3) at RD-147 & 157, which concludes that bund structure was failed due to piping. While the values of exit gradient at RD-25 & 29 were lesser than safe exit gradient and remain safe. SEEP/W results of are shown in Figure 3-6.

Based on FoS calculated from SLOPE/W, stability analysis, stability of bund structures is not the key issue. As the values of FoS for slope stability are more than the critical FoS against sliding.

Table 2: Results of SEEP/W and SLOPE/W

RD	Water Level (ft)	States	Seepage Analysis			Slope Analysis		Remarks
			Safe Exit Gradient	Exit Gradient Obtained	Phreatic line	Safe FoS	Obtained FoS	
RD 147+000	HFL	SS	0.3	0.34	exposed	1.5	1.21	Unsafe
		RDD	0.3	0.13	Within toe	1.3	1.11	
	Extreme Condition	SS	0.3	0.41	Exposed	1.4	1.06	
		RDD	0.3	0.22	Exposed	1.1	1.00	
RD 157+000	HFL	SS	0.3	0.33	exposed	1.5	1.25	
		RDD	0.3	0.20	Within toe	1.3	1.25	
	Extreme Condition	SS	0.3	0.38	Exposed	1.4	1.15	
		RDD	0.3	0.30	exposed	1.1	1.072	
RD 25+000	HFL	SS	0.3	0.18	Within toe	1.5	1.5	Safe
		RDD	0.3	0.056	Within toe	1.3	1.8	
	Extreme Condition	SS	0.3	0.26	Within toe	1.4	1.4	
		RD	0.3	0.072	Within toe	1.1	1.8	
RD 29+000	HFL	SS	0.3	0.17	Within toe	1.5	1.53	
		RDD	0.3	0.002	Within toe	1.3	2.12	
	Extreme Condition	SS	0.3	0.23	Within toe	1.4	1.48	
		RDD	0.3	0.003	Within toe	1.1	2.08	

SS-Steady state, RDD-Rapid draw down

IOP Conf. Series: Materials Science and Engineering **414** (2018) 012035 doi:10.1088/1757-899X/414/1/012035

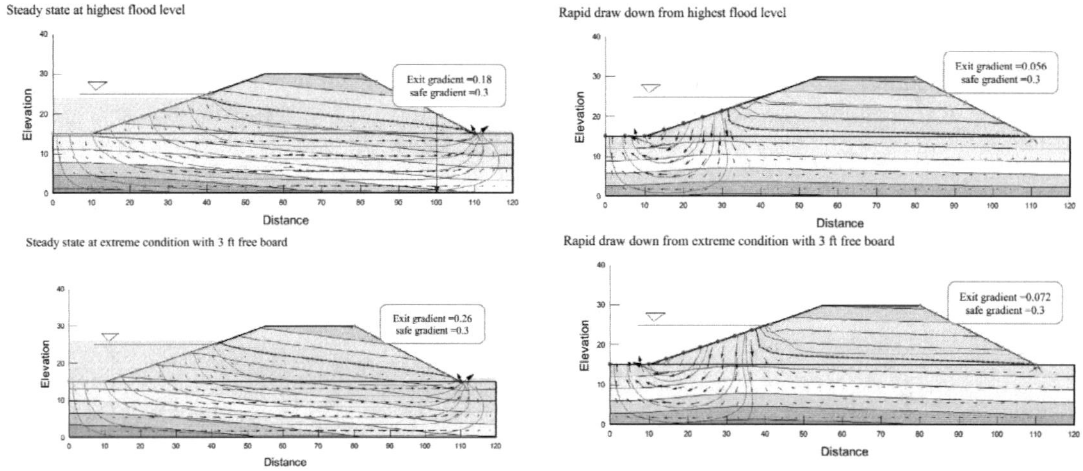

Figure 3. Seepage analysis results for various cases of RD-25

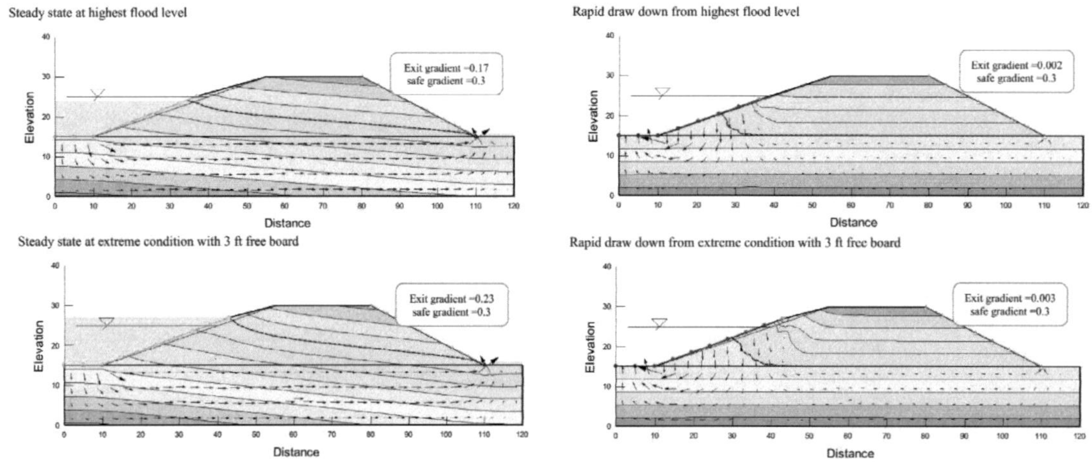

Figure 4. Seepage analysis results for various cases of RD-29

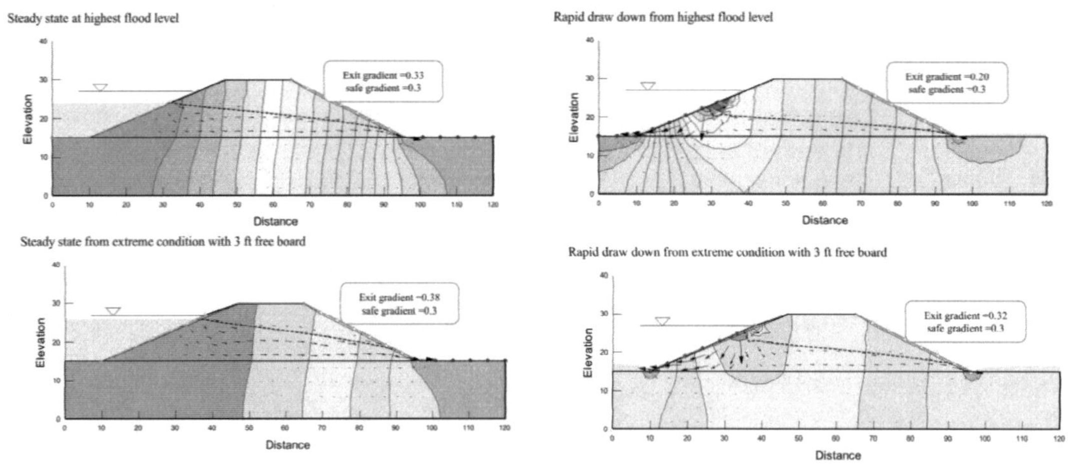

Figure 5. Seepage analysis results for various cases of RD-147

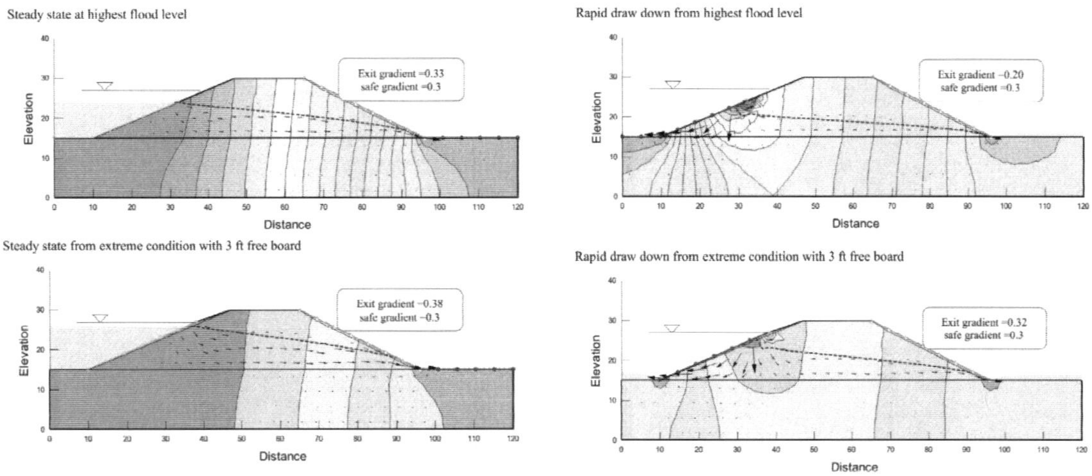

Figure 6. Seepage analysis results for various cases of RD-157

5. Conclusions

In this paper the health of flood bund for the four selected sites along the Chenab river in the Muzzafer Garh district, Punjab is evaluated in term of *REBVI*, safe exit gradient and safety factor against slope failure. Soil parameters for the embankment and foundation of bund were selected based on the extensive laboratory tests performed on collected soil samples from in-situ testing. Based on *REBVI* the calculated value RD-147 & 157 are shown to be susceptible to failure whereas RD-25 & 29 are found to be safe. The selected flood bunds are also numerically simulated using SEEP/W and SLOPE/W under different critical conditions to calculate the seepage and stability FoS respectively. The FoS against the seepage shows that the RD-147 & 157 are unsafe while RD-25-29 are safe against the seepage. The stability of flood bunds can be improved by using clayey material to make flood bunds impervious. As most of the flood bunds are susceptible to piping, it is suggested to install cut-off walls or berms to lengthen the seepage path.

References

[1] Irrigation Department Govt. of Punjab (2014): "Annual flood report 2014" p. 3-5

[2] Irrigation Department Govt. of Punjab (2016); "Geotechnical Investigations for The Health of Flood Bunds in Punjab' IRR-Phy/622.

[3] Dyer, M, R. Utili , S. and Smith, P. (2007): "The Geotechnical Inspection and Assessment of Flood Defense Embankments" Conference Paper: May 2007 https://www.researchgate.net.publications/282330738

[4] Morris, M. Dyer, M. Smith, P. (2007): "Management of Flood Embankments" A Good Practices Review. Research and Development Technical Report FD2411/TR1. (available at: http//scienceresearch.defra.gov.uk/Document.aspx.)

[5] Mondal, M. Gouri, S. B and Pravar K.S (2012): "Vulnerability Analysis of Embankment Breaching", International Journal of Geology, Earth and Environmental Sciences ISSN: 2277-2081.

[6] Bhanbhro, R. Ansari, N, A. Knutsson, S. (2014); "Frequent breaches in irrigation canal in Sindh Pakistan" Journal of Water Resource and Protection, 2014, p. 214-223 Published Online March 2014 in SciRes

[7] Mountassir, G. E., Snche, M., & Romer, E. (2014). An experimental study on the compaction and collapsible behavior of a flood deface embankment fill. Engineering Geology vol. 179, 132-145.

[8] Hughes, P. N., Jones, R. H., Chambers, J. E., Smethurst, J., & Flesjo, K. (2016). "Challenges in monitoring and managing engineered slopes in a changing climate". 3rd European Conference on Unsaturated Soils – "E-UNSAT 2016", (p. 7). Paris.

IOP Conf. Series: Materials Science and Engineering **414** (2018) 012035 doi:10.1088/1757-899X/414/1/012035

[9] Clarke, B. G. (1997) "Ground Exploration" Chp 2. Page 15.
[10] Terzaghi, K (1943), "Theoretical Soil Mechanics", John Wiley and Sons, Inc., New York.
[11] Krahn, J., Stability modeling with SLOPE/W: An engineering methodology. GEOSLOPE/W International Ltd. Calgary, Alberta, Canada, 2004.

Effect of the confining method on the cyclic undrained behaviours of sand

[1]Saeedullah Jan Mandokhail, [3]Khawer Rehman, [1]Zafar Baloch, [4]Shamshear Sadiq, [2]Marjan Gul Kakar

[1]Assistant Professor, Department of Civil Engineering, BUITEMS, Quetta, Pakistan
[2]Lecturer, Department of Civil Engineering, BUITEMS, Quetta, Pakistan
[3]ssistant Professor, Department of Civil Engineering, Ghulam Ishaq Khan Institute of Engineering Sciences and Technology, Swabi, Pakistan
[4]Graduate Student, Department of Civil and Environment Engineering, Hanyang University, Seoul, South Korea

Email: saeed_mandokhail@yahoo.com

Abstract. Cyclic Simple Shear (CSS) test is widely considered to best simulate vertically propagating shear waves and in-situ stress conditions. CSS test is always performed to generate reliable laboratory data for development and calibration of models in Geotechnical Earthquake Engineering design. Soil sample is laterally confined to simulate in-situ stress condition (k0). Stacked rings (SR) and Wire-reinforced membranes (WR) are the two widely used confining methods. Studies have been conducted on the effect of the two confinement methods on the static behavior of soil. In this paper, the effect of the SR and WR confinement methods on the dynamic soil behavior is studied. Undrained cyclic stress controlled tests were performed on loose and medium dense samples of Ottawa sand laterally confined by SR and WR. The comparison of the tests results showed that the dynamic behavior of the two confinement methods is mostly similar.
Keywords: Cyclic simple shear test, stacked rings, wire-reinforced membranes, stress, strain.

1. Introduction

The Cyclic Simple Shear (CSS) test and Cyclic Triaxial (CTX) test have been extensively performed to determine the behavior of soil under cyclic loading condition, [1-3]. CSS test can best simulate the vertically propagating shear wave induced by earthquake loading than the CTX test [4]. In the CSS test, k_0 (in-situ) consolidation is provided by confining the sample against lateral deformation. The lateral confinement is provided by using the stacked rings or NGI type wire-reinforced membrane, as shown in Figure **1**. Both the stacked rings and wire-reinforced membrane methods have been used in the early development of the Simple Shear apparatus [1, 5].

Studies have been conducted on the effect of the two confinement methods on the drained behavior of soil under static loading. Baxter, Bradshaw [6] and McGuire [7] conducted undrained simple shear tests on cohesive soil and reported that the results with the two confinement methods are comparable. Kwan and El Mohtar [8] tested cohesionless soil under drained condition using the two confinement methods and showed an overall agreement between the measured shear strength and volumetric response with the two confinement methods. However, no studies have been conducted to compare the effect of the two confining systems on the cyclic behavior of soil. This paper presents undrained cyclic simple shear test results conducted on sample of clean Ottawa sand using both stacked rings (Teflon-coated aluminum rings) and wire-reinforced membranes, as shown in Figure 1.

(a) (b)

Figure 1. (a) stack rings (b) NGI type wire-reinforced membrane [9]

2. Material and Testing

The Simple Shear equipment used in this study is "Geocomp Shear Trac-II", an automated device capable of performing consolidation, drained and undrained, static and cyclic simple shear tests. The system is of the type developed at NGI in the mid 1960's. The constant volume condition during the shearing is maintained by adjusting the vertical load through a closed loop computer control with the vertical displacement sensor as the feedback. The change in vertical stress is essentially equal to the developed excess pore water pressure in a constant volume *SS* test. Cyclic undrained tests were performed on clean Ottawa sand designated as ASTM C 778 [10] (also known as ASTM C 109). The particle size distribution curve of Ottawa sand is shown in Figure 2 and properties are presented in Table 1.

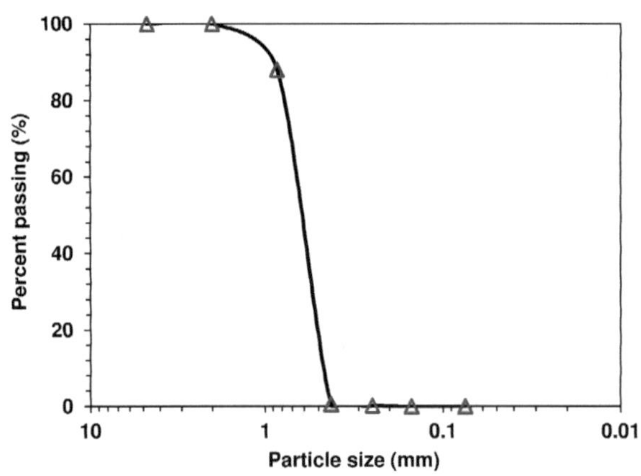

Figure 2. Grain size distribution of Ottawa sand

Table 1. Index properties of Ottawa sand

Soil	e_{max}	e_{min}	G_s	D_{50} (mm)	C_u
Ottawa sand	0.764	0.49	2.65	0.31	1.89

2.1 Sample preparation

Two types of confinement methods SR and WR were used in this study. The installation of the two confinements are shown in Figure 3 and Figure 4. The membrane was installed on the bottom platen and Stack of rings was put around the membrane, as shown in Figure 3 a and b. Split mould shown in Figure 3 c was applied over the stack of rings, as shown in Figure 3 d. The suction was then applied to stretch the membrane tightly against the inner wall of the stack rings. While the wire-reinforced membrane was simply installed on the bottom platen, which do not need the suction to be applied, as shown in Figure 4. The reconstituted samples were made with moist by applying the under compaction method as described by Ladd [11]. The specimen diameter was 6.37 cm and height was 2.4 cm, which fulfils the ASTM D6528-07 [12] requirements. Sand was oven dried and weighted the required amount for the targeted density. The weighted sand was divided into 3 equal parts, mixing each part with 5% of water by weight. The moist soil was directly placed inside the membrane in three layers of equal thickness. Each layer was gently compacted. Top layer was leveled and cape was placed. Split-mould was removed in case the staked of rings was used. The whole assembly was than putted in the shear box for consolidation and shearing, as shown in Figure 5.

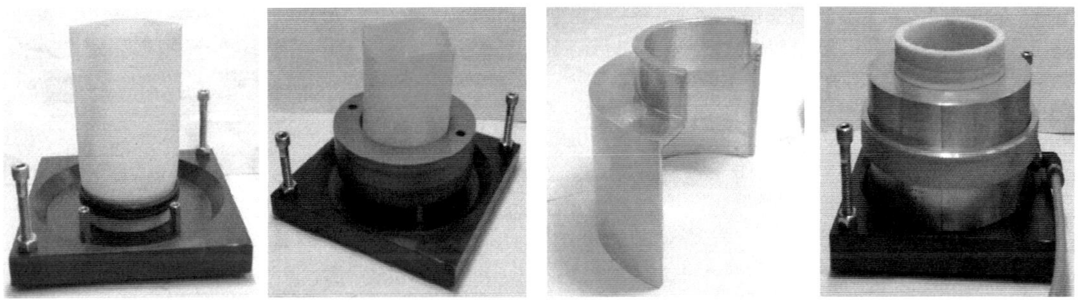

(a) (b) (c) (d)

Figure 3 (a) Membrane installed on the bottom cap (b) stack of rings are placed around membrane (c) One half of split-mould (d) Suction mould is applied

2.2 Consolidation and cyclic shearing

Vertical stress of 100 kPa was applied and the samples were anisotropically consolidate for 20 to 30 minutes. The consolidation was followed by undrained stress controlled cyclic shearing. Uniform sinusoidal loading was applied in the horizontal direction at the rate of 0.1 *Hz*. The cyclic loading was applied to the samples till failure. The tests were performed at different cyclic stress ratios (CSR).

$$CSR = \tau_{cyc} / \sigma'_o \tag{1}$$

where τ_{cyc} is cyclic shear stress and σ'_o normal effective stress.

Figure 4. Installation of wire-reinforced membrane on bottom cap

(a) (b)

Figure 5. Sample is placed in shear box (a) sample tested with SR (b) sample tested with WR

3. Results and discussion

The undrained cyclic behavior of loose and dense Ottawa sand, accomplished with the two confinement methods SR and WM, are shown in Figure 6, Figure 7 and Figure 8, respectively. Where Figure 6 and Figure 7 shows the results of the loose sample subjected to CSR = 0.08 and 0.1 respectively, and the results of dense sample subjected to CSR = 0.3 are shown in Figure 8 for both SR and WM. The effect of the two confinement are analyzed by comparing (a) shear strain versus number of cycles; (b) stress-strain response; (c) excess pore-water pressure versus number of cycles; and (d) Undrained stress path. It can be seen by comparing Figure 6, Figure 7 and Figure 8 that the number of cycles to liquefaction are larger for smaller CSR and vice versa. However, the number of cycles to liquefaction in the two confinement systems are similar. Insignificant discrepancies consists in the stress-strain response, excess pore pressure development and stress path between SR and WR. The overall comparison of the two confinement methods show comparable results.

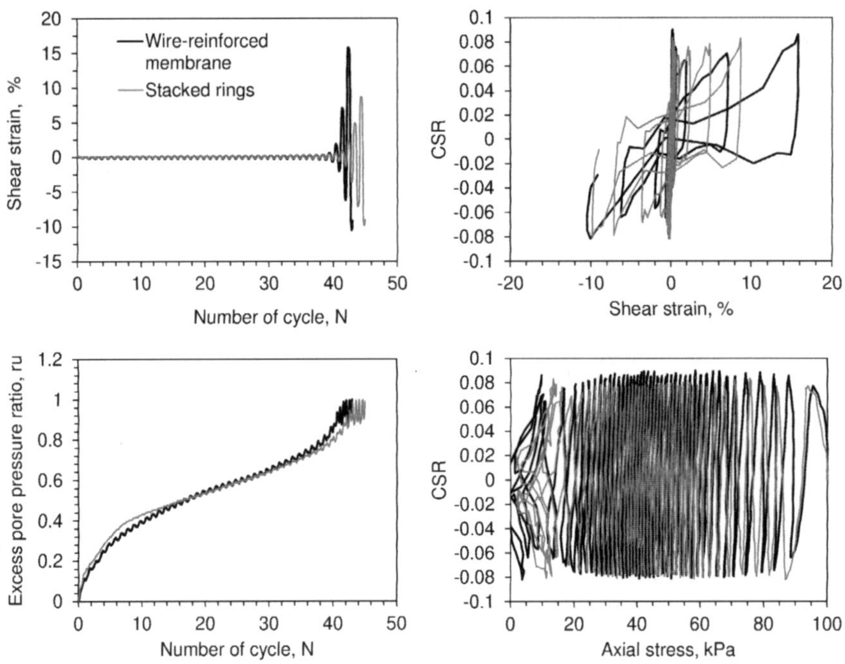

Figure 6. Comparison of test results obtained with wire-reinforced membrane and stack ring, (Loose Ottawa sand, $\sigma'_{vo} = 100\ kPa$, $CSR = 0.08$)

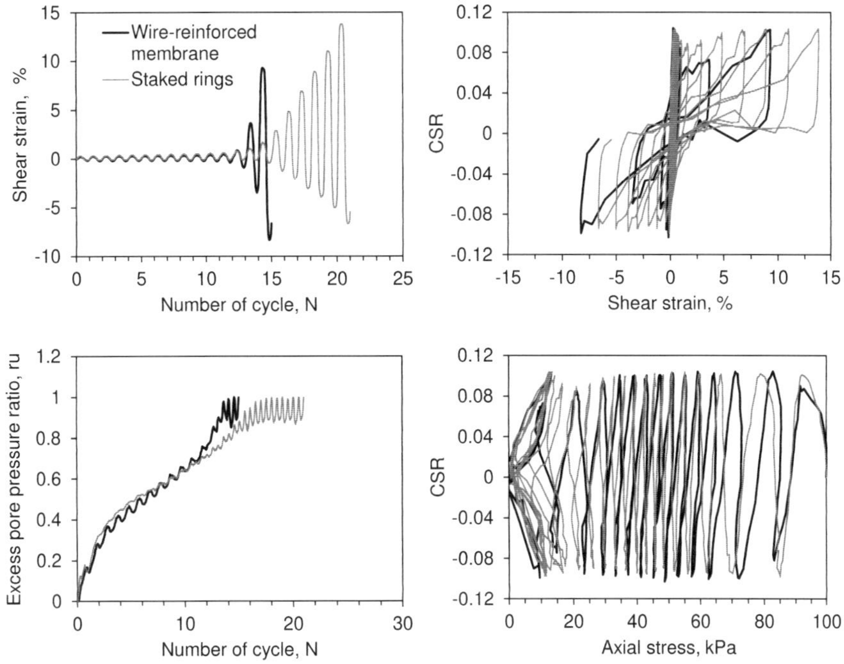

Figure 7. Comparison of test results obtained with wire-reinforced membrane and stack ring, (Loose Ottawa sand, $\sigma'_{vo} = 100\ kPa$, $CSR = 0.1$)

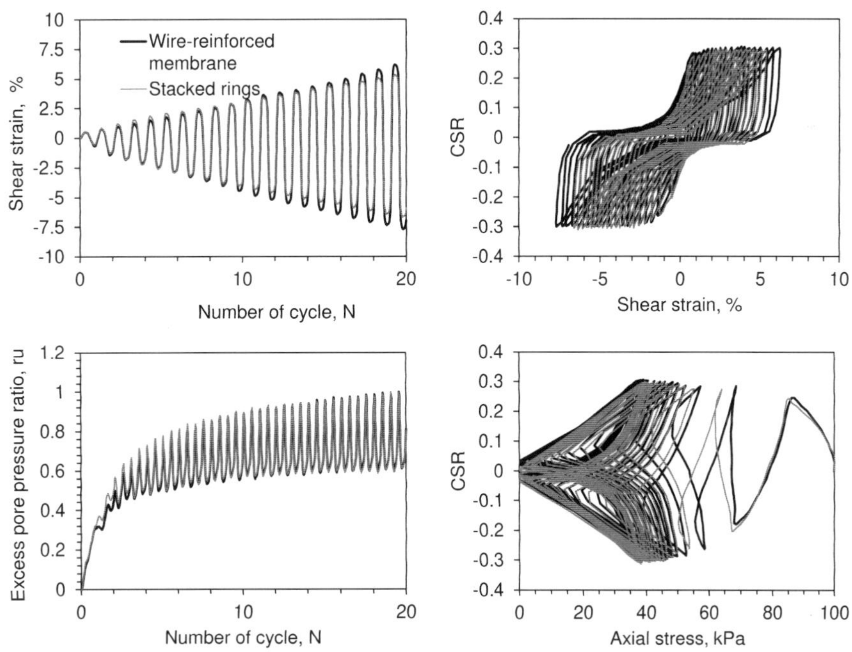

Figure 8. Comparison of test results obtained with wire-reinforced membrane and stack ring, (Dense Ottawa sand, $\sigma'_{vo} = 100\ kPa$, $CSR = 0.3$)

4. Conclusion

In this paper, the comparison between the two widely used lateral confining systems, stacked rings and

IOP Conf. Series: Materials Science and Engineering **414** (2018) 012036 doi:10.1088/1757-899X/414/1/012036

wire-reinforced membrane, in undrained stress control cyclic simple shear testing is presented. Undrained cyclic simple shear tests with the two confinement systems were performed on loose and dense samples, laterally confined by stack of rings and wire-reinforced membrane, of clean Ottawa sand. The comparison of test results shows that the two confinement systems provide comparable results with minor inconsistencies in terms of excess pore pressure generation and stress-strain behavior.

References

[1] Bjerrum, L. and A. Landva, Direct simple-shear tests on a Norwegian quick clay. *Geotechnique*, 1966. **16**(1): p. 1-20.

[2] Finn, W.D.L., S.K. Bhatia, and D.J. Pickering, The cyclic simple shear test. *Soil Mechanics-Transient and Cyclic Loads*, 1982: p. 583-607.

[3] Ishihara, K. and F. Yamazaki, Cyclic simple shear tests on saturated sand in multi-directional loading. *Soils and Foundations*, 1980. **20**(1): p. 45-59.

[4] Boulanger, R.W., et al., A low-compliance bi-directional cyclic simple shear apparatus. *ASTM Geotechnical Testing Journal*, 1993. **16**(1): p. 36-45.

[5] Kjellman, W., Testing the shear strength of clay in Sweden. *Geotechnique*, 1951. **2**(3): p. 225-232.

[6] Baxter, C.D.P., et al., *DSS test results using wire-reinforced membranes and stacked rings*, in *GeoFlorida 2010: Advances in Analysis, Modeling & Design*. 2010. p. 600-607.

[7] McGuire, S.T., *Comparison of direct simple shear confinement methods on clay and silt specimens*. 2011, University of Rhode Island.

[8] Kwan, W.S. and C. El Mohtar. *Comparison between Shear Strength of Dry Sand Measured in CSS Device Using Wire-Reinforced Membranes and Stacked Rings*. in *Geo-Congress 2014: Geo-characterization and Modeling for Sustainability*. 2014. ASCE.

[9] Mandokhail, S.J., *Development of Normalized Liquefaction Resistance Curve for Granular Soil and Application to Effective Stress Site Response Analysis*. 2017, Hanyang University, Seoul: South Korea.

[10] ASTM, Standard specification for standard sand. *ASTM C 778, West Conshohocken, PA*, 1995.

[11] Ladd, R.S., Specimen preparation and liquefaction of sands. *Journal of the Geotechnical Engineering Division, ASCE*, 1974. **100**(GT10): p. 1180-1184.

[12] ASTM, Standard test method for consolidated undrained direct simple shear testing of cohesive soils. *D 6528-07, West Conshohocken, PA.*, 2007.

River Flow Dynamics with Two-dimensional Shallow-water Equations

Khawar Rahman[1] and Saeedullah Jan Mandokhail[2]

[1]Assistant Professor, Department of Civil Engineering, Ghulam Ishaq Khan Institute of Engineering Sciences and Technology, Topi, District Swabi, Pakistan

[2]Chairperson Department of Civil Engineering, BUITEMS, Quetta, Pakistan

khawar.rehman@giki.edu.pk

Abstract. This study proposes a numerical model to simulate unsteady flows in rivers and associated riverine processes. The mathematical model is based on Saint-Venant equations. The governing equations are descritized by Godunov-type finite volume method. Use of total variation diminishing (TVD) scheme and MUSCL-Hancock technique satisfies second-order temporal and spatial accuracy requirements, respectively. Hydrodynamic flux components associated are computed by Harten-Lax-van Leer (HLL) method. A modified method is introduced for approximating flow features along the shear wave. The proposed scheme is used to model an idealized dam-break case, and a case of dam-break in a channel with a 90° bend which results in bore formation and complex two-dimensional flow; the predictions show excellent agreement with available experimental results and analytical solutions.

1. Introduction

River flows are extensively modelled by non-linear shallow-water equations. The similarity of the numerical predictions to the real flows depends upon the structure on depth averaged equations considered to model the problem. Two-dimensional shallow-water equations are widely applied in modeling hydraulic, coastal, and environmental engineering flows due to lower computational cost compared to three-dimensional models [1,2].

To ensure steady solution, the well-balanced condition has to be satisfied; to ensure well-balancing of proposed numerical scheme, the bed slope terms are discretized by revised surface gradient method (RGSM) so that when added with momentum terms, the system remains conserved [3]. A cell centered finite volume method is used for discretization of governing equations. MUSCL-Hancock technique is employed for reconstruction of flow variables on unstructured computational domain. The proposed method performs very well for various benchmark test cases.

2. Numerical Method

For flows in natural rivers, open channels, and floodplain systems over a horizontal plane, the depth averaged shallow-water equation are given as;

$$\frac{\partial h}{\partial t}+\frac{\partial hu_j}{\partial x_j}=0$$
$$\frac{\partial (hu_i)}{\partial t}+\frac{\partial}{\partial x}\left(hu_i^{\,2}+\frac{1}{2}gh^2\right)+\frac{\partial h}{\partial x_j}\left(u_iu_j\right)=-gh\left(\frac{\partial z}{\partial x_i}-\tau_i\right) \tag{1}$$

where i=j=1,2; h is water surface elevation, u and v represent velocity vectors in x and y direction, respectively, z is bed elevation, and τ is the bed friction. The equation is discretized by Godunov type finite volume method with Runge-Kutta scheme for time discretization. Discretization of Equation (1) by finite volume method is expressed as;

$$\frac{\boldsymbol{U}_i^{n+1} - \boldsymbol{U}_i^n}{\Delta t} = -\left(\frac{\partial \boldsymbol{F}_x}{\partial x} + \frac{\partial \boldsymbol{F}_y}{\partial y}\right) + \boldsymbol{S}_i^n \tag{2}$$

where $\boldsymbol{U}$ represents conserved variables in Equation (1), superscript n refers to time level , $\boldsymbol{S}$ is a vector of bed inclination and friction entities and $\boldsymbol{F}$ represent flux vector.

$$\boldsymbol{S} = \boldsymbol{S}_b + \boldsymbol{S}_f = \begin{bmatrix} ghS_{ox} \\ ghS_{oy} \end{bmatrix} + \begin{bmatrix} -ghS_{fx} \\ -ghS_{fy} \end{bmatrix} \tag{3}$$

where S_{ox}, S_{oy} and S_{fx}, S_{fy} are bathymetry slope and frictional terms in the x- and y-axis directions, respectively. Manning's relation is used to compute friction terms.

$$S_{fx} = \frac{\mathrm{n}^2 u\sqrt{u^2 + v^2}}{h^{4/3}}, S_{fy} = \frac{\mathrm{n}^2 v\sqrt{u^2 + v^2}}{h^{4/3}} \tag{4}$$

In Equation (4) 'n' is the Manning's roughness coefficient. Equation (2) is then discretized over the control volume (in this study triangle) to approximate the solution. Fluxes in Equation (2) are approximated HLL scheme (Toro, 2001).

$$\boldsymbol{F}_{HLL} = \begin{cases} \boldsymbol{F}_L & \text{if } 0 \le S_L \\ \boldsymbol{F}^{hll} = \dfrac{S_R F_L - S_L F_R + S_R S_L (U_R - U_L)}{S_R - S_L} & \text{if } S_L \le 0 \le S_R \\ \boldsymbol{F}_R & \text{if } 0 \ge S_R \end{cases} \tag{5}$$

In Equation (5) S_R, and S_L are the right and left propagating waves, respectively. Depth of water and wave speeds needed to compute numerical fluxes by HLL scheme [4]. Small water depth, near wet–dry interfaces cause numerical instabilities, which result in negative water depths or unphysical high velocities; therefore, a threshold water depth of $\varepsilon = 10^{-5}$ is set to avoid instabilities. Since the proposed scheme is time-explicit, CFL condition is used for limiting time-step.

$$\Delta t \le \min\left(\frac{R_i}{2\max_j (U + c)_{ij}}\right) \tag{6}$$

where R_i is the shortest distance between cell center and vertex, $\boldsymbol{U}$ is velocity vector, and c is wave celerity.

3. Results

3.1. Dam-Break in L-Shape Channel

L-shaped channel is considered to test the ability of numerical scheme of modeling complex dam-break flows. This test case has been designed by CADAM team [5]. The initial water depth in the reservoir is 0.58 m while the channel is dry. Manning's roughness coefficient equal to 0.019 is fixed throughout the domain. A thin wall (dam) at x = 2.39 m separates the reservoir from channel.

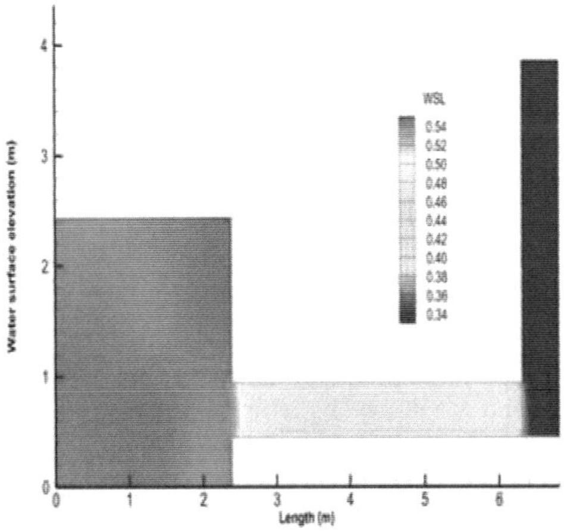

Figure 1. Wave propagation after t = 5 sec of dam-break.

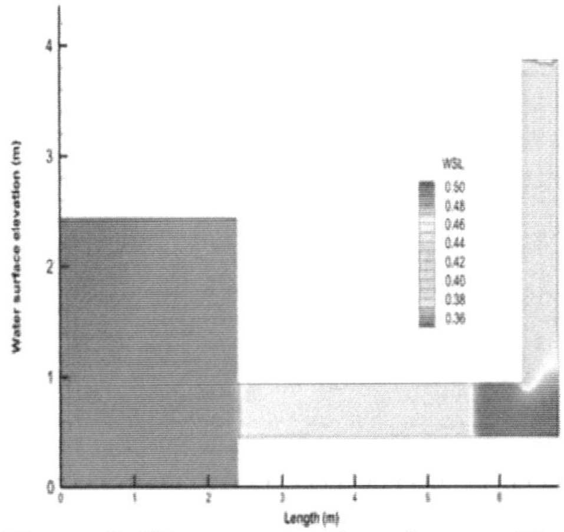

Figure 2. Wave propagation after t = 12 sec of dam-break.

After the removal of dam the wave propagation is depicted in the channel at t = 5 and 12 sec in figure 1 and 2, respectively. The observed pattern of wave propagation and bore reflection from the channel wall at the end agrees very well with the experimental and numerical studies [6] confirming the proposed scheme's ability to model dam-break flows in two dimensions.

3.2. *Tidal flow*

The flow over large step is simulated to test model's ability for simulating tidal flows. A channel of length (L) of 1500 m is considered. The bathymetry of the problem is given as

$$z(x)=\begin{cases}8.0\text{m} & \text{if } |x-750|\leq 187.5\text{m}\\ 0.0 & \text{otherwise}\end{cases} \tag{7}$$

For initial flow conditions, water depth $H(x, 0)$ is set equal to 16 m in the whole domain. For boundary conditions the velocity $u(L,t)$ is set as zero, and water depth is set equal to

$$h(0,t)=H(x,0)+4-4\sin\left\{\pi\left(\frac{4t}{86400}+0.5\right)\right\} \tag{8}$$

The analytical solution for this problem was proposed by [7], and is given as

$$\begin{aligned}h(x,t)&=H(x)+4-4\sin\left\{\pi\left(\frac{4t}{86400}+0.5\right)\right\}\\ u(x,t)&=\frac{(x-L)\pi}{5400h(x,t)}\cos\left\{\pi\left(\frac{4t}{86400}+0.5\right)\right\}\end{aligned} \tag{9}$$

Simulations are run for a total of 35,000 s; the computed water surface elevation and velocity (at 10,800 s, 21,600 s, 32,400 s) are compared with analytical solution, as shown in Figure 3 and 4, respectively. Excellent agreement is observed between computed and analytical solution.

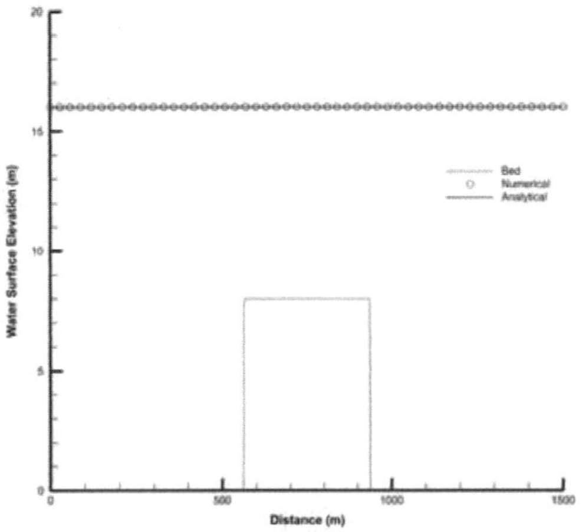

Figure 3. Comparison of water surface elevation with analytical solution.

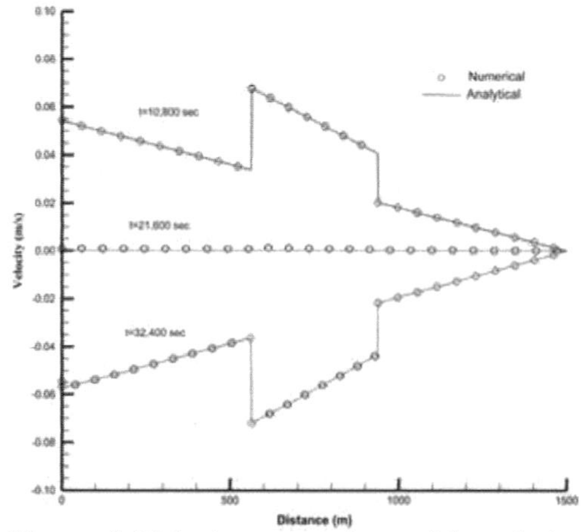

Figure 4. Velocity comparison with analytical solution at t = 10,800, 21,600, and 32,400 s.

4. Conclusions

A two-dimensional depth averaged formulation is proposed in this study to model open channel flows. The model is based on non-linear shallow-water equ0ations with a treatment for wet-dry fronts. The spurious oscillations associated with bathymetry are avoided by employing high resolution reconstruction technique. Proposed scheme is used to model complicated dam-break flows involving shocks making it suitable for wide applications in modeling open-channel flows without sediment. Numerical results agree well with experimental studies and analytical solutions. However, to model the effect of eddies associated with flows undergoing rapid changes in direction and varying bed conditions, the proposed scheme will be coupled with suitable turbulence model which will enable modelling complicated river and estuarine flows.

References

[1] Zhou, J. G., Causon, D. M., Ingram, D. M. & Mingham, C. G. 2002. Numerical solutions of the shallow water equations with discontinuous bed topography. *International Journal for Numerical Methods in Fluids,* 38**,** 769-788

[2] Hou, J., Simons, F., Mahgoub, M. & Hinkelmann, R. 2013. A robust well-balanced model on unstructured grids for shallow water flows with wetting and drying over complex topography. *Computer Methods in Applied Mechanics and Engineering,* 257**,** 126-149.

[3] Kim, D., Cho, Y. & Kim, H. 2008. Well-Balanced Scheme between Flux and Source Terms for Computation of Shallow-Water Equations over Irregular Bathymetry. *Journal of Engineering Mechanics,* 134**,** 277-29

[4] Toro, E. 2001. Shock-capturing methods for free-surface shallòw flows. 2001. *Chichester, etc.: Wiley.*

[5] Morris, M. 1998. 2nd CADAM Meeting Proceedings. *HR Wallingford,* 2

[6] Liu, X., Landry, B. & García, M. 2008. Two-dimensional scour simulations based on coupled model of shallow water equations and sediment transport on unstructured meshes. *Coastal Engineering,* 55**,** 800-810.

[7] Bermudez, A. & Vazquez, M. E. 1994. Upwind methods for hyperbolic conservation laws with source terms. *Computers & Fluids,* 23**,** 1049-1071.

Potential Effect of Sugar Mill waste water as Substrate for Bio-Electricity Generation using Laboratory Scale Double Chamber Microbial Fuel Cell

M Siddique[1], A S Jatoi[2], M H Rajput[3], S A Soomro[3], S Aziz[3], F Mushtaq[1], G Khan[1], M A Abro[3,4], M Najam Khan[1], A K Shah[5] and S K Sami[1]

[1]Chemical Engineering Department, Balochistan University of Information Technology, Engineering and Management Sciences (BUITEMS), Quetta, Pakistan.
[2]Chemical Engineering Department, Dawood University of Engineering and Technology (DUET) Karachi, Pakistan.
[3]Chemical Engineering Department Mehran University of Engineering and Technology (MUET) Jamshoro, Sindh, Pakistan.
[4]College of Chemical Engineering, Beijing University of Chemical Technology, Beijing 100029, P. R. China.
[5]Metullery and Materials Engineering Department, Dawood University of Engineering and Technology (DUET) Karachi, Pakistan.

Email: siddiqnasar786@gmail.com

Abstract: Bio-electrochemical system for power generation getting attention due to utilization of waste material. Based on that study was made to convert sugar industry waste water for bio-electricity generation using double chamber microbial fuel cell. Different organic load in form of substrate concentration and parametric effect were tested to optimize the best condition for electricity generation. From 100g/l to 300g/l with step size 100g/l, for aeration rate from 100-250ml/min with step size 50ml/min, and for as pH from 4.5 to 6.5 with step size 0.5pH. The maximum power generation were observed at pH 6, aeration rate 200ml/min and organic load 200g/l about 820mA. Regarding above results that found favourable condition on environment as well as waste reduction.
Keywords. Sugar mill waste water, bio-electricity, double chamber microbial fuel cell.

1. Introduction
The waste water generated during the sugar manufacturing process of sugarcane contain lots of organic compound in the form of carbohydrates like glucose, sucrose etc. This can be treated as well as used as substrate for the generation of power in Microbial fuel cell. For the treatment of waste water, we design Microbial fuel cells (MFCs). In Microbial fuel cells (MFCs) there are two chambers which are anode and cathode and separated by polymeric proton exchange membrane. To utilize the substrate and convert it into electricity we introduce bacteria in anode chamber. Most MCFs contain aqueous cathode in which dissolved oxygen can be provided to electrode by mean of water bubbled with air. To make MFCs more economical and energy efficient, we investigate the power generation in an air-cathode MFC having carbon electrode during the absence and presence of the polymeric proton exchange membrane (PEM). The waste water and glucose were used as substrate and the biocatalyst were the bacteria which are in the domestic waste water. The power density is much higher than that of the normally reported aqueous cathode MFCs, using a maximum of 262 ± 10 mW / m^2 (6.6 ± 0.3 mW / L, liquid volume). The maximum power density "494 ± 21 mW / m^2" (12.5 ± 0.5 mW / L) is achieved by eliminating PEM. The Coulomb efficiency is 40-55% of PEM, and the efficiency of removing PEM is 9-12%, which indicates that in the absence of PEM, the oxygen in the anode compartment diffuses in large amounts [1]. The chemical energy of the substrate in waste water is converted into electrical energy by utilization of bacteria in Microbial fuel cells (MFCs). In this cell electron are being transfer by the bacteria to the electrode rather than directly to an electron acceptor. Its technical feasibility has recently been tested, and the scientific

community has great enthusiasm for MFC's ability to generate “green power” sources by using household and industrial waste to generate energy. Acetate, lactate, and glucose are used as the feed stocks for MFCs.By using the Microbial fuel cell, we can generate the electricity from the domestic waste water and at the same the biological treatment of waste water has also been done such as removal of chemical oxygen demand; COD. These tests are carried out in the single chamber microbial fuel cell (SCMFC) with eight graphite electrodes (anodes) and a single air cathode. Primary clarified effluent which is taken from the local wastewater treatment plant was used in the system at continuous flow conditions. The waste water COD removed up to 80% while generating the electricity (maximum of 26 mW m^{-2}) in the prototype SCMFC reactor. Power generated was proportional to the hydraulic retention time (3–33 hour) and to the influent wastewater strength (50–220 mg/dm^3 of COD) [2]. Fuel cell and voltammetry were utilized for understanding the mechanism of electron transfer with the addition of Shewanellaputrefaciens.The bacterium's electrochemical activity without any electrochemical mediators were determined by the both methods. By the studies of cyclic voltammetric, electrochemical activities were showed by the Shewanellaputrefaciens anaerobically, but for another condition regarding aerobically there was not enough. The activates measured regarding electrochemical system is nothing but the related closely with electric potential and current generation [3]. From the various operating condition a mediator-less microbial fuel cell was optimized. pH, resistance, electrolyte used, and dissolved oxygen concentration in the cathode compartment are the several factors on which current generation depends. In neutral solution (pH=7) generation of current was maximum. The proton transfer and dissolved oxygen supply limited the cathode reaction at lower resistance than 500 Ω. Under the operating condition, the rate determining factor was resistance at over 500 Ω. Up to some extent proton limitation were reduced by high strength buffer solution. At the DO (dissolve oxygen limited condition the concentration of DO was around 6 mg l^{-1} The fact that oxygen limitation was observed at high DO concentration is believed to be due to the poor oxygen reducing activity of the electrode used, graphite [4].Some of bacterial culture use glucose as carbon source to identify the power generation with respect to glucose utilization. With the utilization of glucose dose 0.5–3 g l^{-1} d^{-1} the power output about 3.6 W m^{-2} and 89% electron recovery in terms of electricity reported by [5].

Lab-scale double chamber MFC was used for power generation. Electric current was produced from the sewage organics using the Saccharomyces cerverciae sp. as a mediator and potassium ferricyanide as an oxidizing agent. The anodic and cathodic solutions were introduced in batch and continuous way the cells were connected in series. The cathode in continuous modes and anode in batch-feed were used to operate the cell. A maximum voltage of 830mV (3.3 mA) was obtained per liter of the sludge under optimize operating parameters such as oxygen flow rate, pH and substrate concentration [6]. Two dehydrogenases, cellobiose dehydrogenase and pyranosyl-capsule heat dehydrogenase, Agaricusblazei, used surface display system of yeast to show the first surface in Saccharomyces cerevisiae. The use of dehydrogenases displayed on the surface of microbial fuel cells produces high power output. The potential midpoint - 28 mV at pH (vs. Ag/ AgCl electrode) = 6.5 represents the dehydrogenase shown on the surface of the microbial fuel cell and the output power of 3.3 μWcm produced - 2 using lactose as its use without The fuel in the anode compartment of the mediator. In microbial fuel cells, pyranose dehydrogenase is exposed on the surface and produces high output efficacy by using different substrates. The maximum power of D-xylose is 3.9 μWcm^{-2}. These results indicate that cellobiose dehydrogenase and pyranose dehydrogenase exposed on the surface can be successfully used in microbial bioelectrochemical systems [7].

Protection and perseverance of our environment and energy crises are the two major problems that have played havoc with our lives. The conversion methods of other renewable energy contain transitional conversion into mechanical power but Microbial fuel cell (MFC) offers an advantage that transmit chemical energy directly into the electrical energy. In this context, the sewage sludge which is the sources of major environmental pollution but also contains high levels of organic compounds is utilize to generate power in the MFC. By using the microbial fuel cell, we converted the chemical energy from the substrate of sewage sludge into the electricity by using mediator which is methylene blue and biocatalyst which is Saccharomyces cerevisiae sp. By varying the agarose's concentration in the salt bridge of MFC we can examined the efficiency of MFC. Concentration of agarose salt which range from 7% to 12% was utilized for analyzes of the best possible concentration was noticed being 10% because it exhibited maximum voltage generation of 0.97V [8].The utilization of the fossil fuels and pollutant producer must be reduced to make the world more feasible and conventional for living. These types of problems can be solved by treating the waste in an appropriate manner, such as sewage sludge. As we know that the sewage sludge is full of organic compound so it can be used as a perfect substrate for the power generation. According to the study it was found that dual compartment MFC with the bio cathode was used to generate the voltage from the sewage sludge (2L). An internal resistance of 36-46 ohm were developed by the bio cathode to obtained maximum voltage (2.5 V) Saccharomyces cerevisiae sp. was used as biocatalyst. Methylene blue (10 ml) was used a mediator and potassium ferricyanide (350 ml) was used as an oxidizing agent for the conversion of sewage sludge into voltage generation [9].The results show that although the substrate has a high impedance, all the general parameters on the 5th day prove to be the maximum, and then the trend is observed after 6 days. The corresponding maximum values of the generated parameters are 0.825 V, 0.0113 μA, 0.009223 μW and 0.000000947 mW/m^2, respectively. It has also been found that the obtained voltage, current, power and power density have similar patterns [10].This work is related with the potential use of sugar mill waste water for electricity generation using microbial fuel cell.

2. Materials and Methods

2.1 Microorganisms

Pure strain of Saccharomyces Cerevisiae was used as a biocatalyst [11]. The work of Saccharomyces Cerevisiae is same as other yeasts do, which is capable of transferring electron from biodegradation of substrate, which are acetate or glucose. Acetate and glucose work as substrate on the behalf of containing carbon source which dissociate into electron and proton [12, 13]. Saccharomyces cerevisiae sp. as a biocatalyst and methylene blue as mediator were used for the conversion of [9] sugar processing water into bio-electricity. This could make direction towards the implication of microorganism for degradation of substrate into electron, which make continuous flow of electron.

2.2 Fabrication and Operation of double chamber MFC

Microbial fuel cell were fabricated with two plastic bottle attach with salt bridge having volume of 3liter of each chamber. Microbial fuel cell made of two chambers with different operation condition, one of them maintains at anaerobic condition which is anode and other chamber cathode maintained at aerobic condition. Figure 01 shows typical diagram of microbial fuel cell which consists of salt bridge with two chambers. In addition with them one pinhole drilled at top of the chamber in order to connect electrode with resistors for maintain the flow of electron from anode to cathode. As proton were concerned, it oxides with the addition of air in cathode chamber. During operation of microbial fuel cell substrate degrade by microbes, which split substrate into electron and proton.

2.3 Salt Bridge

Salt bridge or proton exchange membrane constructed through use of agar salt and different salt like NaCl and KCl for conducting proton from anode chamber by degradation of substrate. The typical composition of salt bridge is 8-13%agarose, 8g NaCl and 150ml of distilled water. The mixture of salts monitored continuously during heating and pouring into PVC pipe. The maximum power output observed by using 10% agarose concentration. Then mixture of salts and agarose were left for cooling in order to making it gel type media. After that PVC pipe attach with anode and cathode chamber for making circuit complete.

Table 1. Dimension of salt bridge, anode and cathode chamber

Items	Length (in)	Dia (in)	Height(in)
Cathodic chamber	-	5	10
Anodic chamber	-	5	10
Salt bridge	1	0.5	-

IOP Conf. Series: Materials Science and Engineering **414** (2018) 012038 doi:10.1088/1757-899X/414/1/012038

2.4 Anode and Cathode

Anode and cathode chamber made with two different conditions based on air supply. In anode chamber degradation of substrate in absence of oxygen and cathode chamber oxidize proton that emit from anode chamber with the use of different air ratio. Carbon cloth was utilized as electrode for conducting electron during conversion of organic matter into electron and proton. The power output was observed through use of multimeter. The total volume of anode and cathode chamber is 800ml with working volume of 500ml.

2.5 Assembling the Fuel Cell

The sample of sugar mill waste water put into anode chamber with different concentration and power were observed after 24 hour due to utilization of substrate and substantially growth in cell for microbial growth. Cathode chamber maintained pH with addition of 10%H_2SO_4 dilute solution for maintains the pH and 10%NaOH for anode chamber. Electrode was connected with lid inserted for receiving electron from anode chamber. The current is generated by moment of electron from anode to cathode chamber. Electric current were measured by multimeter which maintained at mode of fuel cell output voltage.

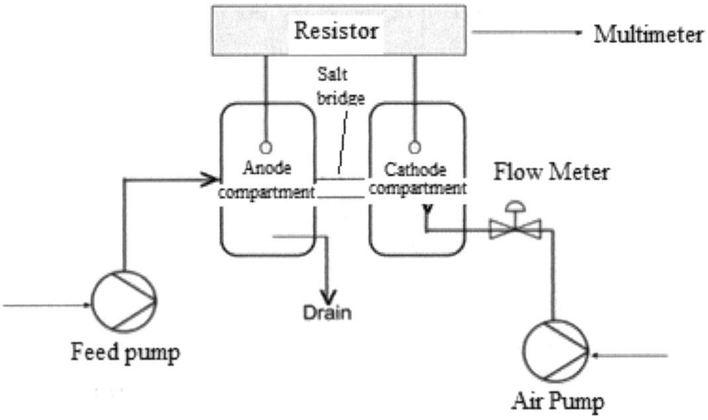

Figure 1. Typical schematic diagram of Microbial Fuel cell

2.6 Electrical parameters and measurements

Electric power was measured during experiments of degradation of sugar mill waste water into electron and proton. Electron flow through external resistors from anode to cathode chamber. The current and voltage measured after 24hour, periodically after 20min.

3. Results and Discussion

Microbial fuel cell for electricity generation gained huge interest in field of renewable energy resources. Microbial fuel cell is a device which utilizes organic matter for electricity generation with the help of electron and proton. Microbial fuel cell has also application in bioremediation for treatment of waste water [14]. The electricity generation was monitored by different operational condition for both season's summer and winter. Anodic and cathodic chamber optimization regarding maximum current and maximum power is about 0.784 mA and 0.645 mA explained by [15] at organic concentration in solutions about 720 mgL^{-1} which produces maximum power density of 768 mWm^{-2} and minimum 556 mWm^{-2}.

Determined voltage generated from Microbial fuel cell and organic concentration of raw material for degrading seems to maintain the Monod type equation. Performance of microbial fuel cell related with molecular weight and structures of eight amino acids [15]. Electricity generation from sugar mill waste water using microbial fuel cell were under study to optimize the parametric effect on that, to measure potential of sugar mill waste water different operational parameter were under study.

Table 2. Current, voltage, power, power and current density at various aeration rate, pH, and substrate

Parameter	Oxygen flow rate (ml/min)				pH value				Substrate % w/v		
	100	150	200	250	5	6	7	8	25	50	75
Current (mA)	0.81	0.89	0.94	0.98	0.79	0.83	0.86	0.89	0.75	0.79	0.83
Voltage (volts)	0.77	0.86	0.95	0.97	0.75	0.79	0.82	0.85	0.72	0.74	0.77
Power (mW)	0.62	0.77	0.89	0.91	0.59	0.66	0.71	0.76	0.54	0.58	0.64
Power density (mW/m^2)	50.43	61.53	71.98	78.42	49.38	54.64	58.77	63.04	45.0	48.72	53.3
Current density (mA/m^2)	67.29	73.98	78.97	82.48	65.83	69.17	71.67	74.17	62.5	65.83	69.2

3.1 Effects of oxygen flowrate on electricity generation

During running of microbial fuel cell different ranges of oxygen flow rate were tested to investigate the optimize condition for electricity generation from sugar mill waste water. Regarding 500 Ω resistance dissolved oxygen supply and proton transfer decrease to cathode reaction [4] also explain the strategies behind dissolved oxygen, on the basis of that aeration rate had significantly fact on current generation. The microbial fuel cell might be used as a biochemical oxygen demand (BOD) sensor [4]. During degradation of organic matter in anode chamber, that splits into electron and proton for electricity generation from 100-250ml/min were analyzed periodically to measure the voltage generation using 50Ω resister with the help of digital multimeter. In figure 2, it is clearly observed that when oxygen flowrate increasing up to the level of 200ml/min voltage generation were also increased up to the 360 mv/l after 24-hourperiod of time for microbial growth.

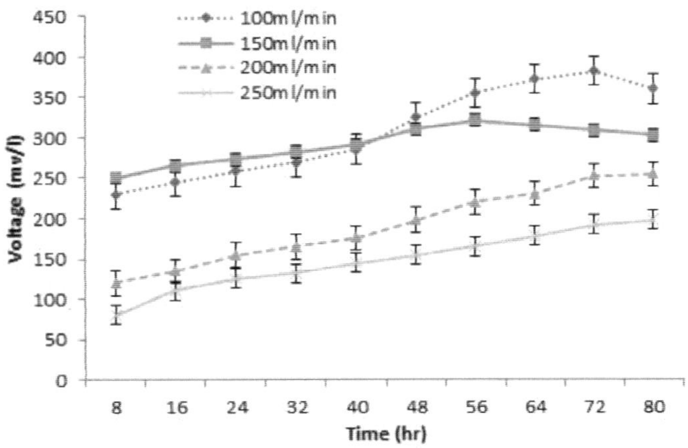

Figure 2. Effects of oxygen flowrate on electricity generation

3.2 Effects of pH on electricity generation

The maximum current and voltage generation were observed at pH ranges between 8 and 10 when compared with lower pH ranges. Bacterial growth inhibited by buffer and changed electrolyte. In microbial fuel cell anode chamber gave favorable results with neutral pH ranges and minimum at lower pH ranges explained by [16]. The highest current was generated at a pH of 6.5 in the anode chamber containing 4% CE. The greater pH difference between the anode and cathode electrolytes favors higher currents and voltages. In the analyzed COD range (100-600 mg / l), a linear relationship was observed between the current and the substrate removed [17]. Microbial growth of saccharomyces cell during running of microbial fuel cell affected by pH due to tolerable pH ranges of microbes specie increased. From 5-8 were tested with step size 01 using buffer solution to maintain the pH ranges according required range of pH under 48 hr running period of microbial fuel cell. The maximum voltage generation from microbial fuel cell by utilizing sugar mill waste water. In Figure 3, effects of pH on electricity generation clearly observed and maximum Voltage generated at pH 8 and about 276mv/l.

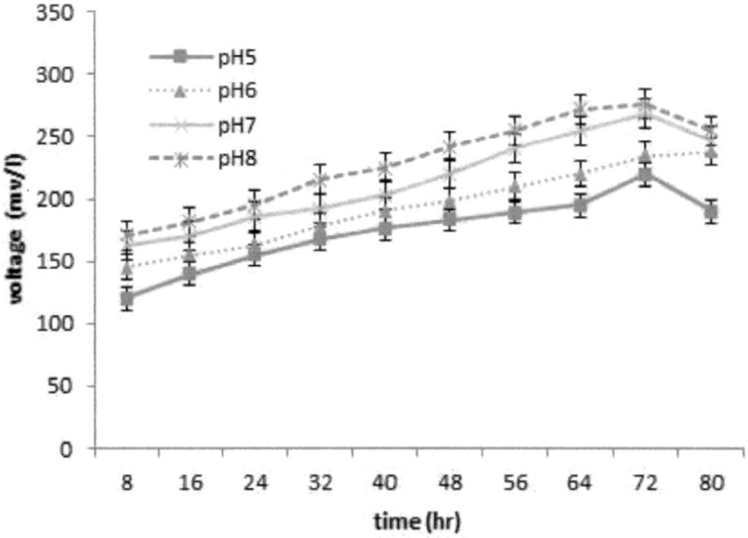

Figure 3. Effects of pH on electricity generation

3.3 Effects of substrate utilization on electricity generation

It was observed that the amount of power generation increased with increasing substrate concentration as shown in Figure 4. Starting from about 25% substrate concentration, the power obtained at this substrate concentration was 0.725V. At a substrate concentration of 75%, power generation increased by 1550mV. Further increases in the concentration up to 100 % resulted in the decrease in power production by more than 100% when it reached the value of 1V. This may be due to various factors such as pH-induced decrease in enzyme activity. This also shows that higher concentrations of substrates can affect the performance of the anode, which can lead to lower simultaneous generation of electricity.

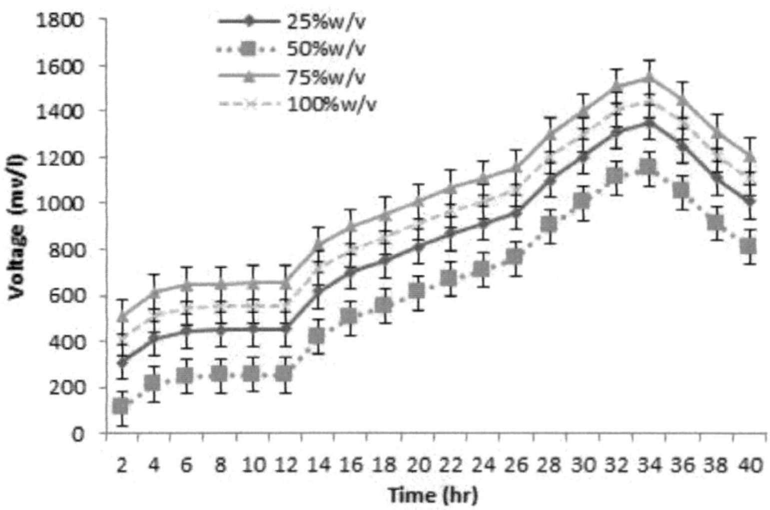

Figure 4. Effects of Substrate on electricity generation

3.4 Effects of substrate utilization on current and power densities

Different substrate concentration was utilized to produced protons and electrons for making complete circuit between anode and cathode chamber. In microbial fuel cell electrodes were provided for acceptance of electron from anode chamber. With the use of bio-catalyst *saccharomyces cerevisiae* substrate degraded into protons and electrons, where proton get oxidation in cathode chamber with the use of aeration rate provides through fish pump. Figure 6 clearly show that when increase in substrate content, there is significant change of power and current density. It is due to increase in organic load present in sugar waste water, because during processing of sugarcane some starch and glucose were remained in waste water. Figure 5 shows the maximum power and current density were obtained at 100w/v% about 58mW/m^2 and 74mA/m^2.

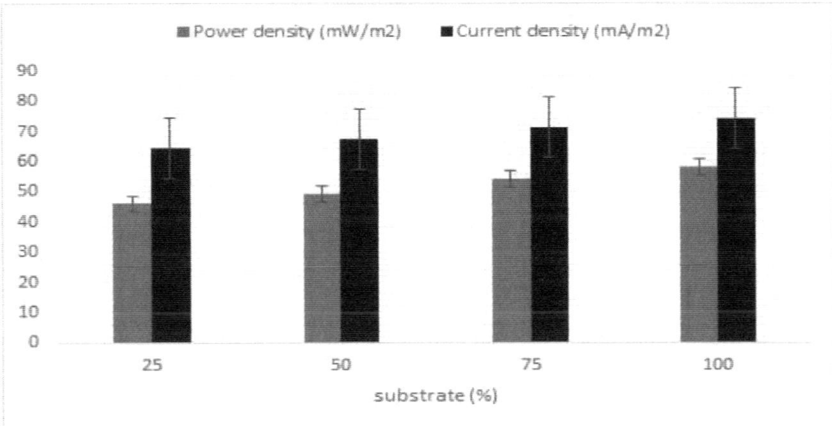

Figure 5. Effects of substrate on current and power density

4. Conclusion

Microbial fuel cell getting importance for treating waste water for power generation. From 100g/l to 300g/l with step size 100g/l, for aeration rate from 100-250ml/min with step size 50ml/min, Maximum electricity generation at 200ml/min about 360 mv/l and for as pH from 4.5 to 6.5 with step size 0.5pH. The maximum power generation was observed at pH 6 about aeration rate 200ml/min and organic load 200g/l about 820mA. Regarding above results that found favorable condition on environment as well as waste reduction.

5. Acknowledgment

This experimental work was carried out in Chemical Engineering Department, Mehran University of Engineering and Technology (MUET), Jamshoro, Sindh, Pakistan.

References

[1] Liu H and Logan B E 2004 Electricity generation using an air-cathode single chamber microbial fuel cell in the presence and absence of a proton exchange membrane Environmental science & technology 38 4040-6

[2] Liu H, Ramnarayanan R and Logan B E 2004 Production of electricity during wastewater treatment using a single chamber microbial fuel cell Environmental science & technology 38 2281-5

[3] Kim H J, Park H S, Hyun M S, Chang I S, Kim M and Kim B H 2002 A mediator-less microbial fuel cell using a metal reducing bacterium, Shewanella putrefaciens Enzyme and Microbial Technology 30 145-52

[4] Gil G-C, Chang I-S, Kim B H, Kim M, Jang J-K, Park H S and Kim H J 2003 Operational parameters affecting the performannce of a mediator-less microbial fuel cell Biosensors and Bioelectronics 18 327-34
[5] Rabaey K, Lissens G, Siciliano S D and Verstraete W 2003 A microbial fuel cell capable of converting glucose to electricity at high rate and efficiency Biotechnology letters 25 1531-5
[6] Aziz S, Memon A R, Shah S F, Soomro S A, Parkash A and Jatoi A S 2013 Electricity Generation From Sewage Sludge Using Environment-Friendly Double Chamber Microbial Fuel Cell Science International 25
[7] Gal I, Schlesinger O, Amir L and Alfonta L 2016 Yeast surface display of dehydrogenases in microbial fuel-cells Bioelectrochemistry 112 53-60
[8] Parkash A, Aziz S, Abro M, Kousar A, Soomro S A and Jatoi A S 2015 Impact of Agarose Concentrations on Electricity Generation using Hostel Sludge Based Duel Chambered Microbial Fuel Cell Science International (Lahore) 27 1057-61
[9] Parkash A, Aziz S and Soomro S A 2015 Utilization of Sewage Sludge for Production of Electricity using Mediated Salt Bridge Based Dual Chamber Microbial Fuel Cell Journal of Bioprocessing & Biotechniques 5 1
[10] Parkash A 2016 Characterization of Generated Voltage, Current, Power and Power Density from Cow Dung Using Double Chambered Microbial Fuel Cell J Phys Chem Biophys 6 2161-0398
[11] Zhao F, Harnisch F, Schröder U, Scholz F, Bogdanoff P and Herrmann I 2005 Application of pyrolysed iron (II) phthalocyanine and CoTMPP based oxygen reduction catalysts as cathode materials in microbial fuel cells Electrochemistry Communications 7 1405-10
[12] You S-J 2005 Identification of denitrifying bacteria diversity in an activated sludge system by using nitrite reductase genes Biotechnology letters 27 1477-82
[13] Bond D R, Holmes D E, Tender L M and Lovley D R 2002 Electrode-reducing microorganisms that harvest energy from marine sediments Science 295 483-5
[14] He C-S, Mu Z-X, Yang H-Y, Wang Y-Z, Mu Y and Yu H-Q 2015 Electron acceptors for energy generation in microbial fuel cells fed with wastewaters: A mini-review Chemosphere 140 12-7
[15] Yang Q, Wang X, Feng Y, Lee H, Liu J, Shi X, Qu Y and Ren N 2012 Electricity generation using eight amino acids by air–cathode microbial fuel cells Fuel 102 478-82
[16] He Z, Huang Y, Manohar A K and Mansfeld F 2008 Effect of electrolyte pH on the rate of the anodic and cathodic reactions in an air-cathode microbial fuel cell Bioelectrochemistry 74 78-82
[17] Jadhav G and Ghangrekar M 2009 Performance of microbial fuel cell subjected to variation in pH, temperature, external load & substrate concentration Bioresource Technology 100 717-23

Effective use of Enzyme Zymase for Enhancement of Ethanol Production Couple with Parametric Effect

M Siddique[1], AS Jatoi[2], MH Rajput[3], MN Khan[1], AN Mengal[4], S Aziz[3], SA Soomro[3], F Mushtaq[1], A Shah[5] and SK Sami[1]
[1]Chemical Engineering Department, Balochistan University of Information Technology, Engineering and Management Sciences (BUITEMS), Quetta, Pakistan.
[2]Chemical Engineering Department, Dawood University of Engineering and Technology (DUET) Karachi, Pakistan.
[3]Chemical Engineering Department, Mehran University of Engineering and Technology (MUET) Jamshoro, Sindh, Pakistan.
[4]Mechanical Engineering Department, Balochistan University of Information Technology, Engineering and Management Sciences (BUITEMS), Quetta, Pakistan.
[5]Metallurgy and Materials Engineering Department, Dawood University of Engineering and Technology (DUET) Karachi, Pakistan.

siddiqnasar786@gmail.com

Abstract. Pakistan has large resources of sugarcane crops coupled with significant amount of molasses generated annually. In this study, the effects of enzyme zymase concentration (2-5 g/l), pH (5.5-6), aeration rate (0.1-0.2 vvm/l) and agitation speed (250-300rpm) were investigated for ethanol production. *Saccharomy cesservisae* was used as a biocatalyst. The maximum ethanol production of 82g/l was obtained at 0.2vvm/l, pH5.5, 300rpm and 35°C at 4g/l of enzyme. This study suggests a novel method of utilizing the enzyme separately for improving ethanol production.
Keywords. Sugarcane molasses, Enzyme zymase, pH, Aeration rate, Agitation speed, ethanol production.

1. Introduction

Natural energy sources such as oil and coal have been consumed at much high rates in recent decades. Due to its impact on the environment and the fact that it may eventually be exhausted, the heavy reliance on these fuels is bound to end. Therefore, alternative fuel resources such as ethanol are becoming more and more important. Bioethanol is one of the most important renewable fuels, helping to reduce the negative environmental impact of the global use of fossil fuels [1]. The traditional ethanol industry produces ethanol from variety of biomass-based waste, which is considered an important option to solve this problem [2]. When organic substrates are fermented, such as agro-industrial residues like sugarcane molasses [3], wheat stillage [4], fish residues [5], pineapple waste [6], carbon sources can be converted into ethanol, a popular well known bio-fuel. Several well-known microorganism for ethanol production were focused, such as *Saccharomyces cerevisiae*, *Leuconostocoenos* and *Zymomonasmobilis*, which are capable of converting sugary material into valuable fuel products [7, 8]. During sugar cane refining, the molasses generated as a by-product has been identified as a potential sugar source that can be converted to value chemical fuels. This is because sugarcane is rich in sucrose and its substrate does not require pretreatment before fermentation. In conventional batch fermentation process, Saccharomyces cerevisiae is used to produce distilled ethanol. However, growth inhibition may be the result of product or substrate concentration which limit its productivity [9]. Ethanol production can be improved by different technologies. These include continuous culture, continuous fermentation of cell recycle, vacuum distillation of cell recycle [10] and immobilization of yeast cells [11]. The maximum yield, theoretical yield (YE/S), volumetric productivity (Q_P) and total sugar consumption of ethanol were reported 19.15 g/l, 46.23%, 2.39 g l^{-1} $h^{-1}$and 96%, respectively [12].

The most useful microorganism for the production of ethanol by alcoholic fermentation is *Saccharomyces cerevisia* [13]. This leads us to define new strategies for intensive production to maximize Q_p and yield. The use of continuous reactors to produce ethanol has been studied to improve economics and performance of fermentation process [14]. Although, high Q_P can be achieved if high cell retention is employed [15]. The general strategy for cell retention involves fixing the microbial cells in a fermenter or separating the membrane from the product stream and then recycling it back to the fermenter [16]. Higher ethanol concentrations were obtained in the case of non-sterilized molasses (47.4-50.6 kgm^{-3} at 35°C and 44.2-48.2 kgm^{-3} at 40°C) when compared to sterilized molasses, where ethanol concentration reported from 35.6 to 46.6 kgm^{-3} at 35°Cand 30.8-44.2 kgm^{-3} at 40°C [16]. The increase in ethanol productivity and to reduce the intensity of work, factors such as bioreactor volume, energy consumption and cellular immobilization of ethanol production have been studied evaluated [17-19]. Various researchers have proposed the idea of immobilizing yeasts and bacteria during fermentation from various agricultural biomass residues. It was included that the completion of process for continued use of the same, and not the organic and inorganic recipients, between the organic matter such as calcium Alginate, polyacrylamide; polyvinyl alcohol [18] and inorganic materials such as alumina [20], silica [21] and zeolite [18, 22]. In raw biomass containing glycosylic radicals readily available for fermentable sugars may be considered ideal substrate for ethanol production. However, pretreatment step of converting the starch and cellulose materials into fermentable sugars is required. Recent studies on alcoholic fermentation have shown that sugarcane and sugar beet can provide high yields of ethanol and renewable substrates. The current work was focused on investigation of optimum doze of enzyme zymase coupled with different parametric effects for ethanol production.

2. Materials and Methods

A viscous by product obtained by refining sugar cane or sugar beets known as molasses or black treacle was collected from Al Abbas Distillery, Sindh. The amount of sugar in molasses depends on method of extraction and the age of the plant. The characterization of molasses is shown in Table 1. The enzyme zymase was previously produced by [23] and extracted from yeast *klyeromyces Marxians* was used. During fermentation process, enzyme loses their activity by different operational parameter. For this purpose, enzyme *zymase* was used separately within the 24 hours of fermentation process. The enzyme zymase was added in order to maintain activity for utilization with sugary material.

Table 1. Characteristic of Molasses

Component	**brix**	**Sucrose**	**Purity**	**TSI**	**TS**	**RS**	**NFS**	**FS**
Average value	90	35.4	39.60	51.9	50.1	14.1	3.2	48

The ethanol production can be inhibited by the different process parameters coupled with the activity of the microorganism. The different process parameters were studied and investigated in order to obtain the optimized dose of zymase to improve ethanol production at a fixed temperature of 35°C. The experimental setup for ethanol production can be seen in Figure 1.

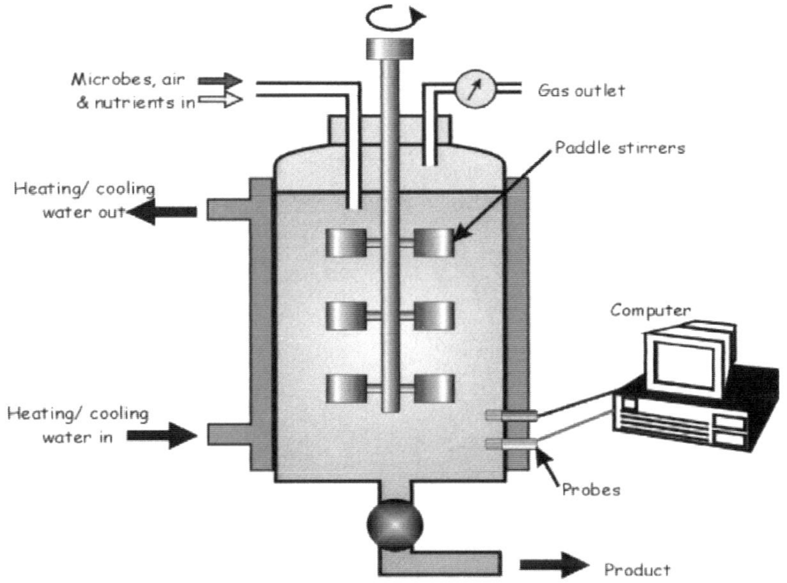

Figure1. Experimental Setup

3. Results and Discussion

3.1 Effects of enzyme zymase on ethanol production

The effects of zymase concentration on biomass production are shown in Figure 2(a). It can be observed that as the time proceeds, cell concentration changes. Maximum biomass concentration was observed with 3g/l of zymase concentration. The minimum biomass production occurred at 2g/land 5g/l because the substrate needed to fill the active side of the enzyme was utilized and thus resulted in inhibiting the growth. With 3 g/l, the production of ethanol was observed maximum. The zymase concentration on sugar utilization can be seen in Figure 2(b). It can be observed that sugar concentration decreases as the time proceed; likewise the parametric condition had important effect. The maximum sugar utilizedwhen3g/l and 5g/l zymase was used. Figure 2(c) highlighted the effects of various dose of zymase on cell growth. It can be observed that with the use of enzyme from 2-5 g/l altered ethanol production.

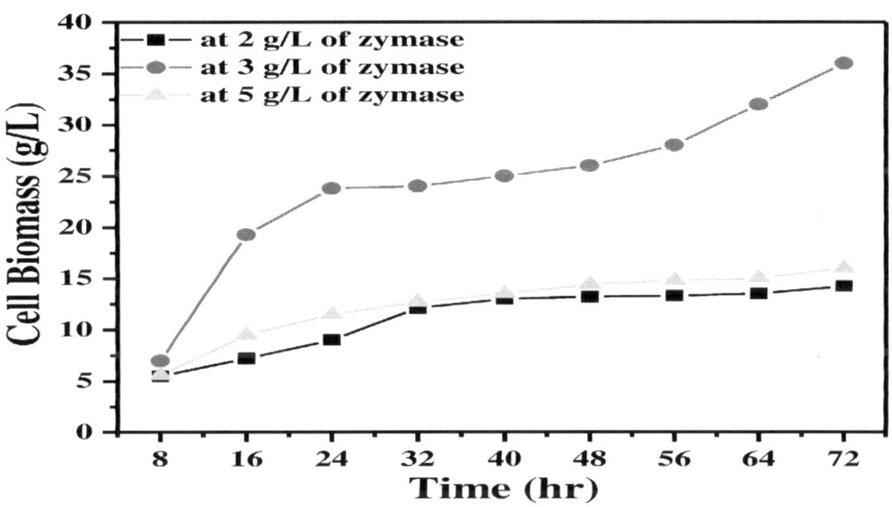

Figure 2 (a). Cell growth at various doses of enzyme

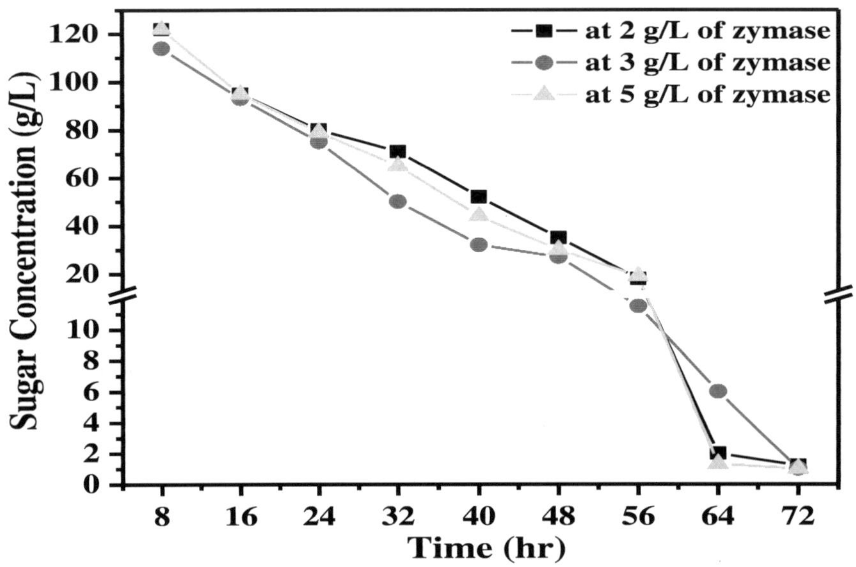

Figure 2 (b). Cell growth at various doses of enzyme

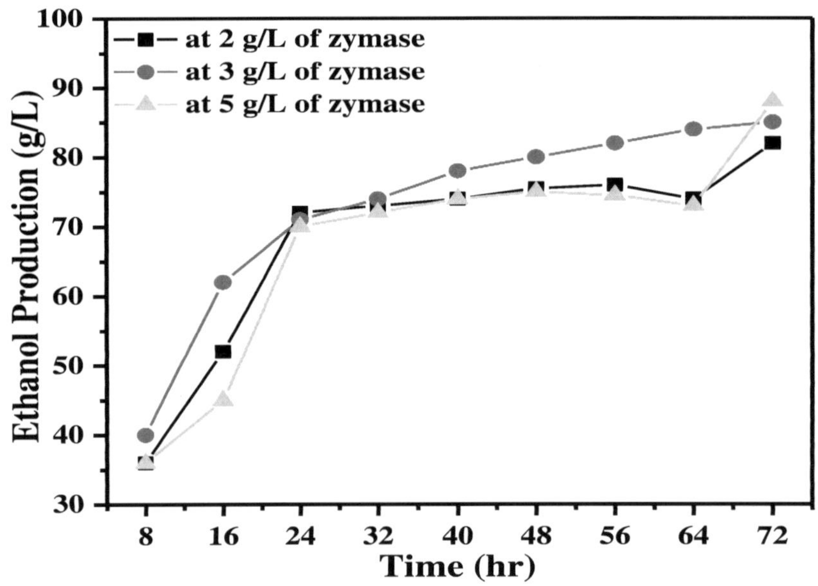

Figure 2 (c). Ethanol production at various doses of enzyme

3.2 Operational parameter on ethanol production

3.2.1 Effects of aeration rate

Figure 3 highlighted the effects of different aeration rates with 3 g/l of enzyme. Aeration rates of 0.1vvm/l to 0.2vvm/l were tested for ethanol production. Proper aeration rate can be provided, which can result in increase in microbial activity, as the microbes would be able to digest more fermentable sugar present in molasses. This could be beneficial for industry as the percentage of un-fermentable sugar can be utilized, otherwise rejected as waste in the spent wash. Effective utilization of the enzyme zymase dosage could also make the sugar consumable.

3.2.2 Effects of pH

Figure 4 presents the effects of pH on substrate utilization and cell biomass on ethanol production. It can be seen; maximum ethanol was produced at 5.5pH. The microbial activity is challenged when pH is changed. pH is supports microbial growth and convert sugars. Changes in pH can also affect enzyme activity. The most favorable pH, the point at which the enzyme is most active is considered optimum pH. Extremely high or very low pH usually results in the loss of complete activity of enzymes. pH is also considered a factor in enzyme stability. As with activity, there is also a region of optimal pH stability for each enzyme. During fermentation process, range of optimal pH is required for microbial activity to ensure maximum ethanol production. This study was carried out to investigate the optimal dose of enzyme for the conversion of un-fermentable sugar into fermentable sugar. The main objective of the enzyme is to hydrolyze the un-fermentable sugar and change it into convertible sugar for ethanol production.

3.2.3 Effects of agitation speed

Figure 5 highlights the effect of different agitation speeds on ethanol production. It was observed that the controlled condition for maximum ethanol occurred at 300rpm.Homogenizing the temperature needs some external source which could result in higher microbial growth to be inhibited with respect to the agitation speed. The agitator is provided to overcome this problem as the homogenize temperature can alter the ethanol production, which can be either increased or decreased. In order to enhance the cell growth, we used different agitator speeds leading to an increase in ethanol production. Different ranges of agitation speeds were used to identify the optimized range for maximum ethanol production.

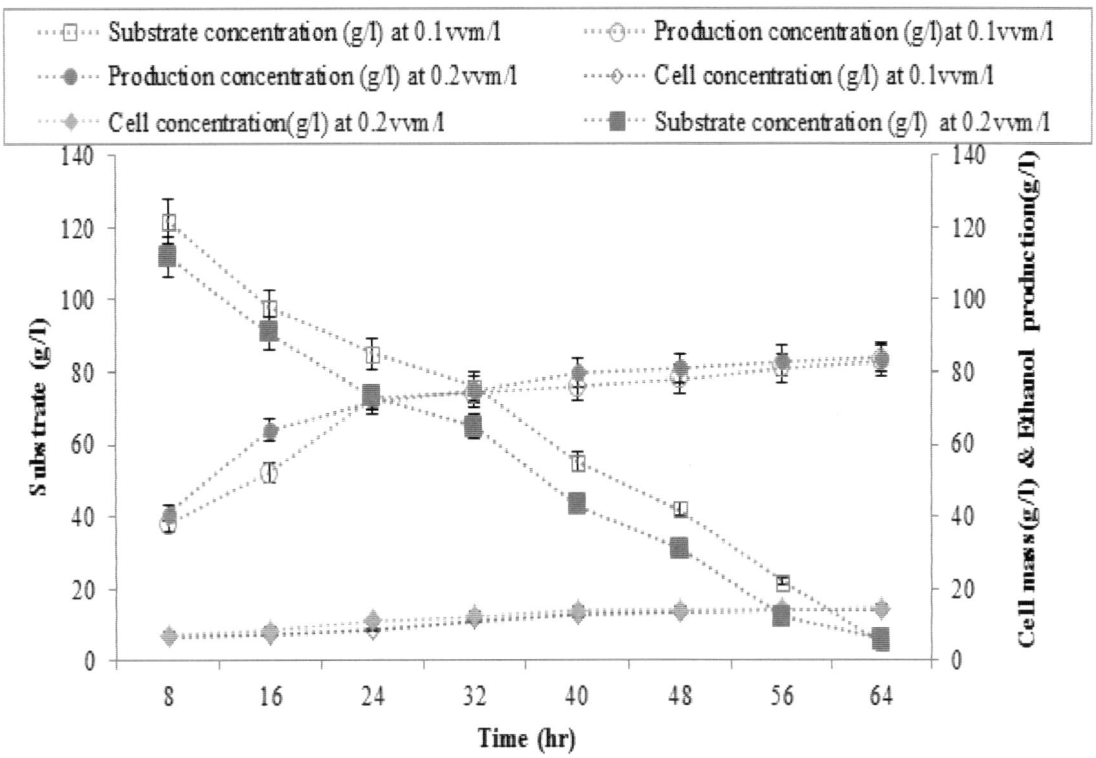

Figure 3. Effect of aeration rate on substrate utilization, cell growth and ethanol production

IOP Conf. Series: Materials Science and Engineering **414** (2018) 012039 doi:10.1088/1757-899X/414/1/012039

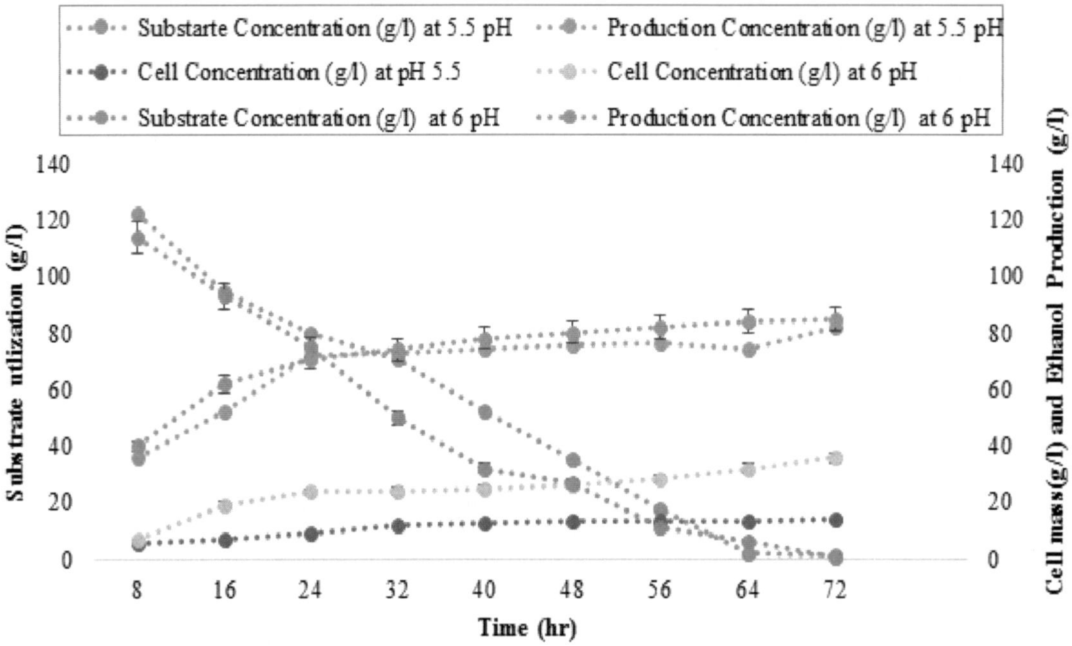

Figure 4. Effects of pH on substrate utilization, cell growth and ethanol production

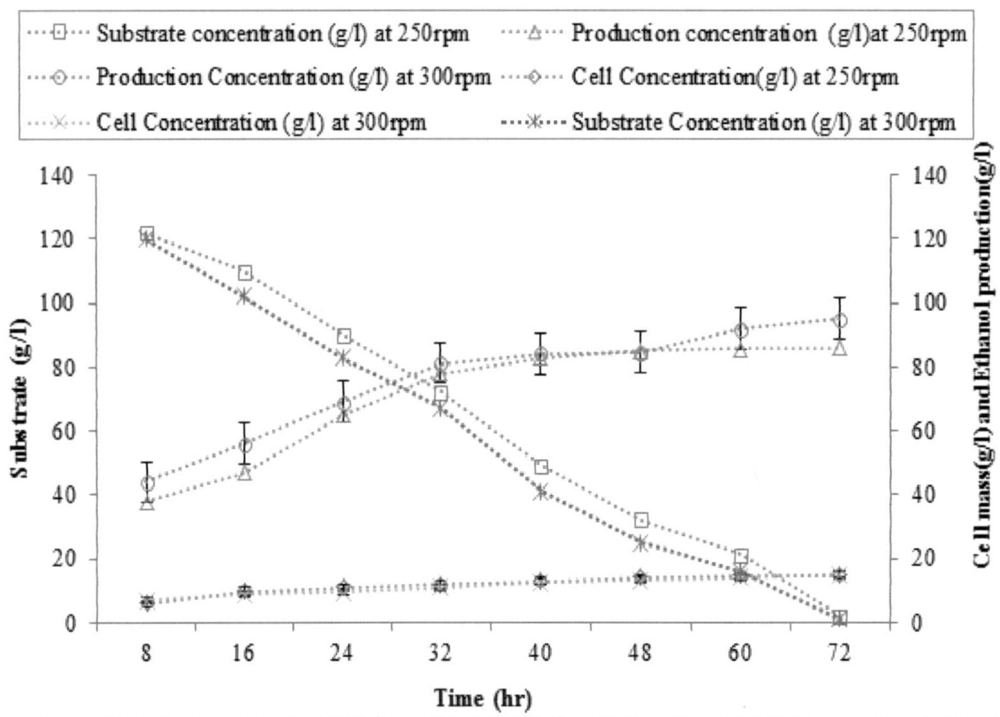

Figure 5. Effect of Agitation intensity on substrate utilization, cell growth and ethanol production

4. Conclusion

Present study was conducted to integrate ethanol production by converting the un-fermentable sugar into fermentable sugar. The enzyme zymase was used as biocatalyst to improve ethanol production various operational and nutritional parameters were investigated and analyzed to determine the optimum ethanol production. The operational parameters included the aeration rate, pH and agitation speed together with the nutritional parameters was studied. The maximum ethanol production of 82g/l was obtained at 0.2vvm/l, pH5.5, 300rpm and 5g/l of enzyme at a fixed temperature of 35°C. This study suggests a novel method of utilizing the enzyme separately for improving ethanol production.

References

[1] Cardona C A and Sánchez Ó J 2007 Fuel ethanol production: process design trends and integration opportunities *Bioresource technology* **98** 2415-57
[2] Saxena R, Adhikari D and Goyal H 2009 Biomass-based energy fuel through biochemical routes: a review *Renewable and Sustainable Energy Reviews* **13** 167-78
[3] Cazetta M, Celligoi M, Buzato J and Scarmino I 2007 Fermentation of molasses by Zymomonas mobilis: Effects of temperature and sugar concentration on ethanol production *Bioresource technology* **98** 2824-8
[4] Davis L, Jeon Y-J, Svenson C, Rogers P, Pearce J and Peiris P 2005 Evaluation of wheat stillage for ethanol production by recombinant Zymomonas mobilis *Biomass and Bioenergy* **29** 49-59
[5] Ruanglek V, Maneewatthana D and Tripetchkul S 2006 Evaluation of Thai agro-industrial wastes for bio-ethanol production by Zymomonas mobilis *Process biochemistry* **41** 1432-7
[6] Tanaka K, Hilary Z D and Ishizaki A 1999 Investigation of the utility of pineapple juice and pineapple waste material as low-cost substrate for ethanol fermentation by Zymomonas mobilis *Journal of bioscience and bioengineering* **87** 642-6
[7] Sitton O and Gaddy J 1980 Ethanol production in an immobilized-cell reactor *Biotechnology and Bioengineering* **22** 1735-48
[8] Spettoli P, Bottacin A, Nuti M P and Zamorani A 1982 Immobilization of Leuconostoc oenos ML 34 in calcium alginate gels and its application to wine technology *American Journal of Enology and Viticulture* **33** 1-5
[9] Baptista C, Cóias J, Oliveira A, Oliveira N, Rocha J, Dempsey M J, Lannigan K C and Benson P S 2006 Natural immobilisation of microorganisms for continuous ethanol production *Enzyme and Microbial technology* **40** 127-31
[10] Cysewski G R and Wilke C R 1978 Process design and economic studies of alternative fermentation methods for the production of ethanol *Biotechnology and Bioengineering* **20** 1421-44
[11] Kolot F 1980 New trends in yeast technology--immobilized cells *Process biochemistry*
[12] Ghorbani F, Younesi H, Sari A E and Najafpour G 2011 Cane molasses fermentation for continuous ethanol production in an immobilized cells reactor by Saccharomyces cerevisiae *Renewable Energy* **36** 503-9
[13] Najafpour G, Younesi H and Ismail K S K 2004 Ethanol fermentation in an immobilized cell reactor using Saccharomyces cerevisiae *Bioresource technology* **92** 251-60
[14] Purwadi R and Taherzadeh M J 2008 The performance of serial bioreactors in rapid continuous production of ethanol from dilute-acid hydrolyzates using immobilized cells *Bioresource technology* **99** 2226-33
[15] Loukatos P, Kanellaki M, Komaitis M, Athanasiadis I and Koutinas A A 2003 A new technological approach proposed for distillate production using immobilized cells *Journal of bioscience and bioengineering* **95** 35-9
[16] Palmqvist E, Galbe M and Hahn-Hägerdal B 1998 Evaluation of cell recycling in continuous fermentation of enzymatic hydrolysates of spruce with Saccharomyces cerevisiae and on-line monitoring of glucose and ethanol *Applied microbiology and biotechnology* **50** 545-51

[17] Kopsahelis N, Nisiotou A, Kourkoutas Y, Panas P, Nychas G J-E and Kanellaki M 2009 Molecular characterization and molasses fermentation performance of a wild yeast strain operating in an extremely wide temperature range *Bioresource technology* **100** 4854-62

[18] Kourkoutas Y, Bekatorou A, Banat I M, Marchant R and Koutinas A 2004 Immobilization technologies and support materials suitable in alcohol beverages production: a review *Food microbiology* **21** 377-97

[19] Marignetti N, Fumi M D, Silva A, Battistotti G, Pomini M and Colagrande O 1997 Model for prediction of immobilizedAcetobacter cell number in calcium alginate gel droplets *Biotechnology techniques* **11** 7-8

[20] Loukatos P, Kiaris M, Ligas I, Bourgos G, Kanellaki M, Komaitis M and Koutinas A 2000 Continuous wine making by γ-alumina-supported biocatalyst *Applied biochemistry and biotechnology* **89** 1-13

[21] Gill I and Ballesteros A 1998 Encapsulation of biologicals within silicate, siloxane, and hybrid sol– gel polymers: an efficient and generic approach *Journal of the American Chemical Society* **120** 8587-98

[22] Hartmann M 2005 Ordered mesoporous materials for bioadsorption and biocatalysis *Chemistry of Materials* **17** 4577-93

[23] Aziz S, Jalal F, Nawaz M, Niaz B, Ali Shah F, Hafeez-ur-Rahman Memon M, Latif F, Nadeem S and Rajoka M I 2011 Hyperproduction and thermal characterization of a novel invertase from a double mutant derivative of Kluyveromyces marxianus *Food Technology and Biotechnology* **49** 465-73

Solar Based Smart Irrigation System Using PID Controller

Shehzar S Sheikh[1], A Javed[2], M Anas[2] and Faisal Ahmed[2]

[1]Department of Electrical Power Engineering, U.S. Pakistan Center for Advanced Studies in Energy (USPCASE), National University of Science and technology (NUST), Islamabad, Pakistan.
[2] Faculty of Information and Communication Technology, Balochistan University of Information Technology, Engineering and Management Sciences, Quetta, Pakistan.

Email: shehzarshahzad@gmail.com

Abstract. An approach to minimize the human interface for the agriculture sector is presented in this paper which provides comfort to the farmer in monitoring and performs automatic irrigation operation. Here, a combination of software integrated microcontroller that is Arduino, four real-time sensors fetching input data from field and a DC motor pump as output are used. A controlling technique Proportional Integral and Derivative (PID) controller is integrated into Arduino. The inputs include four sensors: wind speed, ambient temperature, humidity, and radiation sensors while taking soil moisture as the reference point. These four sensors work at different ratios on the basis of PID and penman model of evapotranspiration. The system is powered by a lead-acid battery which is recharged by a solar panel. The four designated sensors measure field conditions and simultaneously send information through wireless communication to a remote placed monitoring device which has a Liquid Crystal Display (LCD). The complete hardware and software modeling is performed in this paper.
Keywords: Arduino, Irrigation System, Lead Acid Battery, Moisture sensor, Penman Model of Evapotranspiration, Proportional Integral and Derivative (PID), Solar Panel.

1. Introduction

Water is an indispensable element of human life and plants. The Food and Agriculture Organization of United Nations estimates 70% of water consumption is for agriculture, 20% for industry, and 10% for domestic usage [1]. A design should be devised or adopted by such methods to decrease the wastage of water as well as maintain the basic parameters required for the growth of a plant. In literature proposed methods that not only control the moisture level/water level but also control the supply of water in the fields. However, our proposed system has four sensors which work simultaneously to perpetuate the basic parameters of plant and control the water level even if the farmer is not in the field. Moreover, it also gives signals to the farmer either it is an overflow condition or underflow condition.

There are some problems regarding the irrigation system: the specific amount of nutrients and water is applied to the soil, to cope with this issue, a new adaptive smart irrigation system algorithm is applied. In this algorithm, the water is balanced in a control tray by measuring the information provided by the sensors. For this purpose, a controller is designed which is a PI controller. A feedback process is used to measure the current and future situation of the tray then it is controlled by the PI controller. An open loop-controlled technique for irrigation is used by simple relay based ON and OFF mechanism. The communication is performed between the receiving and transmitting end by GPRS/GSM modules [2]. An automatic solar-powered drip by Wireless Sensor Network (WSN) for irrigation is present. Arduino is used for controlling the soil moisture input [3]. A method by Mobile Irrigation Care Unit (MICU) has previously used: a smart wireless monitoring system mounted on a solar-powered vehicle which ensures proper irrigation of each plant [4].

D. Rodriguez et al. [5] presented a work, based on Wireless Sensor Network (WSN) technology which is cost effective and less energy consuming system. It works on multi-sensors simultaneously and senses the water level, moisture in the field and inform the farmer either by a text message or by an alarming unit. If the level of water is less or above than the predefined value, then it alarms the controller to ON/OFF. P. B. Chikankar et al. [6], presents a modified form of PID controller along with an integral function which efficiently demonstrates implicable regulation of soil moisture by supplying the appropriate amount of water to the agricultural field and meet all the requirements of the field. Moreover, it also controls the rapid changes in the environment like rainfall, underwater and makes rapid ON/OFF actions in the controller.

The Artificial Neural Network (ANN) is presented using MATLAB, [7]. The input parameters like air, temperature, soil moisture, radiations, and humidity are modeled, then the amount of water required for irrigation is estimated and afterward associated results are simulated. PID controlled irrigation system using temperature, humidity, and moisture is presented in [8], which is powered by a solar panel. The control tray is designed which is simply an ON/OFF mechanism to control the amount of water but it is not so simple there is a need to develop a controller i.e., PID controller which allows fully automatic control operation and reduces the cost of equipment, [9]. A closed loop controller based on modified PID is used with constrained integral that provides good soil moisture regulation by applying the appropriate amount of water and is fast to environmental change, [10].

An innovative application of PID technique is introduced which uses penman model of evapotranspiration [11] for irrigation purpose. Here modeling of four parameters, as discussed in the above paragraph, for the operation of DC pump controlling using PID Controller algorithm on Arduino provides a comprehensive study of irrigation in the agriculture sector. The penman model is considered the most scientific model for evapotranspiration. A general penman model of evapotranspiration is explained with a moderated correlation.

2. Methodology

The irrigation system by PID techniques which is real-time need of irrigation sector. The researchers combined two different fields of studies by a combination technique model: Irrigation with penman model of evapotranspiration and an Arduino as the most used microcontroller in modern electronics applications with an efficient PID as a controlling algorithm, as shown in figure 1.

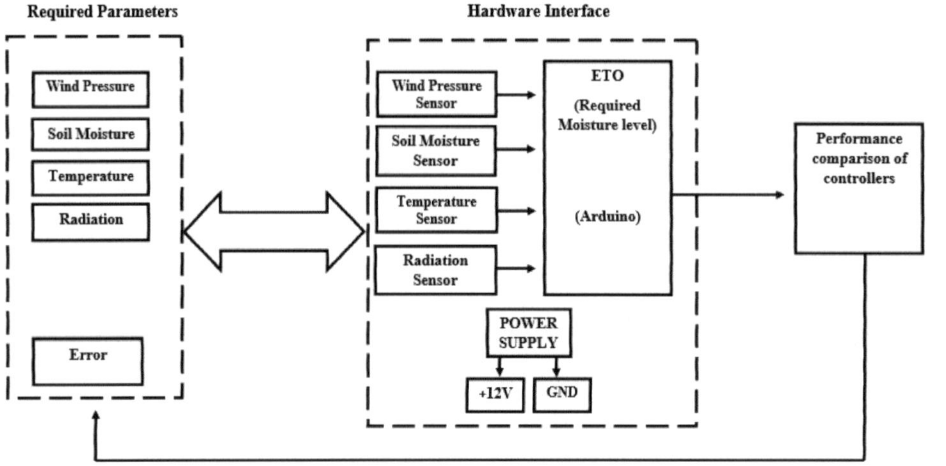

Figure 1. General Configuration of Proposed Methodology.

This work is comparatively efficient with respect to the previously lesser number of input sensing parameters in this field of research. Further, PID Controller is designed on the aggressive and non-aggressive model which efficiently reacts to small variations in sensors values. The system discussed in figure 1 consists of hardware and back-end software interface.

The hardware consists of an Arduino system with four required sensors; namely as, ambient humidity, wind pressure, solar radiation and soil moisture sensors that substantiate to take the required moisture level for the irrigation purpose. The Arduino requires 12 volts DC source for this purpose the complete system is solar powered with a battery backup. The output values of our sensors are transferred to the PID controller to perform the governing decision for the DC pump operation. Hereafter, the performance comparative controlling techniques are used for measuring all the sensors to its required moisture level. The software interface is based on Arduino coding that provides a timely change in the ratios of sensors input to PID controller and set point on soil moisture i.e., in case of rain and long-term corps routine. The sensors unit is integrated with the Arduino in the field which transmits radio signals to the receiver end. The receiver is used for monitoring and displaying output signals and these signals are converted into a required form for observation i.e., analog and digital readings. In this system, the back-end software and hardware have two-way communication with the wireless a device (Radio Frequency based) module which continuously observes and displays transmitter values.

2.1. Irrigation System Hardware

The complete hardware of this work is shown in figure 2, which contains a solar panel, fake sun, electric dimmer switch, transmitter, receiver, DC motor pump, four sensors, two black buckets: one contains the dry soil and other with water, and a Digital Multi-Meter (DMM) for measurements.

Figure 2. Complete Hardware of Irrigation System.

2.2. Arduino Uno Microcontroller

In this system, an open source platform Arduino Uno microcontroller is used, which contains two parts: a physical Printed Circuit Board (PCB) and ATmega328P microcontroller. It has 14 digital inputs and output pins. Six of them can be used as Pulse Width Modulation (PWM) outputs, six pins are for analog inputs, a USB connection, and a power jack. The Arduino software (IDE) 1.0 for programmable coding is used.

2.3. DC Motor Pump

This project includes a DC Motor Pump for irrigating the soil with the help of PWM technique which controls the DC Pump to water the soil up to its requirement while maintaining upper and lower limits. The DC pump is the only output in this project as shown in figure 3 that reacts on the given set of values which includes wind, humidity, irradiation and moisture values. Figure 4 shows the PSpice (Electronic Circuit simulator) diagram of DC pump modeling.

IOP Conf. Series: Materials Science and Engineering **414** (2018) 012040 doi:10.1088/1757-899X/414/1/012040

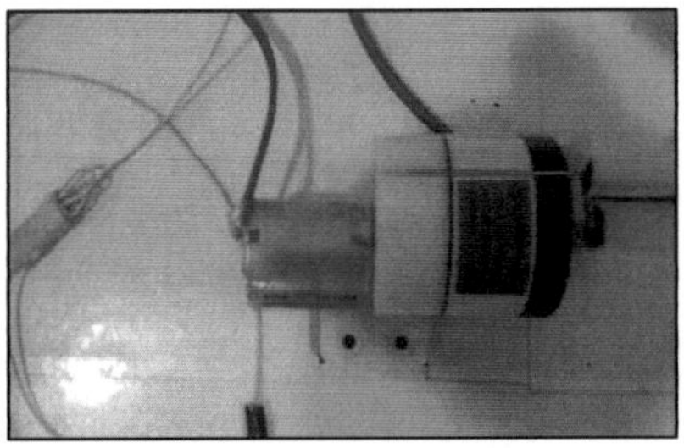

Figure 3. DC Pump used for Irrigation.

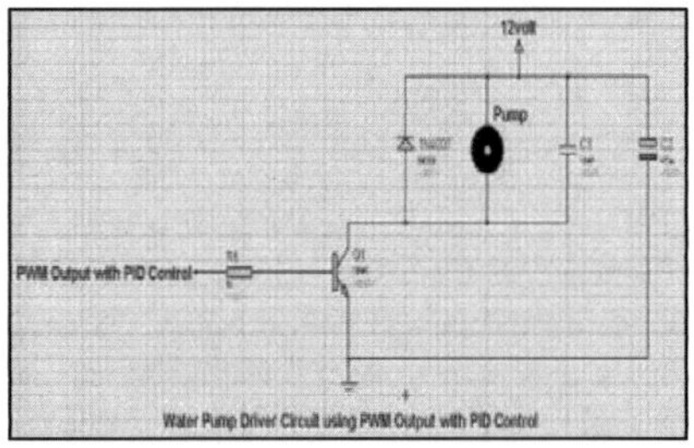

Figure 4. Equivalent model of DC Pump Simulated in PSpice.

2.4. Soil Moisture Sensor

The soil moisture sensor measures soil moisture content in the active root area. The soil is nonconductive in nature; however, due to the presence of water content in soil increases its conductivity. The variation in conductivity of soil depends on moisture which is opposite to resistance (0-5 kilo Ohms). In this way, a soil moisture sensor measures soil conductivity in an analog waveform which is further converted into digital format for microcontroller processing.

2.5. Fake Sun and Solar Panel

The fake sun is to demonstrate a prototype of the sun in figure 5. Here two 100 Watts incandescent bulbs are used which show the visible region of a sun irradiance. In addition to make this setup compact the black surrounding is used to capture heat for our calculation.

Figure 5. Fake Sun setup enclosing two incandescent bulbs and solar panel at top of the bucket.

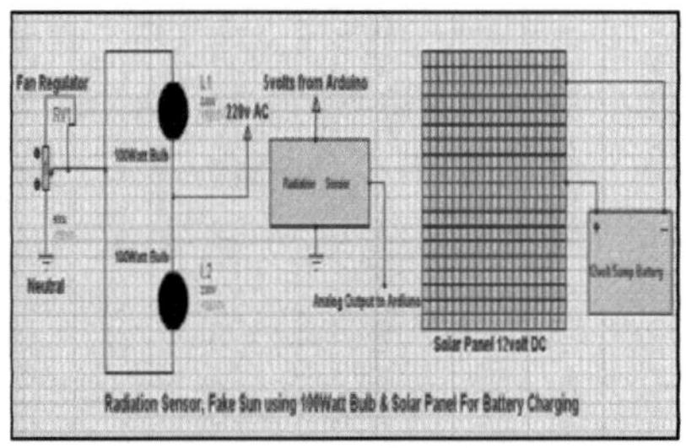

Figure 6. Equivalent model of fake sun and solar panel simulated in PSpice.

In this setup solar panel of 12 volts and 5 amperes is placed at top of the fake sun. The solar panel produces a different amount of output depending upon the sunlight versus time in a day, and to demonstrate this phenomenon of variation in solar irradiance, the researchers used an electric dimmer switch which can vary the electric power of bulbs i.e., the intensity of light can be changed. Figure 6 manifests a solar panel of the monocrystalline structure is used, which is cost effective with 21.3% of efficiency. Here the solar panel is directly charging 12 volts battery that does not need any charge-controller for tracking maximum power output due to two reasons. Firstly, the desire irradiance can be adjusted by electric dimmer switch which set desire intensity of light bulbs. Secondly, battery withdraws a fixed amount of charge for charging versus any variable loads where the continuous tracking is required to harvest the maximum amount of power from solar panels.

IOP Conf. Series: Materials Science and Engineering **414** (2018) 012040 doi:10.1088/1757-899X/414/1/012040

2.6. Lead Acid Battery

The table 1 exhibits lead acid battery and other exiting batteries including Nickel Metal hydride, Sodium Nickel (ZEBRA) and Ultra-Capacitors comparison. This comparison includes the cost in dollars per kilo Watts hours ($/KWh) and self-discharge rate per day (%/day) which explains how much a battery discharge due to internal reactions. In table analysis, it is clear that lead-acid battery is suitable in terms of cost ranging from 100- 150 ($/kWh), and self-discharge of 0.29-0.57(%/day) for our work.

Table 1. Comparison of Lead-Acid Battery vs Other Exiting Batteries in the Market.

Specifications	Lead Acid	Nickel Metal Hydride	Sodium Nickel (Zebra)	Ultra-Capacitor
Cost ($/Kwh)	100-150	150-250	100-200	300-2000
Self-Discharge (%/Day)	0.29-0.57	01-1.43	15	20-40

2.7. Transmitter Module

The complete assembly of the transmitter, which contains Arduino and input parameters (sensors) shown in figure 7. We have set a different ratio criteria for all input parameters on basis of their importance that can be modified as per agriculture needs: crop type, field type and sessional variation. Input sensors requires different voltages for their operation which is shown in table 2.

Table 2. Voltage distribution of sensors.

Input Voltage	Wind Speed	Ambient Temperature	Air Humidity	Soil Moisture	Radiation
12 DC	0-12V	0-5V	0-5V	0-5V	0-5V

This system has 12 Volt DC motor pump which variates its speed using pulse width modulation technique (PWM): a power electronics technique used to control DC-DC voltage conversion with no power losses.

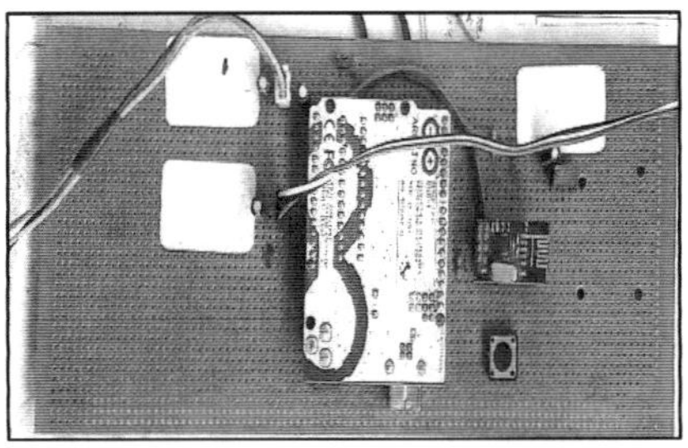

Figure 7. Hardware of transmitter end including and wires corresponding to sensors.

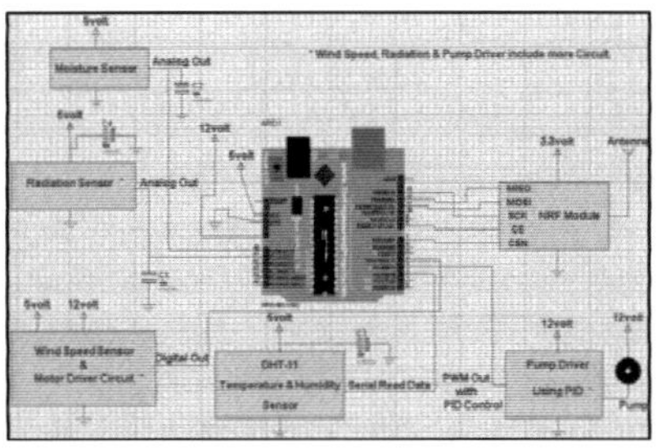

Figure 8. Equivalent model of a transmitter including wind speed, radiation, NRF, DHT-11 and moisture sensors; NRF Module; DC Pump simulated in PSpice.

NRF (Radio frequency based device) is used for communication between transmitter and receiver. This device makes our system out performed previously developed system in terms of cost effectiveness and long distant communication with increased number of sensors. These four input sensors produce analog signals which are translated into digital format to be processed and make decision in terms of activating DC motor pump. furthermore, the transmitter simultaneously send signal to receiver which contains updated values of all parameters: wind, ambient temperature and air humidity, solar radiation and soil moisture variations, keep farmer informed of any variations.

2.8. Receiver Module
We designed a portable equipment for the formers that enables them to monitor field parameters remotely. This module consist of an Arduino, NFR (for wireless communication), Liquid Crystal Display (LCD) and a 9 Volts battery as shown in figure 9 (hardware) and figure 10 (Simulated setup). The receiver module efficiently keeps updating its values which includes: ambient temperature, moisture, humidity, wind speed, and radiation.

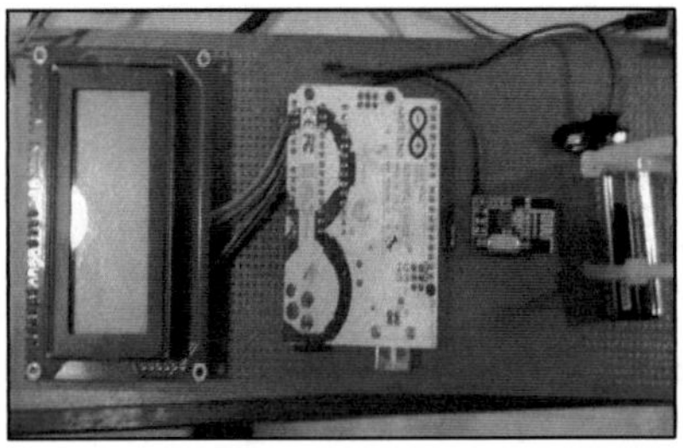

Figure 9. Hardware of Receiver End Including NRF.

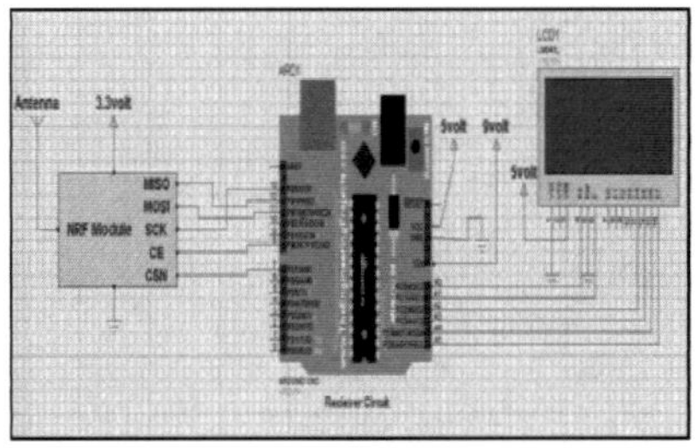

Figure 10. Equivalent Model of a Receiver Including LCD and NRF Module in PSpice.

3. Proportional integral, and Derivative (PID) Controller for The Irrigation System

The PID controller is fast and efficient in terms of their decision-making capability. We integrated PID for a simple application with its pragmatic implementation for irrigation control in agricultural sector, operating a PID algorithm in aggressive and non-aggressive mode for DC motor pump speed controlling and deciding either/whether to turn on or turn off the pump. The pump is automatically controlled by PWM which variates the speed of DC pump in different situations: mainly at starting and just before the end of desired amount of water with precision.

We provided different input ratios to PID algorithm; the ratios are set on the basis of input sensors (air humidity, wind speed, radiation, and soil moisture) importance and other effecting variables as shown in table 3. In Addition, The penman model of evapotranspiration model has four most important parameters which we integrated with PID controller, [11].

Table 3. Sensors Percentage for PID Controller.

Sensors	Wind Speed	Ambient Temperature	Air Humidity	Soil Moisture	Radiation
Percentages	0.3%	1.2%	0.3%	98%	0.2%

4. Modeling of Input Parameters

The process evapotranspiration is influenced by four parameters (ambient temperature, air humidity, wind speed, radiation and soil moisture). The modeling of parameters is remodeled into Arduino coding. The penman model of evapotranspiration discuss parameters:

4.1. Air Humidity: It is modeled as

- A sine wave with an amplitude of 10 %
- Biasing of 60% (constant)
- A frequency of 0.2618 rad/h

4.2. Wind Speed: It is modeled as

- A sine wave with an amplitude of 1 Km/h
- Biasing of 3.5 Km/h (constant)
- A frequency of 0.2618 rad/h

4.3. *Radiation:* It is modeled as maximum possible radiation at earth's surface (Rmax).

- A sine wave with an amplitude of $\frac{2MJ}{m^2}$
- Biasing of 112 MJ/m
- A frequency of 0.2618 rad/h

5. Evapotranspiration Model

We related irrigation with the Penman-Monteith equation (01) of evapotranspiration [11] that has generally been considered and accepted as a scientifically sound formulation for evapotranspiration (ET_O) as reference estimation (comprehensive model for irrigation). This equation is expressed as a combined function of air humidity, radiation, maximum and minimum moisture, vapor pressure and wind speed. The updated method by Food and Agriculture Organization (FAO) in May 1990, of Penman equation is given as under:

$$ET_o = \frac{\left\{0.408\Delta(R_n - G) + \gamma \frac{900}{T + 273} u^2 (e_s - e_a)\right\}}{\Delta + \gamma(1 + 0.34u^2)} \quad (01)$$

Where:
ET_0 = reference evapotranspiration, mm day^{-1}.
Rn = net radiation at the crop surface, MJ m^{-2} d^{-1}.
G = soil heat flux density, MJ m^{-2} d^{-1}.
T = mean daily air temperature at 2 m height, °C.
u^2 = wind speed at 2 m height, ms^{-1}.
e_s = saturation vapor pressure, kPa.
e_a = actual vapor pressure, kPa.
$e_s - e_a$ = saturation vapor pressure deficit, kPa.
Δ = slope of the vapor pressure curve, kPa °C^{-1}.
γ = psychrometric constant, kPa °C^{-1}.

6. Software Modeling and Devices Communication

The work includes four real-time sensors fetching data. The microcontroller is working on software-based programming and tunning each of the parameters (sensors) with respect to system desire conditions. As discussed above, the Arduino (microcontroller) performs analog to digital conversion for data analyzing for PID controller.

Transmitter end consists of sensors with controller input and DC motor pump as the output of the system. To keep a timely monitoring for system performance in all conditions this may include transients in sensors values, self-generated values of PID or any malfunction of the system. In addition, the monitoring gives the first-hand comfort to a farmer for distant monitoring the field parameters. For monitoring parameters and system performance, efficient wired or wireless communication is required. In this paper, wireless communication is carried out by NFR24L01 (Radio frequency model number). Which enables the farmer to monitor all parameters from the distance of 100 meters. LCD screen shows values on receiver end which includes air humidity, ambient temperature, solar radiation, wind speed and soil moisture I n figure 11.

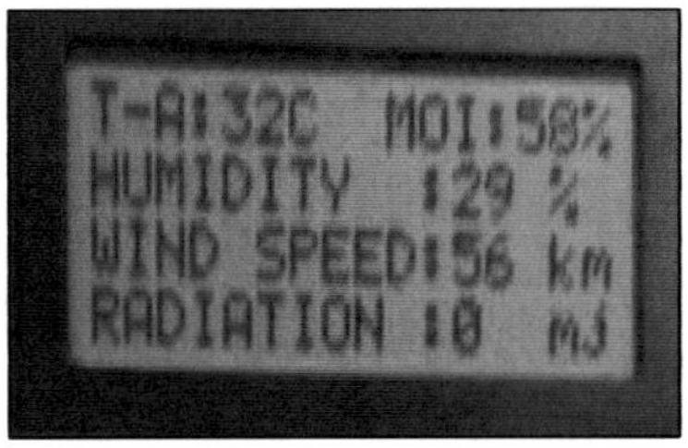

Figure 11. LCD display of receiver end which shows T-A (ambient temperature), MOI (moisture), HUMIDITY (air humidity), wind speed and radiation sensors values.

7. Conclusion

The proposed work has emphasized on the importance and significance of modernization of irrigation in agriculture field while considering all-natural parameters including temperature, air humidity, wind speed, radiation and soil moisture which have scientific importance. The consideration involves modernization sometimes makes system costly; however, in this research all components are not only cost effective but also robust in performance. In addition, this work is a professional and practical combination of hardware and controlling technique; such as, a PID controller. This work is intended to widen the knowledge in parameters consideration for irrigation and simple in system designing. Furthermore, the purpose of this model is to facilitate and help farmers for controlling and managing water wastage, and providing them proper monitoring while excluding human error which is one of the most practical solutions of agriculture sector in Pakistan and other under developing countries of the world.

8. References

[1] *Food and Agriculture Organization of the United Nations [Online]. Available: http://www.fao.org/nr/water/aquastat/water_us e/i%0Andex.s tm, Last accessed: July 6, 2017*

[2] Kumar A, Kumar A and Sharma K P 2017 Smart Irrigation System using IoT : SIS *international journal of Engineering & Technology (IJERT)* **6** 103–7

[3] Sadeesh N, Elavarasi A, Karthikeyan R, Rithesh R and Samundeeswari V 2017 Energy Efficient Solar Powered Smart Crop Watering with Water System using Arduino Controller **3** 303–7

[4] Kumar S, Sethuraman C and Srinivas K 2017 Solar Powered Automatic Drip Irrigation System (SPADIS) Wireless Sensor Network Technology *Int. Res. J. of Eng. Technol* **4** 722–31

[5] Rodriguez D, Reca J, Martinez J, Lopez-Luque R and M. Urrestarazu 2015 Development of a New Control Algorithm for Automatic Irrigation Scheduling in Soilless Culture *Appl. Math. Inf. Sci* **9** 47–56

[6] Chikankar B P, Mehetre D and Das S 2015 *An automatic irrigation system using ZigBee in wireless sensor network* (Int. Conf. Pervasive Comput.) p 1–5

[7] Goodchild S M, Kühn D K, Jenkins D M, Burek J K and Dutton J A 2015 A Method for Precision Closed-loop Irrigation Using a Modified PID Control Algorithm *Sensors & Transducers* **188** 61–68

[8] Ososanya T E, Haghani S, Mahmoud H W and Lakeou S 2015 *Design and Implementation of a Solar-Powered Smar Irrigation System* (122nd ASEE Annual Conference & Exposition)

[9] Kumar N and Shimi M Smart Farming System for Indian Farmers using Arduino based Technology *int. j. of Adv. Research, Ideas and Innovation in Tech.* **3** 105–10

[10] Fareed A, Ullah G M, Hussain A, Zaidi M and M. Baig 2017 Smart Irrigation Care Unit

[11] Zotarelli L, Dukes D M, Romero C C, Migliaccio W K and Morgan T K 2010 Step by step calculation of the Penman-Monteith Evapotranspiration (FAO-56 Method) *Inst. Food* **1–10**

Human Machine Interface Glove Using Piezoresistive Textile Based Sensors

Surjeet Kumar[1], M Junaid Sultan[1], Anayat Ullah[1], Syed Zameer[2], Saqib Siddiqui[2] and Syed Kamran Sami[3]

[1] Department of Electronic Engineering, Balochistan University of Information Technology, Engineering and Management Sciences, Quetta, 87300, Balochistan, Pakistan
[2] Department of Textile Engineering, Balochistan University of Information Technology, Engineering and Management Sciences, Quetta, 87300, Balochistan, Pakistan
[3] Department of Chemical Engineering, Balochistan University of Information Technology, Engineering and Management Sciences, Quetta, 87300, Balochistan, Pakistan

Email: surjeetkhurana9@gmail.com

Abstract. Human machine interface technology is focused upon new ways of interaction between human beings and machines. Gesture recognition gloves are getting increasingly popular as human-machine interface devices. Conventionally, these gloves use electronic sensors to sense different hand gestures. As electronic sensors are bulky and uncomfortable, we propose a glove with textile based piezoresistive yarn sensors. This glove is flexible, more comfortable and cheap as compared to the conventional human machine interface gloves. We have examined that this glove can effectively sense our gestures and can be used for teleoperations, sign language to speech conversion systems and gaming.
Keywords. Piezoresistive, Textile Sensors, Human Machine Interface Glove.

1. Introduction

Machines have been close companions to human beings for decades. The rapid increase in a number of machines means more interactions with us. The recent developments in human machine interface technology are focused on developing more efficient ways of communication between humans and machines. As we use hand gestures extensively in our everyday communication with human beings, these gestures can also be used to communicate with the machines. In order to recognize these gestures and use those to control machines two main approaches are used; the first approach is to use visual based recognition systems that extract gesture data from graphic images; the second approach is to use haptic data recognition glove based systems. The latter glove based systems are better than the visual based systems because these are cheap, portable and require less processing power.

The glove based gesture recognition systems are applicable in design and 3d modeling, manufacturing, healthcare, robotics, device control, sign language recognition etc. For 3D modeling and design, glove based systems are used with computer-aided design software that can be used by engineers, architects, and industrial designers. Glove based systems are a better way to interact with robots with natural hand movements. In medicine and healthcare, electronic gloves are used for the rehabilitation and assessment of hand motion by acquiring gesture data. For sign language recognition and conversion to speech or text, gloves are used to recognize different gestures that are programmed to produce a specific letter or word with the help of a computer program. For data management, glove based systems are used for visual representation of data and images and improve the interaction between users and the data. The

Data glove was used by NASA to interact with the simulated airflow around an aircraft at NASA laboratories [1].

In past various attempts were carried out to measure analog haptic data using gloves and then use it to control machines, the first prototypes of glove based systems were made in the 1980s, such as the MIT-LED glove which was used to track the position of the hand and then show it as a computer graphics animation [2]. The digital data entry glove [3] used touch, proximity, tilt and inertial sensors to measure finger joint flexion and wrist tilting and was able to convert these motions into 96 ASCII letters and numbers. The Data glove [4] commercialized in 1982 was based on plastic tubes and light sensors that were able to measure and store joint angles. The Cyber Glove [5] consists of 22 piezoresistive sensors for measuring finger flexion, including software for virtual modeling. The Human glove [6] uses 20 Hall-effect sensors to measure joint bending angles for motion analysis and medical applications. Most of the aforementioned gloves use electronic sensors mounted on the cloth supports. A common drawback in this approach is a lower level of comfort and obtrusive nature of the glove. Real time implementation of these systems requires a lot of work in this area.

In this work a conductive yarn with piezoresistive properties is integrated into a glove, the yarn is integrated in such a way that it is stretched and compressed with the movement of finger joints. The resistance of this yarn changes as it is stretched or compressed, this change is notable and can be used to monitor hand gestures. Scanning electron microscopy (SEM) is used to study morphology and understand the working of piezoelectric yarn. A portable system was designed to monitor analog data for each gesture. The raw data was processed in MATLAB computation software. A 3d printed robotic hand was controlled with the glove to demonstrate its application in teleoperations.

2. Methods

The piezoresistive yarn was imported from Sparkfun USA, it has a blend of 20% stainless steel and 80% polyester fibers. It was noticed that for a specified length, the resistance of this yarn was decreased as the yarn is stretched. To understand this phenomenon Scanning Electron Microscopy (SEM) was done using COXEM CX-200 SEM to understand the piezoresistive nature of this yarn.

Based on the above-mentioned phenomenon, this yarn was integrated into a glove. As shown in Figure 1(a) the yarn has 3 plies, Figure 1(a-b) show the embroidery design and yarn placement on the glove. The yarn was embroidered over each finger in a glove so that the yarn is stretched or released with the finger movements causing a change in resistance. Snap buttons were used to create electrical contacts with the yarn.

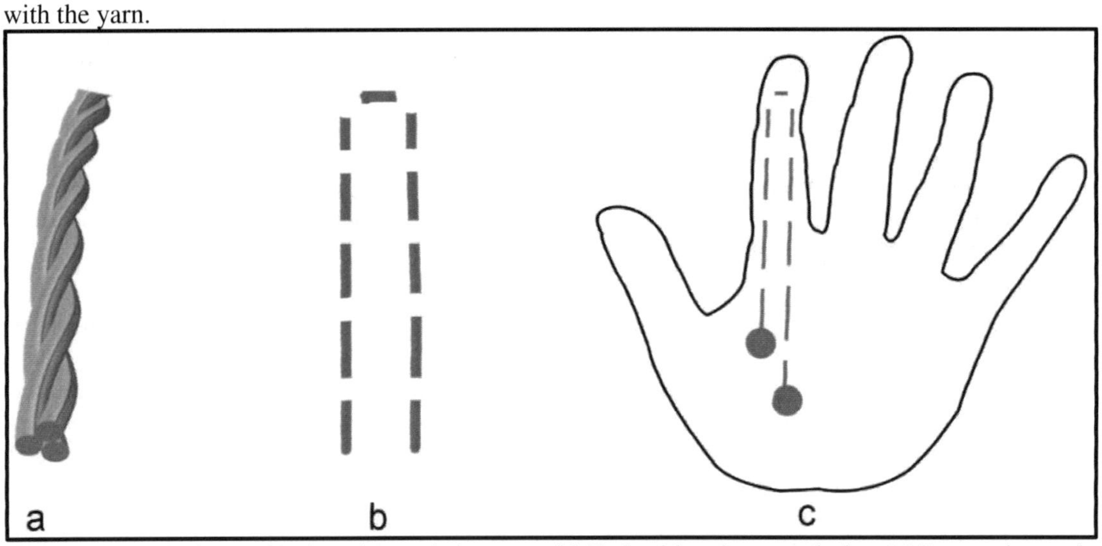

Figure 1: (a) 3 ply conductive piezoresistive yarn (b) Embroidered sensor design (c) Integration and placement of piezoresistive yarn in the glove.

Figure 2 shows the block diagram of the working of entire human machine interface systems employed in this work. Hand gestures are sensed by the piezoresistive yarn sensors by measuring resistance for repeated gestures. This change in resistance is acquired in the form of analog data by an Arduino Nano Board with in-built 8-bit Analog to Digital Converter (ADC). This analog data has random noise, and overshoots occur as a finger is moved rapidly. The raw data needs to be filtered before it can be used to recognize gestures. A five point moving average filter was applied to filter out high frequency component in the signals.

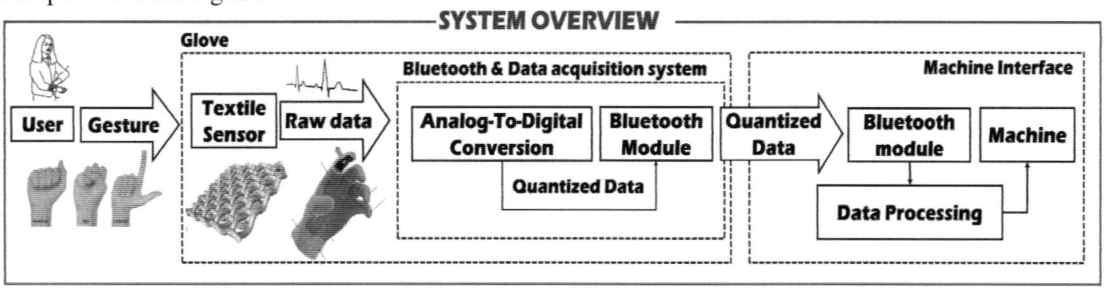

Figure 2: Block Diagram for the gesture recognition glove system.

The Arduino board is controlled with a MATLAB script. Arduino IO package works as a server program that allows Arduino control with MATLAB commands. The acquired data for each finger sensor is stored in a separate MATLAB vector.

A robotic hand was used to demonstrate the application of gesture recognition gloves in teleoperations. This robotic hand was 3d printed using an open source CAD design from InMoov. The robotic hand was assembled, and 5 servo motors were used to move each finger.

3. Results

SEM was used to observe the morphology of the piezoresistive yarn. Figure 3(a) shows the 500 times magnified SEM image of this yarn. The SEM results show that the fiber has a diameter of 18 to 20 micrometers. And Figure 3(b) shows that the interfiber space is around 20 to 40 micrometers. This shows that the yarn has high porosity. As the yarn is stretched the interfiber spaces are reduced and as the steel fibers come in contact with each other the overall resistance of the yarn decreases. And when the yarn is unstretched interfiber spaces are increased again.

IOP Conf. Series: Materials Science and Engineering **414** (2018) 012041 doi:10.1088/1757-899X/414/1/012041

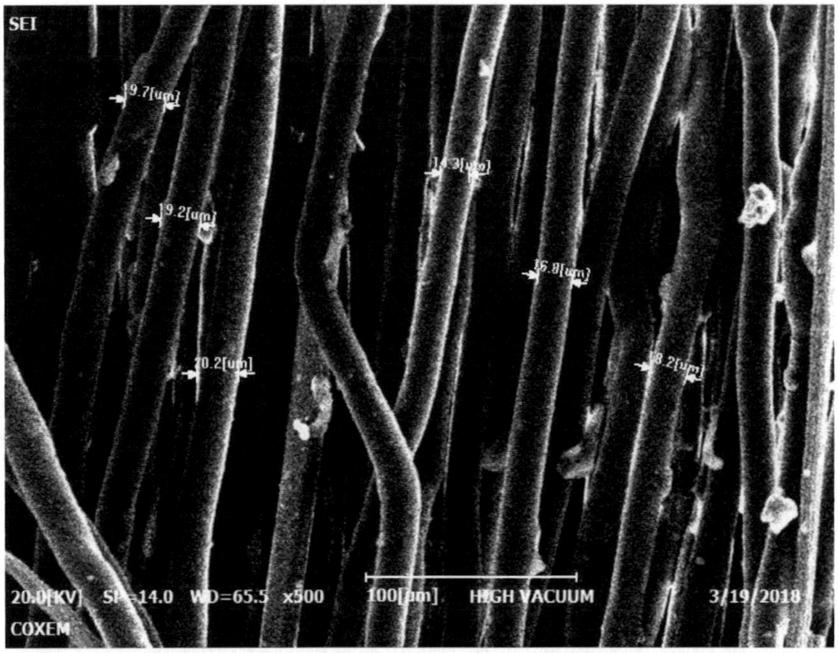

(a)

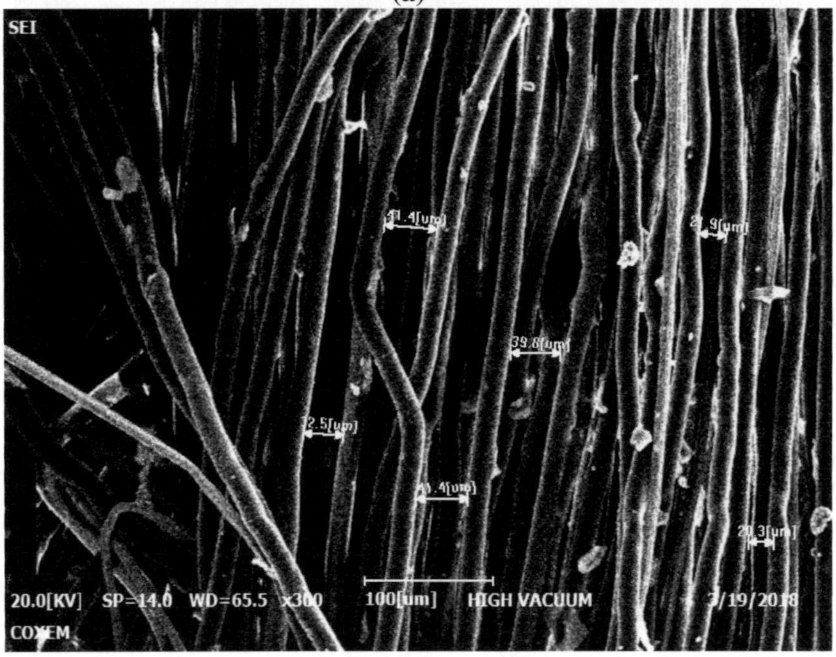

(b)

Figure 3: (a) Scanning Electron Microscope (SEM) image of piezoresistive yarn showing fiber diameter. (b) Scanning Electron Microscope (SEM) image of piezoresistive yarn showing interfiber spcaes.

Figure 4(a) shows the changing resistance of the conductive yarn for continuous bending and releasing of the five fingers, Figure 4(b) shows filtered resistance signal for the yarn. Each of the finger sensors is responsive to movement of the finger. It can be seen that there is a significant change in yarn resistance of the order of K Ohms.

IOP Conf. Series: Materials Science and Engineering **414** (2018) 012041 doi:10.1088/1757-899X/414/1/012041

The above results show that the piezoresistive yarn sensors are sensitive to even minor finger movements, in addition to this the yarn sensor's signal shows overshoots as the finger is moved rapidly. To filter out the

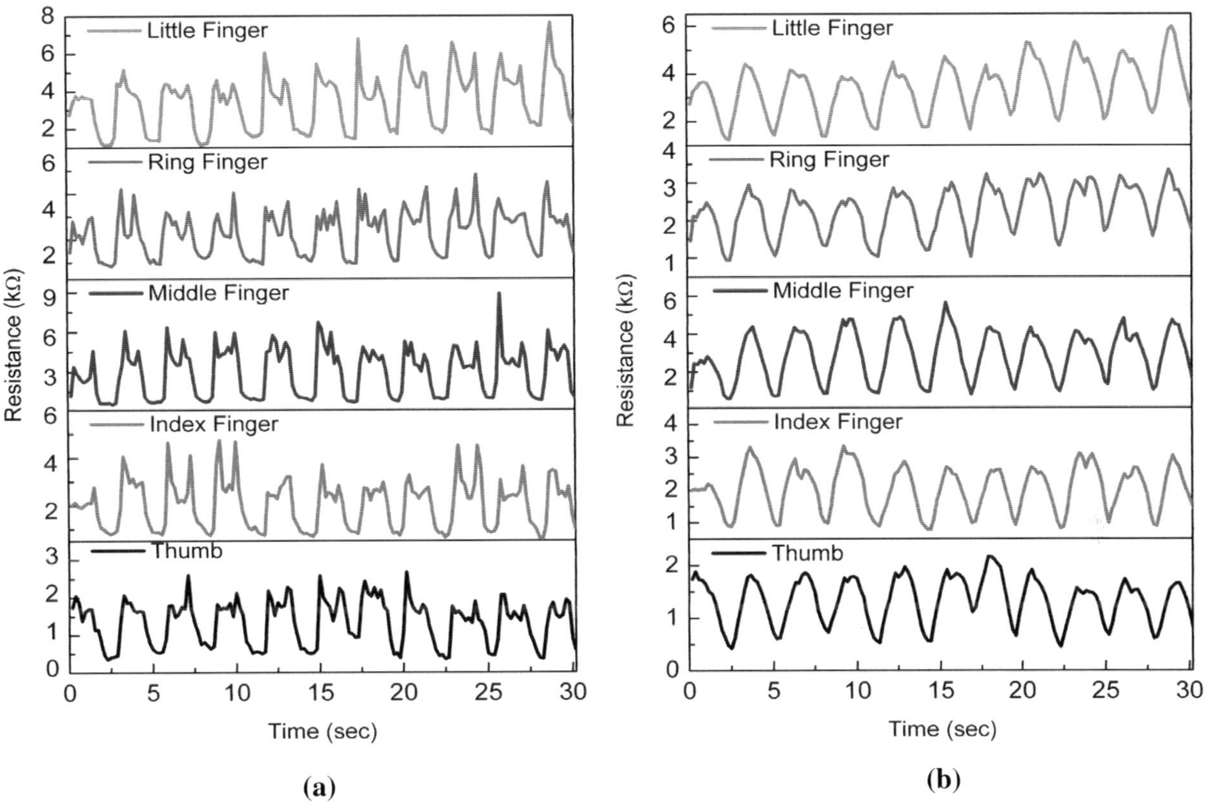

Figure 4: (a) Resistance verses time graph for piezoresistive yarn sensors for the five fingers. (b) Signal output from a moving average filter.

overshoots and fluctuations in the signal additional signal processing techniques are required. The filtered output from the 5 points moving average filter is smooth and can be used to recognize gesture or control other machines effectively. The filtered output was used to control the servo motors that control the InMoov robotic hand. Figure 5 shows the robotic hand and HMI glove; the robotic hand could mimic the movement of the person wearing the glove. This shows that this glove can be easily used in teleoperations.

4. Conclusion

As wearable gesture recognition devices are gaining popularity, a gesture recognition glove is developed with piezoresistive yarn as a sensor. This yarn has 20% stainless steel and 80% polyester. The SEM images show that the fiber has diameters of 18-20 micrometers. And the inter fiber spaces are in the range of 20-40 micrometers in normal condition. As pressure is applied across the two ends of the yarn the inter fiber spaces in the yarn start decreasing. This yarn is integrated into the glove in such

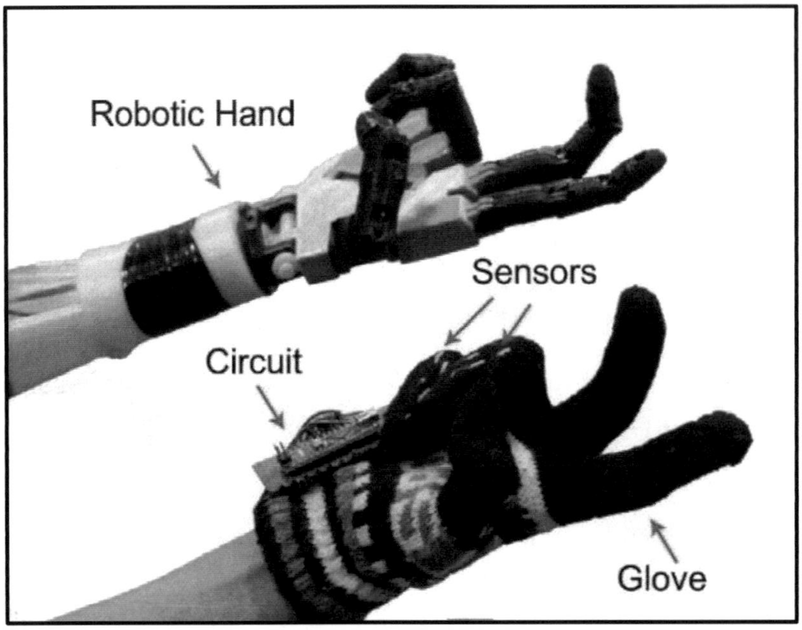

Figure 5: InMoov Robotic Hand controlled with piezoresistive yarn sensors integrated into the glove.

a way that with the movement of each finger a corresponding yarn sensor experiences pressure. The inter fiber spaces decrease as pressure is applied across the yarn and as the stainless steel fibers come in contact with each other the overall resistance of the yarn decreases. An Arduino board with 10 bit ADC is used to measure the resistance of this yarn with the help of a voltage divider circuit. This signal is transmitted to a PC and then it is processed. 5 points moving average filter was applied to filter out the overshoots and random noise signals. Results show that the piezoresistive yarn sensor is sensitive to finger movements. The gesture recognition system was used to control a robotic hand remotely which demonstrates the capability of teleoperations using this type of glove.

5. Acknowledgments

We would like to thank Office of Research Innovation and Commercialization, BUITEMS, for funding this project. We also acknowledge Mr. Ali Raza Shafqat, Mr. Naqeeb Ullah, Mr. Luqman Sarangzai Mr. Umer Iftikhar, Mr. Janzaib Masood, Mr. Imran Khan, Ms. Mutanza Malik for their valuable suggestions.

References

[1] S. Bryson and C. Levit, "The virtual wind tunnel," *IEEE Comput. Graph. Appl.*, **vol. 12**, no. 4, pp. 25–34, Jul. 1992.

[2] D. J. Sturman and D. Zeltzer, "A survey of glove-based input," *IEEE Comput. Graph. Appl.*, **vol.14**, no. 1, pp. 30–39, Jan. 1994.

[3] G. Grimes, "Digital data entry glove interface device," *U.S. Patent 4 414 537*, AT&T Bell Lab,Murray Hill, NJ, USA, 1983.

[4] T. G. Zimmerman, "Optical flex sensor," *U.S. Patent 4 542 291*, Sep. 29, 1982.

[5] P. Coiffet and G. C. Burdea, Virtual Reality Technology. New York, NY, USA: Wiley, 2003.

[6] L. Dipietro, A. M. Sabatini, and P. Dario, "Evaluation of an instrumented glove for hand movement acquisition," J. Rehabil. Res. Dev., **vol. 40**, no. 2, pp. 179–190, 2003.

IOP Conf. Series: Materials Science and Engineering **414** (2018) 012042 doi:10.1088/1757-899X/414/1/012042

Trend analysis in precipitation at individual and regional levels in Baluchistan, Pakistan.

Erum Aamir[1] and Ishitaq Hassan[1]

[1]Civil Department, CUST, Islamabad, Pakistan

E-mail: erum21@hotmail.com

Abstract. Pakistan's largest province by area, is Baluchistan hosts deep sea Gawadar port and is playing a vital role in one belt one road (OBOR) plan. CPEC, a Regional route comes under OBOR which connect Pakistan's Gwadar port to Chinese Xinjiang Province. CPEC Corridor upon construction is expected to revolutionize the infrastructure, economy, trade, finance, demography, environment, culture, socio-economic conditions, of Pakistan. Much of western and central CPEC routes pass through it. However, Baluchistan, due to its topography and rugged terrain, is under constant risk of rainfall driven flash floods. Flash floods are responsible for colossal damages to the infrastructure, especially roads, bridges, trade, communication that ultimately badly affects the humans, economy and the environment. The intensity and frequency of rainfall are increasing due to climate change and flash floods put everything at high risk. In this view, the assessment of changes in rainfall has got significant importance. More or less no appreciable studies have been conducted on the precipitation trend analysis and in Baluchistan. This research is focused on assessment of trend analysis in precipitation of Baluchistan using 40 years (1977-2016) data of 13 stations in Baluchistan. The data has been obtained from the Pakistan Meteorological Department (PMD). Precipitation data sample is checked for normal distribution, abrupt changes, cycles, outliers and missing values. Normal distribution of precipitation data set is done by conducting tests for normality like the Shapiro-Wilk W test, Anderson–Darling, Lilliefors, and Jarque–Bera test. Statistical tests, being the most direct methods of detecting changes in extreme rainfall intensities were adopted. Two nonparametric tests Mann-Kendall (MK) and Spearman's rho (SR) have been used to find the trends in annual and seasonal precipitations. Nonparametric methods are selected as they are less sensitive to data gaps if it exists. Theil Sen slope estimator has been used to compute the magnitude of the trend. The seasonal and regional MK test has also been applied to test the seasonal and regional trends.
Keywords. Precipitation trend analysis, regional levels, Global Climate Risk Index, Baluchistan.

1. Introduction

Pakistan is ranked 18th of most vulnerable nations in 2011 by Global Climate Risk Index. The country is categorized as a 3rd most vulnerable country among 180 countries, impacted losses due to changing weather, [1]. Therefore, it is significant to analyses historical data, its changes, finding out variability in precipitation time series to predict probable future impacts of climate change, [2-4].

Parametric and non-parametric statistical techniques are employed to undertake trend and variability analysis for precipitation time series, [5-7]. The parametric analysis includes linear regression, which needs independent normally disputed data. Nevertheless, the required data scarcely exists for hydrometeorological variables, [8]. On the other hand, the non-parametric analysis is usually employed due to its simplicity as compared to the parametric technique. Generally, Mann Kendall (MK) and Sen's Slope Estimator (SSE) is used for non-parametric analysis, [9, 10]. The MK test has been strongly recommended by the World Meteorological Organization to estimate the trends in hydrometeorological datasets, [3]. The MK test is very important because it does not require any hypothesis for the statistical distribution of the data and can be implemented to the datasets with inconsistent and uneven sampling intervals. Numerous researches have been performed in the recent past using non-parametric analysis to study frequency change of precipitation time series, [11-14].

Few studies have been carried out regarding precipitation and trend analysis in Pakistan. The frequency of extreme precipitation occurrence was studied by Maida and Ghulam, [15]. data gathered from 41 stations and analyzed through the non-parametric Kolmogorov-Smirnov test. The study noted a rise in extreme precipitation events in the country. Investigation of precipitation trends through Dunnett T3 test and Variance Analysis of 30 years showed a decrease in precipitation trends, [16]. Hussain and Lee [14], examined data of 15 stations for the seasonal variability of extreme precipitation events. MK test and simple linear regression was used. Decreasing and increasing seasonal trends in extreme precipitation was identified in southwestern and northeastern part of Pakistan respectively. Abbas et al. [2], explored the change in the extreme in precipitation events in Punjab through MK test using data from five stations. Nevertheless, no considerable trend was identified. Similarly, Ijaz Ahmad et al. [17], carried out a study regarding precipitation behavior in the Swat River Basin. MK and Spearman's rho test was employed on data from 13 stations and a mixture of increasing and decreasing trends were found.

2. Study Area

Balochistan is the biggest province of Pakistan in terms of the area out of the four provinces. It has an area of 347,190 square kilometers comprising nearly 44 % of Pakistan's totals land area, figure-1. It encompasses the southwestern part of the country. Quetta is the provincial capital and largest city having 1 million population. Punjab and the Khyber Pakhtunkhwa lie on the northeast of Balochistan, Sindh to the east and southeast, it has 800 km long coast with the Arabian Sea to the south. Neighboring countries are Iran to the west and Afghanistan to the north and northwest.

The major economic and financial source of growth of Balochistan is Gwadar Port on the Arabian Sea, and renewable resources like natural gas, coal, precious stones, gems, marble, zinc, lead, copper in found Khuzdar area etc.

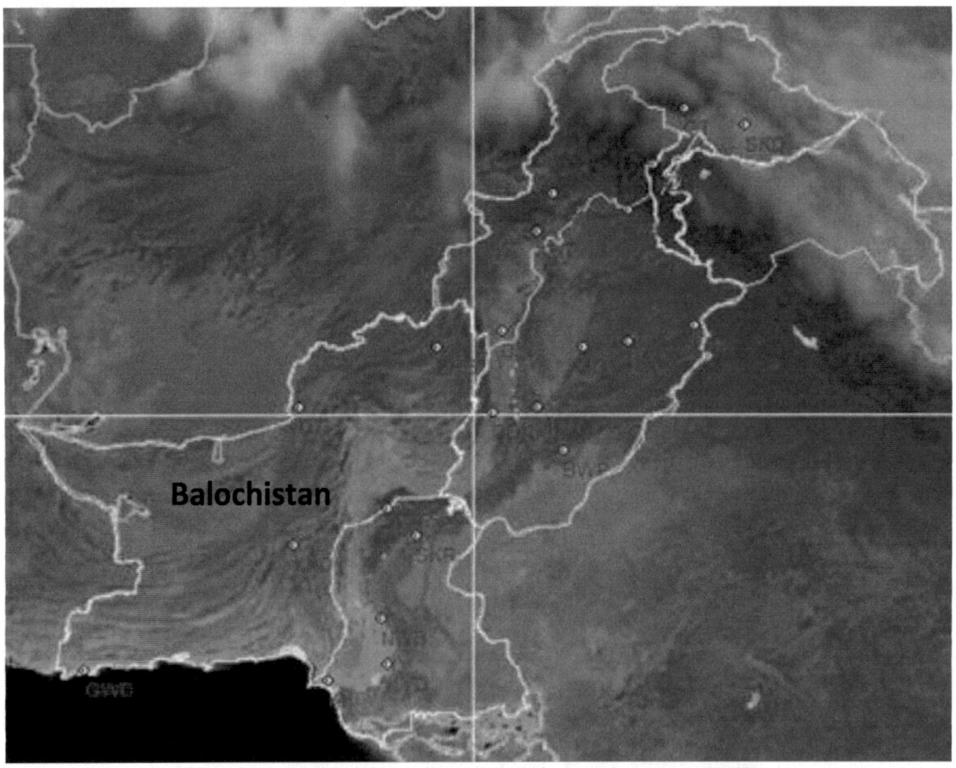

Figure 1. Baluchistan Province, the study area, (PMD).

3. Data Preparation

This research was carried out on monthly and annual precipitation data in mm/day from 15 different stations all over Balochistan. Stations were selected and scrutinized on the basis of 3 parameters (i) 40 years Data from (1977-2016) (ii) Authentic data (iii) Reliable source i.e. Pakistan Meteorological Department (PMD). Locations of all PMD weather stations in Baluchistan are shown in figure 2.

Figure 2. PMD Weather Stations of Baluchistan, (PMD).

Out of the total 15 weather stations in Balochistan, 13 weather stations were selected for the study because of their accurate, consistent, continuous data i.e. 40 years of data. Table no 1 show all 13 stations along with their longitude, latitude, mean sea level, and Mean annual precipitation and Max annual precipitation.

Table 1. Details of Selected Stations

S/n	Stations	Longitude	Latitude	Mean Sea Level	Mean Annual Precipitation	Max. Annual Precipitation
1	Dalbandin	28.88	64.39	843 m	83.30	203.7
2	Jewani	35.77	64.68	57 m	101.39	386.4
3	Kalat	29.05	66.58	2,007 m	207.49	976.9
4	Lasbella	25.87	66.71	46m	181.73	481.6
5	Nokkundi	28.82	62.75	679 m	34.44	187.6
6	Ormara	25.26	64.60	7 m	71.96	505.0
7	Panjgur	26.73	64.14	980 m	96.33	304.3
8	Pasni	25.25	63.41	10 m	102.39	356.2
9	Quetta	30.17	66.97	1,680 m	270.67	949.8
10	Sibbi	29.55	67.88	130 m	181.18	371.0
11	Zohb	31.34	69.46	1,426 m	284.00	495.0
12	Khuzdar	27.18	66.60	1,237 m	267.33	594.7
13	Barakhan	29.89	69.52	1,098 m	422.96	784.0

The precipitation data of Turbat and Gawadar weather stations do not spread over the entire study period from 1977 to 2016 and thus are not considered for the analysis.

3.1 Monthly and Annual Precipitation Data

The data, for this research obtained from PMD, was on monthly basis, which was converted to annual precipitation data for analysis. The Study period is from 1977-2016 (40 years) for the selected 13 stations all over Baluchistan.

3.2 Seasonal Data

To conduct seasonal analysis, the data is prepared by summing three to two months into 5 most common seasons namely winter, spring, summer, monsoon, and autumn. The monthly values were added to prepare the seasonal data. Table-2 illustrates the seasons being considered and the months included in each of the seasons.

Table 2. Seasonal Data

S. No	Seasons	Months
1	Winter	December, January, February
2	Spring	March, April
3	Summer	May, June
4	Monsoon	July, August, September
5	Autumn	October, November

Pakistan has been divided into 5 zones based on temperature by PMD namely zone A to Zone E. Figure-3 shows the zoning of Pakistan on the basis of temperature as defined by PMD. This research follows the temperature zoning of PMD and the study area consisting of Baluchistan province falls into three temperature-based Zones, namely: zone C, zone E primarily and some portion zone D.

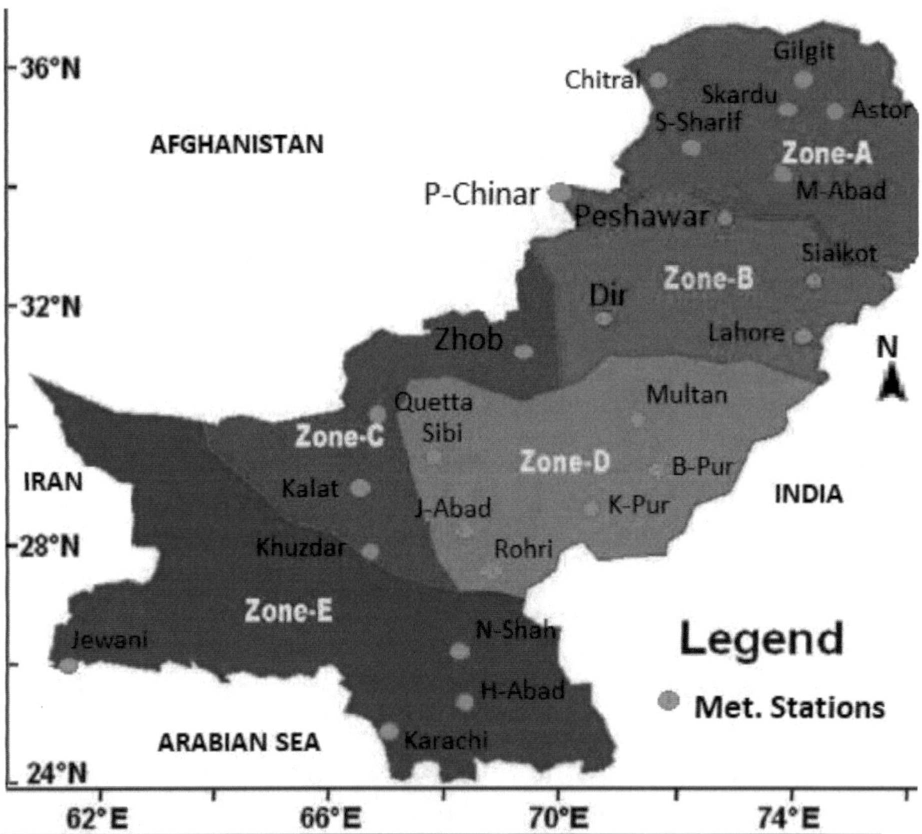

Figure 3: The temperature-based zoning map of Pakistan, [4, 11, 13, 14, 16].

3.3 Regions and Regional Data
For regional trend analysis (Table 9) Baluchistan is divided into four regions: region-1 represent province Balochistan which includes all 13 stations, regions-2 includes 5 stations which fall in zone C, region-3 includes 2 stations which fall in zone D and Region-4 includes remaining 6 stations which falls within zones E. Table-3 provides the details of regions and the stations included in each region.

Table 3. Details of Regions

Region-1	Region-2		Region-3		Region-4	
All 13 stations in Baluchistan	Dalbundin Kalat Quetta Zhob Khuzdar	Zone C	Sibi Barakhan	Zone D	Jewani Lasbella Nokkundi Ormara Panjgur Pasni	Zone E

4. Methodology
The two major methods to evaluate significant trends in precipitation time series data sets are; parametric and nonparametric trend methods. The data should be continuous independent, normally distributed and have the same standard distribution in parametric trend analysis, where as in the nonparametric method; the only condition is that the data should be independent.

Normality tests were used to check whether the time series data for each of the 13 locations are normally distributed or not. Parametric Test such as simple linear regression utilized to calculate the trend slope, its significance and validity to detect the trend in the time series precipitation data. Validity is determined by checking the normal distribution of residuals obtained from simple linear regression.

Test such as simple linear regression utilized to calculate the trend slope, its significance and validity to detect the trend in the time series precipitation data. Validity is determined by checking the normal distribution of residuals obtained from simple linear regression.

Non-parametric test such as Mann-Kendall (MK) test and Spearman's rho (SR) were used to evaluate the trends in the annual and seasonal rainfall. The reasons for adopting these tests is that they are robust and non-sensitive if the data has some gaps and do not follow the normal distribution. The magnitude of the trend estimated by Theil and Sen slope estimator (TSA). The MK and SR were applied on annual and seasonal data to evaluate the trend on annually, monthly and seasonally at each of the 13 selected stations.

The regional MK test was applied to test the regional trends on four regions: Baluchistan as a whole, temperature-based Zone C, D, and E.

4.1 Normality Tests
Precipitation data sample is checked for the characteristics such as Normal Distribution, Abrupt Changes, Cycles, Outliers and Missing Values. Normal distribution of precipitation data set is done by conducting four tests for normality like the Shapiro-Wilk W test, Anderson–Darling, Lilliefors, and Jarque–Bera test. A 2011 study concluded that Shapiro–Wilk has the best power for a given significance, followed closely by Anderson–Darling when comparing. Some research proposes this test has weaknesses. Like, the test has low power for distributions with short tails, especially for bimodal distributions. Some researchers have not included its results in their researches because of its low overall performance.

4.2 Linear Regression Analysis (Parametric Tests)

The linear regression, a parametric test and utilized to estimate linear trends in time series data. The least squares regression utilized to fit "best" straight line. The linear trend observed when slope of regression line statistically different from zero by using t-test. The positive and negative slope indicating increasing and decreasing trends respectively.

Simple linear method of least squares was used to detect the significant trend and for the comparison of trend slope. Simple linear regression calculates (1) the coefficient of variation (R^2), which is an estimate of the degree of co-variation between AR and time (1=perfect correlation, 0=zero correlation) and (2) the p-value representing the appropriateness of rejecting the null hypothesis of no trend with time. If p-values are less than 0.05, then there is a 95 percent probability that data indicate an increasing or decreasing trend, depending on the sign of the slope. The validity of linear regression results depends on whether or not the regression residuals are normally distributed.

4.3 Mann-Kendall (MK) test

The Mann-Kendall (MK) test analysis technique was widely used in detecting trends in climate variables, [18-24]. The MK test is a robust method of detecting a trend in precipitation data with a strength being its application in skewed data. The MK test is a rank-based procedure and has a low sensitivity to abrupt breaks, which may be caused by discontinuous data.

The MK test is a nonparametric test for monotonic trends where data is either consistently increasing or decreasing over time. Therefore, the MK test is not suitable when the cyclic trend observed, where data is alternatively increasing and then decreasing.

In the MK test, it is assumed that the rainfall data (X1, X2,.. ., Xn) represent "n" independent and equally distributed among random variables. The series x of Mann-Kendall statistic S by Mann [10], and Kendall [9], is written as:

$$S = \sum_{i-l}^{n-1} \sum_{j=i+1}^{n} sgn(X_j - X_i)$$

$$sgn(X_j - X_i) = \begin{cases} +1 & (X_j - X_i) > 0 \\ 0 & \text{if } (X_j - X_i) = 0 \\ -1 & (X_j - X_i) < 0 \end{cases}$$

Where i and j are observations of the xi and xj of time series and sign is a function of signum. The S is computed as a variance;

$$Var = \frac{n(n-1)(2n+5) - \sum_{i=1}^{g} t_i(t_i - 1)(2t_i + 5)}{18}$$

Where number of groups g is the tied rank and number of ties t in the group. For a sample size of n>10 or larger, the statistics Z is computed by

$$Z = \begin{cases} \frac{S-1}{\sigma} & for\ S > 0 \\ \frac{S+1}{\sigma} & for\ S < 0 \\ 0 & for\ S = 0 \end{cases}$$

The positive and negative Z values indicate increasing and decreasing trends respectively. The trends are significant if |Z| values are comparatively larger than standard normal deviate $Z_{1-\alpha/2}$ for the value of significance level (α).

When $|Z| > Z_{1-\alpha/2}$ or the p-value is smaller than the significance level (α), the null hypothesis of no trend is rejected. The alternative hypothesis that there is a significance trend in the time series is accepted. The $Z_{1-\alpha/2}$ and p-value are attained from table of standard normal distribution. For significance level of α=0.05 null hypothesis of no trend is rejected while |Z|>1.96.

4.4 Theil and Sen Slope (TSA) Estimator

The Theil and Sen's median slope estimator [25, 26], used for computing magnitude of the trends in time series data. The method as compared to simple linear regression gives a better estimate of trends because it is not affected by outliers. The Theil-Sen approach (TSA), a frequently-used to assess the significant linear trends in time series, employed in research. The TSA is powerful as compared to least-squares method because of relative insensitivity to extreme presentation, [27]. The slope Q between two points of time series x estimated by the relationship;

$$Q = \frac{x_k - x_j}{k - j}, k \neq j$$

The time series x with n point observations, N = n (n - 1)/2 and Q may evaluate. Sen's method used to estimate slope as median N values of Q. The Q* as slope estimator can be calculated by;

$$Q^* = \begin{cases} Q_{(N+1)/2}, N\ odd \\ \dfrac{Q_{N-2} + Q_{(N+2)/2}}{2}, N\ even \end{cases}$$

If the significant trends are 95% confidence intervals by means of non-parametric techniques as defined Salmi, et.al [28]. The quantity α is first calculated as;

$$C_\alpha = Z_{1-\alpha/2}\sqrt{V(S)}$$

The standard normal deviate Z, and α is 0.05. M_1 and M_2 indices calculated by;

$$M_1 = \frac{N - C\alpha}{2}$$
$$M_2 = \frac{N + C\alpha}{2}$$

The limits of confidence M1th and (M2+1) are the major among estimates of Q, with interpolation of non-integer figures of M_1 and M_2.

4.5 Spearman's Rho (SR) test

The SR test is a simple with uniform power for linear and non-linear trends utilized to verify the absence of trends [29, 30]. The null hypothesis (H_0) in the test of time series are independent and identically distributed, while alternative hypothesis (H_1) is that increasing or decreasing trends exist, [31]. The SR test statistic D and the standardized test statistic ZSR are gives as;

$$D = 1 - \frac{6\sum_{i=1}^{n}(R_i - i)^2}{n(n^2 - 1)}$$

The variance associated with SR test statistic D is computed as

$$Z_{SR} = D\sqrt{\frac{n-2}{1-D^2}}$$

Where Ri is the "ith" observation, Xi time series and n is the length in time series. Positive and negative values of Z_{SR} indicate upward trends and downward trends in the time series respectively. When $|Z_{SR}|>$ Z-crit for t(n-2,1-α/2), rejected null hypothesis, the significant trend occurs in time series. The t(n-2,1-α/2) is a critical value of t from the t-student table, for 5% significant level. The value of t(n-2,1-α/2) for n=40, the sample size in this study is 2.024.

4.6 Regional Mann-Kendall (RMK) test

In order to assess trends at a regional scale, the regional MK test was employed by Hirsch, Slack, the Smith and Douglas et al. [32, 33, 34], to quantitatively combine results of the MK test for individual stations and to evaluate the regional trends. In the regional MK test, the Sr of regional data is calculated as:

$$S_r = \sum_{i=1}^{n} S_i$$

Where Sr is Kendall's S for the "ith" station in a region with m stations within the region. If Sr is estimated using independent identically distributed data, Sr is approximately normally distributed for large m with mean equal to 0 and the variance as noted below

$$var\,(S_r) = \sum_{i=1}^{n} Var = \sigma^2$$

$$Z_r = \begin{cases} \dfrac{S_r - 1}{\sigma} & for\ S_r > 0 \\ \dfrac{S_r + 1}{\sigma} & for\ S_r < 0 \\ 0 & for\ S_r = 0 \end{cases}$$

To determine whether to reject or not the null hypothesis of no trend, and test statistic Z_r is assessed against the critical value Zcrit corresponding to the specific significance level α of the test. For the two-tailed test, the critical value is defined as Φ^{-1} (1 – α/2), where Φ is cumulative distribution function of standard normal distribution, [35]. The null hypothesis is rejected and the trend is considered significant statistically if the value of |Zr|≥ Zcrit.

5. Results and Discussions

The MK, SR and RMK tests were run on the time series data with significance levels of 10% and 5% for annual, monthly and seasonal time series whereas with 5% on regional time series and the results are presented in the three sub-sections as discussed below.

The first section discusses the applicability and validity of linear regression test. The second section discusses the presence of significant trends detected through non-parametric tests including MK and SR at each of the selected 13 stations at the local level on the annual, monthly and seasonal time series data. The third section discusses the annual and seasonal trends at the regional level using RMK test.

5.1 Assessing normality of time series data

The normality tests including Shapiro-Wilk W test, Anderson–Darling, Lilliefors, and Jarque–Bera test were run on the annual time series to evaluate the normal distribution of time series data and the results are given in Table-4.

IOP Conf. Series: Materials Science and Engineering **414** (2018) 012042 doi:10.1088/1757-899X/414/1/012042

Table 4. Normality Test on Annual Time Series

S.No	Stations	Shapiro-Wilk	Anderson-Darling	Lilliefors	Jarque-Bera
1	Dalbandin	passed	passed	passed	passed
2	Jewani	failed	failed	failed	failed
3	Kalat	failed	failed	failed	failed
4	Lasbella	failed	failed	failed	failed
5	Nokkundi	failed	failed	failed	failed
6	Ormara	failed	failed	failed	failed
7	Panjgur	failed	failed	failed	failed
8	Pasni	failed	failed	failed	failed
9	Quetta	failed	failed	failed	failed
10	Sibi	passed	passed	passed	passed
11	Zhob	passed	passed	passed	passed
12	Khuzdar	failed	failed	Passed	failed
13	Barakhan	passed	passed	passed	passed

Normality test runs on annual precipitation time series data shows that the time series data of only four stations namely Dalbandin, Sibbi, Zhob, and Barakhan are normally distributed and mostly the data is not normally distributed.

5.2 Trend using Linear Regression and its validity

Parametric tests - Linear regression analysis with least square fit test were run on all the time series data for detecting of the trends and the results are presented in Table-5.

Table 5. Linear Regression on Annual Time Series

S.No	Stations	R^2	Regression Slope	P-value	Result	Residual normality
1	Dalbandin	0.0499	-0.9179	0.166	No trend	passed
2	Jewani	0.07532	-2.0969	0.087	No trend	failed
3	Kalat	0.0015	0.5609	0.811	No trend	failed
4	Lasbella	0.0541	-2.3279	0.149	No trend	failed
5	Nokkundi	0.015	0.403	0.451	No trend	failed
6	Ormara	0.003	0.4423	0.736	No trend	failed
7	Panjgur	0.062	-1.2097	0.121	No trend	failed
8	Pasni	0.0357	-1.3075	0.243	No trend	failed
9	Quetta	0.1264	-4.707	0.024	Decreasing	failed
10	Sibbi	0.038	1.509	0.228	No trend	passed
11	Zhob	0.0534	-1.8367	0.151	No trend	passed
12	Khuzdar	0.0302	-1.7735	0.283	No trend	failed
13	Barakhan	0.017	-1.590	0.419	No trend	passed

Linear Regression failed in all stations except Quetta. The conclusion drawn is that the parametric tests are not suited for trend analysis on the annual precipitation because the data is not normally distributed and doesn't have the same standard distribution. Non-parametric MK and SR tests should be adopted to evaluate the significant trends in the precipitation time series data and TSA should be used for calculating the magnitude of the slope.

5.3 Annual, Monthly, and Seasonal Trends

5.3.1 Annual Trends
MK annual precipitation analysis shows that only two stations namely Panjgur and Quetta out of thirteen has 5% significant trend, which is also decreasing in nature. Sen's Slope result shows decreasing trend in nine (9) stations and increasing trend in two (2) stations and no trend in the remaining two (2) stations, however, Sen's Slope is not recommended for trend analysis, it is used mainly to determine the trend. Results of MK analysis on the annual precipitation data for the thirteen (13) stations are shown in Table-6.

Table 6. MK Test Results Increasing (Decreasing) on Annual Time Series with 5% Significance

S/n	Stations	Mann-Kendall					Sen's nonparametric estimator of slope			
		Kendall's tau	S	Z	Prob (%)	Result	Sen Slope	Lower Limit	Upper Limit	Result
1	Dalbandin	-0.137	-107.00	-1.235	21.25%	no trend	-1.035	-1.258	-0.623	(trend)
2	Jewani	-0.209	-163.00	-1.887	5.75%	no trend	-1.347	-1.75	-0.95	(trend)
3	Kalat	0.068	53.00	0.605	53.69%	no trend	0.814	0.22	1.56	(trend)
4	Lasbella	-0.121	-94.00	-1.083	28.08%	no trend	-1.754	-2.42	-1.06	(trend)
5	Nokkundi	-0.209	-163.00	-1.887	5.75%	no trend	-1.358	-1.75	-0.95	(trend)
6	Ormara	-0.040	-31.00	-0.349	71.79%	no trend	-0.205	-0.53	0.04	no trend
7	Panjgur	-0.236	-184.00	-2.132	3.24%	(trend)	-1.198	-1.38	-0.95	no trend
8	Pasni	-0.122	-95.00	-1.095	26.83%	no trend	-1.039	-1.39	-0.62	(trend)
9	Quetta	-0.241	-188.00	-2.178	2.88%	(trend)	-2.968	-3.50	-2.24	(trend)
10	Sibbi	0.144	112.00	1.293	19.75%	no trend	1.712	1.23	2.42	(trend)
11	Zhob	-0.203	-158.00	-1.829	6.7%	no trend	-2.121	-2.64	-1.65	(trend)
12	Khuzdar	-0.172	-134.00	-1.549	12.19%	no trend	-2.288	-2.81	-1.52	(trend)
13	Barakhan	-0.079	-62.00	-0.710	47.99%	no trend	-1.584	-2.46	-0.66	(trend)

Spearman's rho also showed trends in two (2) stations namely Panjgur and Quetta out of thirteen (13) stations, which validates findings of MK results. The results are presented in Table-7.

Table 7. SR Increasing (Decreasing) Test Results on Annual Time Series with 5% Significance

S.No	Stations	Spearman's Rho		
		Abs.Z_{SR}	**t-crit(df,1-alpha/2)**	**Result**
1	Dalbandin	1.19517	2.024394164	no trend
2	Jewani	1.91667	2.024394164	no trend
3	Kalat	0.71783	2.024394164	no trend
4	Lasbella	0.95980	2.024394164	no trend
5	Nokkundi	1.91667	2.024394164	no trend
6	Ormara	0.17876	2.024394164	no trend
7	Panjgur	2.31238	2.024394164	(trend)
8	Pasni	1.04218	2.024394164	no trend
9	Quetta	2.42614	2.024394164	(trend)
10	Sibbi	1.17868	2.024394164	no trend
11	Zhob	1.92531	2.024394164	no trend
12	Khuzdar	1.63060	2.024394164	no trend
13	Barakhan	0.82972	2.024394164	no trend

5.3.2 Monthly Trends

MK tests were run on the monthly data to evaluate the increasing (decreasing) trends in the monthly time series data for the significance levels of 10% and 5%. Minimum of 16 observations are considered as a threshold for detecting the monthly trends. Increasing (decreasing) trend results are presented in Table-8.

Table 8. MK Test Results Increasing (Decreasing) on Monthly Time series on 10% and 5% Significance

Months	1	2	3	4	5	6	7	8	9	10	11	12	13
	Dalbandin	Jewani	Kalat	Lasbella	Nokkundi	Ormara	Panjgur	Pasni	Quetta	Sibbi	Zhob	Khuzdar	Barakhan
Jan	-10%	-	-	-	-	-	-	-	-5%		-5%	-	-5%
Feb	-	-	-	-	-	-	-	-	-			-	-
Mar	-	-	-	-	-	-	-	-	-10%			-	-
April	-	-	-	-	-	-	-	-	-			-	-
May	-	-	-	-	-	-	-	-	10%	10%		-	-
Jun	-	-	-	-	-	-	-	-	5%	10%	5%	-	-
July	-	-	-10%	-	-	-	-5%	-	-	-	-	-	-
Aug	-	-	-	-	-	-	-	-	-	-	-	-	-
Sept	-	-	-	-	-	-	-	-	-	-	-	-	-
Oct	-	-	-	-	-	-5%	-	-	-	-	-	-	-
Nov	-	-	-	-	-	-	-	-	-	-	-	10%	5%
Dec	- 5%	-5%	-	-	-	-	10%	-	-10%	-	-	-10%	-

5.3.3 Seasonal Trends

Similarly, MK and SR tests were run on prepared seasonal data on the five (5) seasons namely winter, spring, summer, monsoon and autumn for 13 stations. Results of the seasonal trends are further scrutinized with respect to the strength of trend and depth of observations. Minimum of 16 observations are considered as a threshold for detecting the monthly trends. Increasing (decreasing) trend results are presented in Table-9.

Table 9. MK Test Results Increasing (Decreasing) on Monthly Time Series with 10% and 5% Significance

Seasons	1	2	3	4	5	6	7	8	9	10	11	12	13
	Dalbandin	Jewani	Kalat	Lasbella	Nokkundi	Ormara	Panjgur	Pasni	Quetta	Sibbi	Zhob	Khuzdar	Barakhan
Winter	-	-5%	-	-	-	-	-	-	- 5%	-	-	-	-5%
Spring	-	-	-	-	-	-	-	-	-	-	-	-	-
Summer	-	-	-	-	-	-	-	-	10%	10%	-	-	-
Monsoon	-	-	-	-	-	-	-5%	-	-	-	-	-	-
Autumn	-	-	-	-	-	-5%	-	-	-	10%	-	-5%	-

5.3.4 Regional Trends

MK tests were run at the regional level for the four regions i-e province, Zones C, D & E on the annual, seasonal and monthly time series. The results from the regional MK test as a single region (for the four regions) for the study periods are shown in Tables-9. The regional trend analysis for all the cases shows that there is a significant positive or (negative trend) in the AR intensity data at the 10% and 5% significance levels for 4 regions as a single region.

Table 10. RMK Test Results Increasing (Decreasing) on Annual, Seasonal and Monthly Time Series with 10% and 5% Significance

REGIONS	Annual	Winter	Spring	Summer	Monsoon	Autumn	Jan	Feb	Mar	Apr	May	Jun	Jul	Aug	Sep	Oct	Nov	Dec
PROVINCE	-5%	-5%	-5%	-5%	-5%	-	-5%	-	-5%	-	-	5%	-10%	-5%	5%	-5%	-	5%
ZONE C	-5%	-5%	-5%	-	-5%	-	-5%	-	-	-	-	5%	-5%	-10%	10%	-	-	-5%
ZONE D	-	-	-	5%	-	-5%	-5%	-	-	-	-	5%	-5%	-	-	-	-5%	-
ZONE E	-5%	-5%	-5%	-	-5%	-5%	-5%	-10%	-5%	-	-	10%	-	-	-	-5%	-	-5%

6. Conclusion and recommendations

A study is conducted to find out trends of rainfall intensity in Pakistan. The selected study area for the study is Baluchistan-the largest province of Pakistan based on land area. Data for the study is collected from 13 different stations throughout the province. The time span of the data is four decades, from 1977 to 2016. The data is analyzed on the annual, seasonal and monthly basis to observe trends in rainfall intensity. Non-parametric test namely Mann-Kendall (MK) and Spearman's rho (SR) tests for trend analysis is conducted at 5% and 10% significance levels to monthly rainfall time series data on annual, monthly and seasonal basis. Theil and Sen's (TSA) slope approach applied for quantify slope trend. The Regional Mann Kendall test (RMK) test is employed to evaluate trends at a regional scale for the four regions.

The results of MK and SR tests for annual time series on individual stations at local level show that the number of statistically significant negative trends was greater than the statistically significant positive trends, thus, indicating the decreasing trend in rainfall. Similar trends are observed when the MK test is applied to monthly and seasonal time series, which confirms the findings of the MK tests on annual rainfall time series on individual stations. Whereas, in the case of regional tests the number of negative trends are significantly increased that depicts the decrease in precipitation on overall Baluchistan on a large scale.

As discussed earlier in this research paper regarding the significance of the study and the study area, Baluchistan: not only the largest province of Pakistan according to land mass but also a significant portion of CPEC is passing through Baluchistan. Moreover, Gawadar port is also situated in the same province, which will play a pivotal role in socio-economic development of Pakistan. The study will be a baseline study to compare the effect of CPEC and other anthropogenic activity on precipitation trend once it is completed. The analysis and results can be employed as an input for infrastructure planning of the area, which is going to be the financial hub of the country.

It is recommended that the MK test should also be conducted on annual maximum rainfall intensity data and annual maximum rainfall temperature data. Additionally, it is also highly recommended to study the potential effects of large-scale climatic indices to know the long-term precipitation trend of Baluchistan. The author is undertaking this study as a next task. Such in-depth studies will help in identification and adaptation of alternate solutions and mitigation strategies for Baluchistan, Pakistan.

Acknowledgment
The authors would like to thank the PMD for sharing precipitation data, to reviewers for their valuable comments and review to enhance the research excellence.

References
[1] Kreft S, Eckstein D (2013) Who suffers most from extreme weather events? Weather-related loss events in 2012 and 1993 to 2012. Germanwatch, Bonn
[2] Abbas F, Ahmad A, Safeeq M, Ali S, Saleem F, Hammad HM, Farhad W (2014) Changes in precipitation extremes over arid to semiarid and subhumid Punjab, Pakistan. Theor Appl Climatol **116**:671–680
[3] Ahmad I, Tang D, Wang TF, Wang M, Wagan B (2015) Precipitation trends over time using Mann-Kendall and Spearman's rho tests in Swat river basin, Pakistan. *Adv Meteorol* 2015: Article ID 431860
[4] Dahri ZH, Ludwig F, Moors E, Ahmad B, Khan A, Kabat P (2016) An appraisal of precipitation distribution in the high-altitude catchments of the Indus basin. *Sci Total Environ* 548–549:289–306
[5] Voss R, May W, Roeckner E (2002) Enhanced resolution modeling study on anthropogenic climate change: changes in extremes of the hydro- logical cycle. *Int J Climatol* **22**:755–777
[6] Zolina O, Simmer C, Belyaev K, Kapala A, Gulev S (2009) Improving estimates of heavy and extreme precipitation using daily records from European rain gauges. *J Hydrometeorol* **10**:701–716
[7] Schär C, Ban N, Fischer EM, Rajczak J, Schmidli J, Frei C, Giorgi F, Karl, "Percentile. Indices for Assessing Changes in Heavy Precipitation Events." *Climatic Change* 137, no. 1–2 (April 5, 2016): 201–216. As Published http://dx.doi.org/10.1007/s10584-016-1669-2.
[8[Pingale S.M., Khare D., Jat M. K., Adamowski J. 2016. Trend analysis of climatic variables in an arid and semi-arid region of the Ajmer District, Rajasthan, India. *Journal of Water and Land Development.* No. **28** p. 3–18.
[9] Kendall M (1938) A new measure of rank correlation. *Biometrika* 30:81– 89
[10] Mann, H.B., (1945) Non parametric tests against trend. *Econometrica* **13**: 245–259
[11] Khattak MS, Babel MS, Sharif M (2011) Hydro-meteorological trends in the upper Indus river basin in Pakistan. *Clim Res* **46**:103–119
[12] Hanif M, Khan AH, Adnan S (2013) Latitudinal precipitation character- istics and trends in Pakistan. *J Hydrol* **492**:266–272
[13] Hussain MS, Lee S (2013) The regional and the seasonal variability of extreme precipitation trends in Pakistan. *Asia-Pac J Atmos Sci* **49**: 421–441
[14] Hussain MS, Lee S (2014) Long-term variability and changes of the precipitation regime in Pakistan. *Asia-Pac J Atmos Sci* **50**:271–282
[15] Maida Z, Ghulam R (2011) Frequency of extreme temperature and pre- cipitation events in Pakistan 1965-2009. *Sci Int* **23**:313–319

[16] Salma S, Rehman S, Shah MA (2012) Rainfall trends in different climate zones of Pakistan. *Pak J Meteorol* **9**:37–47

[17] Ijaz Ahmad, Deshan Tang, TianFang Wang, MeiWang, and Bakhtawar Wagan (2015) Precipitation Trends over Time Using Mann-Kendall and Spearman's rho Tests in Swat River Basin, Pakistan.

[18] Burn D. H. and Elnur M. A. H., "Detection of hydrologic trends and variability," *Journal of Hydrology*, vol. **255**, no. 1–4, pp. 107–122, 2002.

[19] Liu D., Guo S., Chen X., and Shao Q., "Analysis of trends of annual and seasonal precipitation from 1956 to 2000 in Guangdong Province, China," *Hydrological Sciences Journal*, vol. **57**, no. 2, pp. 358–369, 2012.

[20] Chaouche K., Neppel L., Dieulin C. et al., "Analyses of precipitation, temperature and evapotranspiration in a French Mediterranean region in the context of climate change," Comptes Rendus*: Geoscience*, vol. **342**, no. 3, pp. 234–243, 2010.

[21] Machiwal D. and Jha M. K., "Time series analysis of hydrologic data for water resources planning and management: a review," *Journal of Hydrology and Hydromechanics*, vol. **54**, pp. 237–257, 2009.

[22] Verworn H., Krämer S., Becker M., and Pfister A., (2008) "The impact of climate change on rainfall runoff statistics in the Emscher-Lippe region," in *Proceedings of the 11th International Conference on Urban Drainage,* pp. 1–10, Edinburgh, UK.

[23] Yang X., Xu L., Liu K., Li C., Hu J., and Xia X., (2012) "Trends in temperature and precipitation in the zhangweinan river basin during the last 53 years," *Procedia Environmental Sciences*, vol. **13**, pp. 1966–1974.

[24] Scarpati O. E., Spescha L. B., Lay J. A. F., and Capriolo A. D., (2011) "Soil water surplus in salado river basin and its variability during the last forty years (buenos aires province, argentina)," *Water,* vol. **3,** pp. 132–145.

[25] Theil, H. (1950). A rank-invariant method of linear and polynomial regression analysis, I. Proc. Kon. Ned. Akad. v. Wetensch. A 53, 386-392.

[26] Sen, P. K. (1968). Estimates of the regression coefficient based on Kendall's tau. J. Amer. Statist. *Assoc*., **63,** 1379-1389.

[27] Hirsch, R.M.; Slack, J.R.; Smith, R.A. (1982) Techniques of trend analysis for monthly water quality data. *Water Resour. Res*. 1982, **18**, 107–121.

[28] Salmi, T.; Maatta, A.; Anttila, P.; Ruoho-Airola, T.; Amnell, T. (2002) Detecting Trends of Annual Values of Atmospheric Pollutants by the Mann-Kendall Test and Sen's Slope Estimates, Finnish Meteorological Institute: Helsinki, Finland.

[29] Dahmen ER, Hall MJ (1990) Screening of hydrological data: tests for stationarity and relative consistency. ILRI, The Netherlands, p 58, *Publication #49*

[30] Tonkaz T, Çetin M, Kâzım T (2007) The impact of water resources development projects on water vapor pressure trends in a semi-arid region, Turkey. *Climatic Change* **82**:195–209

[31] Yue S, Pilon P, Cavadias G (2002a) Power of the Mann-Kendall and Spearman'S rho tests for detecting monotonic trends in hydrological series. *J Hydrol* **259**:254–271

[32] Hirsch, R.M., Slack, J.R., and Smith, R.A., (1982). Techniques of trend analysis for monthly water quality data. *Water Resource Research*, **18**, 107–121.

[33] Hirsch, R.M. and Slack, J.R., (1984). A nonparametric trend test for seasonal data with serial dependence. *Water Resource Research*, **20**, 727–732.

[34] Douglas, E.M., Vogel, R.M., and Kroll, C.N., 2000. Trends in floods and low flows in the United States: impact of spatial correlation. *Journal of Hydrology*, **240**, 90–105.

[35] Helsel, D.R. and Hirsch, R.M., 2002. Statistical methods in water resources. Reston, VA: US Geological Survey Reston.

Historical and Future Climatological Drought Projections Over Quetta Valley, Balochistan, Pakistan

Imran Hameed Durrani[1], Shahzada Adnan[2], and Syed Mobasher Aftab[1]

[1]Faculty of Engineering and Architecture, BUITEMS, Quetta, Pakistan
[2]Pakistan Meteorological Department, Islamabad, Pakistan
[3]University of Balochistan, Quetta, Pakistan

E-mail: syed.aftab@buitms.edu.pk

Abstract. The historical climatological records provide eminent evidence that drought is a permanent natural disastrous phenomenon of Balochistan. The precipitation data from 1950 to 2010 characterized that 18 drought events flashed with an interval of 3.3 years when average precipitation deficiency ranges from 20 to >40%. The longest historical drought span over 1945 to 1955 and the most recent dry-spell extended from 1997 to 2003. During the last drought, the average precipitation values decreased from 0-21% in Balochistan, while in Quetta from 40-70%. The historical record depicts that drought study also identifies the future drought frequency and severity with return period over the time. Strong affiliation has been recognized between the soil-moisture and standardized precipitation index, which ultimately helps to identify the spatial behavior of droughts. The Representative Concentration Pathways RCP4.5 and RCP8.5 were downscale at 25 km and 50 km, have been used for future behavior of meteorological parameters. The both RCPs show a positive increasing trend for mean temperature and a negative trend for precipitation for the 21st century. A significant increase in temperature has observed 0.04 °C/y and 0.12 °C/y at RCP4.5 and 8.5 over Quetta Valley respectively. The precipitation values represent no significant change in trend observed at 95% confidence level for the century. A marginally declining trend is identified by the statistical tests in precipitation. The precipitation and mean temperature data have been downscaled for midcentury 2040-2069 and far future 2071-2100 using different Regional Climate Models, namely ERA-40, ECHAM5 and fvGCM to identify association with the observational data of Quetta Valley. The observational data of mean temperature and precipitation shows a strong correlation to the downscaled data of ERA-40 as R^2 = 0.97 and R^2 = 0.47 respectively. The ERA-40 exhibits somewhat underestimate the mean temperature and overestimate the precipitation data. The observational data used to calibrate the downscaled data. The variability in temperature and precipitation has an enormous social and economic impact on the residents of Quetta Valley. The climatological variability devastated the ecosystem, depleted the groundwater resources and exhausted environment. The study outcome assists different stakeholders to predict and device immediate, short and long-term strategies to combat droughts. The Water Resource Managers and Planners may develop preeminent future "Drought Mitigation Policy" in the light of climate change over the Quetta Valley.
Keywords: Drought, Climatological Projections, Standardized Precipitation Index, Regional Climate Models, Quetta.

1. Introduction

The Quetta Sub-basin is a part of Pishin River Basin, which is one of eighteen river basins of Balochistan. The sub-basin extends from Latitude 29° 40′ to 30° 25′ N and longitude 66° 50′ to 67° 20′ E and covers an area of 1,860 km2. Quetta itself is one of the 34 districts and the provincial capital of Balochistan, Pakistan. The city is situated at about 30° 10′ N and 67° 01′ E, it has an elevation of 1,680 m above mean sea level. The valley located in arid zone where groundwater is the major source for domestic, agriculture, commercial and industrial utilization. The aquifers developed in the valley are the only source for urban and rural water supply schemes. A continental and semi-arid climate with great variations from summer to winter temperatures endures in the valley. The recorded highest temperature was 42 °C on 10 July 1998. The lowest temperature -18.3 °C recorded on 8 January 1970. The summer season continues from May-September with an average temperature 24-26 °C. The autumn spans from September-November with temperatures from 12-18 °C. The winter from October-March, temperatures varies 4-5 °C. Spring from April-May with average temperatures 15 °C. Quetta Valley does not have a monsoon season of sustained rainfall. The highest 24-hour rainfall 113 mm recorded on 17 December 2000. The highest monthly rainfall 232.4 mm documented in March 1982. The highest annual rainfall is 949.8 mm recorded in 1982. The snow falls in December-February, very rarely in March.

The imbalance in the precipitation and temperature is an important factor, the evaporation is higher than precipitation and the sustainability of groundwater table cannot be maintained. The climate change also impacts the variability in precipitation, fewer snowfall events and water availability are extremely high during dry years that worsen the conditions. The rural economy and livelihood of communities dependent on groundwater resources which are reliant on precipitation. The groundwater resources severely affected by repetition of droughts of diverse intensities and durations. The seasonal droughts cause severe damages to crop, the livestock and orchards are also vulnerable to persistent droughts. The valley has faced several droughts since 1955 to 2008 due to enormous gaps in regular rainfall patterns and an increase in temperature. A balance in climatic factors is required for sustainability of groundwater resources. The naturally developed gaps due to change in climate rigorously affected the natural restoration process of aquifers and resulted in heavy depletion of the groundwater table in the valley. The valley is surrounded by hills on three sides, it has limitations in terms of expansion. The population density is very high in urban and rural areas with vast urbanization. The periodic droughts, aridity, depleted groundwater resources, degraded rangelands, and rugged terrain is the foremost challenges communities are facing. The location map is presented in figure-1.

Figure 1. Location Map; Study area shown in shaded box, Quetta Valley, Balochistan, Pakistan

2. Drought Climatology

The drought is an inception of a slow disaster and its impact differs from region to region. The meteorological, agricultural, hydrological and socioeconomic are the four major types of droughts. At a time either one or the other type of droughts may observe simultaneously with varying intensities in different regions of Balochistan. The present study based on the analysis of meteorological and hydrological droughts. The impacts are consistent with the environment, weather and soil conditions. The percentage of normal precipitation is used to define drought severity. The precipitation deficiency also utilized to describe drought, which permits a realistic regional evaluation. The results of problems require different definitions of drought, since both the relative and absolute magnitudes of the precipitation deficiency may have obtained from analytical procedures. A single definition of drought is not applicable to the various drought problems encountered in meteorology and hydrology. The selected method should be adaptable for hydrologic application during periods of extended droughts. The irregularities in precipitation are a climate phenomenon that disturbs the hydrologic cycle and causes droughts. The climatology of precipitation, temperature, and atmospheric moisture content helps to assess the precipitation frequency. The correlation between precipitation, temperature, and atmospheric moisture help to assess the severity of droughts.

The atmospheric pattern of Quetta Valley demonstrates that droughts are recurring in rotational patterns that produce sporadically less than average precipitation. The atmospheric pattern during a drought may attain by drought circulation patterns like monthly and yearly averages, air dryness, and variation in temperature gradients. The causes of atmospheric droughts in Quetta Valley may go far beyond to provincial level. The regional nature of atmospheric circulations is the reason that produces sustained periods of less than normal precipitation. The drought physiognomies in Quetta Valley premeditated to ascertain the spatial and temporal distribution of droughts. The drought severity associated with climatological conditions and to ascertain the relationship of precipitation deficiency to depletion of the groundwater table in Quetta Valley. Stress has been placed upon hydrologic application in the approach to the problem and in the presentation of results. The study, based primarily upon monthly precipitation data from the Quetta Meteorological Station for the 60-year sampling period, 1950-2010. The baseline period was taken in the thirty-year period from 1960-1990 for data validation. For the future predictions midcentury from 2040-2069, and late-century from 2071-2100 have been designated.

3. Historical Droughts in Quetta Valley

The standardized precipitation index (SPI) is a standard means for defining and monitoring droughts. The index helps to delineate the rarity of drought at a given time scale for any rainfall station with historic data. The SPI provides the amount of rainfall, indicate the amount in relation to the normal and define whether a climatical station or region is experiencing drought. The drought frequency and historical drought years analyzed by SPI at different timescales of 3, 6, 9, 12 and 24 months [1]. The comparison between soil-moisture and precipitation with 12-SPI was conducted for sixty years during 1950-2010 over Quetta Valley.

The historical drought years along with their intensity were identified over Quetta Valley by using the SPI at time scales of 3, 6, 9, 12 and 24 months. The observational data of precipitation used for historical drought years, frequency, category, and types were identified. The SPI is calculated by using the observational data of precipitation to determine drought frequency, category, intensity, and types. Four different intensity classes, i.e. mild, moderate, severe and extreme is used. Similarly, four different types of drought are calculated as; 3-months SPI depicts the dry period, 6-months SPI meteorological, 9-months SPI agriculture, 12 and 24-months SPI, hydrological and extreme hydrological drought respectively.

The 60 years historical analysis represent that 18 droughts of mild to extreme intensities at the time scale from 3 to 24-SPI occurred over Quetta Valley. The historical drought with intensity at different SPI timescale is presented in table-1. The results show that only four drought years were identified at three months SPI. Two extreme drought events recorded during 1971 and 2000 of 9-24 SPI. The two severe drought events of 24-SPI observed on 1972 and 2001. The maximum twelve drought events of different intensities were experienced in 9 and 24 month's timescales. Quetta receives a somewhat reasonable amount of precipitation only in winter season under the influence of western weather patterns. Consequently, the 12-months timescale is a good indicator to monitor drought over the Quetta Valley. Similarly, the frequency of drought is very high in Quetta Valley as it receives drought after every three years of mild to moderate intensities. The drought severity monitored by correlating the soil-moisture departure and 12-SPI, soil-moisture departure and precipitation departure, precipitation departure and 12-SPI. The temporal comparison assessed between 12-SPI, soil-moisture and precipitation from 1950-2010 over Quetta Valley. It was observed that the highest correlation exists between soil-moisture to 12-SPI with R^2 is 0.78, and 12-SPI to precipitation where R^2 is 0.76, and soil-moisture to precipitation R^2 is 0.67. The results show a good association among the parameters of soil-moisture and precipitation. The two extreme drought years 1971 and 2000 were identified for which the soil-moisture and precipitation departures were -76.7%, -79.0% and -86.5, -35.20 % respectively. The historical data represent a decreasing trend in seasonal as well as monsoon precipitation and a warmer trend in all seasons over Quetta Valley. A temporal comparison was conducted between 12-month SPI, percentage departure of soil-moisture and precipitation from 1950-2010 over Quetta Valley, shown in figure 2.

Table 1. The Historical Droughts with intensity at different SPI Timescale.

Time Scale	Extreme	Severe	Moderate	Mild
3-SPI	-	-	1971, 2000	1999, 2004
6-SPI	-	1971, 2000	2004	1955, 1958, 1969, 1999, 2002, 2006, 2008
9-SPI	1971, 2000	-	1955, 1999, 2002, 2004, 2006	1952, 1953, 1969, 1973, 2008
12-SPI	1971, 2000	-	1955, 1999, 2002, 2006	1953, 1961, 1969, 1973, 2004
24-SPI	2000	1971, 1972, 2001	1970, 2002	1953, 1955, 1969, 1973, 2003, 2004

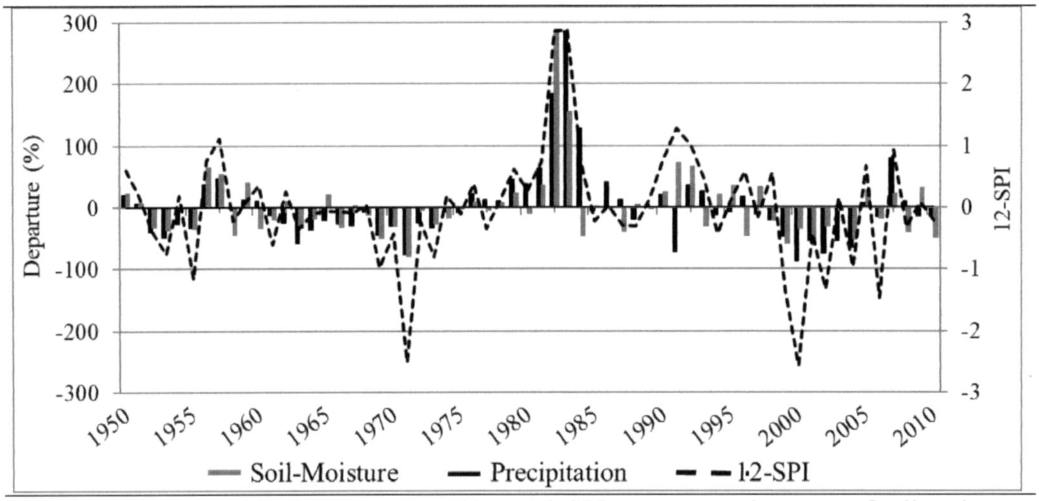

Figure 2. Temporal comparison between 12-month SPI, percentage departure of soil-moisture and precipitation from 1950-2010 over Quetta Valley.

4. Representative Concentration Pathways

The Representative Concentration Pathways (RCPs) are four greenhouse gas concentration trajectories adopted by the Intergovernmental Panel on Climate Change (IPCC), for the fifth assessment report (AR-5), [2], utilized for climate modeling. The four possible climate futures may assess depending on the quantities of greenhouse gas emission in future scenarios. The RCPs are fully integrated scenarios and are consistent sets of projections for components of radiative forcing that serve as input for climate modeling, pattern scaling, and atmospheric modeling. The RCP4.5 and 8.5 were downscale at 25 km and 50 km under the guideline of the IPCC, AR-5, 2014, to determine the future projection of temperature and precipitation for Quetta valley. Observational and Global Precipitation Climatological Centre (GPCC) data of precipitation are used for the period of 1950-2010, [3]. The same period's soil moisture data accessed from the Climate Prediction Centre, Islamabad, 0.5° resolution was used to identify the drought behavior [4].

The future projection of temperature based on RCPs for 4.5 and 8.5 is downscaled for Quetta Valley at 25 km and 50 km spatial resolution. RCP4.5 for 25 km spatial resolution represents an increase in simulated temperature from 2010-2100 from 15-22 °C. A continuous increase in temperature of 0.08 °C simulated for ninety years and the R^2 is 0.885. The RCP8.5 for 25 km spatial resolution characterize an increase in temperature from 17-26 °C in the simulated ninety years from 2010 to 2100. The simulation results represent that temperature at the rate of 0.1 °C/y will increase in ninety years, the R^2 value is 0.55. The future projected model and emission scenarios show that the temperature has a significant increase of 0.04 °C/y and 0.12 °C/y for RCP4.5 and RCP8.5 respectively. The simulated average increase in temperature for the century is 0.08°C/y. The linear relationship of RCP4.5 and 8.5 for 25 km spatial resolution and the observed mean monthly temperature are shown in figure 3a.

The RCP4.5 and 8.5 for 50 km spatial resolution simulated for average yearly temperature from 2010-2100. The simulation represents a constant increase in temperature ranges from 16.5-23.5 °C in ninety years from 2010-2100. The increase in temperature is almost 0.08 °C/y and the R^2 is 0.89. The RCP8.5 for 50 km spatial resolution illustrate the temperature increase from 16-23.5 °C in the simulated ninety years from 2010 to 2100. The simulation results represent that temperatures will increase at 0.08 °C/y in ninety years, the R^2 is 0.55. The linear relationship of RCP4.5 and 8.5 for 50 km spatial resolution and the observed mean monthly temperature are shown in figure 3b. The results show that the mean annual temperature is increasing for both RCP4.5 and 8.5 for 25 and 50 km spatial resolution. The increase is slightly sharp for 50 km resolution. The temperature shows more variation in mid-century from 2041-2070 whereas a sharp increase in the late century from 2071-2100. The future projected model and emission scenarios show that the temperature has a significant increase of 0.04 °C/y and 0.12 °C/y for RCP4.5 and RCP8.5 respectively. The simulated average increase in temperature for the century is 0.08 °C/y.

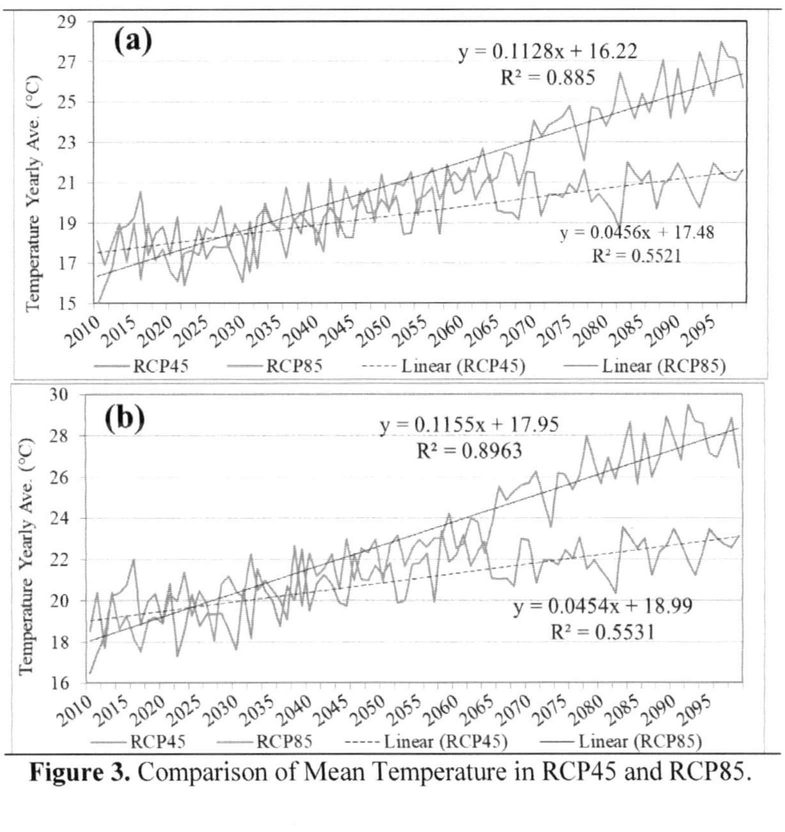

Figure 3. Comparison of Mean Temperature in RCP45 and RCP85.

The future projection of cumulative yearly precipitation established by RCP4.5 and 8.5 downscaled for Quetta Valley at 25 km and 50 km spatial resolution. The RCP4.5 for 25 km spatial resolution characterize simulated yearly precipitation 2-29 mm with exceptional few peaks of the high precipitation range from 30-55mm during 2010-2100. A somewhat same trend in precipitation is simulated for ninety years, represented R^2 as 0.001. The RCP8.5 for 25 km spatial resolution characterize the cumulative yearly precipitation from 1-27 mm/y. A few peaks of high precipitations of 30-81 mm/y simulated for ninety years from 2010 to 2100. The simulation results represent an equitably similar trend of precipitation, the R^2 value is 0.0003. The linear relationship of RCP4.5 and 8.5 for 25 km spatial resolution and the observed yearly precipitation are shown in figure 4a.

The future projection of cumulative yearly precipitation established by RCP4.5 and 8.5 downscaled for Quetta Valley at 50 km spatial resolution. The RCP4.5 for 50 km spatial resolution characterize simulated precipitation 1-30 mm with a few peaks of high precipitation ranges from 32-74 mm during 2010-2100. The simulation represents no prominent change in the precipitation pattern and trend, the R^2 is 7E-05. The RCP8.5 for 50 km spatial resolution characterize the cumulative yearly precipitation from 1-30 mm/y. A few peaks of high precipitations of 35-117 mm/y simulated for ninety years. The simulation results represent a similar trend of precipitation, the R^2 value is 0.0008. The future projection of precipitation based on RCPs for 4.5 and 8.5 is downscaled for Quetta Valley at 50 km spatial resolution are presented in figure 4b.

The RCP4.5 and 8.5 for 25 and 50 km spatial resolution results show that the cumulative yearly precipitation patterns are comparable to observed precipitation values. The outcomes display that the mean annual precipitation shows no change for both RCPs for 4.5 and 8.5. The future projected model and emission scenarios show that the rainfall has no significant increase, a slight decrease in annual precipitation was observed for RCP4.5. The precipitation shows more variation in the late century from 2071-2100.

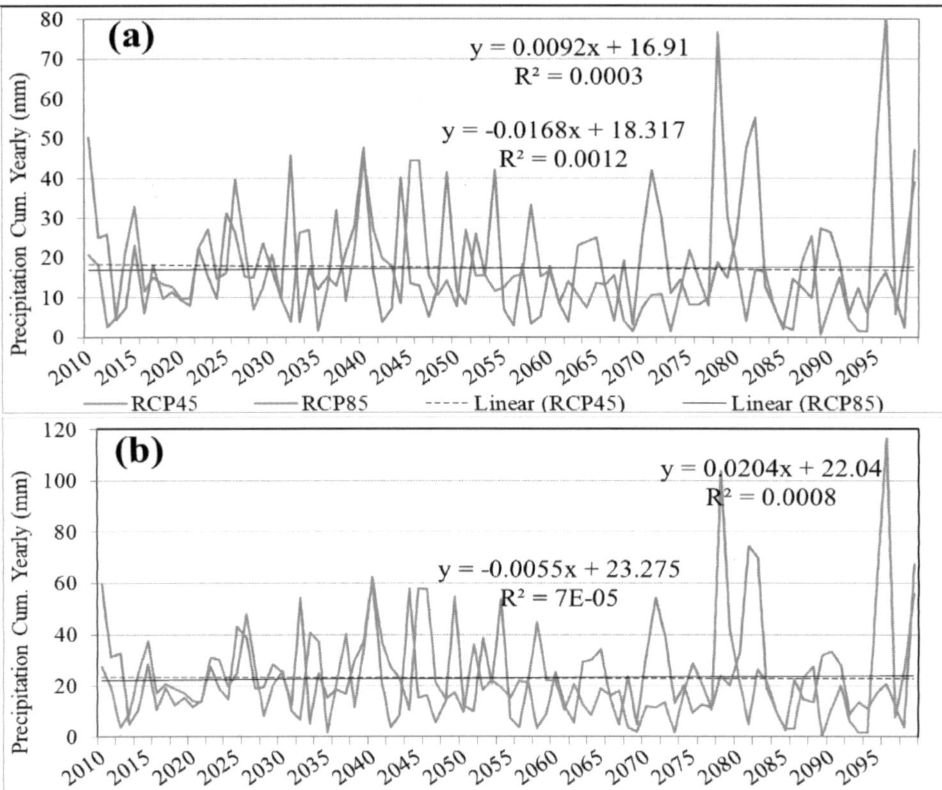

Figure 4. Comparison of annual precipitation in RCP45 and RCP85.

5. Mann-Kendal and Sen's Slope Estimator

The Mann Kendal and Sen's slope estimator are used to determine the significance of the trend and its magnitude, [5], [6]. Man-Kendall test is a nonparametric procedure for finding future drought trends in time series. This test compares the relative magnitudes of data rather than data values themselves. The test is useful because the data need not authorize any particular dissemination. In this test, the data values of time series are compared with all subsequent values. The Mann-Kendall statistics at the start of the test is assumed as zero. If the data value in succeeding time periods is higher than a data value in the previous time-period, it is incremented by one and Vice-versa. The linear trend of precipitation and temperature are calculated using Mann Kendall test, which has been broadly used to analyze the long-term climatological and hydrological time series.

The test statics S, the value of Z is approximately normally distributed and positive value of Z greater than 1.96 represents the significant increasing trend. Whereas, the negative value which is lower than -1.96 donates a significant decreasing trend at 95% significance level with two-tailed tests. Similarly, if the values of Z greater than 1.65 and less than -1.65 shows a significant level with one-tailed test. The result of all increments and decrements provides the final value of Mann-Kendall statistics. A linear trend if present in a time series, the true slope in changes per unit time can be estimated by a nonparametric procedure developed by Sen, [7]. The method assumes a linear trend in the time series. In this method, first the slopes T is calculated. The median of N values of T gives the Sen's estimator of slope Q. A positive value of Q indicates an upward trend and a negative value indicates a downward trend in the time series.

Analysis of temperature and precipitation trends using Mann-Kendall Test and the Sen's Slope Estimator for Quetta Valley were conducted for the 21st Century. The table 2, represents the calculated Z and Q values for temperature and precipitation, 25-50 km resolution for RCP4.5-8.5 from 2010 to 2100. The temperature values showing significance at 95% confidence level for a Z statistic range from 7.69 to 11.19. Similarly, the significance at the 95% confidence level the Q values range from 0.044 to 0.116. The Quetta meteorological station demonstrated a significant trend, the trend is in increasing direction. The increasing trend in temperature might be due to environmental change, fast urbanization heat effect on temperatures and steadily increasing population in Quetta Valley. Studies indicate and reported that temperature increasing trends are dominant nationwide as well as province-wide. The statistical analysis of precipitation obtained by the Mann-Kendall test and the Sen's slope estimator are given in table 2.

The statistical results of precipitation display significance at 95% confidence level under RCP4.5 and RCP8.5 of Quetta Valley for 21st Century. The analysis represents the calculated Z and Q values for precipitation, 25-50km resolution for RCP4.5-8.5 from 2010 to 2100. The precipitation values showing significance at 95% confidence level for Z values range from -0.50 to -1.54. The significance at the 95% confidence level the statistical values of Q vary from -0.025 to -0.91. The precipitation values represent no significant change in trend observed at 95% confidence level for the century by the Mann-Kendall test and the Sen's slope estimator. However, a slightly declining trend is identified by the statistical tests in precipitation.

The difference between parametric or linear regression and non-parametric or Mann-Kendall test and the Sen's Slope estimator methods on the precipitation is small. The temporal pattern of precipitation trend analysis is important for planning and management of water resources. The significance test of the Mann Kendal method represents no significant increase in precipitation. However, a significant increase in temperature has observed 0.04 °C/y and 0.12 °C/y at 95% significance level under RCP4.5 and RCP8.5 over Quetta Valley respectively.

Table 2. Statistical results of temperature and precipitation showing significance at 95% confidence level under RCP4.5 and RCP8.5 of Quetta Valley for the Century.

Parameters	Resolution (km)	Scenarios	Duration		Mann Kendal (Z) values	Sen's Slope (Q) values
Temperature (°C)	25	RCP4.5	2010	2100	7.7	**0.044**[a]
		RCP8.5	2010	2100	11.04	**0.114**
	50	RCP4.5	2010	2100	7.69	**0.044**
		RCP8.5	2010	2100	11.19	**0.116**
Precipitation (mm)	25	RCP4.5	2010	2100	-0.61	-0.021
		RCP8.5	2010	2100	-1.54	-0.062
	50	RCP4.5	2010	2100	-0.5	-0.025
		RCP8.5	2010	2100	-1.54	-0.091

[a]Bold values represent significance at 95% confidence level

6. Regional Climate Modelling

Regional Climate Model (RegCM) is a professional modeling technique for climate-related studies. The Earth Systems Physics, a group of the Abdus Salam International Centre for Theoretical Physics, Trieste, Italy, has prepared Regional Climate Model version 3 (RegCM3), [8]. The RegCM3 interactions on the topic of climate and associated impacts research and modeling. The Regional climate model has the ability to provide more accurate and detailed information after resolving the atmosphere and surface boundary problems. Therefore, the climate change projection based on the Regional Climate model is important in the valley to identify the future vulnerability. The future data is accessed for temperature and precipitation to identify the positive or negative trends. The precipitation data from 1950 to 2010 was used to identify the drought events over the Quetta Valley. The RegCM3 simulations applied to determine the future projection of temperature and precipitation over the valley. The model was run at Global Change Impact Study Centre, Islamabad and the temperature and precipitation data were accessed.

The ECHAM5 is a general circulation model developed by the Max Planck Institute for Meteorology, Germany [9]. The model created by modifying global forecast models to use for climate research. The model configures and resolve climatical issues at different atmospheric settings. The finite-volume General Circulation Model (fvGCM) is a three-dimensional Navier-Stokes solver designed by NASA. The fvGCM is utilizing for quasi-operational weather forecasting. The model designed with innovative algorithms for global high resolution atmospheric modeling. The finite volume transport scheme conserves mass locally and monotonically to ensure proper correlations among the constituents. The velocity preserving horizontal dynamics enhances the simulation of atmospheric oscillations and vortices, which are characteristic of climate and weather phenomena. The atmospheric dynamics are based on the flux form semi Lagrangian advection [10, 11, 12]. The ECHAM5 and fvGCM are used as a driving data with support to the ERA-40 reanalysis data of 50 km resolution under A2 Scenario. The ERA-40 data of the European Center for Medium Range Weather Forecasting (ECMWF) were used in the model simulation. The baseline period was taken for thirty years from 1960-1990 on which data were validated. The future periods designated as midcentury from 2040-2070 and late-century from 2071-2100. The future periods; mid and late-century was calculated for ECHAM, and one period of the late-century for fvGCM.

The comparison of temperature data is conducted between RegCM3 and observational data of Quetta. The modeling results of surface air temperature and precipitation derived from model RegCM3-ERA-40, RegCM3-ECHAM5, RegCM3-fvGCM were analyzed to confirm the validity of numerical modeling. The baseline period is thirty years from 1960-1990 for time series analysis of monthly precipitation. The time series analysis was conducted among observational monthly precipitation data and the RegCM3 model. The simulations were conducted between precipitation and RegCM3-ERA-40, RegCM3-ECHAM5 and RegCM3-fvGCM models. The time series analysis shows considerable variation in the prediction identified by the RegCM3 models during the designated baseline period. The relationship of observed precipitation and RegCM3 models are shown in figure 5.

The time series analysis of the thirty-year period from 1960-1990 and cumulative monthly precipitation with regional climate model ERA-40 were conducted. The time series analysis represents that ERA-40 slightly captured the observed data of precipitation during the baseline period. Throughout the year; based on the cooperation and temporal analysis ERA-40 is slightly coordinated to the observed data. The ERA-40 model's annual mean precipitation is slightly underestimated in comparison to the observational data during all the months except June-September. The overestimation of annual precipitation by the ERA-40 model is the reason for overlaying temperature during July and August. The time series graphical representation of observed and simulated precipitation is given in figure 6.

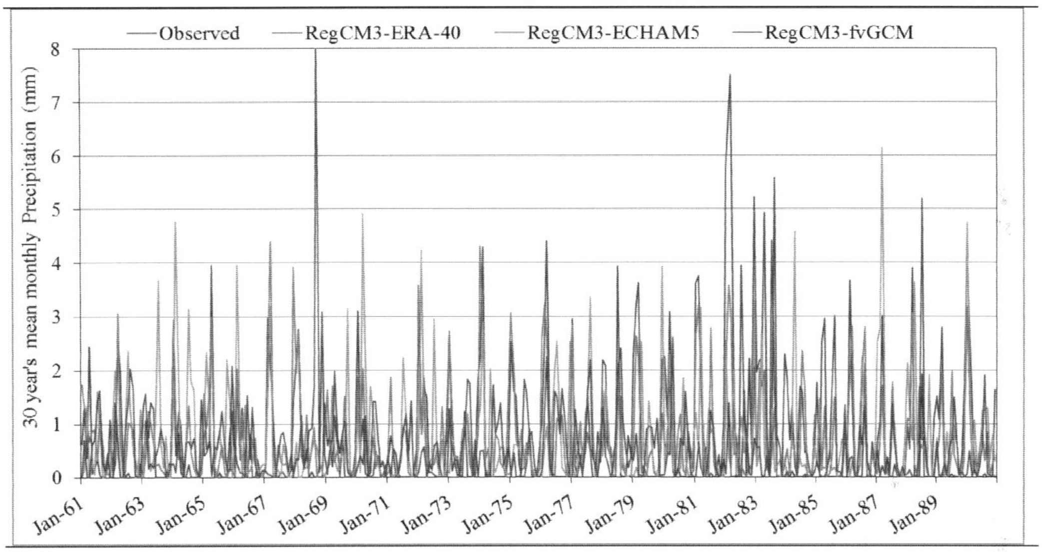

Figure 5. Time series of observed precipitation data and regional climate models of Quetta.

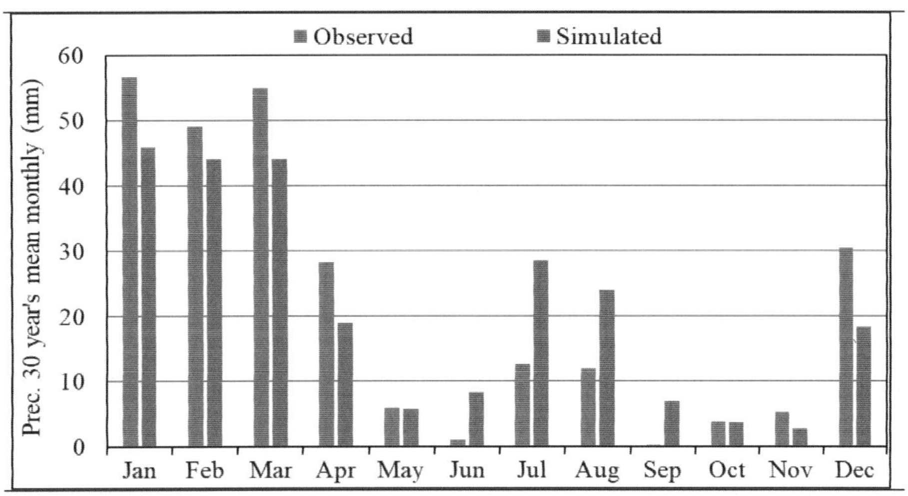

Figure 6. Precipitation (mm) comparison of observed and regional climate model ERA-40 at Quetta

The modeling results of RegCM3-ERA-40, RegCM3-ECHAM5, and RegCM3-fvGCM were analyzed and correlated with observed temperature data. The time series analysis of entire three models is well overlaid the observed temperature data during the baseline period from1960-1990. The observed mean temperature data versus RegCM3-ERA-40, RegCM3-ECHAM5 and RegCM3-fvGCM model results of Quetta are presented in figure 7. Based on the cooperation and temporal analysis ERA-40 is well coordinated to the observed data. The evaluations for ERA-40 temperature have been performed, the data reanalysis is a useful resource for depictions of seasonal variations. The ERA-40 model annual mean temperature is slightly overestimated in comparison to the observational data during all months except July and August. The annual mean temperature comparison of observed and regional climate model ERA-40 is shown in figure 8.

The time series comparisons of the monthly mean temperature of both regional climate models ECHAM5 and fvGCM are conducted for the current century of Quetta. The result of ECHAM5 shows a high annual variation from 3.4 °C on January 2041 to 33.3 °C on June 2069. Depicting the January and June as the coldest and hottest months during 2040-2069 respectively. The lowest temperature 5.78 °C is observed in January 2085 and highest 36.24 °C in June 2091, whereas the fvGCM identifies the lowest temperature for January 2096 as 5.22 °C and highest since June 2096 as 34.71 °C. These results depict that the temperature would rise almost 2°C in Quetta at the end of the century that will increase the evapotranspiration, crop water requirements, and put a pressure on water resources. The time series comparison of temperature ECHMA5 and fvGCM for the Century presented in figure 9.

The time series comparisons of monthly precipitation of both regional climate models ECHMA5 and fvGCM are conducted for the 21st Century of Quetta. The result of ECHAM5 shows a high temporal variability in precipitation. The results depict precipitation remained zero in most of the months of May and November whereas the highest rainfall of 7.03 mm/day is observed on July 2063 at Quetta. Similarly, the highest precipitation is observed during March 2088 as 7.01 mm/day. Whereas, the fvGCM identifies the highest precipitation for March 2100 as 3.94 mm/day. Both models identified precipitation variation during different months. No major increase in precipitation is observed by the two models which are an alarming sign for socioeconomic sectors for the residents of Quetta Valley. The time series comparison of precipitation for both models ECHMA5 and fvGCM of the Century in Quetta is shown in figure 10.

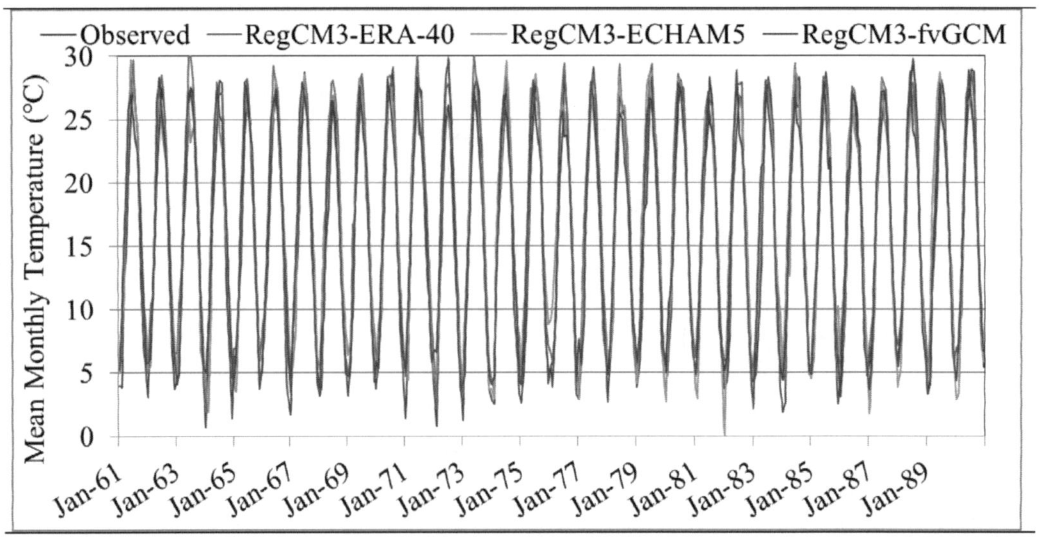

Figure 7. Time series for the mean monthly temperature of observed and regional climate models, Quetta Valley.

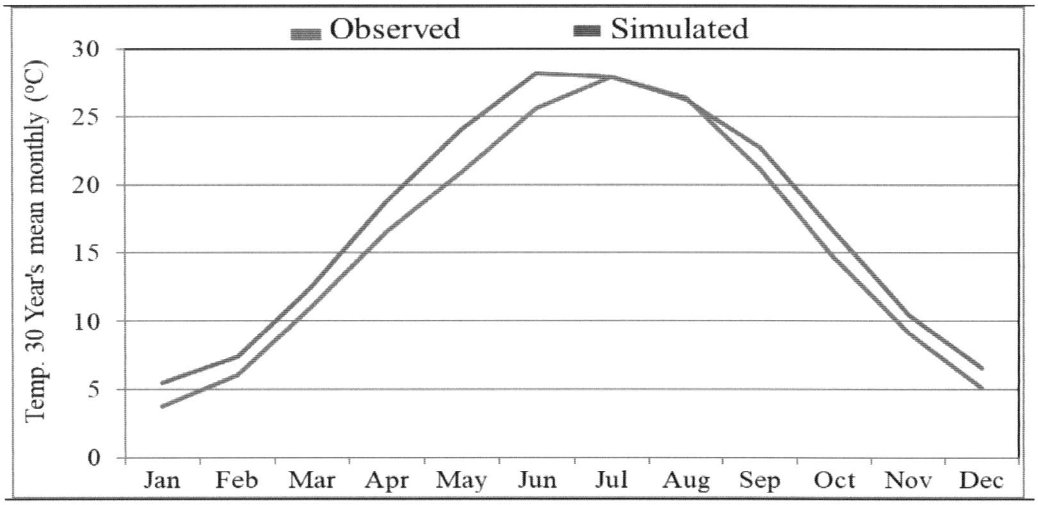

Figure 8. Monthly mean temperature comparison of observed and RegCM3-ERA-40.

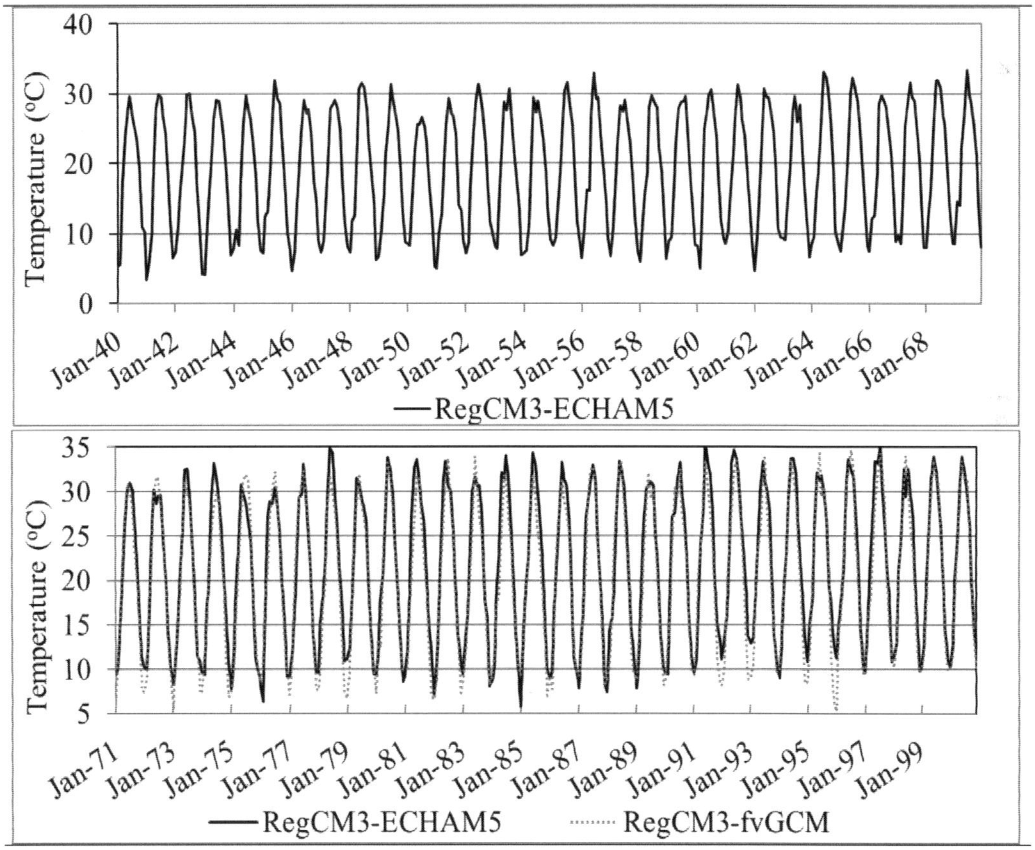

Figure 9. Time series comparison of temperature ECHMA5 and fvGCM for the 21ST Century of Quetta.

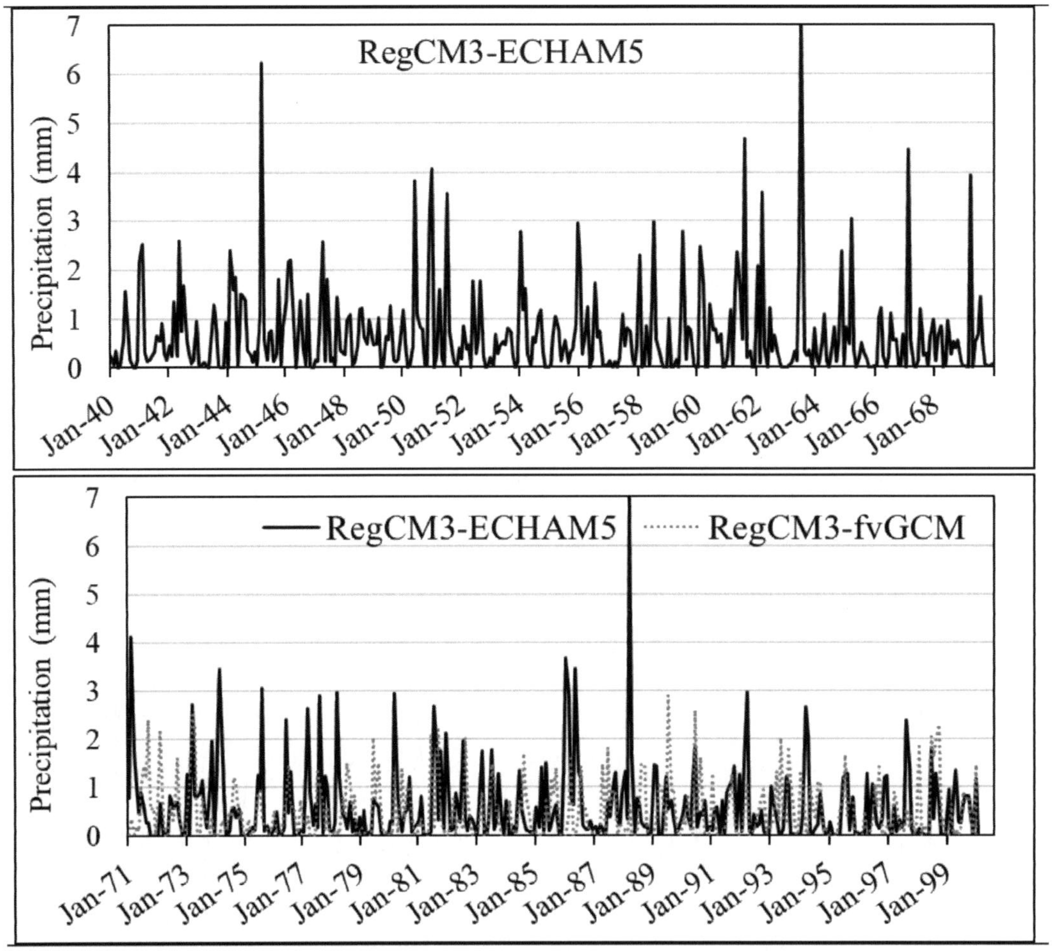

Figure 10. Time series comparison of precipitation ECHMA5 and fvGCM for the 21st Century of Quetta.

The comparison of RegCM3 model with observed monthly temperature data ERA-40, ECHAM5 and fcGCM was conducted, shown in figure 11. The values of R^2 ranges from 0.896 to 0.972 represent that the entire three models show good correlation to observational data. The ERA-40 shows maximum R^2 is 0.972, signify a rational correlation between observed and simulated temperature values.

A comparison between annual observational precipitation data of Quetta and RegCM3 model was conducted. The modeling results of precipitation derived from model RegCM3-ERA-40, RegCM3-ECHAM5, RegCM3-fvGCM were analyzed to confirm the validity of simulation results. The RegCM3-ERA-40 model with observational monthly precipitation data represents a reasonable correlation as R^2 is 0.47.

The RegCM3-ECHAM5 and RegCM3-fvGCM models represent a rational, linear relationship with observed monthly precipitation. The value of R^2 ranges from 0.03-0.002 for RegCM3-ECHAM5 and RegCM3-fvGCM respectively. The ERA-40 shows somewhat good correlation R^2 = 0.471 to observational data in comparison to the rest of the two models i.e. ECHM5 and fvGCM. The comparison of three modeling results with observed monthly precipitation data of Quetta is presented in figure 12.

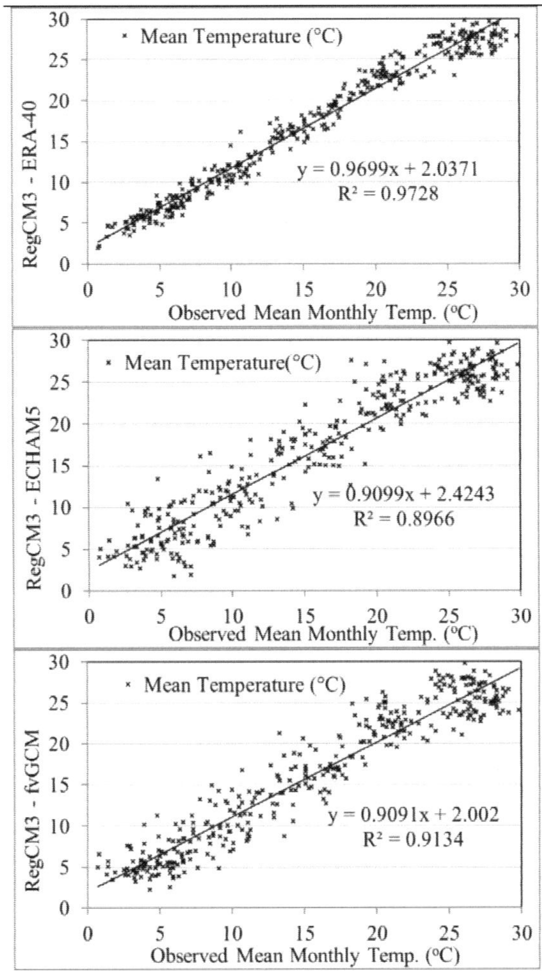

Figure 11. Comparison of RegCM3 model with observed monthly temperature data.

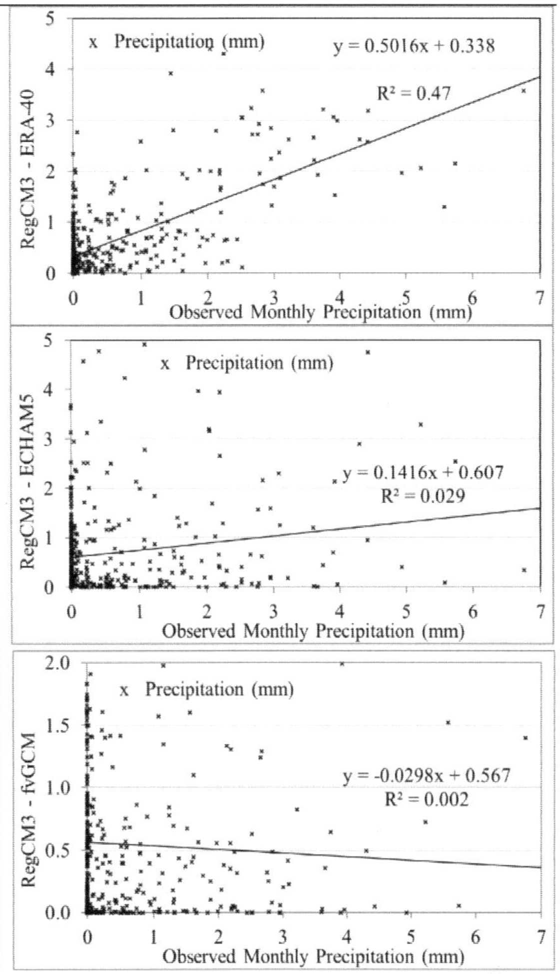

Figure 12. Comparison of RegCM3 model with observed monthly precipitation data.

7. Conclusions

The 60 years historical analysis represent 18 droughts of mild to extreme intensities from 3 to 24-SPI happened as an average after every three years over Quetta Valley. Five severe drought events were identified at six and 24-SPI. Four extreme drought events recorded during 1971 and 2000 at 9 to12-SPI. Only one extreme drought event of 24-SPI observed in 2000.

The temperature projections indicate an increase of 0.04 °C/y and 0.12 °C/y for RCP4.5 and RCP8.5 respectively. The average increase in temperature approximately 0.08 °C/y for the next century. The increasing trend in temperature might be due to environmental change, fast urbanization, and steadily increasing population in Quetta Valley.

The future projection of precipitation based on RCPs, downscale at 25 and 50 km spatial resolution results shows that the cumulative yearly precipitation patterns are comparable to observed precipitation values. The future projected model and emission scenarios demonstrate that no significant increase in precipitation is forecasted.

The drought frequencies will be a major problem in the future for Quetta Valley. The stakeholders like Climate Change Experts, Hydrologists, Agronomist and Water Resource Managers should prepare a viable drought mitigation policy. As per drought severity immediate, the short, midterm and long-term measure should be a significant component of drought mitigation policy.

The severity of droughts directly related to the amount of precipitation. The planning, management, and conservation of surface and groundwater resources should be integrated with drought mitigation policy. An integrated weather and water resources data management for drought modeling, prediction and early warning system are imperative.

To mitigate drought impacts, the structural or physical measures embrace the selection of appropriate crop types and construction of water storage, delay action, and check dams. The community awareness, commitment and ownership, knowledge dissemination, capacity building and operational practices are among non-structural measures.

The sustainable development should be an integral part of the drought mitigation policy. The sustainable development may comprise water, food and environmental security, biodiversity, forests, rangelands, livestock, and watershed management. Implementation of policy with legal context will help sustainable development of groundwater resources for public supply, industry, and agriculture in Quetta Valley.

References

[1] McKee, T.B. Doeskin, N.J. and Kleist, J., 1993. The relationship of drought frequency and duration to time scales. In *Proceedings of the 8th Conference on Applied Climatology. American Meteorological Society*. **17(22)**, Boston, MA, 179–184.

[2] IPCC, Climate Change 2014. *Synthesis Report*. The contribution of Working Groups I, II and III to the Fifth Assessment Report of the Intergovernmental Panel on Climate Change [Core Writing Team, R.K. Pachauri and L.A. Meyer (eds.)]. IPCC, Geneva, Switzerland, 151 pp.

[3] Becker, A. Finger, P. Meyer-Christoffer, A. Rudolf, B. Schamm, K. Schneider, U. and Ziese, M., 2013. A description of the global land-surface precipitation data products of the Global Precipitation Climatology Centre with sample applications including centennial (trend) analysis from 1901-present. *Earth System Science Data*, **5**, 71–99.

[4] Fan, Y. and Vanden Dool, H., 2004. The CPC global monthly soil moisture data set at 0.5-degree resolution for 1948–present. *Journal of. Geophysical. Research*, **109,** D10102, doi:10.1029/2003JD004345.

[5] Mann, H.B., 1945. Nonparametric tests against trend. *Econometrica*. Journal of the Econometric Society, 245-259.

[6] Kendall, M.G., 1975. *Rank Correlation Methods, 4th edition*. Charles Griffin, London, U.K.

[7] Sen, P.K., 1968. Estimates of the regression coefficient based on Kendall's tau. *Journal of the American Statistical Association,* **63**, 1379-1389.

[8] Nellie Elguindi, Xunqiang Bi, Filippo Giorgi, Badrinath Nagarajan, Jeremy Pal, Fabien Solmon, Sara Rauscher, and Ashraf Zakey Trieste, 2007. *RegCM Version 3.1 User's Guide.* Italy.

[9] E. Roeckner, G. Bäuml, L. Bonaventura, R. Brokopf, M. Esch M. Giorgetta, S. Hagemann, I. Kirchner, L. Kornblueh E. Manzini, A. Rhodin, U. Schlese, U. Schulzweida, A. Tompkins, 2003. *The atmospheric general circulation model ECHAM5*. Hamburg, Germany.

[10] Lin, S.-J., and R. B. Rood, 1996. Multidimensional flux form semi-lagrangian transport schemes, Mon. *Wea. Rev*., **124,** 2046 – 2070.

[11] Lin, S.-J., and R. B. Rood, 1997. An explicit flux-form semi Lagrangian shallow-water model on the sphere, Q. J. Roy. *Meteor. Soc*., **123**, 2477 – 2498.

[12] Lin, S.-J., and R. B. Rood, 2004. A "vertically Lagrangian" finite-volume dynamical core for global models. *Monthly Weather Review*, **132**. 2293-2307.

IOP Conf. Series: Materials Science and Engineering **414** (2018) 012044 doi:10.1088/1757-899X/414/1/012044

Impact and Ratio of Lead in Ambient Air from Vehicular Emission in Quetta Valley, Pakistan.

Hussain M[1], F Akhtar[1], S Shah Khan[2]

[1]Department of Environment Management and Policy, Balochistan University of Information Technology, Engineering & Management Sciences (BUITEMS), Takatu Campus, Quetta, Pakistan
[2]Department of Physics, General Muhammad Musa Post Graduate College, Quetta, Pakistan

E-mail: manzoor.hussain@buitms.edu.pk

Abstract. Air pollution is the fourth leading fatal health risk factor consider worldwide, resulting in 4.8 million premature deaths in 2013. Lead in Ambient Air remains a serious problem in many developing and industrializing countries, as well as in some developed countries. Toxic effects of lead is affecting many people especially in motorized, urban environments. Pakistan is the most urbanized country in South Asia and deteriorating Ambient Air conditions particularly in large urban cities are damaging the people's health and quality of life and contributes to environmental degradation. Monitoring of Air Pollution was conducted in Quetta Valley. The seasonal Annual average concentration of Lead in Ambient Air was calculated. The ratio of vehicular emission from exhaust pipe of vehicles was also calculated in Ambient Air. The results were compared with the WHO standards and found four times higher than WHO Standard i.e. 0.5-1 μg. The concentration was found higher in spring and found lower in summer. The results were compared with the major cities of world and within Pakistan. The paper also highlighted the impacts of Lead pollution in Quetta Valley.
Key Words: lead toxicities, Ambient Air, Urban environments, Vehicular emission

1. Introduction

The problem of air pollution is a dominant factor in major cities of the world [1]. Many of the urban centers in the developing countries are beset with the environmental problem especially air pollution due to rapid industrial growth, the population explosion, resource depletion, alteration of the chemical composition of the atmosphere and the cycling of carbon, nitrogen and phosphorous are factors largely responsible for changing the basic physiology of the earth [2]. The WHO criteria pollutants that are responsible for the deterioration of air quality are Oxide of Nitrogen (NOx) Ozone, Oxides of Sulphur (SOx), Carbon Monoxide, Hydrocarbon (HC), Particulate Matters (PM), and Lead (Pb).

Lead is a versatile element due to its low melting point, softness, high density, malleability, ability to form alloys with other metals and resistance to corrosion. There are over 200 naturally occurring ores containing lead. The atomic configuration of lead has the ability to blend with silicate glass to form the colorant. Due to the bright colors it is used in the paints, cosmetics, hair dying and eye shadows. The toxic properties of these compounds have been exploited in the form of pesticides, preservatives and tragically as adulterants especially in wines [3]. Lead is also used in petrol additives. Lead is used to sheathing for electrical cables, shield ionizing radiation, in sound proofing, and soldering. To glaze the tiles the compounds of lead as a poly-vinyl-chloride stabilizer in plastics. The largest use of lead is in the batteries. Therefor the widespread use of lead make lead a strategic and critical element on earth. Now a day the global lead consumption has increase four times and cross the limits of 4 million metric tons. Estimate shows that Lead in atmosphere have exceed 200 times due to anthropogenic sources, about 880 million pounds per year from petrol additives, smelting of lead, iron, zinc and copper and coal combustion. The primary source of lead industrial emissions is still combustion of leaded petrol

[4]. Lead is added to petrol as the organic liquids tetra ethyl lead and tetra methyl lead. About 75% of the lead is emitted from the exhaust pipe of vehicles.

Contemporary researches indicate that vehicular exhaust is one of the major or primary source of lead pollution in urban areas. The mean particle size diameter of lead in aerosols in urban air has been reported to be 0.2-0.4 μm [5]. As the particle size is small, the dispersion of exhaust lead can be considered in terms of the well-established theory of diffusion of gases and smokes. Dispersion is determined by wind and turbulence, since the sedimentation velocity of the particles is smaller compared with turbulent air velocities. In spite of the great improvements in most developed countries due to reduced use of leaded fuels, still road emissions of lead remain a persistent air quality problem [6]. Lead and its public health impacts are drawing increase concern from the environment point of view. Objective of this research work was to collect data on the concentration of lead from different sources. Analysis the data to find ratio concentration between ambient air and vehicular emission of lead.

2. Methodology

2.1. Research Location

Quetta is a funnel shape valley and the capital city of Balochistan, Pakistan and lies between 66015/ 00// to 670 15/ 00// Easting and 300 00/ 00// to 300 25/ 00// Northing in the survey of Pakistan Topo-sheets 34 N/3 & 4 and 34 J/15 & 16. The main valley of Quetta or Shal proper (comprising an area of 1720 sq. km), which unites the two sides at the toe of the horse-shoe, is a parallelogram about sixteen miles by eight [7]. The general character of the city is mountainous. In Quetta city, the three prime locations are selected according to the guidelines provided by the United Nation Environmental Program (UNEP) for the collection of air data i.e. every city has residential, commercial and industrial area. These are illustrated in the map and are briefly defined as below: -

2.2. Monitoring Points

2.2.1Prince Road Choak (Site-1)

Prince Road is an area connecting all the major roads, crossing and commercial sites in Quetta City for example Saleem Complex, Petal Road, and is near the heart of commercial site such as Liaquat Road, Mission road, Quadari Bazar and Alamdar road, etc. Within half a mile of diameter there are also located all the important buildings such as Civil Hospital, Banks, High Court Building, Girls College, State Bank Building, Police Head Quarter Building, Govt. Schools etc. There are all sorts of shops in the vicinity such as cloths, handy craft, shoes, electronics, dry fruits shops, etc. The rush hour of shopping starts from 9:30 A.M. to 8:00 P.M. There are about approximately 10,000 shops in the vicinity.

2.2.2. Satellite town / Bus stand (Site-2)

Satellite Town is a residential area and mobile source. It is densely populated. There are about 10,000 houses and 1000 shops in that area. Adjacent to satellite town there lies Main Bus Stand, about 1000 buses move in and out of the city through this bus stand. Thousands of people travel up and down the country through this bus stand. Within half a mile of diameter there lies residential site of University of Balochistan, University of Balochistan, Phustoonabad, Baloch Colony, etc.

2.2.3. Sirki Road (Site-3)

Sirki road is a small Industrial state situated into the middle of the city. The Industries that are located within vicinity are Chiltan Ghee Mill, Small Industrial Wood Work Center, Soaps Industries, Ice Plants, Marble Cutting Plants etc. Adjacent to Sirki Road there are some residential area also such as Govalmandi choak, Jan Mohammad Road, Faqeer Mohammad Road, etc.

2.2.4. Methodology for Lead in Ambient Air:

Particulate Matter was collected by using high volume sampler, series model 302, Sierra Anderson, USA. The samples were collected at three different sites on Watman 41 filter paper, 20 x 25 cm. the pore size of the filter paper is 0.45 micron. Weighing the filter paper before and after the sampling at constant humidity will give the net weight of the particulate matter in the atmosphere at the particular site. The sampling time was 10 AM to 9 AM next day. The filter paper than analysis in the atomic absorption spectrometer and value of Lead is found.

2.3. Methodology for Vehicular Emission:

2.3.1. Bladder Method

100 vehicles (20 numbers were auto cycle, 20 numbers were Auto Rickshaws, 20 numbers were motor car, 20 numbers were Wagons and 20 numbers were Bus/Trucks) were selected on different streets in Quetta Valley. Smoke Particles were collected by tying the bladder on the exhaust pipes of the vehicles. The bladders were weight before and after filling the smoke. After collecting the smoke, the bladder were tightly closed for about 24 hour and then air was released in such a way that no smoke particle escape from the bladder. The bladder contents were analysis for lead on atomic absorption spectrometer.

2.3.2. Traffic count

Traffic density at each location was determined using digital photography. Table and graph give daily average of vehicles counted at monitoring sites.

2.3.3. Standards for Lead

The World Health Organization (WHO), European Council (EC), United States Environmental Protection Administration (USEPA) and Russian limit/standards are summarized in table 1.

Table 1. Standards for Lead Source: [24]

Pollutants	WHO	EC	USEPA	Russia
Lead	0.5-1 μg/m^3	2.0 μg/m^3	1.5 μg/m^3	0.3 μg/m^3

3. Results

The concentration of TSP was experimentally measured in 2013-2014 using high volume sampler at three sites in Quetta valley. The average concentration of particulate matter in ambient air was found 545 μg /m3, which is more than four times the WHO standards, i.e. 70 μg /m3. The measurement shows that out of 545 μg /m3 of particulate matter, the lead concentration in ambient air is 4.058 μg /m3. The lead concentration in ambient air is more than four times the WHO standards, i.e.0.5-1.0 μg /m3. The highest concentration of lead was recorded at site-1 i.e. 5.8 μg /m3, in the season of spring. This is the main crossing of Prince Road and Jinnah road, the hub of the city. The traffic density at that point was maximum, congestion and traffic jam occurred frequently. The lowest lead concentration was recorded at site-2 i.e.2.31 μg /m3, in summer. Overall the lead concentration at site-2 and site-3 are approximately equal and site-1 has highest concentration of lead in ambient air. Annual mean seasonal average concentration for 2013-2014 was found 4.058 μg/m3 in ambient air.

The vehicular exhaust is one of the major sources of lead [Pb] contributor in Ambient Air in Quetta city, which are produced from motorcycles, Auto rickshaws, cars, buses, trucks, tractors, pickups etc. Smoke particles were obtained by bladder method from the exhaust of various types of vehicles. The concentration found in the smoke ranged from 1.5 μg to 4.5 μg. The average concentration of lead emitted from vehicles is found 3.25 μg. The percentage ratio of lead in ambient air from vehicles is 80.09 % and other sources contribute only 19.91 %. The results are summarized in table below:

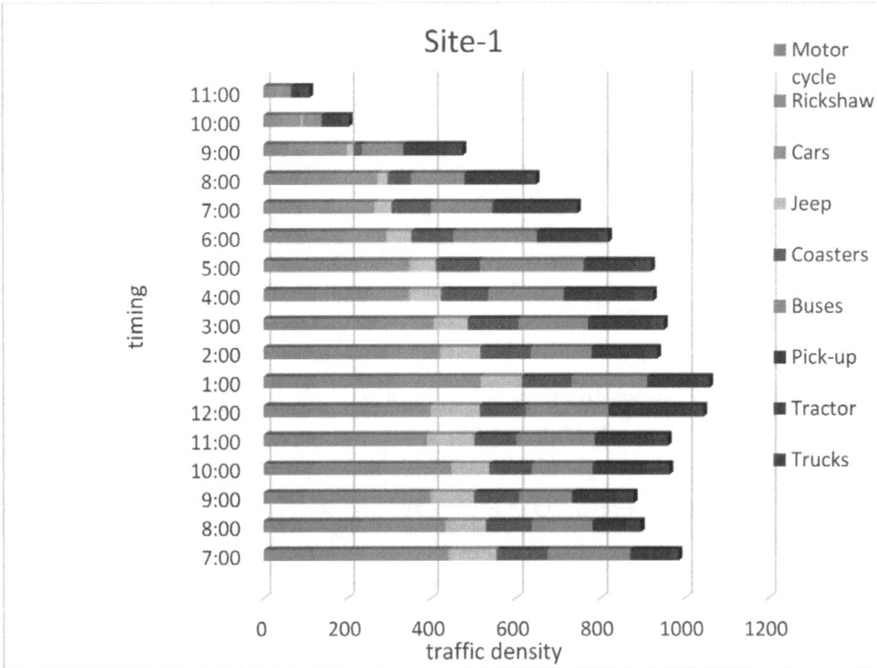

Figure 1. The percentage ratio of lead in ambient air from vehicles Source: Compiled

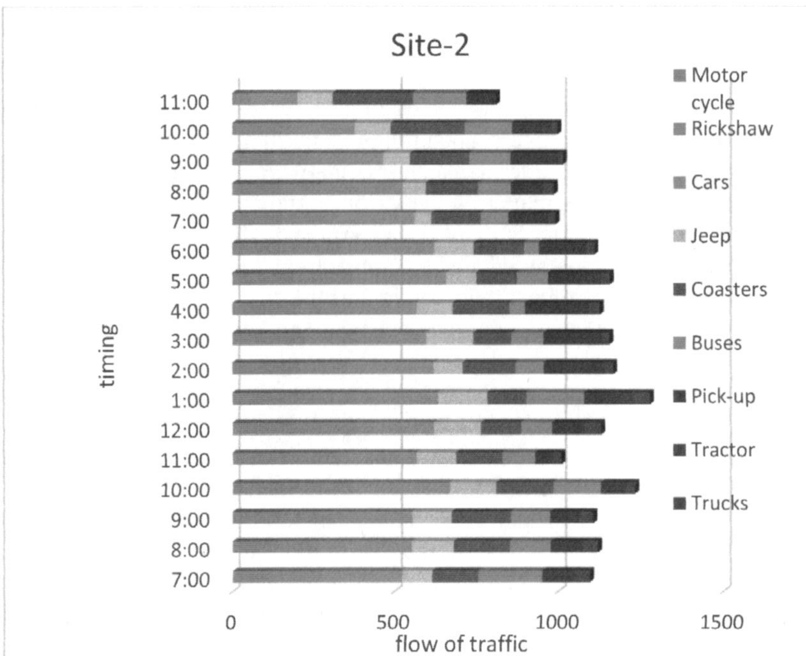

Figure 2. The percentage ratio of lead in ambient air from vehicles Source: Compiled

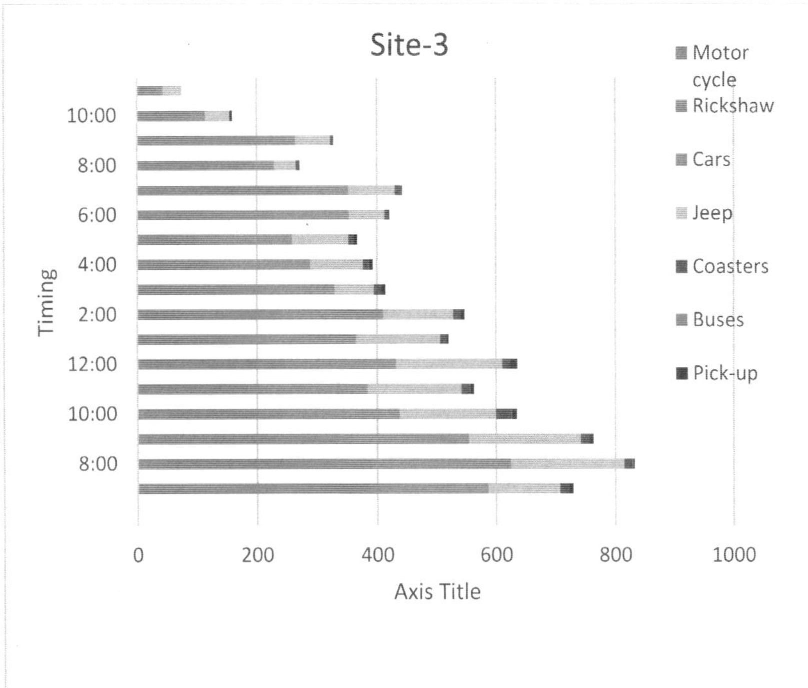

Figure 3. The percentage ratio of lead in ambient air from vehicles Source: Compiled

Table 2. The table below shows average lead emission from vehicles Source: Compiled.

Mode of Transportation	Pb Concentration
Auto Rickshaw	3.0 μg
Auto cycle	1.5 μg
Cars/Jeeps	4.0 μg
Buses/Truck	4.5 μg
Total	13.0 μg
Average	3.25 μg

Table 3. Average Traffic Flow Per Hour Source: Compiled.

Sites	Average Traffic Flow per Hour
Site--1	793
Site-2	1076
Site-3	524

Table 4. Seasonal Ambient Air Lead concentration in Quetta Valley Source: Compiled.

Seasons	Site-1	Site-2	Site-3	Seasonal Average
Autumn	5.2 μg/m^3	4.11 μg/m^3	4.24 μg/m^3	4.5 μg/m^3
Winter	4.71 μg/m^3	3.44 μg/m^3	3.56 μg/m^3	3.90 μg/m^3
Summer	3.81 μg/m^3	2.31 μg/m^3	2.54 μg/m^3	2.89 μg/m^3
Spring	5.8 μg/m^3	4.7 μg/m^3	4.28 μg/m^3	4.93 μg/m^3
Std. Deviation	0.72	0.88	0.70	
Total	19.52 μg/m^3	14.56 μg/m^3	14.62 μg/m^3	16.22 μg/m^3
Average at site	4.88 μg/m^3	3.64 μg/m^3	3.65 μg/m^3	4.05
Annual Average	**4.058 μg/m^3**			

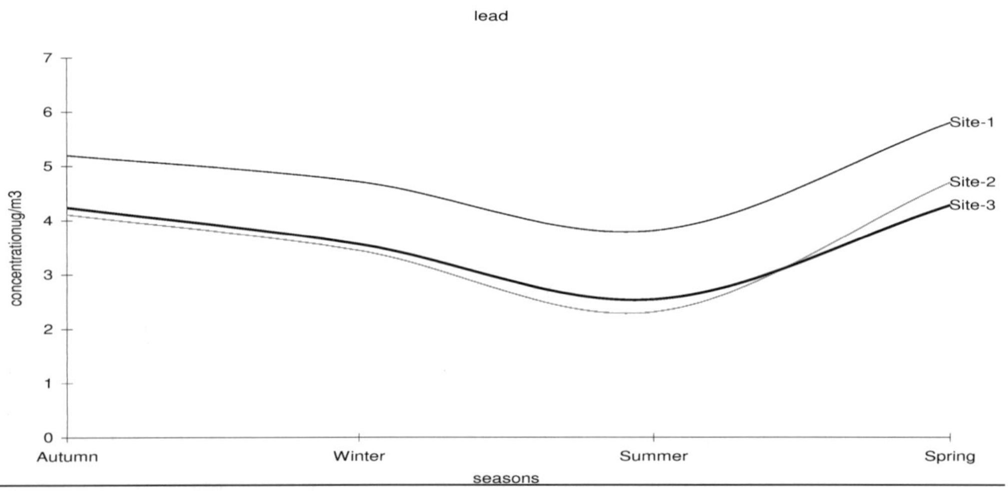

Figure 4. Seasonal Variation of Lead in Quetta City

4. Discussion

Recent researches indicate that vehicular exhaust is one of the major or primary source of toxic air pollutants and contribute significantly to emission inventories especially in urban areas [6]. The pollutants which are cause of concern with respect to transport are CO, HC, NOx, PM and Lead [8]. The ground level vehicles use different types of fuels such as gas fueled, petrol fueled, diesels fueled, solar and bio-product fueled. The physical and chemical physiognomies of natural gas, petrol and diesel are dissimilar in regions around the world, like lead, benzene, Sulphur content [9,10] making it difficult for the findings in one site to be generalized to other locations. This complex situation is further complicated by the effect of different meteorological conditions, different percentage of heavy and light polluter vehicles (more motorcycles in the developing world), design of roads (graded or non-graded), driving behaviors, different maintenance level of the vehicles as well as quality of and control measures for vehicles, and at the end most important exposure profiles of the people [11].

Technologically advanced republics have fought for unpolluted air quality by regulating all major and minor sources of air pollution standards. In these countries industrial and vehicular emissions have reduced significantly. For example, due to new motor vehicles emission standards in 1988, new vehicles are designed, which are 90 % cleaner than those manufacture in 1970 but unfortunately in developing countries these standards are not applied or even in some countries there exists no standards to cope with the pollution problem. As the vehicles fleet continue to grow, vehicular emissions and associated products enter the atmosphere and make changes in the atmosphere. Random acceleration and deceleration due to non-motorized vehicles in developing countries are becoming increasingly important contributors to nearly every major air pollution problem facing the world [6]. In metropolitan cities, where more than 70 % of the population confined the levels of vehicle's pollution frequently exceeded the guideline of WHO standards [1].

Pakistan is one of the unfortunate developing country (EPA laws are not applied properly), now facing the Pollution problem from verity of ways [12, 13]. One of source of pollution in Pakistan is from vehicular emissions [14]. Pakistan had 0.85 million cars on road in 1982, from 1991 to 2012, the number of motorcycles and scooters grew more than 450%, and motor cars, close to 650% and an average annual growth rate in excess of 8.5%. The economic survey of Pakistan pinpoint that vehicular and industrial emission are the main causes of poor air quality in Pakistan and world bank report says that urban air pollution is causing estimated US $ 369 million to the national economy of Pakistan [1].

Airborne Pb is associated with the density of motor vehicles traffic using fuel and the concentration of Pb additive in the fuel. Most of the petrol/gasoline in Pakistan contains 1.5 to 2.0 g/liter of tetra alkyl lead as anti-knocking agent in automobiles. Additionally, so called lead free gasoline available in Pakistan also contains lead quantity ranging from 0.60 to 0.95 g/liter. Lead concentration in all the major cities of Pakistan shows alarming level of lead in ambient air [15, 16]. The lead concentration in all the major cities is more than three time to the standards of WHO and double to the USEPA and EC standards and multiple fold higher from the standard of Russia. In Pakistan a vehicle produces 20 times more HC, 25 times Carbon, 8 times more Lead and 3.6 of NOx in gram per KM compared with an average vehicle in US.

Table 5. Lead concentration in major cities of Pakistan, [16].

City	Pb Concentration
Quetta	4.058 $\mu g/m^3$
Karachi	3.954 $\mu g/m^3$
Peshawar	3.915 $\mu g/m^3$
Lahore	3.51 $\mu g/m^3$
Rawal Pindi	3.35 $\mu g/m^3$

Due to reluctance of the Government to implement the Environmental Protection Agency's laws passed by the Federal Government in 1997. It is feared that an estimated Outdoor air pollution alone caused more than 80,000 hospital admissions per year [1]. The main reason for the deteriorating air quality in metropolis cities is becoming worse because of haphazard industrialization and rapid increase in the number of vehicles. Growing lead concentration in ambient air is also associated with the lead in blood [1, 17]. Lead and its public health impacts are drawing increase concern from the environment point of view [4]. Higher lead level in blood are also recorded in Karachi, Islamabad, Lahore and other major cities of Pakistan [18, 19].The lead concentration in all the major cities shows more than three time to the standards of WHO and double to the USEPA and EC standards and multiple fold higher from the standard of Russia while the acceptable blood level is ≥ 5 µg/d L [20], which results in many dangerous diseases such as blood cancer [21]. Similar studies were carried out by Kamal 1996, and Zaidi, 1997, for Quetta Valley and found high correlation between the distribution of lead deposited on roadside trees and the occurrence of high blood pressure, ear, nose and throat related illness, fatigue, gastrointestinal diseases and cancer among people living in those areas. Beside that lead in air have fatal effect on trees and plants [22].

Quetta is the provisional capital of Balochistan with a vehicular density more than 100000 and is increasing at a rate of 12 % per year. Measurement in Quetta city shows that organic lead vapors accounts for 80.09% of total lead. The concentration is found higher near filling stations; road crossing, congested and traffic jam places and busy streets/road. Quetta city has heterogeneous traffic flow, a substantial part of total traffic is non- motorized which enhance severe congestion and pollution problems especially on road intersection. The popular mode of transport is two stroke engine/motorcycles and Rickshaws and reliance on private cars. Journeying through by buses or public transport especially is very low. A slow and reckless driving habit of the Quetta's population is also one of the main causes that burns more fuel and in return high emission of lead from vehicles.

Quetta is a boarder city therefore smuggle oil/gasoline is easily available in the city. The smuggled oil/gasoline is not of high quality and has high contents of lead and Sulphur. The situation gets worse when adulteration/impurities/ mixing are added to gain little profit. Because of smuggled oil/gasoline decline in sell is recorded at government approved Petrol Pumps and this is severely affecting the economy of the country. In Quetta majority of the vehicles are more than five to six years old because government of Pakistan gave subsidies tax rate to allow import of reconditioned automobiles more than five-year-old. In addition to this, different kinds of second hand engines are imported/ smuggled from different part of the world whose economical life is over.

Government had introduced compressed natural gas (CNG) and liquid Petroleum gas (LPG) used as alternate fuel in automobiles. For example, Introduction of four stroke Rickshaws in Quetta Valley to cope with the problem of air pollution is an effort by the Government. According to the Government these Rickshaws are environmentally friendly and are able to provide door to door services in the streets geometry of the congested and narrow roads of Quetta city. These Rickshaws which are built locally and are without catalytic converter therefore they also add lead concentration in the ambient air. [13, 23, 25, 26]. Due to some of the above reasons and non-compliance of adequate air quality standards, Quetta is becoming one of the top-most Lead polluted city in the world. Lead concentration in major cities of the World is given in table 6.

Table 6. Lead concentration in major cities of the World, Source: [16].

City	Pb Concentration	City	Pb Concentration
Quetta	4.058 μg /m^3	Calcutta	0.73 μg /m^3
Mexico	3.83 μg /m^3	Delhi	0.67 μg /m^3
Bombay	3.60 μg /m^3	Jakarta	1.86 μg /m^3
Dhaka	3.12 μg /m^3	Manila	1.3 μg /m^3
Sydney	3.33 μg /m^3	London	0.229 μg /m^3
Santiago	2.30 μg /m^3	Moscow	0.07 μg /m^3
Los Angeles	0.70 μg /m^3	New York	0.12 μg /m^3
Kyoto	0.40 μg /m^3	New Jersey	0.41 μg /m^3

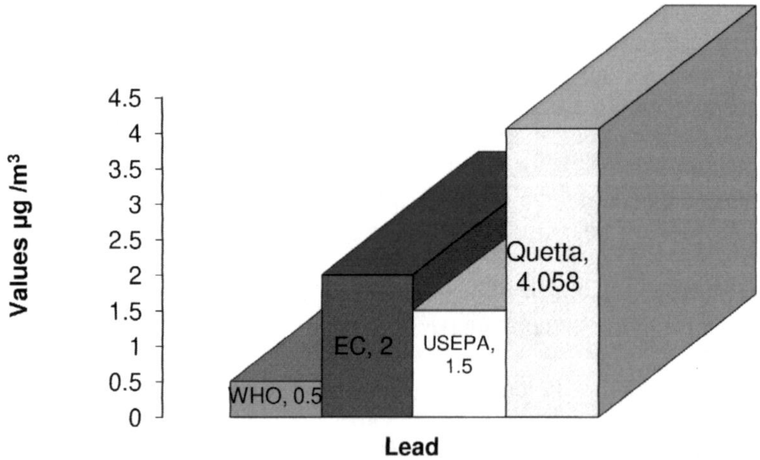

Figure 5. Standards Comparison of Lead Level with Quetta City Source: Compiled

5. Conclusions

The study analysis that Quetta is becoming one of the topmost Lead polluted city in the world. It is also well-established fact that Lead in Petrol/gasoline is the major source of Lead pollution in ambient air. At present there is no legislature regulation to check and control the content of Lead in Petrol/Gasoline in Pakistan. Establishment of proper rule & regulation and strict enforcement of these regulations are needed to control the Lead and other toxic pollutants in air for the safety and health of the general public.

6. References

[1] World Bank and Institute for Health Metrics and Evaluation Report, 2016.*The Cost of Air Pollution: Strengthening the Economic Case for Action.* The World Bank, 1818 H Street NW, Washington, DC 20433, USA; fax: 202-522-2625; e-mail: pub rights@ worldbank.org.

[2] Xi Anglue, Luke p, Nacher, 2006A Review of traffic related pollution exposure assessment studies in the developing world. *Environmental International*, **32**-pp106-120.

[3] Nriagu, J.O. 1990."Global metal pollution. Poisoning the biosphere?" *Environment,* **32**, 7, 7-11,

[4] Abdullah, M. S. 2013.Toxic effects of lead in humans: an overview. Global Advanced Research Journal of Environmental Science and Toxicology *(GARJEST).* **2**(6) pp. 157-162

[5] Frank J. Kelly and Julia C. F. 2015. Air pollution and public health: emerging hazards and improved understanding of risk, *Environ Geochem Health*_**37**(4): 631–649.

[6] Mellouki A, George C, Chai F, Mu Y, Chen J, Li H. 2016 Sources, chemistry, impacts and regulations of complex air pollution: Preface. *J Environ Sci* (China) **40**:1–2.

[7] Shahid N. khan, Muhammad Y, Mohsin A. K. & Hussain S. A. 1986.Environmental geology, mineral, raw material and aggregate resources of Quetta valley, Balochistan, Pakistan P- 1-2 *GSP Rec.* Vol. **LXXVI-** Environmental Geology.

[8] Marr R.A and Harley C. 2002 Modelling the effect of weekday weekend differences in motor vehicle emissions on photochemical pollution in Central California, *Environmental Science and Technology* **36**, pp. 4099–4106, 2002

[9] Khan S. S. Sabir H. U., Shahab N., and Akhta F 2004. "Concentration of Air borne Particulate Matter (PM_{10}) in Quetta valley", *Research Journal, University of Balochistan*, **2**, No-2, pp-159-166.

[10] Khan S. S. Sabir H. U., Shahab N., and Akhta F 2005 "Seasonal variation of Total Suspended Particles (TSP) in Quetta (Pakistan)", Research Journal, Balochistan study centre, *University of Balochistan*, **xiv-xv,** pp-37-43.

[11] Gwillian K 2003 Urban transport in developing countries. *Transp. Rev*, **23**(2): 197-216

[12] Ghauri B., Manzar, S. Mirza, 1994.An Assessment of air quality in Karachi Pakistan, *Environ Monit Assess,* **32** pp 37-5.

[13] Ghauri B., Lodhi A., and Mansha M. 2007. Development of Baseline (Air Quality) Data in Pakistan. *Environmental Monitoring and Assessment* **127**: 237–52

[14] Shams ZI. 1998 Health and environment: lead pollution in Karachi is a serious health hazard. Karachi, Pakistan, *University of Karachi Environics Institute of Environmental Studies.*

[15] Zafar I. S. and Mirza A. A.2004, Lead in particulate deposits and in leaves of roadside plants, Karachi, Pakistan *The Environmentalist* 0251-1088 (Print) 1573-2991 (Online) pp63-67 Springer Link,

[16] Ghauri, B. 2010. Institutional Analysis of Air Quality Management in Urban Pakistan. Study commissioned by the World Bank, Washington, DC. Clean air initiative .org/portal / system/files/.../AQM_Draft_Final_Report.pdf.

[17] Calderon-G. L,mora T. A, Fordham, V. Chung CJ. G. Rodriguez A. A. 2003 Respiratory damage in children exposed to urban pollution Pediatric pulmonol, pp 148-61.

[18] JICA 2000, Report Investigation of air and water quality. Lahore, Rawalpindi, Islamabad. Japan International Corporation Agency/Pakistan Environmental Protection Agency, Islamabad.

[19] Akbar,S. F.M. Kakar, Sami, M. Sarwar,G. and Khan M 2004Dust fall, Smoke Particles & Photochemical Smog in Quetta City, *R. J. U. of Balochistan*, **2** (2) pp-12.

[20] Ul-Haq Arain N.,M.A.,N. Badar, Rasheed M. and Haque Z.2011.Drinking water: a major source of lead exposure in Karachi, Pakistan Eastern Mediterranean Health. *Journal La Revue de Santé de la Méditerranée orientale EMHJ* **Vol. 17** No. 11 882

[21] Yamamoto SS, Phalkey R, Malik AA. 2014 A systematic review of air pollution as a risk factor for cardiovascular disease in South Asia: Limited evidence from India and Pakistan. *Int J Hyg Environ Health.* **217**:133–44.

[22] Keller,B., Faciano, A., Tsega, A., and Ehrlich, J.2017.Epidemiologic Characteristics of Children with Blood Lead Levels ≥45 µg/dL..*The Journal of Pediatrics***180**: 229-34.

[23] Sami M., Waseem A. Akbar S. 2006 Quantitative estimation of dust fall and smoke particles in Quetta Valley, *j. Zhejiang Univ,Science* B 7970 524-547

[24] U.S., EPA, 2003 Latest findings on National Air Quality 2001 Status and Trends, EPA 454/K-02-001, *Research Triangle Park*, NC, 2003

[25] Nriagu, J.O. 1990. "Global metal pollution. Poisoning the biosphere?" *Environment*, **32,** 7, 7-11

[26] Talib H., Izhar u. H. M. Ali K. study of environmental Pollutants in and around the city of Lahore, determination of Lead in Blood of various Population groups, T*he science of Total Environment*, **99**, pp 137-143, 1990

Simulation of Proton Exchange Membrane Fuel Cell by using ANSYS Fluent

Asifa Awan[1], Mahmood Saleem[1] and Abdul Basit[2]

[1]Institute of Chemical Engineering, University of the Punjab, Lahore, Pakistan
[2]Department of Chemical Engineering, University Technology PETRONAS, Malaysia

Email: msaleem.icet@pu.edu.pk

Abstract. Proton exchange membrane fuel cells (PEMFC) are attractive alternative source of electricity. The current study involves the computational fluid dynamics simulations of PEMFC under isothermal and non-isothermal conditions to investigate the performance of fuel cell. Effect of pressure and temperature on fuel cell performance is studied under non-isothermal conditions. PEMFC is modeled at 323 K and 1 atm under isothermal conditions whereas under non-isothermal conditions, the simulation is run on 353 K and 3 atm. The results show that the current density increases with increase in operating pressure of PEMFC and vice-versa with operating temperature.
Keywords. Simulation, proton exchange membrane, fuel cell, ANSYS Fluent

1. Introduction

Fuel cell converts chemical energy into electrical energy. It runs continuously until the reactants run out. The ever growing efficiency and very low emissions of fuel cells will ensure their commercial success in coming years [1]. PEMFC's operate at low temperatures (below 100 ^{o}C) which is easy to operate and handle. Recent research work on PEMFC is generally focused on its size, cost, performance and durability. Computer simulations are getting more attention of researchers as low cost option which produces reliable results.

PEMFC performance depends upon various factors like: Physical parameters (i.e. ref. current density, conc.), operating parameters (i.e. operating temperature, pressure), geometrical parameters, advanced parameters like contact resistance. Siegel C. et.al reviewed PEM fuel cell simulations performed on different softwares [2]. Researchers [3] validated PEMFC model for investigating convergence criteria. According to authors [4], incomplete and unreliable data is the main issue for obtaining correct tests.There is a lot of work going on flow fields to get better performance of PEMFC. Computational model developed for a 5 cm^2 fuel cell with serpentine and parallel flow fields are considered in experimentation and in modeling [5].

Simulation has been performed by modifying the design of flow field which decrease the pressure drop in PEMFC. Modifications are based on theory of injection engine to reduce pressure drop [6]. A comprehensive non isothermal, 3-D model is established to investigate the performance of PEMFC with straight and serpentine flow fields [7]. Aspect ratio of flow fields also has effect on fuel cell performance [8]. The local transport phenomena and cell performance is performed for parallel and integrated flow fields [9]. Effect of single (1-s), double (2-s) and triple (3-s) serpentine flow fields is also investigated. It is reported that effects of flow fields variations are negligible below 0.7 operating voltage [10]. A stepped flow field is proposed to improve cell performance [11]. Researchers made the comparison between the conventional straight gas flow channel and a novel wave-like channel [12]. Different shapes and lengths of parallel flow field affect cell performance [13].Water management is one of the main issues in PEM fuel cell [14].

Percolation theory was coupled with 3-D PEMFC model to investigate the effects of water flooding on gas diffusion layer [15]. Using different height and length tapered flow fields due to which velocity increases and water flooding does not produce, it also enhances rib-convection among adjacent channels [16]. Muthukumar M et.al analyzed the effects of landing to channel (LXC) on the generation of power and current density [17]. Researchers [18] focus on the relation between flow velocity and optimum channel width. PEM permits only proton ions and PEM's activity depends upon the humidity on surface of membrane [19]. Researchers made the comparison between 3 phenomenological membrane models and the results indicates that springer model and Nguyen and white model over predict drying of the membrane while the fuller and new man model provides the best match with experimental data [20]. In present study the transport equations are used after [7].

Table 1. PEMFC Dimensions [7].

Cell dimensions	Units	Values
Gas channel length	mm	10
Height of gas channel	mm	1
Width of the gas channel	mm	1
Width of the cell	mm	2
Thickness of catalyst layer	mm	0.014
Thickness of gas diffusion layer	mm	0.0254
Thickness of current collector	mm	2.5
Thickness of membrane	mm	0.051
Overall cell height	mm	5.1295

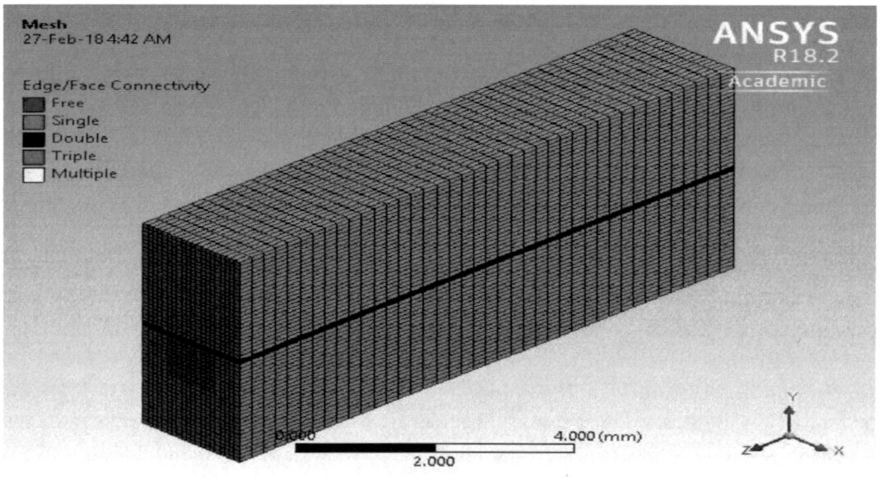

Figure 1. 3-D view of single channel PEMFC

2. Methodology

The simulated 3-D PEMFC consists of single straight channel shown in figure 1. The dimensions are given in table [1]. Boundary conditions and physical parameters used in the simulation are shown in tables 2 and 3 respectively. The model is developed with the following assumptions:

- Flow is laminar.
- Incompressible fluids.
- Steady state system exists.
- Inlet gases follow Ideal gas law.
- The catalyst layer, membrane and gas diffusion layer are Isotropic materials.

- For isothermal modeling temperature is constant.
- For non-isothermal modeling heat flux is constant.

Table 2. Boundary conditions (BC).

BC types	Location	Parameters	Non-Isothermal	Isothermal	Units
Velocity inlet	Inlet anode flow channel face	Velocity inlet	2	0.3	m/s
		Mass fraction of H_2	0.3	0.3	-
		Mass fraction of H_2O	0.7	0.7	-
	Inlet Cathode flow channel face	Velocity inlet	2	0.5	m/s
		Mass fraction of O_2	0.14	0.212	-
		Mass fraction of H_2O	0.2	.079	-
Pressure Outlet	Outlet anode flow channel face	Anode outlet gas pressure	0	0	Pa
		Temperature	Default	323	K
	Outlet cathode flow channel face	Cathode outlet gas pressure	0	0	Pa
		Temperature	Default	323	K
Wall	The terminal and upper anode current collector face	Specified electric potential	0 (only terminal)	0	Volts
	The terminal and lower cathode current collector face	Specified electric potential	0.5-0.9 (only terminal)	0.4-0.9	
	All Outer cell faces	Thermal condition constant Temperature	-	323	K
		Heat flux	0	-	

Table 3. Parameters values.

Parameters	**Isothermal**	**Non-isothermal**	**Units**
Cell operating temperature	323	353	K
Cell operating pressure	1	3	Atm
Open-circuit voltage	1.07	1.05	Volt
Anode Reference exchange current density	10,000	30	A/m^2
Cathode Reference exchange current density	20	0.004	A/m^2
CL Electric conductivity	Default	100	1/ohm-m
Current collector Electric conductivity	4000	Default	1/ohm-m
GDL Electric conductivity	300	Default	1/ohm-m
Anode exchange coefficient	1	0.5	-
Cathode exchange coefficient	1	2	-
Reference concentration of anode	1	0.04	-
Reference concentration of cathode	1	0.00086	
CL Porosity	0.112	0.6	-
GDL Porosity	0.6	0.6	-

ANSYS meshing is used to generate hexahedral mesh with 59200 cells whereas Fluent is used as a solver. Joule heating, electrochemistry and Butler-Volmer are used as model equations for

PEMFC simulation. SIMPLE algorithm is used for solving equations. The least square cell base is chosen under spatial discretization.

3. Results and Discussion

The simulation of single channel PEMFC is performed according to the condition given in tables 2 and 3. Results are presented in figure 2 and 3. The models are validated by using data of Lee Wang et.al. [21] and the results are in good agreement with the experimental data except for the low current density region under non-isothermal conditions and high current density region under isothermal conditions. As the redox reaction in the fuel cell increases, the current density also increases. At high current density, production of water increases in catalyst layer and gas diffusion layer (GDL), due to which effective porosity of catalyst layer and GDL decreases and material resistance increases. The assumption of the simulation that water is being produced in the form of mist does not remain valid due to this phenomenon at high current densities as the large quantities of liquid water fill the pores of catalyst and membrane, thereby increasing transport resistance [5].

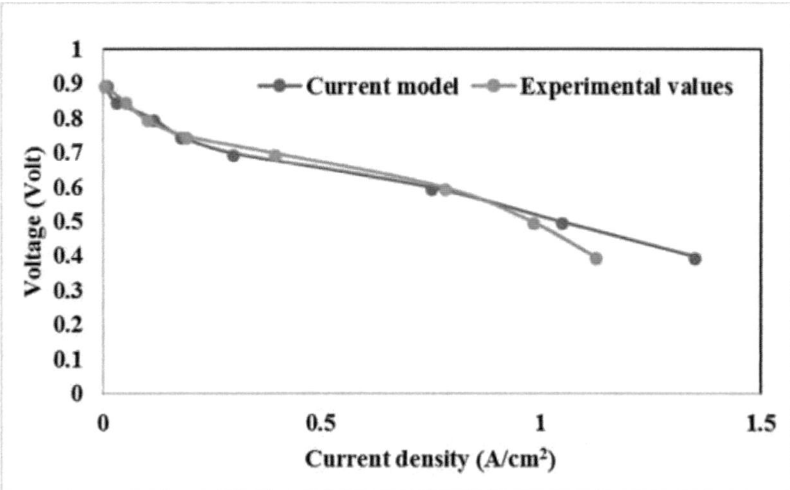

Figure 2. Comparison of simulation and experimental results under isothermal conditions.

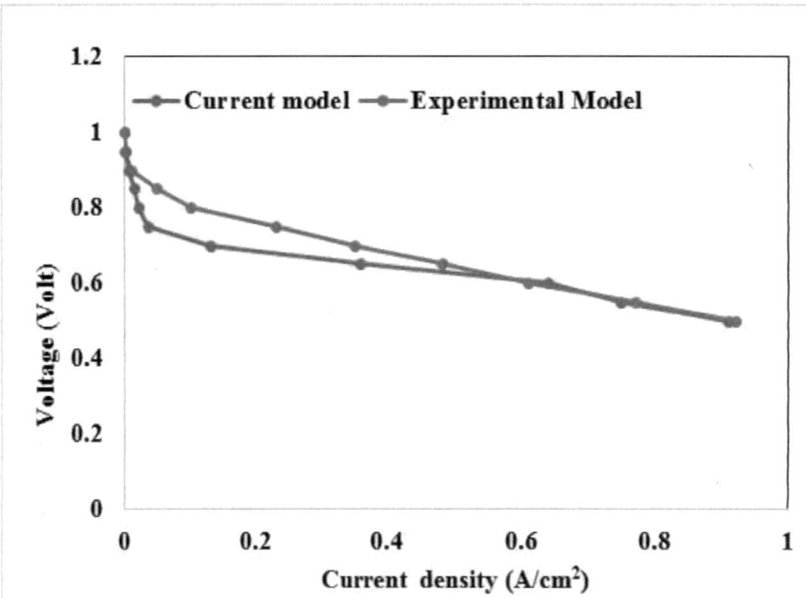

Figure 3. Comparison of simulation and experimental results under non-isothermal condition.

The most affecting reactant at higher current density is oxygen, as shown in figure 4 (a). The mass fraction of oxygen is very high in cathode channel at 353 K. Under the current collector, oxygen is more depleted in MEA region. This might be due to the limited oxygen diffusivity and also due to the difficulty of water removal in cathode side. As vapor saturation pressure depends on temperature, the relative humidity decreases with the increase in temperature. The membrane might be dehydrated thereby slowing the protonic ions permeation and leads to ohmic losses. Another reason of high mass fraction of oxygen might be slow protonic ions generation in membrane as compared to oxygen diffusion into GDL. On the other hand, it can be clearly seen that at high current density there is low change in hydrogen value as compared to oxygen as shown in figure 4 (b). Because hydrogen (small molecule) has better diffusivity than oxygen and it shows better transport at high current density even if the porosity is low. Usually there is no blockage at porous region of hydrogen side because flooding is not taking place on anode side. Pressure decreases from inlet to outlet as shown in pressure contours in figure 4(c). Temperature increases from inlet to outlet as shown in temperature contours in figure 4(d) due to reduction reaction and ohmic losses.

The current densities over-potential near the CCL/PEM interfaces are so big that the maximum current density generated in cathode region is produced on this interface. At this current density, starvation region is clearly shown in figure 5 under the current collector region; this might be due to oxygen diffusivity problem [22].

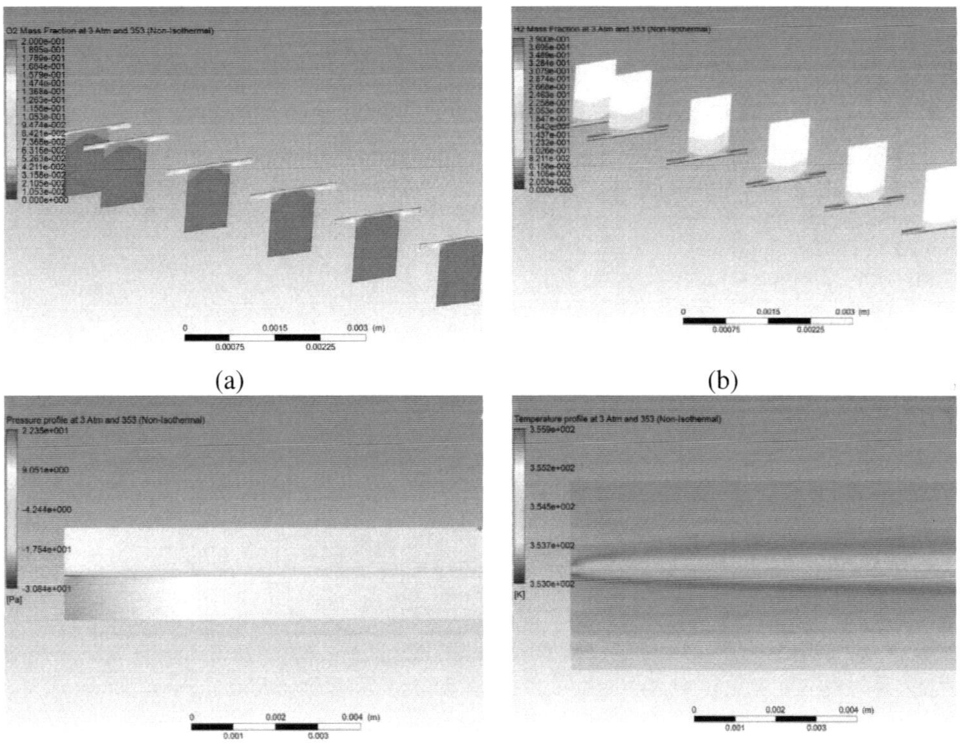

(a) (b)

(c) (d)

Figure 4. Mass fraction distribution under non-isothermal conditions at 0.6 volts for (a) O_2 (b) H_2 (c) Static pressure (d) Temperature.

Figure 5. Current Distribution at 0.6 volts non- isothermal.

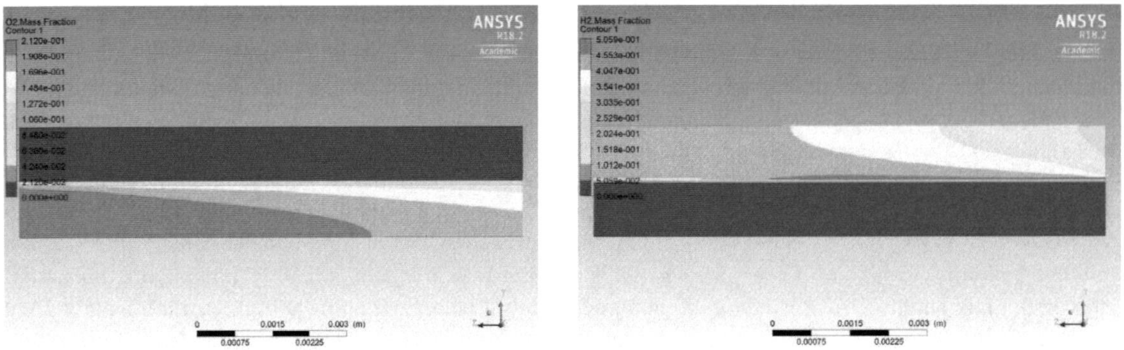

(a) (b)

Figure 6. Mass fraction distribution under isothermal conditions at 0.6 volts for (a) O_2 (b) H_2

Oxygen mass fraction decreases from inlet to outlet because oxygen is consumed in the reaction on its way to the outlet. Current density is greater at the hydrogen inlet as compared to oxygen inlet since the reaction starts at hydrogen inlet as shown in figure 6.

It is shown in figure 4 and figure 6 that no gases are present in MEA region of fuel cell. This means that at anode side, the hydrogen ions enter into the catalyst layer while on cathode side, oxygen ions enter into the catalyst layer instead of gas molecules [23].

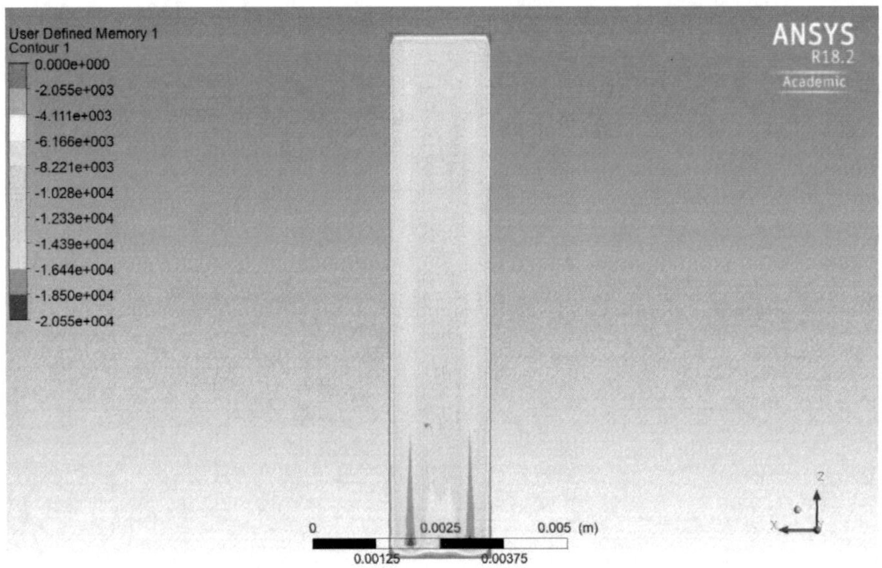

Figure 7. Current distribution at 0.6 volts (isothermal).

Current density value is low at collectors rib because there is depletion of oxygen in those region which is shown in figure 7. Drying of anode side might be major cause of decrease in cell performance. Multi-phase model should be used to observe this phenomenon in detail [5].

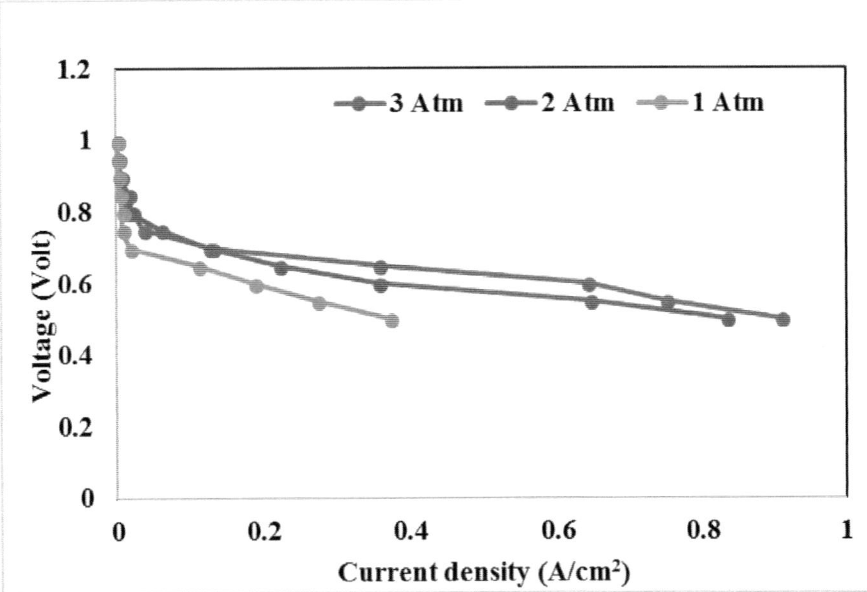

Figure 8. Effects of Pressure on voltage and current density at 253 K.

Polarization curves of different pressures at 353 K temperature are shown in figure 8. The pressures of both anode and cathode sides were kept the same. The performance of the fuel cell improves with the increase of pressure. The overall polarization curves shift positively as the pressure increases. With the increase in pressure, partial pressure of the reactant gases increases [21].

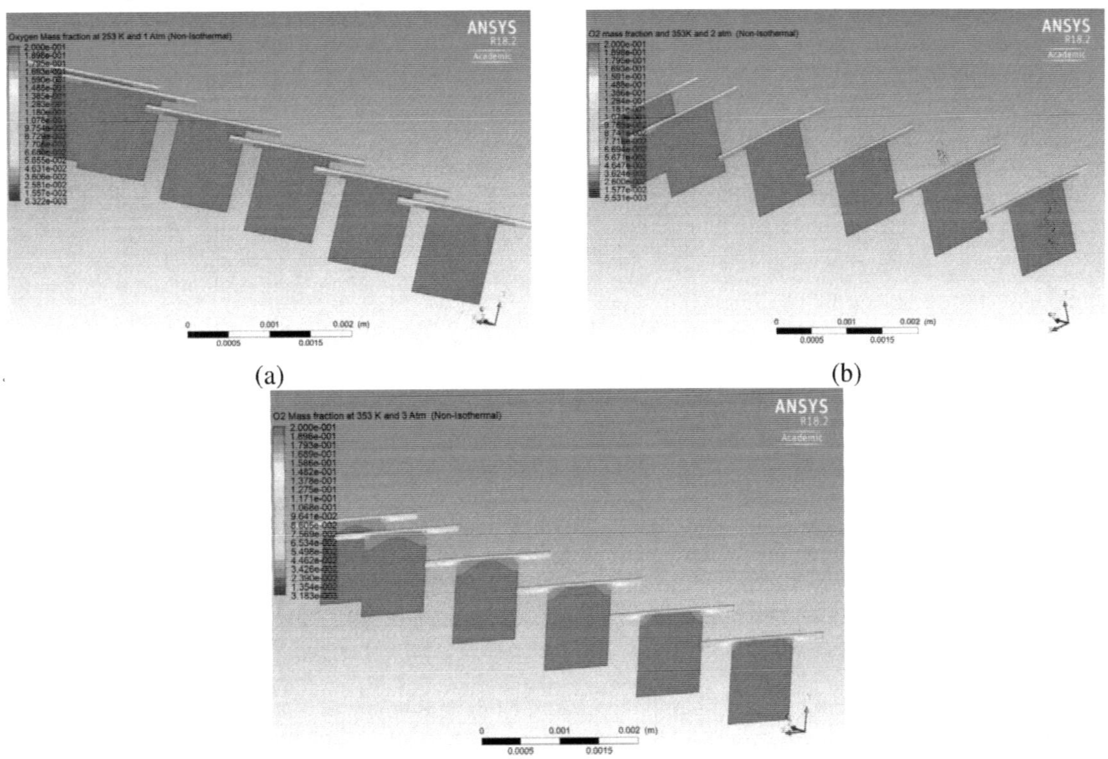

(a) (b)

(c)

Figure 9. Mass fraction of oxygen at (a) 1 atm (b) 2 atm (c) 3 atm.

The mass fraction of oxygen decreases in PEMFC as the operating pressure is increased as shown in figure 9. The decreased mass fraction is due to increase in reaction rate which is favored by increased operation pressure.

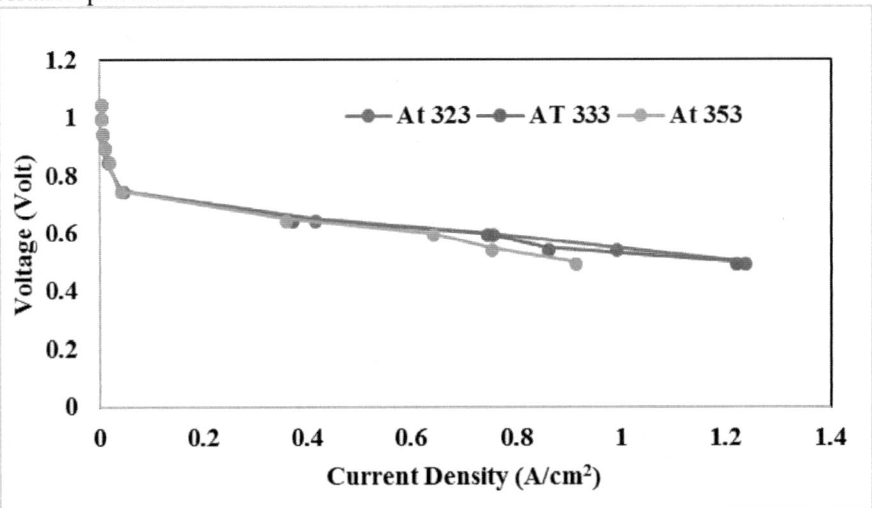

Figure 10. Effect of Temperature on voltage and current density.

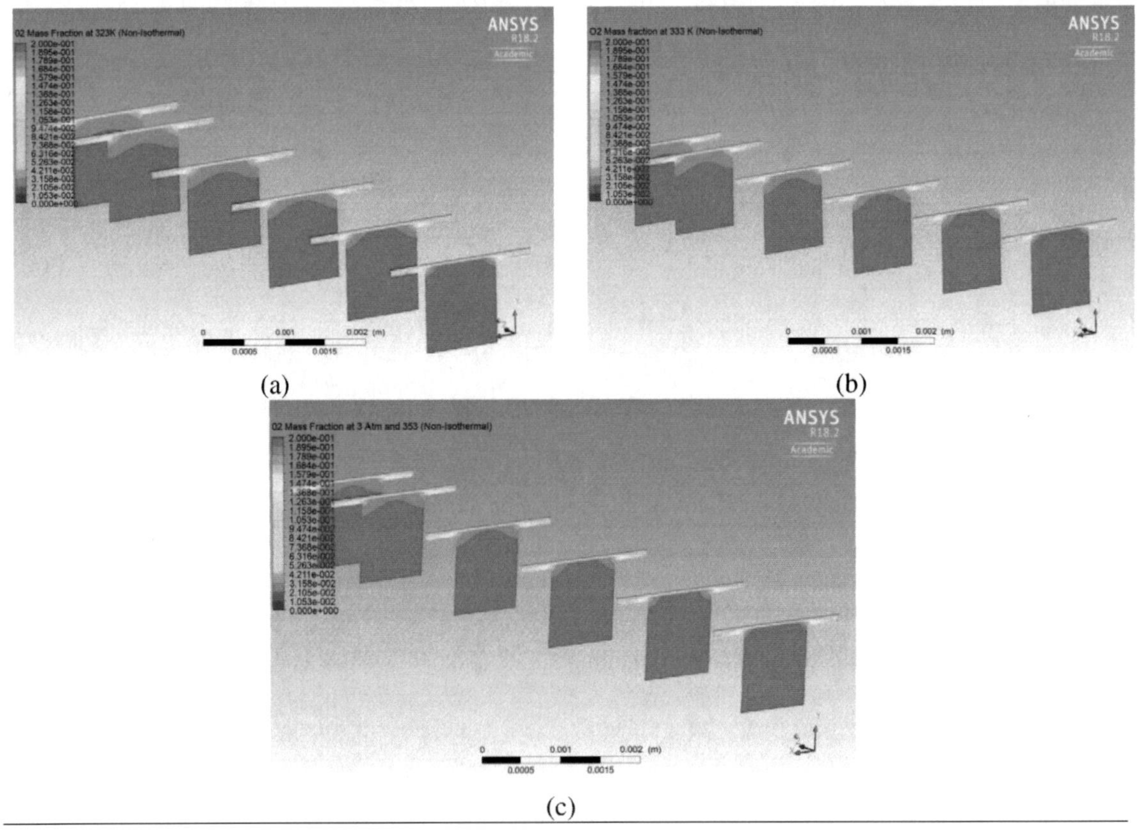

Figure 11. Mass Fraction of O_2 at 323K (a), 333K (b) and 353K (c).

Polarization curve at different temperatures are shown in figure 10. Oxygen mass fraction is shown in figure 11. Oxygen mass fraction distribution is often considered the criteria of fuel cell performance because the diffusivity of hydrogen is high as compared to oxygen. As the temperature is increased, humidification decreases and leaves the membrane less hydrated [24]. Oxygen mass fraction is slightly high at 353 K and 333 K as compared to 323K as shown in figure 11.

4. Conclusions

Simulation of single straight channel, proton exchange Membrane fuel cell is carried out under isothermal and non-isothermal conditions. Non-isothermal flow design is co current flow and isothermal model is counter flow design. It is observed that current was being produced at inlet of hydrogen. No oxygen and hydrogen molecules can enter into catalyst layer and membrane. Models predictions are compared with published experimental data which are in good agreement. The deviations observed might be due to single phase assumption in the study. Temperature and pressure are one of very important parameters which play main role in cell performance. The results show that cell performance under non-isothermal conditions is better at higher pressures.

List of Symbols

CL	Catalyst layer	i_a, i_c	Transfer current density for anode and cathode
GDL	Gas diffusion layer	i_c^{ref}	Ref. exchange current density per surface area Cathode
CC	Current collector	ζ_a	Specific active surface area at anode
F	Faraday constant	ζ_c	Specific active surface area at cathode
M	molecular weight (kg/mol)	i_a^{ref} , i_c^{ref}	Ref. exchange current density per surface area of Anode
$\boldsymbol{\varepsilon}$	porosity	$\emptyset_{sol}$	Electric potential
$\boldsymbol{\rho}$	density	$\emptyset_{mem}$	Protonic potential
$\vec{\mathbf{u}}$	velocity vector	$\sigma_{sol}, \sigma_{mem}$	Electrode conductivity, Ionic conductivity
$\vec{\boldsymbol{\tau}}$	shear tensor	λ_{eff}	Effective thermal conductivity in a porous media
$\boldsymbol{\mu}$	Viscosity	$\vec{J}_k$	diffusional mass flux vector of specie k
$\mathbf{k_i}$	Permeability of GDL and CL	λ_S	Thermal conductivity of solid porous media
$\vec{\boldsymbol{\tau}}_{\mathbf{eff}}$	Effective stress tensor	$\eta_{a,c}$	Over potential at anode and cathode
$\mathbf{h_k}$	Enthalpy of species k	$D_{k,j}^{eff}$	Effective diffusional coefficent
$\mathbf{Y_k}$	mass fraction y species	p_0	Reference pressure
T	Temperature	p	pressure
$\mathbf{T_0}$	Reference temperature		

References

[1] Barnett, B.M. and Teagan, W.P 1992 The role of fuel cells in our energy future Journal of Power Sources 37 15-31

[2] Siegel, C., 2008 Review of computational heat and mass transfer modeling in polymer-electrolyte-membrane (PEM) fuel cells Energy 33 1331–1352

[3] Arvay A, Ahmed A, Peng X-H and Kannan AM 2012 Feb Convergence criteria establishment for 3D simulation of proton exchange membrane fuel cell Int J Hydrog Energy 37 2482–9

[4] Bednarek T and Tsotridis G. 2017 Mar Issues associated with modelling of proton exchange membrane fuel cell by computational fluid dynamics J Power Sources 343 550–63
[5] Iranzo A, Muñoz M, Rosa F and Pino J. 2010 Oct Numerical model for the performance prediction of a PEM fuel cell Model results and experimental validation. Int J Hydrog Energy 35 11533–50
[6] Wilberforce T, El-Hassan Z, Khatib FN, Al Makky A, Mooney J and Barouaji A, et al. 2017 Oct Development of Bi-polar plate design of PEM fuel cell using CFD techniques Int J Hydrog Energy 42 25663–85
[7] Hashemi F, Rowshanzamir S and Rezakazemi M. 2012 Feb CFD simulation of PEM fuel cell performance: Effect of straight and serpentine flow fields Math Comput Model 55 1540–57
[8] Cooper NJ, Santamaria AD, Becton MK and Park JW. 2017 Mar Investigation of the performance improvement in decreasing aspect ratio interdigitated flow field PEMFCs Energy Convers Manag 136 307–17
[9] Wang X-D, Duan Y-Y, Yan W-M and Peng X-F. 2008 Jun Effects of flow channel geometry on cell performance for PEM fuel cells with parallel and interdigitated flow fields Electrochimica Acta 53 5334–43
[10] Velisala V and Srinivasulu GN. 2018 Mar Numerical Simulation and Experimental Comparison of Single, Double and Triple Serpentine Flow Channel Configuration on Performance of a PEM Fuel Cell Arab J Sci Eng. 43 1225–34
[11] Min C-H. 2009 Jan Performance of a proton exchange membrane fuel cell with a stepped flow field design J Power Sources 186 370–6
[12] Kuo J-K, Yen T-H and Chen C-K. 2008 Feb Three-dimensional numerical analysis of PEM fuel cells with straight and wave-like gas flow fields channels J Power Sources 177 96–103
[13] Barakat E, Ahmed K, Ahmed M, Abdel-Rahman AK and Ali AHH 2013 Int. Conf. & Exhibition on Clean Energy (Ottawa, Ontario, Canada) vol 1 p 268
[14] Le AD and Zhou B. 2008 Jul A general model of proton exchange membrane fuel cell J Power Sources 182 197–222
[15] Dawes JE, Hanspal NS, Family OA and Turan A. 2009 Jun Three-dimensional CFD modelling of PEM fuelcells: An investigation into effects of water flooding ChemEng Sci. 64 2781-94
[16] Wang C, Zhang Q, Lu J, Shen S, Yan X and Zhu F et al. 2017 Sep Effect of height/width-tapered flow fields on the cell performance of polymer electrolyte membrane fuel cells Int J Hydrog Energy 42 23107–17
[17] Muthukumar M, Karthikeyan P, Vairavel M, Loganathan C, Praveenkumar S and Kumar APS. 2014 Numerical Studies on PEM Fuel Cell with Different Landing to Channel Width of Flow Channel Procedia Eng. 97 1534–42
[18] ZareNezhad B and Sabzemeidani MM. 2015 Predicting the effect of cell geometry and fluid velocity on pem fuelcell performance by CFD simulation J Chem Technol Metall. 50 176–182
[19] Shimpalee S, Dutta S, Lee WK and Van Zee JW. 1999 Int Mechanical Engineering Congress & Exposition (Nashville, Tennessee, November 14 - 19, 1999) pp 367–374
[20] Kamarajugadda S and Mazumder S. 2008 Jul On the implementation of membrane models in computational fluid dynamics calculations of polymer electrolyte membrane fuel cells Comput Chem Eng. 32 1650–60
[21] Wang L. 2003 Nov A parametric study of PEM fuel cell performances Int J Hydrog Energy 28 1263–72
[22] Baca CM, Travis R and Bang M.2008 Mar Three-dimensional, single-phase, non-isothermal CFD model of a PEM fuel cell J Power Sources 178 269–81
[23] Cheddie DF and Munroe NDH. 2006 Sep Three dimensional modeling of high temperature PEM fuel cells J Power Sources 160 215–23
[24] Kazemi Esfeh H, Azarafza A and A. Hamid MK. 2017 On the computational fluid dynamics of PEM fuel cells (PEMFCs): an investigation on mesh independence analysis RSC Adv. 7 32893–902

China-Pakistan Economic Corridor; Prospects and Challenges for Balochistan, Pakistan.

Muhammad Ahmed Farooqui[1] and Syed Mobasher Aftab[2]

[1]Office of Academics and Research, COMSATS University Islamabad, Lahore Campus, Defence Road, Off Raiwind Road, Lahore, Pakistan
[2]Balochistan University of Information Technology, Engineering and Management Sciences, Quetta, Pakistan

Email: farooqui@ciitlahore.edu.pk

ABSTRACT
Balochistan is rich in natural resource but due to the extremely harsh geographical environment and under-developed civil infrastructure (including roads, energy and water supplies), these natural resources could not have been tapped adequately in the past. With the launch of China-Pakistan Economic Corridor (C-PEC), there are ample opportunities for the exploitation of these resources that are expected to pave the way for unprecedented economic activities in Balochistan. From the perspective of natural resources these activities include, but not limited to, development and management of water resource, mining of coal and minerals, exploration of unexplored hydrocarbons, exploitation of dimension stones and construction material, and the establishment of related and subsidiary industries. All these activities require human resource development, better health, and educational facilities especially institution for technical and skill development of masses and most importantly the sustainable development of water resources and reservoirs for drinking, municipal and industrial purposes. In this paper key natural resources of Balochistan are identified in the context of their sustainable use under the C-PEC induced development projects. These include water, minerals, hydrocarbons, shale-gas, and coal. Suggestions are presented for a way forward to take maximum advantage of these resources for revenue generation and for the ultimate benefit to the local communities in Balochistan. A large number of mega projects including development of deep-sea Gwadar port, power generation plants, highway construction, and the establishment of special economic zones have been part of the C-PEC and are already being undertaken. However, it appears that small and medium-sized local industries and production units have not been given due attention. It is proposed here that the local people through local bodies may be identified and matched with relevant Chinese counterparts to develop joint ventures. Most of the areas of production and development identified in this paper are those whose products are already being exported to China through Karachi port. Hence the demand for these products in China is already established and need not any further survey. The products proposed in this paper are considered viable as all kinds of transport for C-PEC shall be passing through a large part of Balochistan where these resources occur naturally.
Keywords. China-Pakistan Economic Corridor, natural resources, local communities, Balochistan.

1. Introduction

In the seventy years history of Pakistan, the initiation and launching of the China-Pakistan Economic Corridor (C-PEC) is a significant economic activity that has been truly termed as s "game changer" [1]. Since the fall of Dhaka that led to the creation of Bangladesh in 1971, Pakistan has been continuously struggling to regain its position on the economic front. One after the other several internal and external conflicts and issues that surrounded Pakistan unwillingly, including, Soviet invasion in Afghanistan, Iran-Iraq war, Kuwait war, Iraq war, continuous turbulence in Afghanistan for the last four decades, intermittent escalation of tension with India on Kashmir issue, nuclear tests by India and Pakistan, political instability in Pakistan due to military rule , and continuous, mostly uncontrolled, population growth, have been the major causes that did not allow the planners to frame and implement any long-term sustainable economic policy that would have helped Pakistan regain its economic status [2, 3, 4, 5, 6, 7].

The One Belt One Road (OBOR) Initiative and the launching of C-PEC in 2013, when the official MoU was signed between China and Pakistan, has emerged as a strong ray of hope for Pakistan to regain its lost economic strength. The C-PEC aimed to connect western regions of China and Pakistan through integrated investments in energy, trade, and communication. The C-PEC is going to enhance trade activities through Pakistan between China and the Middle East, Africa and Central Asia. The corridor will be a strategic game-changer in the region and would certainly bring long-term opportunities for Pakistan to become a richer and stronger economy. This paper provides an overview of the present and future opportunities for economic growth of Pakistan, especially Balochistan for which comprehensive long-term planning in urgently and essentially required. The paper also identifies the focus areas for effective economic development and the available natural resources in Balochistan for a profitable business. Such businesses are recommended to be initiated as joint ventures of local people of Balochistan and Chinese companies so that the real benefit of development percolated down to the deprived communities.

2. Focus Areas of Technology for Pakistan

According to the latest biennial report of the US National Science Foundation and the National Science Board, China has gained the status of second-largest R&D spender in the world, accounting for 21% of the of approximately $2 trillion, which is next to the United States (26%) [8]. The rate of increase in R&D spending is the point of particular interest. R&D budget in China grew at the rate of 18 percent a year between 2000 and 2015, that is more than four times faster than the United States' rate of four percent. It is almost definite that by the end of 2018, China will be the largest spender on R&D in the world. This should be of interest to not only Pakistani policymakers, but also to the policymakers of the neighboring countries of China. Pakistan is required to train as many people, especially youth, from its rapidly growing population as it can, using both indigenous and foreign institutions. In the context of the latest growing anti-immigration and anti-Muslims sentiments in the United States, China may be an attractive alternative for Pakistani youth.

Increased funding for R&D by the Chinese government has resulted in the increase in the trained technical workforce that ultimately led to a dramatic increase in technical output [8]. The number of science and engineering graduates increased in China increased from 359,000 in 2000 to 1.65 million in 2014, almost a five-fold increase as shown in Table-1 [8]. The number of technology-based research papers being published by Chinese scholars is increasing at an unprecedented rate [9]. Chinese researchers and technologists are heavily involved in technology development and have made impressive contributions in artificial intelligence, telecommunications, robotics, electric cars and renewable energy [10]. These are the areas in which Pakistan also needs to develop expertise and human resource with the help of China.

It is a general perception in Pakistan that most of the C-PEC projects are in two sectors: energy and highways. But in reality, C-PEC projects are mostly about technology. New technologies are bound to be embedded in the C-PEC projects. With fast-paced technology developments, it is evident that in near future only technologically sound nations will be able to survive honorably. The future economic growth of a nation will depend on the confluent use and applications of big data, artificial intelligence, and connectivity; hence Pakistan is required to take initiatives in these focus areas. Therefore, it is imperative for Pakistan to initiate academic and research programs with the help of Chinese institutions to train as many Pakistani youths as possible in the five focus areas of; 1) artificial intelligence, 2) robotics, 3) renewable energy, 4) agriculture and 5) biotechnology [10].

Table 1. A comparison of the R&D spending and Technical Workforce in China and USA [8].

	USA	China
R&D Expenditure in 2017	\$520 Billion	\$420 Billion
R&D Global share in 2017	26%	21%
R&D Rate of Increase	4% per year	18% per year
R&D Expected expenditure in 2019	\$562 Billion	\$588 Billion
Technical Workforce 2000	4,83,000	3,95,000
Technical Workforce 2014	7,42,000	1.65 Million
Technical Workforce 2017 (Estimated)	≈800,000	≈2.1 Million

3. Major Areas of Current Activities in Pakistan

At present a major portion of the funds under C-PEC are allocated to the projects in the following seven areas;

1. Construction of highways and railway tracks
2. Development of Gwadar port
3. Energy generation projects
4. Establishment of industrial zones along the C-PEC route (Fig.1)
5. Laying down of pipelines for oil and gas transmission
6. Cyber networking and telecommunication
7. Water supply projects for industrial and municipal use (mostly in Balochistan)

All projects require technology development and human resource development. Therefore, it is imperative for Pakistan to create and strengthen proper infrastructure and facilities at R&D organizations and universities in Pakistan so that the future demand of technologies technologists, scientists and engineers are met. The list of C-PEC projects and their current statuses are tabulated in tables 2 to 6, [11]. The energy and power generation sector comprised of 18 projects, the overall progress varies from 15 to 75% while a few are completed, table 2. The transportation and communication infrastructure including eight projects, the overall progress varies between 15-70%, table 3. Six projects are related to industrial cooperation and ten projects to Gwadar Port with nominal progress on all of them, table 4 and 5. There are ten Special Economic Zones have been planned to establish in different parts of the country, the feasibility studies and technical evaluation is under progress on all the projects, table 6, [11].

Considering the present rate of development work under C-PEC, it has been estimated that the GDP of Pakistan is expected to experience an annual increase of about 2.5 percent by the year 2030 whereas ca.700,000 jobs are expected to be created in next 10 years. Most of the jobs shall be in the Energy, Communication (Highways and Railways) and Port Development sectors [12, 13].

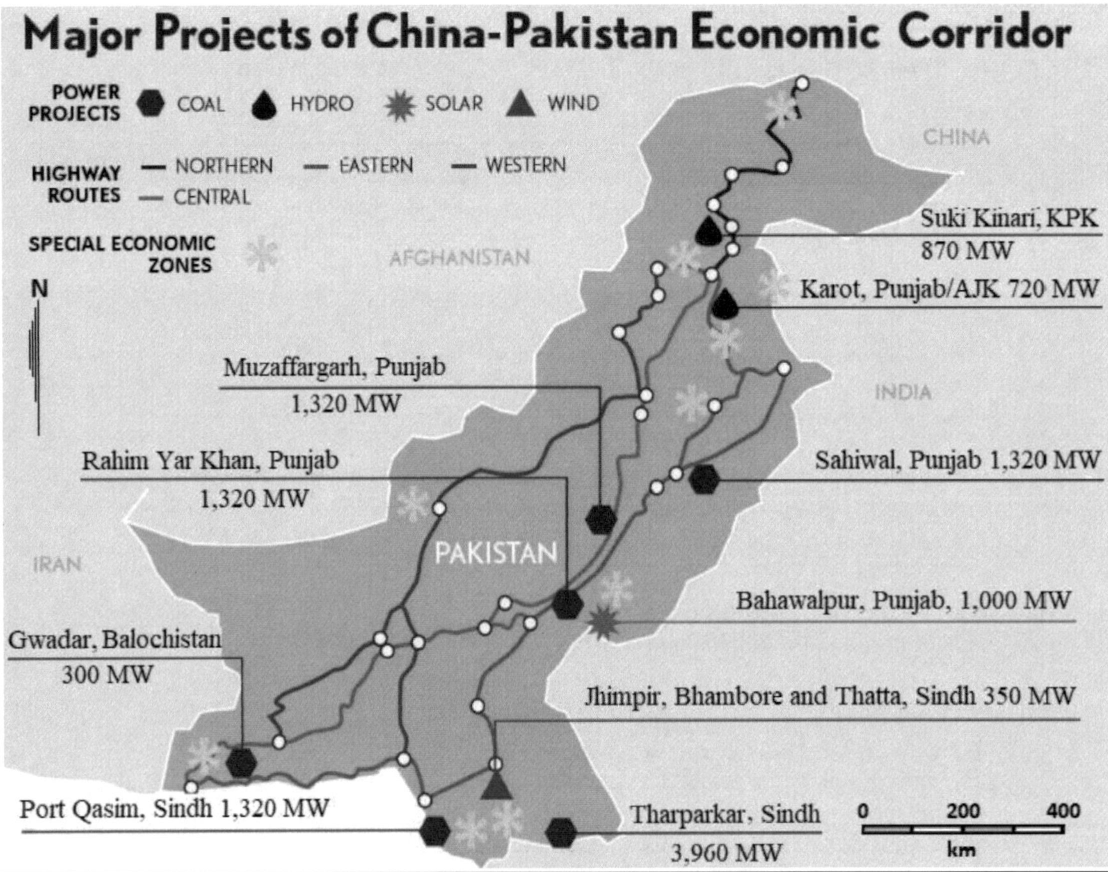

Figure 1. Map showing major projects of China-Pakistan Economic Corridor [13, 14, 15, 16].

4. C-PEC and the Local Communities in Balochistan

The rural communities in Balochistan take pride in their traditions, culture and the natural environment. They are well aware of the natural resources, agriculture, forests, rangelands, watersheds, biodiversity and mineral deposits of the province, however, the harsh living conditions like aridity, depleting groundwater resources, periodic droughts, overgrazed and degraded rangelands, low productive livestock, and rugged terrain are the major hurdles in their advancement. The government departments, both provincial and federal, have been largely unable to mitigate desertification, deforestation, de-vegetation of rangelands and recharging of the groundwater resources.

The province needs an integrated developmental approach in which all stakeholders including government departments, politicians, civil society leaders, community elders and local NGOs are to be brought on one page to take maximum advantage of C-PEC for community development. The western route of the C-PEC has excellent potential for the developmental projects. For a sustainable development-harmonized with local communities, the developmental activities should involve local communities through a consultative process. It's imperative to bring the rural communities into the mainstream by transforming and empowering them in decision making. The empowerment of communities in developmental activities may initiate from project identification to implementation and from operation and maintenance to project's ownership, [17].

The rural communities are dependent on their surrounding environment and natural resources. The small-scale farming, orchards, livestock grazing, limited mining and fishing in coastal areas may improve the community's economic conditions. In some cases, these said economic activities and occupations have already been molded by the natural climate and availability of surface and groundwater resources [18]. The sustainable management, conservation, and protection of water resources is an important factor for the physical, social and economic endurance of rural communities [18].

The impact of climate-induced changes in Balochistan are obvious but are not scientifically understood. The relevant government departments, R&D organizations, and universities may get benefit from the Chinese experience through C-PEC to address and resolve these issues. One of the major challenges is the protection and conservation of groundwater resources [19].

5. Focus Areas for Balochistan

Despite being resource-rich, contributing more than 40 percent to the country's energy requirement in the form of electricity, natural gas, and coal, Balochistan is still the most underdeveloped province of Pakistan. There are multiple reasons and justifications for the slow development of Balochistan and each one has counter-arguments [2, 5, 20]. But the bottom line fact remains unchanged that people of Balochistan are lagging behind in the race towards better socioeconomic conditions and lifestyle.

After a long wait of more than seventy years, the C-PEC is the first ray of hope for the people of Balochistan to prosper. In order to take advantage of this unique opportunity, it is necessary to move forward without indulging into the unending discussions on the background and history of the political, geographical, tribal and social conflicts that kept Balochistan away from the mainstream of development path in the past. With this context, five major focus areas of natural resources are proposed here for initiating new projects table-7.

The focus areas projects are comprised of minerals, coal deposits, and construction materials, which are widely exposed along the new trade corridor. The proved oil and gas reserves are present in many parts of the province, even oil and gas seeps at the surface are known from a long time. The renewable energy resources including solar, wind and geothermal potential are extensively present. New projects are required to initiate the development of these natural resources.

The management of the water resources is the sweltering issue of the province. The climate change has deteriorated the surface and groundwater balance. The different categories of droughts have become a permanent feature in one or the other part of the province. The groundwater table is depleting at an alarming rate in most of the river basins. The development of new water resources and management techniques are needed to apply for a sustainable supply of water.

IOP Conf. Series: Materials Science and Engineering **414** (2018) 012046 doi:10.1088/1757-899X/414/1/012046

Table 2. Energy and Power Generation Projects [11]

No	Project Name	Status	Progress
1	Two 660MW Coal-Fired Power Plants at Port Qasim	Under construction	100%
2	Two 660MW Coal-Fired Power Plants at Sahiwal	Two units inaugurated	100%
3	Four 330MW Engro Thar Coal-fired Power Plant	Under construction	60%
4	50MW (Dawood) Wind Farm, Gharo	Ready for operation	100%
5	900MW Solar Power Park Bahawalpur	Energization achieved	100%
6	100MW Jhimpir Wind Energy Farm Jhimpir	Under construction	100%
7	50MW Sachal Wind Energy Farm, Sachal	Under construction	100%
8	720MW Karot Hydro-Power Project,	Financial close achieved	70%
9	873MW Suki Kinari Hydropower Project	Under construction	65%
10	Two 60MW Coal-Fired Power Plant Rahimyar Khan	Feasibility stage	15%
11	Two 660MW Coal-Fired Power Plant Hub (Habco)	Under construction	50%
12	300 MW Power Plant Gwadar	Feasibility stage	60%
13	Faisalabad-Matiari-Lahore Transmission Line	Negotiation in process	15%
14	Two 660MW Power Plants Gadani Lasbela,	Feasibility stage	15%
15	1100MW Hydro-Power Station Kohala	Negotiation in process	15%
16	300MW Power Plant at Salt Range Mine	Feasibility stage	15%
17	Two 660MW Coal-Fired Power Plants, Thar Coal Mine	Feasibility stage	15%
18	Two 660MW Coal-Fired Power Plants Muzaffargarh	Feasibility stage	15%

Table 3. Transportation and Communication Infrastructure - Roads and Highways [11].

No	Project Name	Status (April 2018)	Progress
1	120 km KKH Phase-II, Havelian- Thakot:	Under construction	70%
2	392 km Karachi-Lahore Motorway (S-M-S)	Under construction	70%
3	Establishment of Havelian Dry port	Feasibility completed	100%
4	Upgradation of Multan-Lahore Section, 339 km; Hyderabad-Multan Section, 749 km; and Kemari-Hyderabad Section, 182 km.	Agreement under discussion	40%
5	Construction of New Havelian Dry Port	Feasibility completed.	25%
6	110 km Khuzdar-Basima Highway (N-30)	Feasibility completed.	25%
7	280 km KKH Phase III (Raikot-Thakot Section)	Feasibility completed. Procedural formalities to be completed soon	25%
8	533 km D.I.Khan-Quetta Highway (N-50)	Feasibility stage	15%

Table 4. Industrial Cooperation Projects [11].

No	Project Name	Status	Progress
1	700 km Gwadar-Nawabshah LNG Terminal and Pipeline	Ready to be inaugurated	40%
2	Haier & Ruba Economic Zone Phase-II	Feasibility stage	15%
3	Laying of Optical Fiber Cable Rawalpindi to Khunjrab	Under construction, Expected Completion by Dec 2018	70%
4	DTMB Demonstration Project	MOU Signed	5%
5	Lahore Orange Line Metro Train	Under construction	70%
6	Promotion and commercialization of TD-LTE in Pakistan	Feasibility stage	15%

Table 5. Gwadar Port and Peripheral Projects [11].

No	Project Name	Status	Progress
1	19 km Gwadar Eastbay Expressway	Framework Agreement Signed	60%
2	New International Airport Gwadar	Framework Agreement Signed	40%
3	Free Economic Zone Gwadar	1st phase completed	60%
4	Smart Port City Master Plan Gwadar	Contract negotiations underway	20%
5	Expansion of Multi-purpose Terminal	Feasibility stage	15%
6	Fresh Water Supply, Wastewater Treatment Plants for Gwadar City	Feasibility stage	15%
8	China-Pakistan Faqeer Primary School Gwadar	Completed.	100%
9	Pak-China Friendship Hospital Gwadar	Feasibility study underway	10%
10	Pak-China Friendship Technical and Vocational College Gwadar	Feasibility study underway	10%

Table 6. Special Economic Zones [11].

No	Name	Type of Industry	Progress
1	Rashkai Economic Zone, M-1, Nowshera, KPK	Fruit/Food/Packaging/Textile Stitching/Knitting	Feasibility study is technically evaluated by China
2	Special Economic Zone Dhabeji Sindh	To be determined during feasibility stage	Feasibility study is technically evaluated by China
3	Bostan Industrial Zone Balochistan	Fruit Processing, Agriculture machinery, Pharmaceutical, Motor Bikes Assembly, Chromite /Ceramic, Ice and Cold storage, Electric Appliance	Feasibility study is technically evaluated by China
4	Allama Iqbal Industrial City (M3), Faisalabad, Punjab	Textile, Steel, Pharmaceuticals, Engineering, Chemicals / Plastics, Food Processing, Agriculture Implements	Feasibility study is technically evaluated by China
5	ICT Model Industrial Zone, Islamabad	Steel, Food Processing, Pharmaceutical & Chemicals, Printing and Packaging, Light Engineering etc.	Feasibility study is technically evaluated by China
7	Industrial Park at Port Qasim, Sindh	Steel, Auto & allied industries, Pharmaceutical, Chemical, Printing and Packaging, Garments.	Feasibility study is technically evaluated by China
8	Special Economic Zone at Mirpur, AJK	Mix industry	Feasibility study is technically evaluated by China
9	Mohmand Marble City FATA-KPK	Marble / Granite, Mineral Processing Industry	Feasibility study is technically evaluated by China
10	Moqpondass Special Economic Zone, Gilgit-Baltistan	Marble / Granite, precious stones, Fruit Processing, Steel Industry, Mineral Processing and Leather Industry	Feasibility study is technically evaluated by China

IOP Conf. Series: Materials Science and Engineering **414** (2018) 012046 doi:10.1088/1757-899X/414/1/012046

6. Human Resource Development

Education and training of indigenous human resource are integral parts of all developmental activities. Human resource in all academic disciplines including engineering, medical, computer, social, business, and natural sciences will be required for C-PEC projects in Balochistan. Higher Education Institutions (HEIs) in Balochistan are to be provided special funds to develop state of the art laboratories either independently or in collaboration with Chinese Universities.

In Pakistan, the skilled labors constitute about 6% of the entire labor-force [21]. The technical and vocational trade institutions are far less to meet the present and future demands in a broader context of C-PEC. In Balochistan, the situation of primary, secondary and the technical and vocational training colleges and the institution is very grim. The number of technical institutions, students enrollment capacity and the number of pass-outs figures are very low as compared to other provinces of Pakistan. Establishment of technical and vocational institutions are required on an emergency basis to fulfill the present and future demands in diverse technical and professional trades. Chinese language institutions are also required to be established for the skilled workforce to learn the Chinese language for a better work-place understanding with their Chinese counterparts.

7. Opportunities for Joint Ventures in Balochistan

A large number of high-cost projects like Industrial Zones, Power Generation Plants, Road & Highways Construction, Special Economic Zones etc. have been part of the C-PEC and are already in progress. However, small production units and business enterprises require more attention owing to its potential to contribute effectively. It can be anticipated that with the opening of commercial transport on C-PEC, a large number of heavy-duty trucks and trawlers will be carrying goods from China to Gwadar port. After getting unloaded these trucks and trawlers will have no or little goods available to take back to China. The traveling of unloaded large vehicles shall be a huge waste of time, fuel, and transport machinery. In order to save these losses, local industrial and production units may be developed by providing facilitation to local people and their relevant Chinese counterparts. It is proposed here that the local people may be identified and matched with relevant Chinese counterparts to develop joint ventures. Most of the areas of production and development identified in this paper are those whose products are already being exported to China through Karachi Port. Hence the demand for these products in China is already established and need not any further survey in China.

Very little facilitation has been provided to match the local business community with their prospective Chinese counterparts to initiate joint ventures in Balochistan. One such meeting was organized (22nd August 2016) by Ministry of Industries and Production (Islamabad) in which business community belonging to Quetta Chamber of Commerce and Industry (QCCI) were invited. The ultimate result of this meeting is yet to come out, however, it is opinionated here that the business community of QCCI has very little to contribute at least at the current stage. This is because Quetta is part of the Western Route of the C-PEC, whereas, according to latest news appearing in press and media, most of the development work on C-PEC is currently concentrated on the Eastern Route of C-PEC.

Geographically, the Balochistan part of the Eastern Route starts from Shahdad Kot in Sindh, crosses Kirther Mountains towards the west and passes through towns of Karkh, Khuzdar, Basima, Nag, Panjgoor, Turbat and ends at Gwadar. An alternate route is from Khuzdar to Wadh, Bela, Winder, and then connects to Coastal Highway leading to Ormara, Pasni and ultimately Gwadar, figure 1. In either case, the town of Khuzdar is common on both routes. Therefore, the products proposed in this paper are considered viable as all kinds of transport for C-PEC shall be passing through Khuzdar. Additionally, the identified products are being produced either along the C-PEC route in Balochistan or in its immediate vicinity.

Table 7. Major natural resources for development projects in Balochistan

No	Natural Resource	Types of deposits	Location in Balochistan
1	Minerals and Coal deposits	• Iron Copper and Gold	Chaghi Mountains.
		• Barite, Lead, and Zinc	Khuzdar and Bela districts.
		• Chromium and Iron	Muslim Bagh and Zhob districts, and Raskoh.
		• Iron, copper, chromium gold, Rare Elements	Raskoh Mountains.
		• Coal deposits	Quetta, Bolan, Kohlu, Barkhan (Chamalang).
2	Construction Material	• Aggregates	All districts of Balochistan.
		• Limestone and sandstone	Sulaiman and Kirther Mountains.
		• Dimension stones: Granite, Limestone, Onyx Marble and sandstone	Sulaiman and, Kirther Mountains, Chaghi Mountains and Raskoh Mountains.
3	Oil and Gas Reserves	• Petroleum	Kharan Basin and Offshore Makran.
		• Gas	Offshore Makran.
		• Shale Gas	Coastal Makran Ranges and Offshore Makran.
4	Energy Resources	• Solar Energy	All Balochistan.
		• Geothermal Energy	Murri-Bugt Hills, Chaghi Hills, Bolan district and Makran Coast (mud volcanoes).
		• Bioenergy	All Balochistan.
		• Ocean Energy	Coastal Makran.
		• Hydrogen Energy	All Balochistan.
5	Water Resources	• Surface Water (perennial)	Nag, Hoshab, Panjgoor (Rakhshan River), Bolan, Harnai, Zarghoon, and Zhob.
		• Rain Water (non-perennial)	All Balochistan; Natural dam sites are available where rainwater can be stored both for groundwater recharge and water supply.
		• Groundwater	Fractured hard rocks, conglomerates and alluvium available all along the C-PEC route in Balochistan; Specific aquifers are a) rocks of Binga and Diz units of Khojak Formation in Central Balochistan (Panjgur and Turbat districts) and b) rocks of Talar and Hinglaj units of Makran Group in Makran Coastal Range between Ormara and Jiwani.
		• Seawater	Along Makran Coast; desalination of sea water

7.1 Dimension Stones

Huge deposits of dimension stones of international quality are available in Balochistan. These stones include limestone, granite, marble, sandstone, and onyx to name a few. From geological prospective Kirther and the Sulaiman Mountains in Balochistan are the natural geological environments where multiple types of high-quality natural stones are available. Such rocks are rarely available in the northern parts of Pakistan and even in Western China (the Tibbet Plateau) because of the intense geological activities (tectonically compressed) due to which most of the rocks are highly crushed or broken into pieces. Large-sized un-fractured dimension stones of same color, texture, and strength which are commercially required, are generally not available in the areas where rocks have gone under high compression. There are some local areas from where these dimension stones are already being mined/quarried for export to China through Karachi Seaport. Some of the locations in Balochistan, mostly along the C-PEC route, from where these stones are being produced, are as under:

1. Saruna, District Khuzdar
2. Karkh, District Khuzdar
3. Nal, District Khuzdar
4. Basima, District Qalat.
5. Duraji, District Lasbela
6. Loi, District Lasbela
7. Gajri, District Lasbela
8. Shah-e-Noorani, District Lasbela
9. Yak Mach, District Chaghi
10. Dalbandin, District Chaghi
11. Nokundi, District Chaghi
12. Different localities in District Loralai, Ziarat, Quetta and Zhob

In these localities, the mining facilities and infrastructure are very primitive with old machinery. Their production can easily be increased by introducing more mechanized mining. Chinese mining companies can play a very productive role by developing joint ventures with local miners to boost the production.

7.2 *Minerals and Mining*

Balochistan is a host of large deposits of metallic and nonmetallic minerals. These include ores of Gold, Copper, Iron, Chromium, Barium, Magnesium. Aluminum, Tungsten, Lead, Zinc, Antimony and many more. The minerals named Chromite, Pyrite, Barite, Magnesite, Galena, and Sphalerite, are already being mined in various localities of Khuzdar, Bela, Chaghi and Muslim Bagh areas of Balochistan. With the exceptions of few, all these ores are being exported to China through Karachi port for further refining and processing. The transport of these ores may be diverted to C-PEC and the production may be enhanced with the help of Chinese mining companies. Some of the major mineral deposits and their locations are listed below for ready reference;

1. Copper-molybdenum deposits of Dasht-e-kain, Chaghi.
2. Copper deposits of other areas of Chaghi District.
3. Iron Ore reserves of Dalbandin and Nokundi District Chaghi.
4. Sulphur reserves of Koh-i-Sultan Nokundi.
5. Chromite deposits of Wad, District Khuzdar.
6. Chromite deposits of Kharan.
7. Laterite deposits of Ziarat District.
8. Antimony deposits of Arambi Village District Qila Abdullah.
9. Chromite deposits of Muslim Bagh District Qilla Saifullah.
10. Magnesite deposits of Muslim Bagh.
11. Chromite deposits of Zhob District.
12. Lead-Zinc Deposits of Lasbela District.
13. Quartzite deposit of Gajri village Lasbela

7.3 *Fisheries/Fish Farming*
Makran Coast and its seawater are regarded as one of the cleanest natural environment having no impact of industrial pollution. Hence it is the habitat of freshest and cleanest marine life. Fish Farming has huge potential all along Makran Coast especially in the towns of Gwadar, Pasni, and Ormara. Local people along the coast are already engaged in traditional fishing activities with minimal resources and training. Fish/Sea Food processing and preservation facilities may be established with the help of local people and their Chinese counterparts. These fish and other raw seafood may then be exported to China via C-PEC.

7.4 *Agriculture*
Climatically, Balochistan comes under desert to semi-desert environment. Water is scarce hence largescale agriculture products may not be expected. However, different types of high quality "dates", the product of desert environment, are enormously available in a vast area between Khuzdar and Gwadar. These dates can be easily processed and preserved for export to China or Central Asian countries (through C-PEC) where dates are not naturally grown.

7.5 *Livestock*
Balochistan has huge potential for the growth of livestock, especially sheep, lamb, and goat. Local people are already involved in traditional livestock business but do not have means for mechanized livestock farming. Small and medium-sized processing plants may be installed along C-PEC or near Khuzdar for the slaughtered cattle and their processing for preservation for ultimate export to China through C-PEC route.

7.6 *Cement Manufacturing Plant*
All the raw material for a cement manufacturing plant are extensively available all the way from Mughal Kot (Zhob) in the north to Surab (near Khuzdar) in the south. The four basic ingredients of cement (Calcium, Silicon, Aluminum, and Iron) are extracted from limestone and shales which are abundantly available in the rocks exposed along the road between Quetta and Zhob. The most viable location for the Cement Plant is the Bostan Industrial Zone (near Quetta) that falls almost halfway on western route of C-PEC between Zhob and Gwadar providing equal access to both northern and southern localities in Balochistan.

7.7 *Ferro-Chrome Beneficiation*
Chromium is an essential ingredient for the production of steel. Muslim Bagh has the second largest chromite (ore of chromium) deposit of the World from where mining is being done for the last 100 years. Currently, the raw chromite ore from Muslim Bagh is being transported all the way to Karachi for its onward export to China at a very low cost. This raw Chromite ore is processed at China to convert it into Ferro-Chrome which is then imported back in Pakistan, at a very high cost, for use in the iron/steel industry (mostly in Pakistan Steel Mills). A Ferro-Chrome plant will not only reduce the transport cost, it will substantially save foreign exchange also. All industrial units in C-PEC Projects will certainly need the iron and steel in one form or other that can be met from the products of Ferro-chrome plant. The most viable location for this industry is near Muslim Bagh.

8. Conclusions
China Pakistan Economic Corridor (C-PEC) has opened up vast opportunities for the socio-economic development of Pakistan. Pakistan needs to focus on key areas of artificial intelligence, robotics, renewable energy, agriculture, and biotechnology to develop its human resource. Trained technical human resource in these areas will be essential for a sustainable development with Chinese investment. Similarly, Balochistan being a resource-rich but highly deprived and underdeveloped province, also needs to focus on seven key areas to take maximum benefits from C-PEC induced trade and transport projects. These are

minerals and coal exploration, construction materials, Oil and Gas exploration, Energy generation including renewable energy, Water, including surface water, groundwater, and seawater, and Human Resource Development. Additionally, Balochistan has the potential offer joint industrial ventures with Chinese counterparts in the production and marketing of dimension stones, mining of mineral deposits, fish farming, agriculture products that are not available in China, farming of Livestock, cement manufacturing, and Ferro-Chrome manufacturing industry.

9. Acknowledgment

The travel grant provided to the authors by the Higher Education Commission through the Balochistan University of Information Technology and Management Sciences, Quetta (Pakistan) is acknowledged and appreciated. The grant was meant to attend and present this paper as Invited Speaker at the *1st International Conference on Advances in Engineering & Technology* held at Quetta on 2-3 April 2018.

References

[1] Naqvi, Baqar Abbas. 2017. CPEC – A Game Changer; *Spearhead Research* 5 July 2017; https:// spearheadresearch.org/index.php/ researchopinions/cpec-a-game-changer (accessed 15 March 2018).
[2] Saleem, Ali Haider 2017. CPEC and Balochistan: *Prospects of Socio-political Stability Strategic Studies* p118-132. Avai. at http://issi.org.pk/wp-content/uploads/2018/01/7-SS_Ali_Haider_Saleem_No-4_2017.pdf.
[3] Workman, Daniel, 2017. "*China's Top Trading Partners*", January 28, 2017.www. world stop exports.com.
[4] Mullen, Rani D. and Poplin, Cody, 2015. "The New Great Game: A Battle for Access and Influence in the Indo-Pacific," *Foreign Affairs*, September 29, 2015.www.foreignaffairs.com
[5] Ahmed, Mansoor. 2017. New Great Game and the CPEC in Balochistan: Opportunities and Challenges: *Pakistan Journal of History and Culture*, Vol. XXXVIII, No. 1, p 83-108
[6] Ramay, Shakeel Ahmad. 2017. China Pakistan Economic Corridor; A Chinese Dream Being Materialized Through Pakistan, *Strategic Development Policy Institute*, Islamabad, 8p. www.sdpi.Org
[7] Tong, Liang 2014. CPEC Industrial Zones & China-Pak. Capacity Cooperation, *Strategic Studies,* p.174-184
[8] National Science Board. 2018. Science and Engineering Indicators 2018. NSB-2018-1. *Alexandria, VA: National Science Foundation.* Available at https://www.nsf.gov/statistics/indicators/. (19 May 2018)
[9] Han, Xueying., Appelbaum, R.P., 2018. China's science, technology, engineering, and mathematics (STEM) research environment: A snapshot. *PLOS ONE* | https://doi.org/10.1371/journal.pone.0195347 April 2018
[10] Burki, Shahid Javed. 2018. Technological development due to CPEC; Opinion Editorial *Tribune* 29-01-2018
[11] Pakistan-China Institute, 2018. http://www.C-PECinfo.com/pci-C-PEC (Accessed 22 March 2018*)*
[12] China Pakistan Economic Corridor (CPEC), *Board of Investment, Government of Pakistan*, www.boi.gov.pk.
[13] CPEC, Directorate General of Public Relations, *Government of Balochistan.*
[14] China Pakistan Economic Corridor, 2018. *CPEC*, www.cpec.gov.pk. Government of Pakistan, Islamabad.
[15] Daniel S, Markey and James West, 2016. *Behind China's Gambit in Pakistan*. Council of Foreign Relations, The New Geopolitics of China, India and Pakistan. www.ForeignAffairs.com.
[16] Bengali, Kaiser 2015, China-Pakistan Economic Corridor; The Route Controversy, *Chief Minister's Policy Reform Unit Government of Balochistan* 24p. www.cmpru.gob.pk (Accessed 29 May 2017)
[17] Techno Consult International, et. al., 2010. Ground Water Management Action Plan for Pishin Lora Basin & Implementation of Pilot Program in Quetta Sub-basin. Balochistan, Pakistan. *Balochistan Small Scale Irrigation Project, Government of Balochistan.*
[18] Planning and Development Department, Government of Balochistan, Quetta. IUCN-The World Conservation Union, 2000. *The Balochistan Conservation Strategy*. Quetta, Pakistan.
[19] Pakistan Poverty Alleviation Fund, Government of Pakistan, 2013. Balochistan Strategy, a roadmap to inclusive socio-economic development of the province. Islamabad, Pakistan.
[20] Abid, M. and Ashfaq, A. 2015. CPEC: Challenge and Opportunities for Pakistan; *Pak. Vision* V16 No.2.
[21] NAVTTC, 2018. *National Vocational & Technical Training Commission* (NAVTTC), Pakistan. In: Government of Pakistan. Islamabad, Pakistan

Assessment of Fluoride in drinking water and its correlation with different physicochemical parameters in the selected areas of Quetta, Pakistan

N Panezai[1], M Ellahi[1], and H Ullah[1]

1Department of Chemistry, Faculty of Arts & Basic Sciences, Balochistan University of Information Technology, Engineering and Management Sciences (BUITEMS), Quetta 87100 Pakistan.

E-mail: mujtaba.ellahi@gmail.com

Abstract. Fluoride is found naturally in water, soil, plants, and animals in trace amount. The level of fluoride in drinking water has to be controlled in order to get rid of dental and skeletal fluorosis. A study conducted on children of Quetta city aged 12-year from randomly selected areas showed that dental fluorosis has been found in 63% of children (the data is available) which is a very high percentage. Hence, the focus of this research is mainly to assess the contents of fluoride and its correlation with other physicochemical parameters of drinking water quality in Quetta city. By applying an even sampling plan with the samples distribution taken from tube wells of different sites of Quetta city, where the drinking water quality is known to have been deteriorated. The technique for the analysis of fluoride in groundwater sample is the colorimetric SPADNS method. The study shows that the Concentration of fluoride varies from 0.32 ppm to 3.0 ppm (Hazara town to the western by-pass area of Quetta) as compared with World Health Organization (WHO) guidelines of drinking water for fluoride (i.e., 1.5 ppm). A total number of 13 drinking water samples out of 20 were found unfit for human consumption due to increase in fluoride concentration. The statistical correlation results of drinking water show that solubility of fluoride depends on the level of pH. The salts of, K^+, Na^+, Ca^{2+}, SO_4^{2-} and Cl^- ions take a major role in attaining the favourable pH for the dissolution of fluoride-containing compounds in drinking water. However, fluoride correlation with other physicochemical parameters has not been studied in drinking water of Quetta city. Therefore, this aspect of research was studied accordingly. Furthermore, it is observed that the prevalence of fluoride-related health problems in the study area is due to high fluoride concentration in drinking water, which is expected from rocks with fluoride-bearing minerals of Quetta valley. On the account of the results, de-fluoridation of the identified sampling sites, and regular monitoring of drinking water is recommended at government level to avert further fluorosis risk.
Key words: Concentration, drinking water, defluoridation, fluoride level, fluorosis risk

1. Introduction

Our planet consists of 71% of water resources and among this only 1% of the water on the Earth's surface is available to be used by humans, and 99% of this usable quantity is located under groundwater [1]. Underground water contains some quantity of dissolved salts with different concentration of ions, which cause different health issues all over the world, besides other ions, fluoride is one of those ions which in high concentration causes health problem [2].

As in drinking water fluoride is present naturally due to geological composition of bedrock containing minerals like fluorapatite, fluorspar, cryolite, hornblende, fluorite, biotite and hydroxyapatite found in local rock and sediment [3]. Due to climatic factors, adsorption and leaching directly affect the migration and exchange of fluoride from soil to water [4]. Water is the convenient source of fluorides for everybody, so the importance of this element in the human body cannot be ignored, as it is mandatory for durable teeth and bones [5]. The preferred optimum level of fluoride which is recommended by WHO is 0.8-1.2 mg L^{-1}, [6] as it supports public health and avoids any sort of tooth decay, as 1.5 mg L^{-1} is the guideline value for a nominal chance for dental fluorosis [7]. When the fluoride level exceeds 1.5mgL^{-1}, the threat of dental fluorosis increases. People frequently exposed to fluoride, suffer from neurological symptoms including a headache, excessive tiredness, sleepiness, indisposition, and giddiness [5]. The concentration of fluoride in drinking water is assessed only through chemical analysis. The commonly used technique for the analysis of fluoride in water sample is colorimetric SPADNS (sodium 2(para sulfo phenyl azo)-l, 8-dihydroxy-3, 6-naphthalene disulfonate) method. Fluoride concentration can also be measured by ion chromatography. It can also be detected through portable Ion-selective electrodes, which is used either in the laboratory or in the field [2].

Reports of high concentration of fluoride in groundwater have been studied throughout the world such as USA, Africa, Russia, Middle East and Pakistan. [08] One such research was conducted in North America suggests that dental fluorosis ranges between 35% to 60% in fluoridated communities [9]. Different studies with regards to water quality have also been conducted in Pakistan. The water quality of Mastung, Mangocher, and Pringabad in Baluchistan have been studied with special focus on the concentration of fluoride and other water quality parameters [1]. A study shows that highest fluoride contamination (22%) has been detected in Balochistan province followed by 19% in Punjab province. [10] Another study conducted on 349 children aged 12-year from 14 schools of Quetta showed that dental caries was found in 81 children (23.2%) with mean DMFT 0.61. Boys had 1.6 times more chance to have dental caries than girls. Dental fluorosis was found in 63.6% of children with the majority of moderate degree (50.5%) [11]. The previous studies have also focused on fluoride toxicity, in a research conducted on drinking water samples of Quetta,81 samples were collected from different sites and 14% were found with increased level of fluoride [8]. Another study conducted on the contamination of fluoride in drinking water and tea samples, in which the concentration of fluoride was detected between 0.61 to 2.55 ppm [12]. However, there has been no study on the statistical correlation of fluoride with other parameters of Quetta city, hence the study has focused on the fluoride correlation and their analysis of physio chemical parameters that distinguish it from other studies. The water samples collected from tube wells in selected areas of Quetta city, specifically those areas where the population was suffering from dental fluorosis have been examined. In this regard concentration of fluoride and other physicochemical parameters have been analyzed. Besides water analysis, a statistical correlation between fluoride and other physicochemical parameters were examined. The statistical correlation results in drinking water showed that the solubility of fluoride depends on the level of pH. The salts of, K^+, Na^+, Ca^{2+}, SO_4^{2-} and Cl^- ions play a major role in attaining the favorable pH for the dissolution of fluoride-containing compounds in drinking water. The findings of this study explain that the correlation assessment of fluoride with various physicochemical parameters in water helps the population of mentioned areas to know the quality of water, especially the concentration of fluoride in their drinking water as shown in figure 1.

2. Materials and Methods

2.1 The Area of Study

The study area for this research is Quetta City. The selection of the sites was mostly based on the physical symptoms of dental fluorosis of the population of concerned area. The drinking water samples were collected by maintaining the distance of 5 to 10 km^2. EPA sampling procedure was followed throughout the sampling. A total of 20 samples were collected from selected sites of groundwater source such as tube wells during November-December 2017. The water samples were kept at 4±1 °C in sterilized polyethene bottles of 1.5-liter capacity properly washed by distilled water followed by sample water.

Analysis of the aesthetic and physicochemical parameters was according to standard methods applied to the analysis of drinking water quality. The pH of the water samples was detected by pH meter (Orion 720A). Total dissolved solids(TDS) were measured by Jenway 4320 TDS meter. The concentration of fluoride in these samples were estimated by SPADNS methods recommended by (APHA_AWWA_WEF,1992) due to its sensibility and efficiency [11]. Magnesium and Calcium ions were determined with the help of UV spectrophotometer. Potassium and sodium ions were estimated by using flame photometer (Model 360, Sherwood Scientific Ltd). Chloride of the samples was measured by the Argentometric method. Bicarbonate and sulphate ions were analyzed according to the standard (APHA_AWWA_WEF,1992) [12]. The results of the samples were compared with WHO and National water quality standards.

2.2 Statistical Analysis

In Statistics the relationship between two continuous variables is evaluated through correlation analysis. Statistical correlation analysis was performed to examine any potential relationship between fluoride and other physicochemical parameters of water samples by using IBM SPSS.

3. Results and Discussion

In total 20 samples of drinking water, 07 samples (35% of the total) were having the content of fluoride within the permissible limit of WHO (1.5mgL^{-1}). The remaining 13 samples (65% of the total) were not suitable for drinking purpose, as the amount of fluoride of these water samples were detected between 1.60 to 3.00 mg/L as shown in figure 1. A study conducted on drinking water samples of Quetta, the concentration of fluoride was ranged between 0.61 to 2.55 ppm [8].

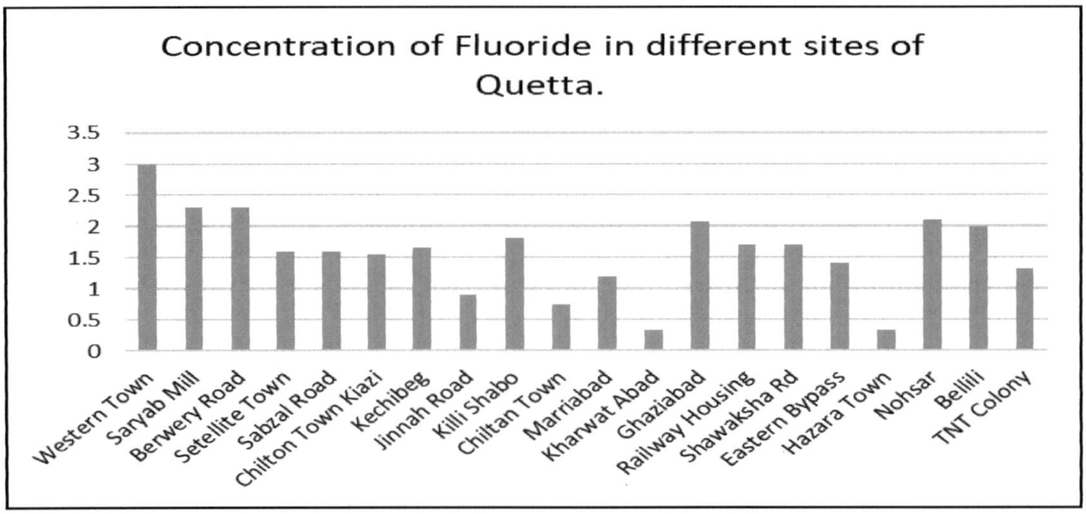

Figure 1. The concentration of fluoride in different sites of Quetta

The fluoride concentration in drinking water, which exceeded the permissible limit, caused adverse effects on health. The deteriorating dental health condition may vary from mild dental fluorosis to crippling skeletal fluorosis as the amount and intake of fluoride increases. Major symptoms other than dental and skeletal fluorosis are depression, weakness of muscles and nervousness may appear as a result of high fluoride intake.

The high content of fluoride in groundwater of Quetta valley is due to metamorphic and igneous rocks such as gneisses and granite [13]. The high amount of fluoride in this area is mainly due to the dissolution of large deposits of fluoride (the estimated total reserves is about 100,000 t) in Balochistan [14]. The fluoride contamination cannot be removed by simply boiling the water or passing through charcoal- based water filtration plants; therefore, reverse osmosis and distillation techniques are suggested for the elimination of drinking water with high fluoride contents.

A study conducted on tap water of Quetta and the results showed that pH of drinking water samples is within the range of WHO standards, which is 6.5-8.5. [15,16,17]. In present study, out of 20 samples the pH values ranged from 7.0 to 8.5, in which one sample acquired the higher acceptable pH value (8.5). For pH of drinking water, no health-based guidelines value has been estimated, as it does not impart any direct effect on human health. However, water samples with pH value near to upper or lower permissible limits are not acceptable. The relationship between fluoride and pH has been depicted in figure 3. In order to control and check the quality of water, it should be monitored periodically. TDS (total dissolved solids) of samples is the totality of Ca^{2+}, Na^{+}, K^{+}, and Mg^{2+} cations and CO_3^{2-}, SO_4^{2-}, Cl^{-}, NO_3^{-}, and HCO_3^{-} anions as the major inorganic constituents [17]. WHO has recommended 500 and 2000mgL^{-1} as the upper and lower permissible levels, respectively. [18] The taste of drinking water is affected as follows: excellent, < 300 mg L^{-1}; good, 300–600 mg L^{-1}; fair, 600–900 mg L^{-1}; poor, 900–1200 mg L^{-1}; and unacceptable, > 1200 mg L^{-1} [19]. A research based on water quality of 16 different points of Quetta city showed that TDS and hardness of the tap water was within the range of WHO limit [15]. Whereas in this study, 19 samples of Quetta have the permissible TDS level (333–1200 mg L^{-1}) and 1 sample has the objectionable TDS >1200mg L^{-1}.

Very little concentration of TDS makes water intolerable for drinking purpose, owing to lack of some useful minerals and flat taste. The lower acceptable limit may be 80mg L^{-1}, as there is no lower recommended limit.

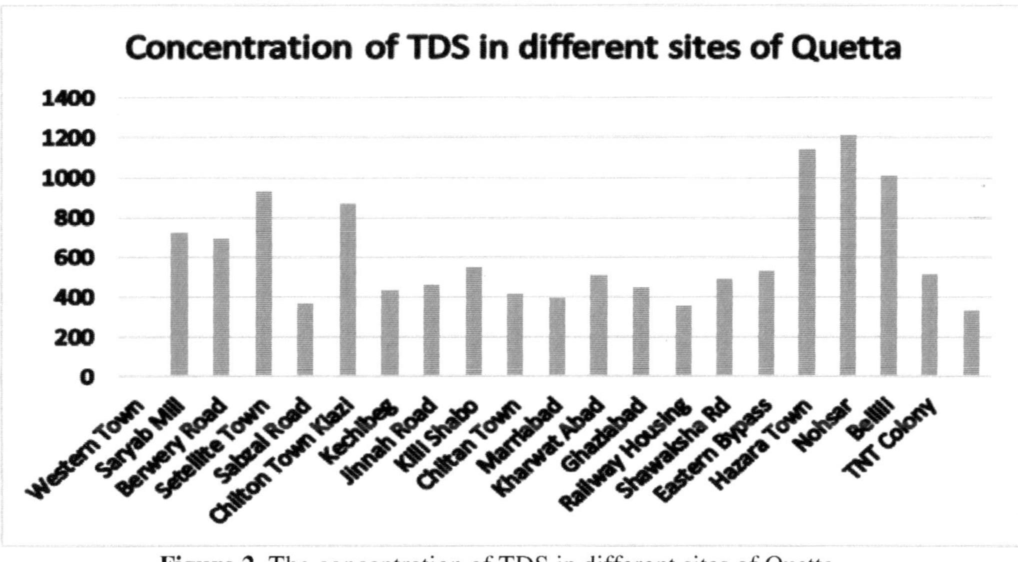

Figure 2. The concentration of TDS in different sites of Quetta.

Ground water with higher TDS value (>500 mg L^{-1}) may be unpleasant because it causes objectionable taste and layers of scales inside the water pipes, household appliances, boilers and heaters. TDS in different sites of Quetta City is shown in figure 2. The scales are caused by some constituents of TDS, like Ca^{2+}, CO_3^{2-}, Cl^{-}, SO_4^{2-}, Mg^{2+} ions.

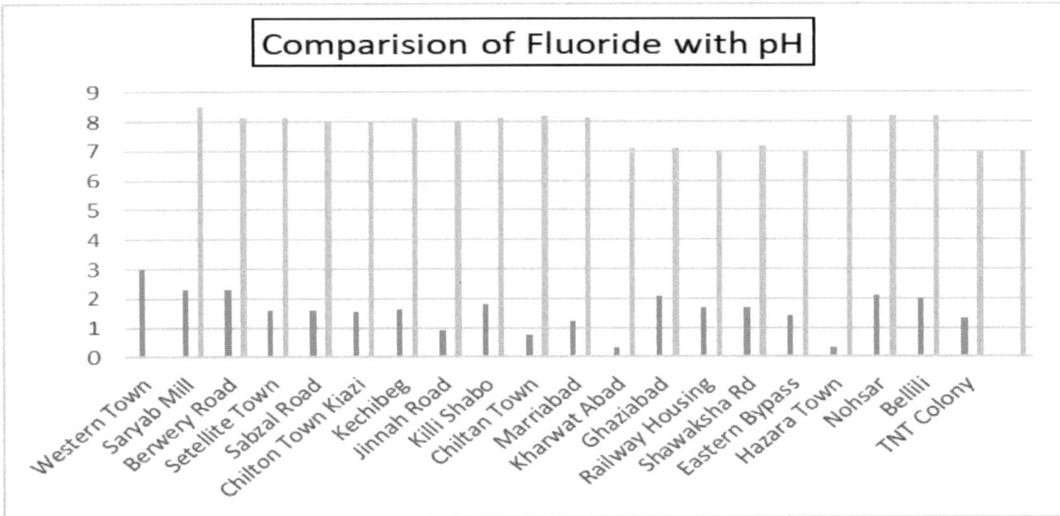

Figure 3. Comparison of fluoride with pH

Note: pH __Fluoride

The hardness of water is caused by the higher concentration of Calcium and magnesium. The data obtained from analysis of water samples showed that only one sample exceeded the maximum permissible limit (75 mg L^{-1}), set by National Standards of Water Quality [16]. Figure 2 shows the concentration of TDS in different sites of Quetta City. The high amount of Ca^{2+} in water can be decreased with the help of treatments like, chemical softening, electro dialysis, ion exchange, and reverse osmosis [7]. Magnesium in drinking water is relatively non-hazardous to humans at low concentration but at high concentration, it causes diuretic and laxative effects on human. The results of the analysis showed that five samples acceded the maximum acceptable level of 50mg L^{-1}[17]. The presence of higher concentration of Ca^{2+} and Mg^{2+} reflects the hardness of water and shows that this water is unfit for drinking.

High intake of sodium is harmful for the patients suffering from renal, cardiac and circulatory diseases. WHO has not recommended any maximum limit for Na^+; though, the possible objectionable limit for the taste is 200 mg L^{-1}. Only 02 drinking water samples were detected with a high concentration of sodium. Water can be purified by demineralization and Ion exchange process. It is evident that the presence of Ca^{2+} and Na^+ cations cause high TDS value in the studied area.

The major source of chloride (Cl^-) is sodium chloride (NaCl) added in the food while drinking water supplies comparatively less concentration. Chloride concentration is associated with Na^+ ions. If the concentration of chloride exceeds 250 mg L^{-1} then it can be detectable through its taste. The high intake of chloride causes kidney and heart diseases [16]. Another experimental work performed on drinking water of Quetta revealed that the concentration of chloride and sulphate ranged from 5.00ppm to 80ppm and 20ppm to 90ppm, respectively [20]. While, in current study only 1 sample was identified with a concentration of Chloride more than 250 mg L^{-1}.

On human health, no toxic effects have been studied for the concentration of SO_4^{2-} in drinking water. However, the concentration between1000-1200 mgL^{-1} may give purgative effect, which may not lead to, diarrhoea, dehydration or weight loss. According to WHO, the lowermost taste detectable limit of $SO4^{2-}$ as a sodium salt is 250mgL^{-1} [17]. The results showed that 4 samples were identified with a higher concentration of Sulphate (>250 mg L^{-1}). The occurrence of high Sulphates may cause corrosion and gives taste to drink water. The contribution of Bicarbonate (HCO_3^-) and carbonate $(CO_3)^{2-}$ to the alkalinity of water have been reported [18]. The amount of HCO_3^-in water samples have been detected in the range of 110 to 320mgL^{-1}.

3.1 Statistical Correlation analysis

The result of each area has been put under the statistical Pearson correlation analysis. The correlation between pH and TDS with fluoride is strongly positive providing 0.841 and 0.775 Pearson correlation coefficient respectively, which is the highly significant result with P value 0.000. This result indicates that the increase in pH and TDS causes to increase in the concentration of fluoride in drinking water samples. These results are formed from the correlation outputs as shown in table 1. It shows that high concentration of fluoride is found in samples which have more concentration of TDS and high pH value. So, the amount of F^- takes part in increasing the concentration of TDS. In the same way, K^+, Na^+, Ca^{2+}, Cl^- and SO_4^{2-} are recognized to have a positive correlation with the F^-.

Table 1. The Pearson correlation of pH and TDS with fluoride.

Parameters	Correlation		P-value	N
	pH	Fluoride		
pH	1	0.841	0.000	20
Fluoride	0.841	1	0.000	20
	Fluoride	TDS		
Fluoride	1	0.775	0.000	20
TDS	0.775	1	0.000	20

4. Conclusions

The results of the analysis showed that most inhabitants of the Quetta city utilize contaminated groundwater for drinking purpose. With the exception of 35 % of drinking water samples, other remaining samples were recognized as polluted and not fit for human drinking purpose. The risk of dental fluorosis and dental caries was found for the remaining 65% of the sites due to the higher concentration of fluoride. The increasing pH and TDS of the groundwater were due to the contribution of different salts of F^-, Cl^-, K^+, Na^+, and SO_4^{2-}. Statistical correlation between fluoride, pH and TDS is strongly positive and highly significant. From correlation analysis, it is found that fluoride concentration in drinking water depends on the level of pH and concentration of TDS. The salts of K^+, Na^+, Ca^{2+}, SO_4^{2-} and Cl^- play a major role in attaining the favourable pH for the dissolution of fluoride-containing compounds in drinking water. Hence it is concluded that rocks present in Quetta valley, are possibly the reason for the increased concentration of fluoride in groundwater. Moreover, the groundwater may come from the strata of the same surroundings. Nevertheless, there has been no inclusive health survey conducted for dental fluorosis to address the problems in the past and there is a very minor difference in the concentration of fluoride that imparts a beneficial or hazardous effect on human health, therefore, there should be regular monitoring of drinking water scheme to maintain the water quality. Hence, the results reveal that in the selected area of Quetta city the assessment of fluoride in drinking water and its correlation with various physicochemical parameter has been studied successfully. Furthermore, there is a need for remedial measures for the removal of fluoride in drinking water on governmental scale.

Acknowledgment

The authors give appreciation to BUITEMS for the analysis performed at Chemical laboratory and WRRC (Water research resources center) Quetta.

References

[1] Lewis A. R, Robert M. C and Walter G. Modeling Chlorine Residuals in Drinking Water Distribution Systems 1994 *j. Environ Eng.***120** 803-20

[2] Chandio A.T, Khan M.N and Sarwar A 2015 Fluoride estimation and its correlation with other physicochemical parameters in drinking water of some areas of Balochistan, Pakistan *J. Environ Monit Assess* **187** 531

[3] Datta D.K, Gupta L.P and Subramanian V 2000 Dissolved fluoride in lower Gangae Brahma putrae Meghna river system in Bengal Basin, Bangladesh *J.Environ Geol* **39** 1163-68.
[4] Wang W, Li R.A, Tan J, Luo K, Yang L, Li H and Li Y 2002 Adsorption and leaching of fluoride in soils of China *Geographical Sciences and Natural Resources Research* 122 Research report Fluoride **35** 122-29.
[5] Jabeen U, Fahmid S, Zameer T and Qureshi S 2016 Determination of Fluoride Concentration in Drinking Water and its Comparison with Mineral Water Samples in Quetta, Pakistan *J. Acad. J. Sci. Res.* **04** 081-84.
[6] WHO 2006 Fluoride in drinking water London IWA.H.Fluoride **7** 32-35.
[7] EPA (United States Environmental Protection Agency) Edition of the drinking water standards and health advisories 2015 Office of water U.S. Environmental Protection Agency Washington, DC.
[8] Tahir M.A and Rasheed H 2013 fluoride in the drinking water in pakistan and the possible risk of crippling fuorisis. *J. Drink. Water Eng. Sci* .**6** 17-23.
[9] Clark D C1994 Trendes in prevelence of dental fluorosis in North America *J.community Dentistry and oral epidomology* .**22** 148-55.
[10] Sami H, Vichayanrat T and Satitvipawee P 2016 Caries with Dental Fluorosis and Oral Health Behaviour Among 12-Year School Children in Moderate-Fluoride Drinking Water Community in Quetta, Pakistan. *J.Coll Physicians Surg Pak* **26** 744-47.

[11] APHA, AWWA & WEF Standard methods for examination of water and wastewater 1992 **18th** ed. American Public Health Association, Washington, DC.

[12] Afzal S, Durrani S, Malghani A.K, Khan M, Sajjad M and Tariq N 2016 Concentrations of Fluoride in Drinking Water and Tea Samples and Associations with Dental Fluorosis *Pakistan Journal of Nutrition* **15** 85-88
[13] WHO 2006 Fluoride in drinking water London:IWA.H.Fluoride 7 32-35.
[14] Mineral profile of Balochistan 2015.
[15] Khan S.S, Tareen H, Jabeen U, Mengal F, Masood Z, Ahmed S, Bibi S, Riaz S, Rizwan S, Mandokhail F, Irum U and Mengal R 2016 Quality Assessment of Drinking Water from the Different Colonies of Quetta City, Pakistan according to WHO Standards *BFAIJ* **07** 699-02.
[16] World Health Organization Guidelines of drinking water quality recommendations: 1, Geneva. recommended standards 1963.
[17] National Standards for Drinking Water Quality Pakistan Environment Protection Agency Ministry of Environment GovernmentofPakistan 2008.

[18] WHO , Guidelines for Drinking Water Quality. In: Health Criteria and Other Supporting Information, second ed. World Health Organization, Geneva 1984 **2**.

[19] Olayanju G. M, Ayuk M. A, and Adelusi 2011 Geoelectrical mapping of the groundwater regime around the Federal Polytechnic Ado-Ekiti, South Western Nigeria. *J. Geol. Min. Res.* **08** 201–10.

[20] Khattak M I 2011 Study of common inorganic anions in water samples of Quetta city by technique of ion chromatography *J. Sci Int (Lahore)* 23 135-41.

Electromagnetic shielding effectiveness of high loft activated carbon web prepared by using acrylic waste

M S Naeem[1], S Z Hassan[2], Z Javed[1], B Ramzan[1], A Rasheed[1], Z Ahmad[1], S Javed[2]

[1] Faculty of Textile Engineering, National Textile University, Faisalabad, Pakistan.
[2]Faculty of Engineering, Balochistan University of Information Technology, Engineering and Management Sciences, Faculty of Engineering, Pakistan.
[3]Department of Microbiology and Molecular Genetics, Faculty of Life Sciences, Punjab University Lahore, Pakistan

E-mail: salman@ntu.edu.pk

Abstract: This work represents a novel method for the formation of activated carbon web using charcoal through physical activation in high temperature muffle furnace. The acrylic fibrous waste was separated from bath mats and transferred to carding machine which was transferred to needle punching machine for obtaining a compact structure of non-woven. The acrylic web was first stabilized at 250 °C with a heating rate of 50 °Chr^{-1}. The stabilized web was again raised to high temperature (700 °C, 900 °C and 1100 °C) using charcoal for single stage carbonization and activation. The impact of temperatures on the characteristics of carbon webs was analyzed in terms of yield, dusting, flexibility, XRD (x-ray diffraction analysis) and EDX (energy dispersive x-ray analysis). The EM (electromagnetic) shielding of prepared webs was analyzed by using wave guide method which measures shielding effectiveness at 2.45 GHz (high frequency region). It was found that shielding effectiveness increased from 2.21 dB to 23.71 dB as the carbonization temperature was increased from 700 °C to 1100 °C. Not only the carbonization temperature but the impact of thickness of carbon webs on shielding effectiveness was also observed. The shielding effectiveness increased when the thickness of carbon webs increased because of internal reflections and more absorption of electro-magnetic radiations.
Keywords. Electromagnetic, shielding effect, activated carbon, acrylic waste

1. Introduction

In the current smart age researchers and scientists are trying to explore alternate flexible materials having good electrical properties. Electrically conducting textile materials are widely used in different fields like military applications, smart materials, actuators, sensors and for EM (electromagnetic) Shielding applications. There are different techniques available for the formation of conducting textiles like coating of conductive particles (carbon nanotubes and carbon particles), metallization, insertion of metallic yarns etc. [1]. There are different standards available for shielding effectiveness but it is preferred that for effective shielding the resistivity should be less than 10^2 ohm.cm^{-2} however mostly synthetic fibres possess electrical resistivity around 10^{14}-10^{15} ohm.cm^{-2} [2]. Different electrical instruments emit electromagnetic radiations with different frequencies. These radiations are not only a source of noise for other electrical instruments but also pose serious health hazards to humans as well [3,4]. Reflection, absorption and internal reflections are the three basic mechanisms that cause effective shielding of different materials [5, 6]. The phenomenon of internal reflections causes shielding effectiveness in bulk materials, however highly conducting materials having less thickness like metals block electromagnetic radiations through reflections. However high density and weight, corrosive nature, cost and less flexibility are main factors that inhibit the use of metals for electromagnetic shielding.

More focus is now shifting towards carbon structures and graphene materials which are a favourable alternative for replacing metals for effective shielding applications. Further shielding effectiveness by using reflection is not a preferred choice since it can disturb the working of other instruments, hence the concept of eco-friendly EMI shielding became popular. This can be achieved by using light weight material having good absorption and weak reflection so that it can absorb the incoming electromagnetic radiations for which carbon foam and carbon non-woven having porous structures with higher values of surface and volume conductivity [7]. This work explained an affordable and simple method for developing porous structure of carbon having good conductive and EMI shielding application.

2. Material and methodology

The raw material (acrylic) used in this study was provided by Grund Industries which is located in Czech Republic. The acrylic was given in the form of discarded bath mats having the monomer of AN (acrylonitrile) around 88 %. The salient properties of acrylic fibres can be seen from table 1.

Table 1: Physical characteristics of acrylic fibrous waste

Tenacity (cN/tex)	21.22
Fineness (tex)	97
Shrinkage (%)	2.1
Elongation (%)	47

Acrylic fibers which were separated from bath mats were converted into acrylic web by using lab carding machine. This structure was further compacted by using needle punching machine. Needle punching machine helped to entangle the fibers in order to get the compact structure. A fixed size of acrylic web having 12 inches' length and width was cut and washed in order to remove impurities. This web was later dried in oven at 100 °C for 2 hours. The dried web was transferred to high temperature muffle furnace for stabilization. The stabilization was performed at 250 °C hr^{-1} with no holding time and having a heating rate of 35 °C hr^{-1}. These stabilized webs were carbonized and activated simultaneously at 1100 °C, 900 °C and at 700 °C having heating rate (300 °C hr^{-1}) without any holding time (by holding time it means that for how much time the sample was placed in furnace at high temperature).

2.1. Physical characterization of prepared carbon webs

The physical characterization of prepared carbon webs was done in order to determine yield, flexibility, dusting and shrinkage. The shrinkage of carbon webs was determined by measuring original length of acrylic web and final length of carbonized web.For the determination of yield of resulting carbon web the following formula is used. The cantilever bending principle was used for the determination of stiffness of carbon web.

$$\text{Yield} = \frac{\text{Weight of activated carbon web}}{\text{Weight of acrylic web}} \times 100 \quad (1)$$

2.1.1. X-ray diffraction (XRD) analysis

The technique of x-ray diffraction analysis was performed which helped to determine degree of crystallinity in prepared carbon webs at different carbonization temperatures (700 °C, 900 °C and 1100 °C). The degree of crystallinity can be calculated by using equation 2.

$$I_c = 1 - \frac{I_1}{I_2} \quad (2)$$

Where, I_1 is the intensity at minimum peak and I_2 is the intensity at maximum peak [8].

2.1.2. Energy dispersive x-ray (EDX) analysis
Energy dispersive x-ray analysis was conducted to determine the relative proportion of different elements in prepared activated carbon webs at different temperatures.

2.1.3. Electromagnetic shielding effectiveness of activated carbon web
The shielding effectiveness of prepared carbon webs was determined by using waveguide instrument. This instrument is able to measure shielding effectiveness at a particular wave length which is 2.45 GHz. The EM shielding can be determined by the help of equation 3.

$$SE = 10\log\frac{P_t}{P_i} \tag{3}$$

Here P_t and P_i are power densities with the presence and absence of carbon webs.

3. Results
3.1. Effect of carbonization temperature on carbon webs
When the carbonization temperature was increased then at high temperature more carbon reacted with the available oxygen and caused a reduction in the yield. This reduction of yield also caused adverse impacts on the shrinkage, dusting anf flexible tendency of activated carbon webs as shown in table 2.

Table 2. Physical properties of AC webs at different temperatures

Temperature (°C)	Yield (%)	Shrinkage	Flexibility	Dusting
700	61.27	Good	Average	Average
900	57.12	Good	Average	Poor
1100	45.11	Average	Average	Poor

3.2. Energy dispersive x-ray (EDX) analysis
The content of different elements in prepared carbon webs can be determined EDX analysis. This technique helped to determine and compare the percentage of different elements in carbon webs prepared at high temperatures. Table 3 shows the relative proportion of different elements in different carbon webs.

Table 3. impact of different carbonization temperature on elemental composition

Element	App concentration	Intensity	Atomic percentage
	700 °C		
C K	0.26	2.12	82.22
O K	0.01	0.761	17.78
	900 °C		
C K	0.37	2.12	87.12
O K	0.02	0.760	12.88
	1100 °C		
C K	0.18	2.10	93.67
O K	0.01	0.744	5.43
Ca K	0.00	0.902	0.90

From table it is clear that carbon web prepared at highest temperature showed higher content of carbon 93.67% along with 5.43 % oxygen content, but as the temperature for carbonization was reduced the carbon content reduced to 83.33 % at 700° C. The rise in carbon content is because of more elimination of elements other than carbon like nitrogen, sulfur and hydrogen because of decomposition at higher temperature [8].

3.3. X-ray diffraction (XRD) analysis

By increase carbonization temperature the degree of crystallinity also kept on increasing which can be determined by using x-ray diffraction analysis. The x-ray diffraction patterns of carbon webs prepared at different carbonization temperatures can be seen from figure 1. Initially the degree of crystallinity was 79.21 % at 700 °C which increased to 83.7 % and 89.41 % when temperature for carbonization was increased to 900 °C and 1100 °C respectively. This rise of crystallinity is because of more alignment of carbon chains causing a reduction in spacing of these planes as well as removal of disorganized carbon.

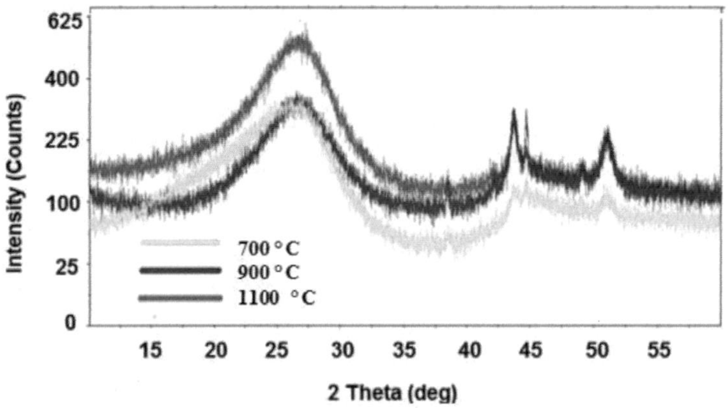

Figure 1. Effect of carbonization temperature on degree of crystallinity of AC webs

3.4. Impact of temperature on conductivity and EMI shielding of AC webs

The surface conductivities kept on increasing while increasing carbonization temperature. It was found that electrical conductivity increased from $0.35 \times 10^{-2}\ \Omega^{-1}$ to $3.81 \times 10^{1}\ \Omega^{-1}$ when the temperature for carbonization was increased from 700 °C to 1100 °C respectively. More conductivity was exhibited by carbon web at 1100 °C due to more graphitization (parallel orientation of chains and less spacing of chains) as can be seen from peaks of XRD spectra [9].

Figure 2 shows shielding effectiveness for electromagnetic radiations by using carbon webs prepared at different temperatures. The electromagnetic shielding effectiveness was determined by the help of wave guide method. It is clear from figure that shielding effectiveness increased from 1.9 dB to 18.88 dB and finally to 23.71 dB at carbonization temperatures of 700 °C, 900 °C and 1100 °C, respectively.

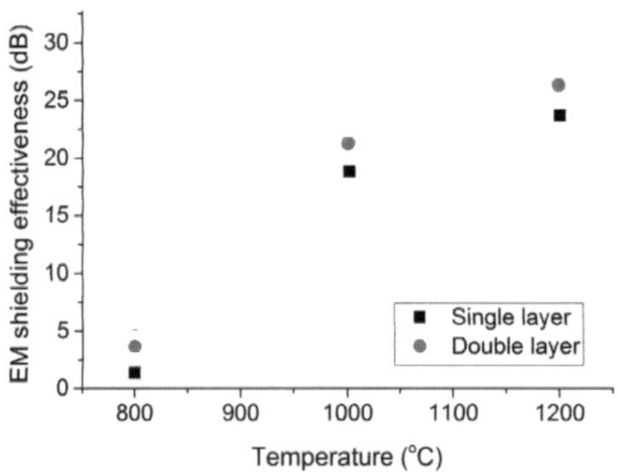

Figure 2, Effect of carbonization temperature on EM shielding

When the temperature for carbonization was less i.e. at 700 °C, there was not so much difference in shielding effectiveness between carbon web and acrylic web. However, as the carbonization temperature was increased, sudden increase in shielding effectiveness was observed due to development of carbon/graphite structure which is clear from XRD pattern shown in figure 1. More crystallinity at higher temperature is responsible for maximum parallel orientation of chains which enables easy movement of electrons and causing higher conductivity due to which shielding effectiveness also increased at higher temperature. As far as thickness of carbon webs in concerned, shielding effectiveness increased by increasing thickness of webs because of higher internal multiple reflections.

4. Conclusions

The focus of this study was the development of porous and continuous structure of activated carbon web with good flexibility and higher carbon content. The novelty of this method is the single stage carbonization and activation using charcoal. Initially the webs were stabilized at lower temperature with slow heating rate for ensuring proper stabilization. Later the stabilized webs were carbonized at 700 °C, 900 °C and 1100 °C with a heating rate of 300 °C. It was found that at higher temperature although a reduction in the yield was observed but at the same time rise in carbon content and degree of crystallinity was also observed. These two are the main reason for higher electrical conductivity at high carbonization temperature. These webs were also checked for EM shielding application by using wave guide method. It was found that shielding effectiveness not only increased by increasing carbon content but also by increasing the thickness of webs. At high temperature the carbon web showed higher electrical conductivity which is the main reason for the rise in shielding effectiveness. Likewise, as the thickness of webs increased their absorption and internal reflections also increased which helped in lowering the impact of electromagnetic radiations.

5. References

[1] Daniela N 2012 *Fibres Text. East. Eur.* **20** 53-56.
[2] Chen C 2007 *J. Mater. Process Technol.* **192** 549-554.
[3] Safarova V, Tunak M and Militký J 2015 *Text. Res. J.* **85** 673-686.
[4] Safarova Vand Militký J 2014 *Text. Res. J.* **84** 1255-1267.
[5] Sano E and Akiba E 2014 *Carbon N Y.* **78** 463-468.
[6] Arjmand M and Sundararaj U 2015 *Compos Sci. Technol.* **118** 257-263.
[7] Li Y, Shen X, Pei Y. Zhang D, Yi 2016 *Carbon N Y* **100** 375-385.
[8] Ma Y, Yin X and Li Q 2013 *Trans. Nonferrous Met. Soc.* **23** 1652-1660.
[9] Jonathan Y and Chen 2016 *Activated Carbon Fiber and Textiles* (Woodhead).

Effect of material and stitching on tensile properties of woven fabrics

Z Ahmad[1,2], B K Sirkova[1,2], S Ahmad[1], M S Naeem[1] and S Z Hassan[3]
[1] National Textile University, Faculty of Textile Engineering, Faisalabad, Pakistan.
[2] Technical University of Liberec, Faculty of Textile Engineering, Liberec, Czech Republic.
[3] Balochistan University of Information Technology, Engineering and Management Sciences, Faculty of Engineering, Quetta, Pakistan.
E-mail: zuhaibahmad.ntu@gmail.com

Abstract: The strongest growth potential for high strength fabrics and advanced composites is in high-performance applications such as aerospace, maritime, automotive and construction industries. The optimization of mechanical properties of these materials is very important in this regard. Therefore, this study has been performed where we have analyzed the effect of material and stitching on tensile properties of woven structures. For this purpose, double layer woven fabric structures have been prepared on sample weaving loom using Basalt and Glass yarns. Their tensile testing has been performed to analyze the impact of fabric parameters of tensile load. ANOVA (analysis of variance) results showed that stitch distance has insignificant impact on the strength of fabric while the effect of material and stitching on peak load was statistical significance.
Keywords: Woven structure, material, tensile load, stitching, binding points.

1. Introduction

Recent years have seen a considerable spread in polymer composite structural materials, which allows the development of alternatives that fulfill technical requirements and methods for improving their mechanical performance [1]. Carbon fibers and glass fibers are mostly used for composites making in conditions of high loads [2]. Basalt fibers are more commonly used because of additional benefits like reduction of cost, improved chemical and physical characteristics [3] and for the reinforcements of thermosetting polymers as compared to the conventional Glass or carbon fibers [4] [5] [6]. Basalt is a natural product from volcanic rock. Basalt rock is melted into fibres in a way that fibres with the diameter of 7–13 μm are formed from the melt at a temperature above 1500 °C by the means of a centrifugal blowing process [2].

Most of the investigators worked on composites and single-layer woven fabrics, like Liu et al prepared polymer composites reinforced by basalt fabric and glass fabrics and concluded that there is no significant difference in their compression strength, shear strength, flexure strength and tensile strength [7]. Whereas, Lopresto et al investigated the Basalt and E-Glass reinforced plastic laminates were compared showing a high performance of the Basalt material in terms of mechanical properties [8]. Babu and Chairman evaluated Glass and Basalt fabric reinforced composites for analyzing their abrasive wear performance and mechanical properties [9]. Similarly, the significance of strain-rate effect on material properties of unidirectional Glass, Basalt, Carbon and plain-woven Aramid fabrics have been investigated by Yao and Zhu [10].

It is necessary to understand and evaluate the mechanical properties of woven structure before being used in composites. So, in this research work, the possibility of Basalt material as reinforcement material has been compared with E-Glass. The comparison has been done on the basis of impact of material, stich distance and the effect of stitching (between layers) has been determined on tensile strength and toughness of fabric.

2. Material and methodology

The continuous multifilament yarns of Glass 68*2 tex and Basalt 66*2 tex were used as raw material. The Basalt and Glass structures were produced on the sample weaving loom. The yarn count for warp and weft was kept the same while woven samples were produced in accordance with experimental design as can be seen from Table 1 [11]. The basic structures for stitched and unstitched fabrics have been shown in Figure 1. The conditioning of yarns before testing and fabric production have been achieved for one day. The Vibrodyn-400 instrument was used to measure the fineness and tenacity of Basalt and Glass fibers, in accordance with EN ISO 1973:1995.

Table 1. Experimental plan for construction of Basalt and Glass fabrics

Sample Code	Ends/ cm	Picks /cm	Design	Sample Code	Ends/ cm	Picks/ cm	Design
B1	16.2	18.2	S.D.= 0.5 x 0.5	**G1**	16.2	18	S.D.= 0.5 x 0.5
B2	16.2	18.2	S.D.= 1x1	**G2**	16.2	18.2	S.D.= 1x1
B3	16.2	18.2	S.D.= 1x1.5	**G3**	16.2	18.2	S.D.= 1x1.5
B4	16.2	18	S.D.= 1x2	**G4**	16.2	18	S.D.= 1x2
B5	16.2	17	unstitched	**G5**	16.2	17	unstitched

*B = Basalt fabric, G = Glass fabric, S.D. = stitching distance (cm)

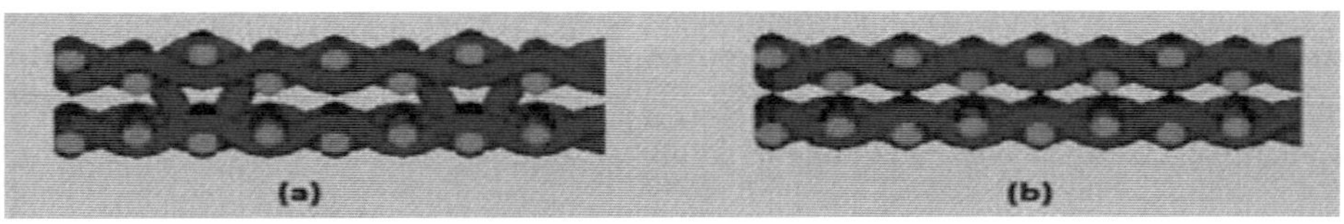

Figure 1. Two-layer fabric (a) stitched, (b) unstitched.

Similarly, TIRA 2300 instrument was used to measure the tensile properties of Basalt and Glass yarns accordance to ASTM D885. The average load-elongation curve for each yarn type was obtained by using Matlab software. Whereas the image analysis has been performed in accordance with IS 22-103-01/01 was used to determine the yarn diameter [12]. The twist tester of MesdanLab was used to measure the yarn twist for both materials in accordance to ASTM D1422. Fiber specification and characteristics of yarns for Glass and Basalt can be seen from Table 2.

The Testometric M350-10CT machine as shown in Figure 2 has been used to measure the tensile properties of all woven fabrics following EN ISO 13934-1. The load-elongation curve data for each of woven fabric (five samples) were obtained. Later the average curve for each type of fabric sample was obtained using Matlab software. The statistical significance was calculated by using ANOVA (at α= 0.05) for tensile strength of fabrics.

Table 2: Characteristics of Basalt and Glass fibers and yarns

Parameters	Basalt fiber	Glass fiber	Parameters	Basalt Yarn	Glass Yarn
Fineness (dtex)	1.72	1.56	Fineness (tex)	132	138
Diameter (um)	10	9	Diameter (mm)	0.44	0.42
Tenacity (cN/tex)	96.97	111.2	Tenacity (cN/tex)	46	40.18
Breaking strain (%)	3.66	4.23	Breaking strain (%)	2.96	2.7
Density (g/cm^3)	2.8	2.56	Twist per meter	94	148

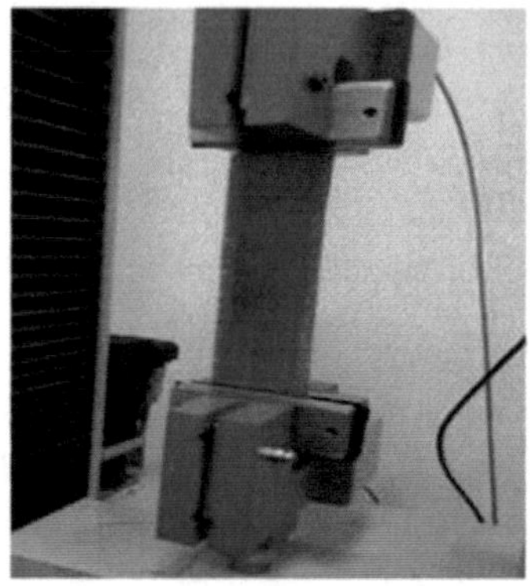

Figure 2. Tensile testing of sample fabrics

3. Results and Discussions

When a woven fabric is subjected to a tensile loading then it goes through the four regions in a load-elongation curve, which has been explained here in Figure 3. The viscoelastic nature of both materials can be observed here as a low slope can be observed in the starting phase which is the first region. Later, in the second region it rises steeply. In first region which is also called the crimp region, the crimp is decreased in one set of yarn during tensioning while it increases in other set of yarn, it is also called the crimp interchange. Hence a large increase in elongation at a low load level in the crimp region can be observed in the curve of load elongation. In the second region, when more extension is added in the fabric by the application of force, the extensions in fibers as well as in yarn also starts. The slope of curve in comparison to previous region increases when straightened yarns bear more load which is termed as elastic region. As far as the non-linear part of load elongation curve is concerned, it can be seen from the third region (which is non-linear part of failure region) before tensile strength is reached. It is because of random breakage of filaments present in bundles of yarn which is also prior to their localized failure. In the last stage which is the post peak region, a swift decrease in the load beyond the tensile strength can be observed, which links to the increasing yarn failure [13].

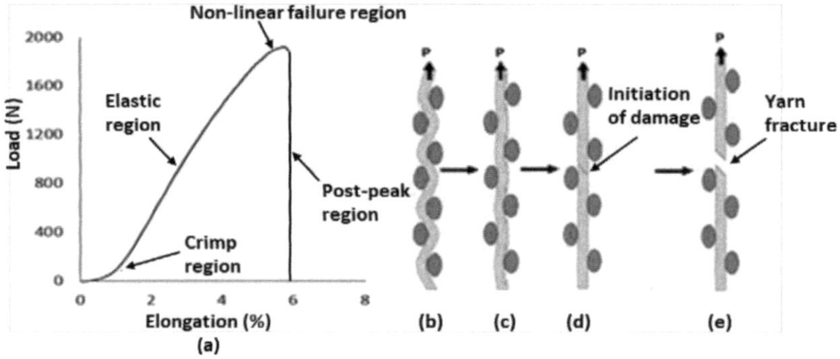

Figure 3. (a) Load-elongation curve during tensile loading, (b) crimp region, (c) elastic region, (d) non-linear failure region and (e) post-peak region (adapted and reproduced) [10]

Figure 4 shows the curve for load-elongation for two-layer Basalt woven fabrics in both (warp and weft) directions. The impact of stitch distance and stitching on curve of load elongation can also be seen from figure. The curves of load elongation for all stitched fabrics (B1-B4) are close to each other in both directions which clearly shows that stitch distance has no effect on tensile load. All two layer stitched fabrics bear similar load but a little high deformation (elongation %) in both directions, when we compare the all stitched fabrics (B1-B4) with the curve of unstitched fabric (B5). From these results it can be concluded that stitching points cause a little influence on elongation % but remain ineffective during breaking load. This is due to the fact as the stitching points connects the two-layer and the rate of stitch points increase the overall crimp in the yarns, which permits them to deform (extend) more before breakage along both directions. It is also not preferable to have more number of stitch points in a woven fabric as in case of (B1), the woven fabric sample exhibits the low tensile load values along both warp and weft direction. The stitched woven fabrics have been found less rigid as compared to unstitched woven fabric. In Table 3, the statistically significant of stitching between two layers of fabric can be observed which is found significant.

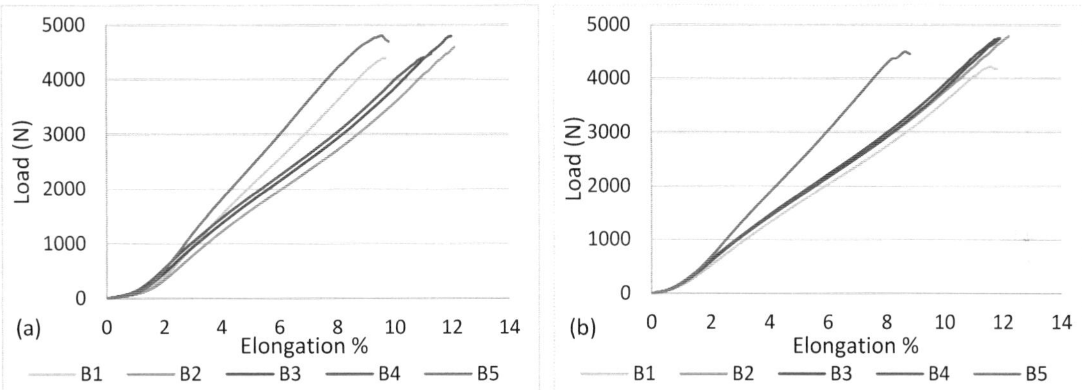

Figure 4. Average Load vs elongation curves for Basalt fabrics along (a) weft and (b) warp.

Similarly, the impact of stitching and stitch length between two layers of fabric can be analyzed in Glass woven fabric samples can be seen from figure 5. It can be analyzed that no sufficient difference exists in tensile strength when we compared two-layer stitched fabrics (G1-G4), stitch distance also has insignificant impact in both directions. Also, the unstitched fabric (G5) has similar tensile strength and less elongation % along both directions as compared to stitched fabrics (G1-G4).

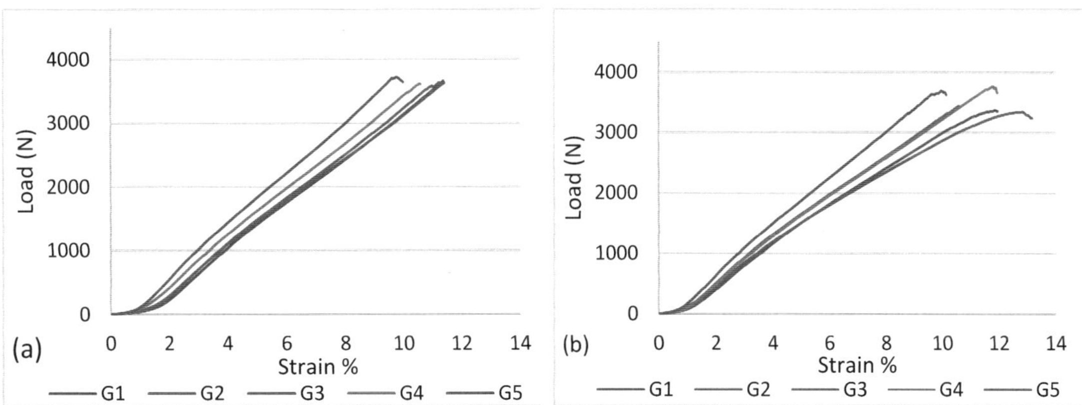

Figure 5. Average Load vs elongation curves for Glass fabrics along (a) weft and (b) warp.

WOR (work of rupture) is defined as the energy required to break a material or total work done to break that material. It is also a measure of toughness of material (area under the load-elongation curve) and been calculate for all fabric samples using equation 1. A small section of force extension curve is "*dl*" and within this small section the force "*F*" is constant as shown in

Figure **6**, while this force increases the sample in length by an amount *dl* [14].

The WOR of a material is proportional to its areal weight (GSM) for fabrics and to their original sample size (area). So, in order to compare the sample fabrics their specific work of rupture (SWOR) has been calculated using equation 2 and shown in Figure 7.

$$Work\ of\ rupture = \int_{0}^{break} F.dl \quad (1)$$

$$\text{Specific work of rupture} = \text{SWOR} = \frac{\text{Work of rupture}}{\text{Areal weight x Initial Area}} \quad (2)$$

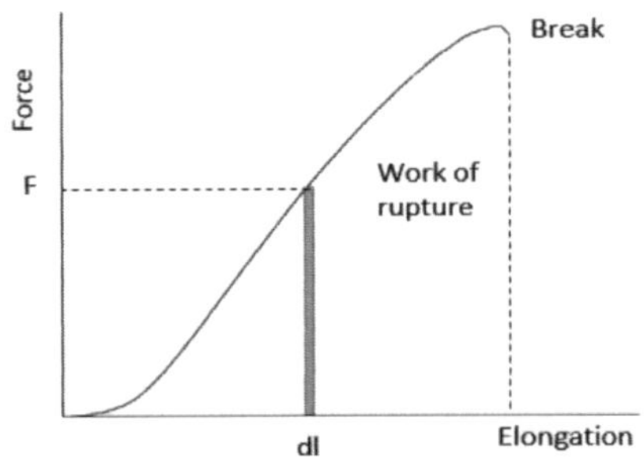

Figure 6. Work of rupture in force extension curve

It can be observed in Figure 7 that all Basalt fabrics exhibits higher toughness as compared to Glass fabrics along both directions. Basalt yarn exhibits higher tenacity than Glass yarns (table 2), similarly their load-elongation values are higher, consequently it leads to higher toughness in Basalt woven fabrics. When the unstitched fabric samples (B5) and (G5) having same density and cross-sectional area as stitched fabric samples (B1-B4) and (G1-G4), are compared with their respective material to analyze the impact of stitching on toughness of fabric then it can be observed that the addition of stitching points increases the toughness as it combines the two-layers and allow them to have high strain values before breakage along both directions of applied load. Whereas the impact of stitch distance is not so noticeable in stitched fabrics (B1-B4) and (G1-G4) along both directions.

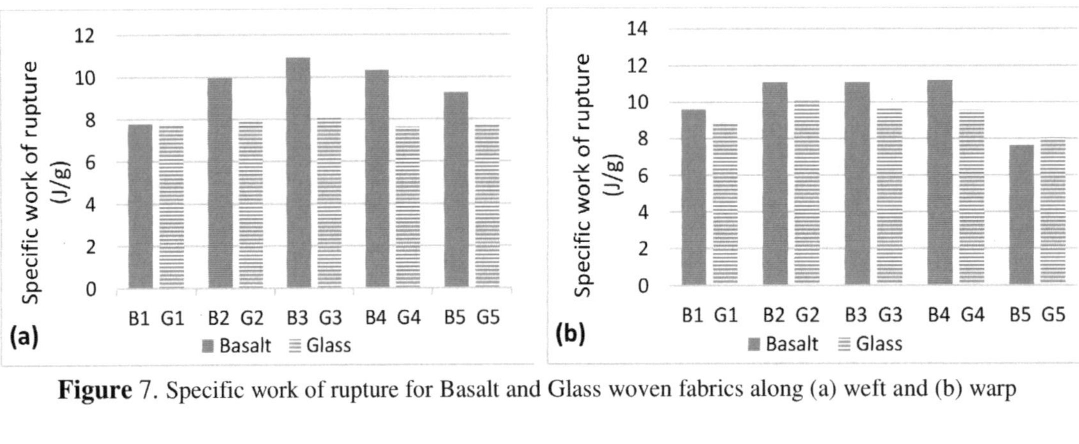

Figure 7. Specific work of rupture for Basalt and Glass woven fabrics along (a) weft and (b) warp

Figure 8 shows the effect of material which is very important for reinforcement materials and in case of Basalt as an alternate material to the Glass structures. The peak load values of woven structures of both materials can be compared along both directions. All the stitched and unstitched structures of Glass material exhibit low peak load values as compared to the Basalt woven structures along both warp and weft directions. As we can see in Table 2, the Basalt yarns have more tenacity as compared with Glass yarn and the similar behavior can be observed in both kind of fabrics as well. The variation of the load in all stitched fabrics has already been observed in load-elongation curve data and here in Figure 8 the peak load analysis also shows insignificant results for both Basalt and Glass structures. Moreover, in Table 3, the ANOVA results explains the statistical significance of material on peak load of fabric as P value is less than 0.05.

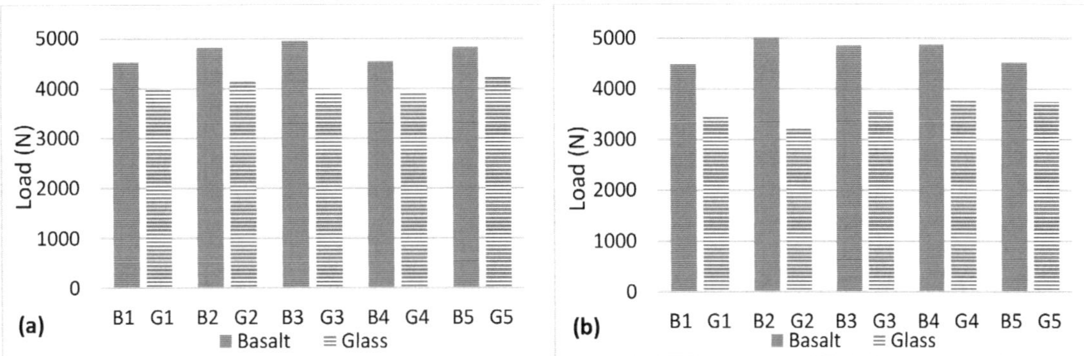

Figure 8. Peak load of Basalt and Glass woven fabric samples along (a) weft and (b) warp

The material effect applied load direction (warp, weft), stitching distance, stitching between two layers of fabric and their interactions on fabric strength can be observed in Table 3. It shows the statistically significant effect different parameters on peak load of woven fabric if the P value is less than 0.05.

Table 3. ANOVA results for peak load (N)

Source	Two-layer stitched Fabrics Significance	Two-layer unstitched fabric Significance
Material	0.000	0.000
Direction (weft, warp)	0.002	0.000
Stitching distance	0.091	-
Stitched vs Unstitched	-	0.020
Material * Direction	0.000	0.000
Material * Stitching distance	0.033	-
Stitching distance * Direction	0.066	-
Material*stitched vs unstitched	-	0.001

4. Conclusions

The following conclusions can be made from this study.

- The values of load-elongation curves for Basalt and Glass stitched woven fabrics are close to each other and the effect of stitch distance between two layers of fabric has been found statistically insignificant as well along warp and weft direction.

- The two-layer unstitched fabric (B5) and (G5) have also the same amount of peak load as compared to their relevant stitched structures (B1-B4) and (G1-G4), with slightly less amount of extension along warp and weft direction. It explains that the stitching or connection points between two layers of woven fabric affect little on extension percentage only and not on peak load.
- The toughness of Basalt fabrics is more as compared to Glass woven fabrics, similarly it is concluded that the effect of material on tensile strength of fabric is prominent along both direction of applied load.
- Two-way ANOVA results concluded that the effect of stitch distance between two layers of fabric is insignificant, whereas the material effect, applied load direction (warp, weft) and stitching between two layers of fabric is statistical significance.
- The Basalt material can be a good alternative to E-Glass as it has shown good material properties regarding tensile for stitched and unstitched two-layer woven fabrics.

5. References

[1] Goldsworthy WB 2000 *Compos. Technol.* **8** 15-16.
[2] Singha K 2012 *International Journal of Textile Science* **1** 19-28.
[3] Czigány T 2005 *Polymer Composites.* ***8*** 309-328.
[4] Deák T and Czigány T 2009 *Textile Research Journal* **79** 645-651.
[5] Botev M, Betchev D and Bikiaris P 1999 *Journal of Applied Polymer Science* **74** 523-531.
[6] "BASALT FIBER & COMPOSITE MATERIALS TECHNOLOGY DEVELOPMENT," [Online]. Available: http://basaltfm.com/eng/. [Accessed 10 12 2016].
[7] Liu Q, Shaw MT and Parnas RS 2006 *Polymer composites* **27** 41-48.
[8] Loprest V, Leone C and Lorio ID 2011 *Composites Part B: Engineering* **42** 717-723.
[9] Babu K and Palani S 2013 *Journal of Applied Polymer Science* **130** 120-130.
[10] Yao Y, Zhu D, Zhang H, Li G and Yiming Y 2016 *Journal of Materials in Civil Engineering,* **28** 333-341.
[11] Ahmad Z and Sirková BK 2018 *The Journal of The Textile Institute* **109** 686-694.
[12] T. R. Center, *Internal Standards, Section B – Textile Materials & Design of Textile Products. IS 22-103-01/01,* Liberec: Faculty of Textiles, Technical University of Liberec, 2004.
[13] HU J 2004 Structure and Mechanics of Woven Fabrics, North America (Woodhead Publishing).
[14] Saville B 1999 *Physical testing of textiles* (Cambridge: England).

Enhancement of anti-microbial activity by natural finishes prepared from herbal spices and wastage peel of fruits applied on textile substrate

Syed Qutaba Bin Tariq[1], Qasim Siddiqui[1], Abdul Malik Rehan[1].

[1]Department of Textile Engineering, Balochistan University of Information Technology, Engineering & Management Sciences (BUITEMS), Quetta, Pakistan

Email. Qasim_siddiqui81@hotmail.com

Abstract: Fabrics with antibacterial assets have become essential to organize and manage the infestation by microbes, and to reduce the formation of odor. The fabrics with antimicrobial finishes are highly hygienic in all dimensions particular, when consumed by human beings. In order to evaluate antimicrobial activity, we have prepared natural and organic extracts from herbs and wastage of fruits like Pomegranate. The fabric samples were tested for antimicrobial activity against bacterial strains like Staphylococcus, E.coli, under qualitative analysis method AATCC 147. The results indicated that the cotton fabric show a better microbial resistance against the above-mentioned strains by both two natural finishes on untreated and treated substrate. As per qualitative analysis, the fabric treated with extract showed best reduction against Staphylococcus by analyzing antimicrobial activity.
Keywords. Textile fabrics, Natural Extract, Fruit peel, antibacterial activity, Staphylococcus Aureus, Escherichia coli.

1. Introduction

Antibacterial fabrics are important not only in medical applications but also in terms of daily life usage [1]. The cotton fabric has ability to use mostly on medical terms fabric so need anti-microbial finish for its cellulosic structure [2,3].

The application of antimicrobial finishing to textiles can prevent bacterial growth and antibacterial textile production has become increasingly prominent for hygienic and medical applications [4,5]. The environmental concerns with commercial anti-microbial finishes has silver chloride, benzimidaole and Triclosan which very hygienic for human. Mostly anti-microbial products like socks and undergarments always contact with human body organs, so they must be abolish the hygiene factor [6].

Pakistan is an agriculture country and has increased growth rate of Pomegranatefruit, on seasonal life lot of peel have been generated. The peel of fruits is a rich source phenols, quinone's, flavonoids, lectines, polyacetylenes which showed good antimicrobial activity. Many polymethoxylated flavones have several important bioactivities, which are very rare in other plants [7,8].

Pomegranate and herbal species is an important medicinal plant of the family Rutaceae. It is cultivated mainly for its alkaloids, which have anticancer activities and the antibacterial potential in crude extracts of different parts. flavonoids have a large spectrum of biological activity including antibacterial, antifungal, anti-diabetic, anticancer and antiviral activities, [8].

In addition the fiber of fruits also contains bioactive compounds, such as polyphenols, the most important being vitamin C (or ascorbic acid), and they certainly prevent and cure vitamin C deficiency-the cause of scurvy, [4]. Antimicrobial activity of the peel extract is directly concerned with the components that they contain [7]. The studies showed that essential oils, protopine and corydaline alkaloids, lactone, polyacetylenes, acyclic sesquiterpenes, hypericin and pseudohypericin compounds are effective toward various bacteria, [8]:

The purpose of the present study was to determine the antimicrobial activity of fabrics functionalized with fruit extract (Pomegranate) and (Aloe Vera) against above bacteria's. Since it is important to maintain the antibacterial activity of fabrics. Overall research methodology is shown at figure 1.

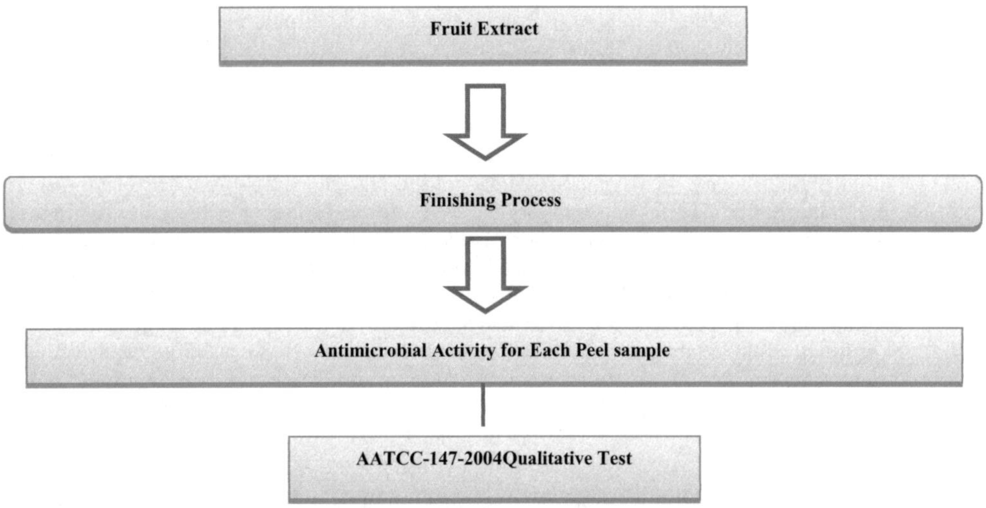

Figure 1. Schematic Diagram of research methodology

2. Material and methods

2.1 Preparation of Fabrics

The materials used for this research study are 100% cotton bleached fabric, obtained from (GulAhmed textile Karachi). The fabrics were passed through rinsing process in a washing machine at 40°Cfor 45-50 minutes using wetting agent, after washed fabric were tumble dried. The specification for fabric & water used is shown in Table 1 & 2.

Table 1. Tested Fabric Specification

Fabric Composition	Weave type	Area weight (g/m^2)	Warp yarn count (tex)	Weft yarn count (tex)	CIE Whiteness index (WI)	Absorbency (Sec)
Bleached 100% cotton	Plain	170.0	45	40	80.5	1.0
Bleached 100% cotton	Plain	115.8	30	30	66.5	0.8

IOP Conf. Series: Materials Science and Engineering **414** (2018) 012050 doi:10.1088/1757-899X/414/1/012050

Table 2. Tested water specification

The quality of water	
pH	7.0
Total hardness (ppm)	45.0
Total dissolve solids (ppm)	142.0

2.2 Preparation of natural extract

2.2.1. Fruit

Fruit collected from local market must sure are fresh and not effected bacteria before then it will change the properties of microbes during application and testing. Fruit were washed with the help of tap water followed by sterilized distilled water, air dried and peeled off further the peels were dried in sun, packed in envelops for drying in hot air oven at 60°C for 3 days and used as raw material for the extraction of antimicrobial compounds [9,10]

2.3 Dispersion of prepared finish and application parameters

2.3.1. Direct application methods

We took 1g, 5g, and 10g of each fruit waste powder (Pomegranate) & (Aloe Vera) in 1000 mlseparate measuring cylinders. Then added 100 ml water in each measuring cylinders. And shake each cylinder continuously for 10 minutes. After 10 minutes we observed that finish was not completely soluble in water. After 30 minutes the finish was more soluble. But still some particle left which abolish by filter.

Table 3. Finishes application parameters for natural fruits extract.

Extract source	Finishing Concentration%	Pickup %	Processing temp & time	
			Drying	Curing
Pomegranate & Aloe Vera Rind	1g, 5g, 10g /100 ml	80	60°C, 10 min	150°C, 3min

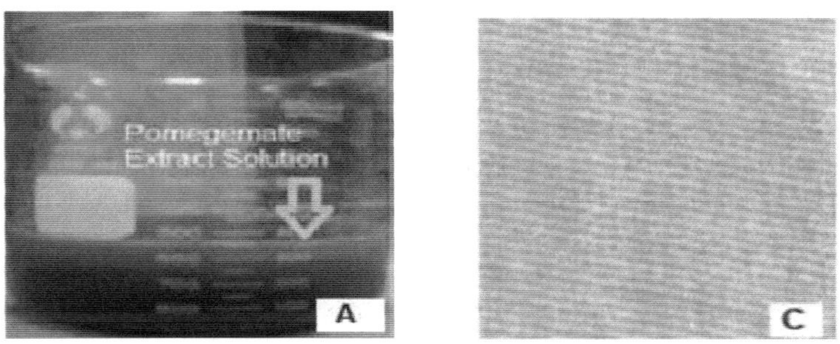

Figure 2. (a) Fruit peel extracts solution. **(c)** Fruit solution treated sample

IOP Conf. Series: Materials Science and Engineering **414** (2018) 012050 doi:10.1088/1757-899X/414/1/012050

2.4 AATCC Test Method 147-2004 (Parallel streak method)
Using a 4 mm inoculating loop, one loop full of the diluted inoculums was transferred to the surface of TSA plates by making five streaks approximately 60 mm in length, spaced 10 mm apart covering the central area of a standard Petri plates without refilling of loop. Test specimens were cut with a rectangular die (25x50 mm) and were placed to inoculate TSA transversely across the five inoculums streaks. Petri plates were incubated for 18 – 24 hr at 37°. Incubated plates were examined for interruption of growth along the streaks of inoculums beneath the specimen and for a clean zone of inhibition along a streak on either side of the test specimen was calculated using the following equation:

$$W= (T - D)/2$$

Where, W is width of clear zone of inhibition in mm, T is total diameter test specimen and clear zone in mm, D is diameter of the test specimen in m3.

3. Results & Discussion
To test activity of antimicrobial, we have above fruit finish sources and check on four types of bacteria activity on different conditions,
a, Extract concentration,

Binders are the mechanism used to keep the color on the fabric when using pigments for printing textiles. The choice of binders will always depend upon the final fastness requirements as well as the cost requirements of the process. Almost all the binders used in textile pigment printing are the addition polymerization products. The binder film in pigment print is a three-dimensional structure, the third dimension is rather less important than the other two. The binder is a film-forming sub-stance made up of long chain macromolecules, which when applied to the textile together with the pigment, produces a three-dimensionally network. The links are formed during some suitable fixing process, which usually consists of dry heat and change in pH value, bringing about either self-cross linking or reaction with other suit-able cross-linking agents. Elasticity and improved adhesion of the film to the substrate is achieved by cross linking. The cross-linking reaction must produce covalent bonds, which are insensitive to hydrolyzing agents (washing liquor, perspiration, industrial atmosphere). The reaction should be activated in dry hot air by curing process. Its formation takes place in two stages: flocculation (or coagulation) and coalescence. During the first stage of film formation, water and surfactants are removed from the binder by;

Binder-CH-ORHO-Cellulose Binder-CH-O-Cellulose + ROH

On the above-mentioned results lemon show the better result in all bacteria (Gram positive & Gram negative) for the reason of this unsaturated fatty acids compose 64.1% of total linoleic acid (21.71%) Compounds have anti-oxidant &anti-microbial activity. Lemon (citrus compositions), as results lemon shows the good number of microbial reduction and will be used as competitive as such commercial anti-microbial finishes in field of finishing. After noticeable for washing results on recent textile industry development and issues, we had selected the Quantitative test by agar nutrients EN-ISO-20645-2004 which also show inhibition zone both test are use same number of standardized microbes, we had tested all fruit extract concentration and show the common results (Table 5, 6 ,7) show the result on highest extract concentration.

In all three concentrations resist the bacteria and show the good anti-microbial activity. But lack of information is that its only show the resist zone not mentions the specific number. Due to up gradation of finishing, we need to improve the testing and need specific number, so we have to choose a different test related to exactly numbers. AATCC- 100-2004 has shown the results in number by scientific machine, limited germs and test the both un-wash & wash sample. But limitation of colony counter device, and we know the high extract factor we chose only 10gm related samples and check the anti-microbial activity. If we discuss the concentration and the number of cycles we follow the domestic washing rules on laundry detergent which mostly used in commercial and home laundering, so we gradually increase the washing cycles up to 20 and check the results which show the increase cycle mean decrease in anti-microbial activity. The best results show the high concentration from all extract, but due to cost affected we need to use the concentration as per our need our as per demand of finishing.

Table 4. Results of Antimicrobial Activity by AATCC-147-2004 (Qualitative)

	S. Aureus (AATCC 6538)	**E.coli (AATCC4352)**
Fabrics	**Inhibition Zone(mm)**	
Fruit Extract Treated Swatches	3.41 ± 0.04mm[a]	0 mm and No growth of microbe[b]
Untreated Swatches (Control Sample)	0mm and No growth reduction	0 mm and No growth reduction

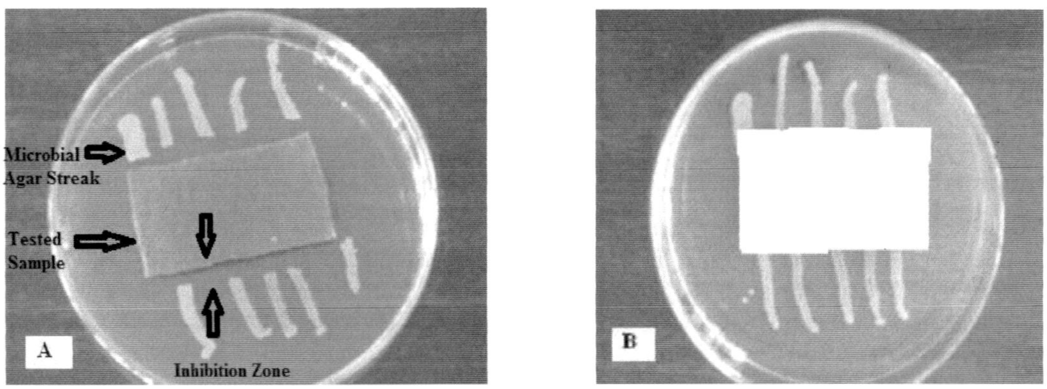

Figure – 3. **(a).** Anti-microbial Activity by agar streak Test, **(b).** Control Sample

4. Conclusion

The binding of cotton fabric using plant natural products was found to exhibit anti-bacterial properties. The fruits rind of fruit as an efficient waste biomass, exhibit efficient anti-microbial agent for the preparation of anti-microbial finish of medical cloths Pomegranate and herbal species extract shown better anti-microbial activity against all microbes. The choice of binders will always depend upon the final fastness requirements as well as the cost requirements of the process. Almost all the binders used in textile are the addition polymerization products. The binder film in pigment print is a three-dimensional structure, the third dimension is rather less important than the other two. The alkaline cross linking agent proved to be better in performance as compared to the acidic binder with third dimension. But as we mention our subject title that we need to use natural finishes so we try to keep away the formaldehyde binders a or other acrylic binders which can be cause the natural extract worst and our product and work can be gone to astray way so we chose the sodium bi carbonate and the acetic acid which both need to fix the dyes on cotton fabric when contact with different dyes alos these both chemical are not harmful and gradually used in industrial way. Both chemical exhaust after unknown time as product will be eco-friendly. Our work on improvement of anti-microbial activity achieved by natural way.

References

[1] T., Ramachandran. Antimicrobial Textile an overview. *Journal of the Institution of Engineers*. (2004). **84**(2): 42–47.

[2] Jantas. R., &Górna., K. Antibacterial finishing of cotton fabrics. *Fibers and Textiles in Eastern Europe*. (2006). 14(1): 55

[3] Gouda., M. Enhancing flame-resistance and antibacterial properties of cotton fabric. *Journal of Industrial Textiles*. (2006). 36(2):167-177

[4] Jakimiak. B., Röhm-Rodowald. E., Staniszewska. M., Cieślak M., Malinowska. G., &Kaleta., A. *Microbiological assessment of efficiency of antibacterial modified textiles*.

[5] Roczniki Państwowego Zakła-du-Higieny. (2006). **57**(2): 84-177.

[6] Renaud. F., N. R. Dore. J., &Freney. H., J. Evaluation of antibacterial properties of a textile product with antimicrobial finish in a hospital environment. *Journal of Industrial Textiles*. (2006). 36(1): 89-94.

[7] Uddin., F. Environment concern in Antimicrobial finishing of Textiles. *International Journal of Textile Science*. (2014). 15-20.

[8] Mahesh. S., Manjunatha. R. A. H., &Vijaya. K., G. *Studies on antimicrobial textile finishes using certain plant natural products. International conference on advances in biotechnology and pharmaceutical sciences Bangkok*. (2011). 253-258.

[9] Ahmad. M. M., Z. Salim-ur-Rehman., F. M. Iqbal-Anjum., & J.I., Sultan. Genetic variability to essential oil composition in four citrus Fruit Species. *Pak, J, Bot.* (2006). 319-324.

[10] AATCC Test Method 100 (2004). Antibacterial finishes on textile materials: Assessment of AATCC Technical Manual, American Association of Textile Chemists and Colorists, *Research Triangle Park, NC*. 150-151

[11] ENISO 20645:2004. Determination of antibacterial activity-agar diffusion plate test. *Technical Committee CEN/TC* **248**. (2004)